Clean Energy in South-East Asia

Other related titles:

You may also like

- PBPO075 | Biswarup Das | Power Distribution Automation | 2016
- PBPO098 | Antonio Moreno-Munoz | Large Scale Grid Integration of Renewable Energy Sources | 2017
- PBPO159 | Oliver Probst, Sergio Castellanos and Rodrigo Palacios | Transforming the Grid Towards Fully Renewable Energy | 2020
- PBPO193 | Surajit Chattopadhyay and Arabinda Das | Overhead Electric Power Lines: Theory and practice | 2021
- PBPO251 | David S-K. Ting and Jacqueline A. Stagner | Clean Energy for Low-Income Communities: Technology, deployment and challenges | 2024

We also publish a wide range of books on the following topics:
Computing and Networks
Control, Robotics and Sensors
Electrical Regulations
Electromagnetics and Radar
Energy Engineering
Healthcare Technologies
History and Management of Technology
IET Codes and Guidance
Materials, Circuits and Devices
Model Forms
Nanomaterials and Nanotechnologies
Optics, Photonics and Lasers
Production, Design and Manufacturing
Security
Telecommunications
Transportation

All books are available in print via https://shop.theiet.org or as eBooks via our Digital Library https://digital-library.theiet.org.

IET ENERGY ENGINEERING SERIES 266

Clean Energy in South-East Asia

Ngo Dang Luu, Nguyen Hung, Nguyen Dinh Long, Le Anh Duc and Le Anh Tuan

The Institution of Engineering and Technology

About the IET

This book is published by the Institution of Engineering and Technology (The IET).

We inspire, inform and influence the global engineering community to engineer a better world. As a diverse home across engineering and technology, we share knowledge that helps make better sense of the world, to accelerate innovation and solve the global challenges that matter.

The IET is a not-for-profit organisation. The surplus we make from our books is used to support activities and products for the engineering community and promote the positive role of science, engineering and technology in the world. This includes education resources and outreach, scholarships and awards, events and courses, publications, professional development and mentoring, and advocacy to governments.

To discover more about the IET please visit https://www.theiet.org/.

About IET books

The IET publishes books across many engineering and technology disciplines. Our authors and editors offer fresh perspectives from universities and industry. Within our subject areas, we have several book series steered by editorial boards made up of leading subject experts.

We peer review each book at the proposal stage to ensure the quality and relevance of our publications.

Get involved

If you are interested in becoming an author, editor, series advisor, or peer reviewer please visit https://www.theiet.org/publishing/publishing-with-iet-books/ or contact author_support@theiet.org.

Discovering our electronic content

All of our books are available online via the IET's Digital Library. Our Digital Library is the home of technical documents, eBooks, conference publications, real-life case studies and journal articles. To find out more, please visit https://digital-library.theiet.org.

In collaboration with the United Nations and the International Publishers Association, the IET is a Signatory member of the SDG Publishers Compact. The Compact aims to accelerate progress to achieve the Sustainable Development Goals (SDGs) by 2030. Signatories aspire to develop sustainable practices and act as champions of the SDGs during the Decade of Action (2020–30), publishing books and journals that will help inform, develop, and inspire action in that direction.

In line with our sustainable goals, our UK printing partner has FSC accreditation, which is reducing our environmental impact to the planet. We use a print-on-demand model to further reduce our carbon footprint.

Published by The Institution of Engineering and Technology, London, United Kingdom

The Institution of Engineering and Technology (the "**Publisher**") is registered as a Charity in England & Wales (no. 211014) and Scotland (no. SC038698).

Copyright © The Institution of Engineering and Technology and its licensors 2025

First published 2025

All intellectual property rights (including copyright) in and to this publication are owned by the Publisher and/or its licensors. All such rights are hereby reserved by their owners and are protected under the Copyright, Designs and Patents Act 1988 ("**CDPA**"), the Berne Convention and the Universal Copyright Convention.

With the exception of:

   (i) any use of the publication solely to the extent as permitted under:

      a. the CDPA (including fair dealing for the purposes of research, private study, criticism or review); or
      b. the terms of a licence granted by the Copyright Licensing Agency ("**CLA**") (only applicable where the publication is represented by the CLA); and/or

   (ii) any use of those parts of the publication which are identified within this publication as being reproduced by the Publisher under a Creative Commons licence, Open Government Licence or other open source licence (if any) in accordance with the terms of such licence, no part of this publication, including any article, illustration, trade mark or other content whatsoever, may be used, reproduced, stored in a retrieval system, distributed or transmitted in any form or by any means (including electronically) without the prior permission in writing of the Publisher and/or its licensors (as applicable).

The commission of any unauthorised activity may give rise to civil or criminal liability.

Please visit https://digital-library.theiet.org/copyrights-and-permissions for information regarding seeking permission to reuse material from this and/or other publications published by the Publisher. Enquiries relating to the use, including any distribution, of this publication (or any part thereof) should be sent to the Publisher at the address below:

The Institution of Engineering and Technology
Futures Place,
Kings Way, Stevenage,
Herts, SG1 2UA, United Kingdom

www.theiet.org

Whilst the Publisher and/or its licensors believe that the information and guidance given in this publication is correct, an individual must rely upon their own skill and judgement when performing any action or omitting to perform any action as a result of any statement, opinion or view expressed in the publication and neither the Publisher nor its licensors assume and hereby expressly disclaim any and all liability to anyone for any loss or damage caused by any action or omission of an action made in reliance on the publication and/or any error or omission in the publication, whether or not such an error or omission is the result of negligence or any other cause. Without limiting or otherwise affecting the generality of this statement and the disclaimer, whilst all URLs cited in the publication are correct at the time of press, the Publisher has no responsibility for the persistence or accuracy of URLs for external or third-party internet websites and does not guarantee that any content on such websites is, or will remain, accurate or appropriate.

Whilst every reasonable effort has been undertaken by the Publisher and its licensors to acknowledge copyright on material reproduced, if there has been an oversight, please contact the Publisher and we will endeavour to correct this upon a reprint.

Trade mark notice: Product or corporate names referred to within this publication may be trade marks or registered trade marks and are used only for identification and explanation without intent to infringe.

Where an author and/or contributor is identified in this publication by name, such author and/or contributor asserts their moral right under the CPDA to be identified as the author and/or contributor of this work.

British Library Cataloguing in Publication Data
A catalogue record for this product is available from the British Library

ISBN 978-1-83724-015-9 (hardback)
ISBN 978-1-83724-016-6 (PDF)

Typeset in India by MPS Limited

Cover image: Floating PV System in a Wetland Area, courtesy by authors

Contents

About the authors xvii

1 **Wind energy development in Vietnam** 1
  1.1 Wind energy potential in Vietnam, wind energy studies and projects in Vietnam, and issues of safety and environmental protection 1
    1.1.1 Wind energy studies in Vietnam 4
    1.1.2 Structural safety of wind turbines and environmental issues 6
  1.2 Wind energy technology, operating principles, equivalent circuits, models, and wind turbine generator controllers 9
    1.2.1 History of wind turbines development 9
    1.2.2 Types of generators used in wind power systems 10
  1.3 Typical connection parameters for wind power projects selection of connection circuit parameters 14
    1.3.1 Connection distance 14
    1.3.2 Conductor cross-sect 15
    1.3.3 Connection power capacity 15
    1.3.4 Connection voltage 16
  1.4 Calculation of the wind power generation: wind speed 16
  1.5 Survey of installation and operating costs for wind power projects globally and in Vietnam 19
    1.5.1 Onshore wind power projects 19
    1.5.2 Offshore wind power projects 20
  1.6 Optimizing power calculation through nonlinear control and sensor fault monitoring for wind turbines 20
    1.6.1 Introduction 21
    1.6.2 Observer-based controller and fault-tolerant control design 21
    1.6.3 Observer design for generalized high-level disturbances 22
    1.6.4 Discussion 26
  1.7 Chapter summary and new research opportunities 27
  Further reading 28

2 **Inexhaustible energy sources: the future of solar energy in energy scenarios** 31
  2.1 Potential and development prospects of solar energy in Vietnam: solar radiation coefficients in potential localities and some typical solar energy projects in Vietnam 31

	2.1.1	Potential and development prospects of solar energy in Vietnam	31
	2.1.2	Notable solar energy projects in Vietnam	32
2.2	Scientific and technological achievements in solar energy: traditional solar energy technology, next-generation solar energy technology, and key components in solar energy system structures		34
	2.2.1	Traditional solar panel technology	34
	2.2.2	Structure and technical specifications of main components in solar power systems	42
2.3	Efficient solar power system models: stand-alone and grid-tied solar power systems		42
	2.3.1	Stand-alone solar power systems	42
	2.3.2	Grid-tied solar power system	43
	2.3.3	Hybrid grid-tied solar power systems	44
	2.3.4	Rooftop solar power systems for self-generation and self-consumption: a new direction for connection to the national grid in Vietnam	45
2.4	Solar power with grid-tied rooftop systems: a power source for smart homes, automatic navigation solar power systems, and agricultural irrigation systems		47
2.5	Summary and future research directions		55
	Further reading		56

3 Technologies and solutions for smart energy systems with the presence of renewable energy — **59**

3.1	Introduction to the application of artificial intelligence and data analysis for fault detection and prediction in multi-source energy systems		59
3.2	Application of artificial intelligence for calculating and estimating the inertia constants of the power system		59
	3.2.1	Overview of power system inertia	59
	3.2.2	Development of renewable energy and its impact on the inertia of the power system in Vietnam	60
	3.2.3	Current status	62
3.3	Initiative content		62
	3.3.1	Research on methodology for calculating and measuring PS inertia	62
3.4	Evaluation of harmonic analysis and equipment loss calculation using specialized software and effective selection of power system protection devices		66
	3.4.1	Study of heat generation and vibration in transformers caused by voltage and current harmonics	66
	3.4.2	Protection equipment check and selection at the workshop	69
3.5	Some energy storage technologies for renewable energy systems and load forecasting in solar and wind power systems		71

		3.5.1 Energy storage technologies	71
		3.5.2 Research on the application of seasonal time series models for monthly electricity production forecasting	73
	3.6	Introduction to high-voltage direct current transmission and its principles and benefits compared to traditional alternating current transmission	86
		3.6.1 Recent achievements in transmission technology	86
		3.6.2 Key technical requirements for high-voltage direct current transmission	88
	3.7	Potential integration of HVDC into existing power grid systems and real-world HVDC transmission projects	88
		3.7.1 Assumptions used in the calculations	88
		3.7.2 Technical assumptions	88
		3.7.3 Economic assumptions	89
	3.8	Calculation of current cost adjustments when input factors change	92
		3.8.1 Power system simulation in PSS/E	92
		3.8.2 Transmission distance of 270 km	93
		3.8.3 Calculation of present value	94
		3.8.4 Transmission distance of 450 km	99
	3.9	Potential DC transmission projects	101
	3.10	Impact of renewable energy sources on the power quality of the grid and discussion on reducing energy losses and power losses, improving voltage quality and reliability	102
		3.10.1 Building a model to calculate power supply reliability and assess the impact of wind power plants on the distribution grid	102
		3.10.2 Analyzing the impact of wind power plants on the power grid and statistical analysis of wind turbine failures at the Tuy Phong Wind Power Plant—Binh Thuan	103
		Further reading	105
4	**Study and application of smart grid**		**107**
	4.1	Introduction to smart grid and its importance in improving performance and network management	107
		4.1.1 Required input data for DAS application	109
		4.1.2 Software requirements	109
	4.2	Technology and advances in smart grid	110
		4.2.1 Level 1: Mini-SCADA	110
		4.2.2 Level 2: DAS	110
		4.2.3 Level 3: DMS model	111
		4.2.4 Definition and function of DMS	111
		4.2.5 Operating principle	111
		4.2.6 Application of distribution grid automation in enhancing power supply reliability at EVNCPC	111

Clean energy in South-East Asia

4.3	Application of artificial intelligence and automation in smart grids on cloud computing platforms	113
	4.3.1 Project effectiveness analysis	113
	4.3.2 Application of artificial intelligence and automation in smart grids on cloud platforms	120
	4.3.3 Fault location process and issues in power transmission lines in Vietnam	123
	4.3.4 Some protection solutions for PV sources	126
	4.3.5 Distance protection (F21) for grid-connected PV sources	128
	4.3.6 Application of STATCOM for dynamic voltage stabilization in Power Systems	141
4.4	Challenges and opportunities in implementing smart grids	156
	4.4.1 Introduction to smart grids	157
	4.4.2 Key challenges in implementation	157
	4.4.3 Benefits and opportunities	159
	Further reading	160

5 Study and application of microgrid systems — 163

5.1	Definition and objectives of microgrid systems	163
5.2	Key components of a microgrid system	163
5.3	Application of microgrid technology in renewable energy sources and cloud-based systems	165
	5.3.1 Global research on wind–diesel hybrid power systems	165
	5.3.2 Research status in Vietnam	167
	5.3.3 Problem statement	169
	5.3.4 Standalone wind power generation system	169
	5.3.5 Wind–diesel hybrid power system	170
	5.3.6 Cloud computing	170
5.4	Specific applications and examples of microgrid power grid systems deployment	173
	5.4.1 Wind–diesel hybrid power system in Phu Quy	173
	5.4.2 Control of wind–diesel hybrid power system	173
	5.4.3 Operation of the wind–diesel hybrid power system	174
	5.4.4 Results from steady-state mode survey	178
	5.4.5 Analysis and proposed solutions to enhance stability	179
	5.4.6 Application of a general control algorithm for the wind–diesel hybrid power generation system on Phu Quy Island	180
	5.4.7 Comparison with reality and other solutions	180
	5.4.8 Operational constraints	181
5.5	Development of a microgrid model for Phu Quoc Island	183
	5.5.1 Overview of the electrical grid model for Phu Quoc Island	183
	5.5.2 Wind turbine model	184
	5.5.3 Power balance	185
	5.5.4 Building the power grid model in MATLAB	186

Contents   xi

5.6	Smart energy management algorithm	187
	5.6.1 Building the MATLAB-PLC communication module	187
	5.6.2 Energy management system algorithm	188
	5.6.3 Electric vehicle charging algorithm	189
5.7	Simulation results	190
	5.7.1 Energy management system test	190
	5.7.2 Electric vehicle charging scenario simulation	190
5.8	Study on the impact of wind power plants on the reliability of the distribution grid	193
	5.8.1 Application of reliability indices calculation for the independent grid of Ly Son Island—Quang Ngai with the integration of wind power	193
	5.8.2 Summary	207
	Further reading	**209**

6 Research and application of IoT technology in sustainable energy development   **213**

6.1	Introduction to wireless sensor systems based on remote management and their role in renewable energy development	213
	6.1.1 IoT applications in solar power monitoring	214
	6.1.2 Studies on monitoring and analyzing the operational status of photovoltaic panels	214
	6.1.3 Studies on the application of MLT in monitoring and analyzing the operational status of photovoltaic panels	215
	6.1.4 Research objective	216
	6.1.5 Research content	216
6.2	Types of sensors and communication technologies for monitoring and managing renewable energy systems	218
	6.2.1 Sensors in renewable energy	218
	6.2.2 Communication technologies in renewable energy management	219
	6.2.3 Benefits of sensor and communication systems in renewable energy management	219
6.3	Application of wireless technology and sensor systems in monitoring and predicting the performance of renewable energy sources and some practical examples related to wireless sensor systems in the renewable energy sector	220
	6.3.1 Failure cases in photovoltaic panels	221
	6.3.2 Internet of Things	222
	6.3.3 Machine learning model for fault classification	225
	6.3.4 Machine learning model training results	229
	6.3.5 Development of an IoT-based solar energy monitoring system	233
	6.3.6 Mobile application	237
	6.3.7 Mobile application and computer software interface	238

xii   *Clean energy in South-East Asia*

	6.3.8	Integration of machine learning models into the IoT system	242
	6.3.9	Experimental results	243
	6.3.10	Achievements	247
	6.3.11	Challenges and limitations	247
	6.3.12	Recommendations and development directions	247
	Further reading		248

7 Impacts of policies and mechanisms on renewable energy development — **251**

- 7.1 Overview of Vietnam's power system: structure and challenges in ensuring adequate, stable, and safe electricity supply — 251
  - 7.1.1 Structure of Vietnam's power system — 251
  - 7.1.2 Solar power — 251
  - 7.1.3 Challenges in ensuring electricity supply — 252
  - 7.1.4 Strategic solutions — 252
- 7.2 Vietnam's regulations on grid connection for wind and solar power projects: capacity, voltage, and frequency requirements — 252
  - 7.2.1 Regulations on connecting wind and solar power projects to the grid — 252
  - 7.2.2 Simulation of the impact of renewable energy systems on the operating parameters of the neighboring grid at the connection point — 261
  - 7.2.3 PSCAD software — 277
  - 7.2.4 Simulation of voltage drop and short circuit at the rooftop solar power plant in Long An Province — 286
- 7.3 Vietnamese electricity market: structure of the electricity market and electricity transmission pricing model — 289
  - 7.3.1 Problem statement — 289
  - 7.3.2 Vietnam's wholesale electricity market — 290
  - 7.3.3 Current transmission pricing method in Vietnam — 298
  - 7.3.4 Transmission electricity price model in Vietnam's competitive wholesale electricity market — 303
  - 7.3.5 Analysis of simulation results — 318
- 7.4 Renewable energy subsidies and the development of FiT in Vietnam: calculation methods for subsidies and FiT mechanism for renewable energy producers — 318
  - 7.4.1 Definition of FiT — 318
  - 7.4.2 FiT pricing method — 319
  - 7.4.3 Results — 323
  - 7.4.4 FiT pricing policy — 325
  - 7.4.5 Challenges and solutions — 326
  - 7.4.6 Benefits of competitive bidding — 326
  - 7.4.7 Policy recommendations — 327

	7.5	Subsidy mechanism: limitations, analysis of subsidy mechanisms, and recommendations to promote renewable energy development in Vietnam	327
		7.5.1 Introduction	327
		7.5.2 Existing subsidy mechanisms and their importance	327
	7.6	Impact of renewable energy development on environmental, economic, and social policies in Vietnam	329
		7.6.1 Introduction	329
		7.6.2 Environmental impact	330
		7.6.3 Economic impact	330
		7.6.4 Social impact	331
	7.7	Solutions for renewable energy development in Vietnam: supportive policies, market mechanisms, and technological solutions	331
	7.8	Summary and future research directions	333
		Further reading	334
8	**Investment procedures for renewable energy projects: a case study in Vietnam**	**337**
	8.1	Investment policies, mechanisms, and competition among power producers in Vietnam	337
		8.1.1 Key policy recommendations	338
		8.1.2 Strengthening the independence of ERAV	338
		8.1.3 Enhancing national load dispatch center (A0) independence	338
		8.1.4 Promoting private investment in transmission infrastructure	339
		8.1.5 Establishing a transparent competitive bidding framework	339
		8.1.6 Solutions for creating a level playing field in the clean energy infrastructure sector in Vietnam	339
	8.2	Aspects of implementing wind and solar energy projects in Vietnam: capital mobilization policies and electricity pricing	341
		8.2.1 Promoting the development of independent renewable energy systems	341
		8.2.2 Characteristics and policy requirements for renewable energy development	341
		8.2.3 Mechanisms and policies to support investment	341
		8.2.4 Preferential policies	342
		8.2.5 Developing human resources	342
		8.2.6 Environmental protection policies	342
		8.2.7 Factors affecting household investment intentions in the rooftop solar energy sector	342
	8.3	Procedure for building and developing renewable energy projects in Vietnam	345
	8.4	Technical criteria in the construction and development of renewable energy projects	346

xiv   *Clean energy in South-East Asia*

        8.4.1 Site selection, survey, land clearance, solar radiation calculation, and climate conditions   346
        8.4.2 Application of fuzzy logic algorithm for supplier selection of solar equipment   347
        8.4.3 Power calculation   347
        8.4.4 Calculating the connection to the electrical system   348
        8.4.5 System testing and project handover   349
  8.5 Estimating the economic and financial indicators of the project and identifying funding sources for solar and wind energy systems   349
        8.5.1 Calculating the cost and benefit components of rooftop grid-connected solar systems   351
        8.5.2 Calculating the economic effectiveness of a rooftop grid-connected solar system   351
        8.5.3 Calculating the financial and economic effectiveness of rooftop grid-connected solar systems   353
        8.5.4 Calculation of costs and benefits of grid-connected rooftop solar power systems   354
        8.5.5 Calculation of the financial economic efficiency of wind power systems   354
        8.5.6 Calculation of the costs and benefits of grid-connected wind power systems   355
        8.5.7 Calculating the effectiveness of the project   356
        8.5.8 A combined approach between grid-connected solar energy and independent investment optimization: a case study in Vietnam   356
        8.5.9 Hybrid Solar-BESS system based on business models   362
        8.5.10 Some experiments: case studies in Vietnam   363
  8.6 Practical example for some solar and wind power projects in Vietnam   367
        8.6.1 Overview of the project characteristics   367
        8.6.2 Legal basis for the estimate   367
        8.6.3 Estimate content   368
        8.6.4 Estimated cost   369
        8.6.5 Project information summary   372
        8.6.6 Preliminary energy estimates   375
        8.6.7 Conclusion and recommendations   381
  Further reading   382

9 Circular economic solutions—effective management of "solar photovoltaic panel waste"   385
  9.1 Introduction to circular economy scenarios for renewable energy development: case studies in Vietnam   385
        9.1.1 Overview of circular economy and renewable energy   385
        9.1.2 Circular economy scenarios in renewable energy development   385

	9.1.3 Case studies in Vietnam	386
	9.1.4 Conclusion	386
9.2	Circular economy approach after the end-of-life of materials in solar and wind energy equipment	387
	9.2.1 Challenges of solar and wind energy equipment after end-of-life	387
	9.2.2 Circular economy approaches after end-of-life use	387
	9.2.3 Case studies in Vietnam and globally	388
9.3	Solar photovoltaic panels: understanding the structure of solar panels to open up research on solar panel waste treatment	389
	9.3.1 Structure of a solar photovoltaic panel	389
	9.3.2 Challenges in solar panel waste treatment	390
	9.3.3 Solar photovoltaic panels: researching the structure for waste treatment	390
9.4	"Solar panel waste": introducing the issue of solar panel waste and environmental challenges in the renewable energy sector	391
	9.4.1 Solar energy development worldwide – current status and forecast	392
9.5	Solutions for solar panel waste: current solutions and future proposals	395
	9.5.1 Solar energy waste recycling technology	395
	9.5.2 General cycle of solar energy waste recycling technology	395
	9.5.3 Study on the waste flow of solar panels from solar power plants	396
	9.5.4 Solar panel recycling technology worldwide	398
	9.5.5 Experimental research on processing, recycling, and material recovery from crystalline silicon PV modules in Vietnam	403
	9.5.6 Chemical method: using chemicals to separate materials in solar modules	410
9.6	Approaches to managing "waste from renewable energy" for a circular economy and sustainable development	411
	9.6.1 Managing waste from renewable energy	411
	9.6.2 Developing recycling and processing technologies	412
	9.6.3 Encouraging the adoption of the circular economy model	412
	9.6.4 Use of recycled materials in production	412
	9.6.5 Creating value from waste	413
	Further reading	413

10 Social impacts and trends in renewable energy development in developing countries: solar energy, wind energy, and beyond   **417**

10.1	Environmental impact assessment of solar and wind energy projects in Vietnam	417
	10.1.1 Overview	417
	10.1.2 Environmental impact assessment for solar energy projects	419

xvi *Clean energy in South-East Asia*

10.1.3 Environmental impacts of wind power projects	421
10.1.4 Mitigation measures	424
10.2 Raising environmental and social awareness in renewable energy projects	424
10.2.1 The importance of environmental and social awareness	425
10.2.2 Steps for implementation and impact assessment	425
10.2.3 Strengthening community and stakeholder involvement	425
10.2.4 Training and capacity building for stakeholders	425
10.2.5 Government policies and commitments	425
10.3 Methods for environmental impact assessment of renewable energy sources to identify effective mitigation and control solutions	426
10.4 Example of environmental impact assessment in central Vietnam	427
10.4.1 Project introduction	427
10.4.2 Assessment and forecasting of impacts during the project preparation phase	428
10.4.3 Impact on air quality	430
10.4.4 Impact on water quality	430
10.4.5 Impact assessment and forecasting during the construction phase	431
10.4.6 Non-waste-related sources of impact	431
10.4.7 Impact assessment and forecasting during the operational phase	432
10.4.8 Recommendations	434
10.4.9 Commitments	435
10.5 IoT solutions and smart lighting system, wireless charging stations for vehicles, and electric ferries using renewable energy	435
10.5.1 Flowchart for central control unit algorithm	437
10.5.2 System test results	439
10.5.3 Development directions	440
10.5.4 Wireless charging station for electric vehicles using renewable energy	441
10.5.5 Overview of the transportation sector in Vietnam	442
10.5.6 Application of adaptive control research in energy management systems at Vietnamese ports	454
10.6 Green hydrogen from wind and solar power via water electrolysis: an advanced solution for sustainable energy	456
10.6.1 Current hydrogen production methods	457
10.6.2 The role of hydrogen in renewable energy	459
10.6.3 Groundbreaking ceremony for the first green hydrogen production Plant in Tra Vinh—March 30, 2023	461
10.6.4 Conclusion	462
Further reading	462

Index   **465**

About the authors

Ngo Dang Luu is the chairman of Anh Minh Global Co. Ltd, Vietnam. He earned his PhD in Science in the Federal Republic of Germany and has over 20 years of experience in the electricity and energy industry in Vietnam. He specializes in technical design, grid interconnection, battery energy storage, and automation applications for the maintenance and operation of solar and wind projects, as well as power system integration. He is also a leading expert in renewable energy optimization, electrical systems, smart grids, microgrids, circular economy, and electric vehicle charging systems.

Nguyen Hung is the vice director of the HUTECH Institute of Engineering, Ho Chi Minh City University of Technology (HUTECH), Vietnam. His research interests span various fields, including robust and nonlinear control, mobile manipulator robot control, power system stability analysis and control, and artificial intelligence applications in renewable energy. In addition, he serves as a reviewer for numerous international journals and conferences.

Nguyen Dinh Long is an assistant professor in the Department of Engineering, Dong Nai University, Vietnam. He received his PhD in electronics and electrical engineering from Queen's University Belfast, UK, where he was also a research fellow. His research interests include convex optimization theory and applications for wireless communications, the Internet of Things, digital transformation, and real-time optimization for interdisciplinary studies. He is currently serving as a reviewer for several IEEE journals, international journals, and conferences.

Le Anh Duc is the rector of Dong Nai University, Vietnam. He has previously held the position of deputy director at the Youth Employment Service Center and deputy head of the Vocational Training Department. He has also served as rector of Long Thanh-Nhon Trach Vocational College and Dong Nai College of High Technology.

Le Anh Tuan has been working at Can Tho University, Vietnam since 1982 and is currently a senior lecturer in the Department of Environmental and Natural Resources. He previously served as the vice director of the Institute for Climate Change Research at Can Tho University. Dr Tuan has many years of teaching and research experience in areas such as water resource planning and management, environmental engineering, climate change and natural disasters, hydrology, meteorology, and renewable energy.

Chapter 1
Wind energy development in Vietnam

1.1 Wind energy potential in Vietnam, wind energy studies and projects in Vietnam, and issues of safety and environmental protection

Vietnam is located near the equator, between 8° and 23° North latitude, and lies within the tropical monsoon climate zone. This climate creates two distinct wind seasons: the Northeast Monsoon and the Southwest Monsoon. The average wind speed in coastal areas ranges from 4.5 to 6 m/s at an altitude of 10–12 m. In remote island regions, the wind speed is even higher, reaching from 6 to 8 m/s, offering significant potential for wind energy development. This potential has been identified through wind measurement data provided by the Department of Meteorology and Hydrology.

The regions with the highest wind energy density in Vietnam are presented in Table 1.1, highlighting areas suitable for wind energy projects.

In Table 1.1, wind speed is measured at altitudes of 10–12 m. However, large-scale wind turbines, with capacities ranging from a few hundred of kilowatts to tens of megawatts, are typically installed at heights of 50–90 m. Currently, there is no data available on wind speed at these higher altitudes. Therefore, some units have conducted wind measurements at heights of 50–90 m at specific survey sites before preparing feasibility reports for wind energy projects.

To promote the development of wind energy in line with the government's goals, the Ministry of Industry and Trade issued Document No. 4308/BCT-TCNL on May 17, 2013. This document requires the 24 provinces and cities with potential for wind energy development to develop provincial-level wind energy master plans. To date, 11 out of the 24 provinces have completed the development of these plans and obtained approval from the Ministry of Industry and Trade. According to these plans, the total estimated wind energy capacity is projected to reach approximately 2,511 MW by 2020 and increase to about 15,380.9 MW by 2030, as shown in Table 1.2.

In addition to the projects approved according to the provincial wind energy development plans, many wind energy projects have been submitted by the provincial People's Committees (PPCs) for inclusion in separate planning documents. As of March 2020, approximately 28 projects with a total capacity of 2,214 MW have been approved for inclusion in national and local power development plans

Table 1.1 Selected regions with the highest wind energy potential in Vietnam

No.	Region	Average wind speed V_{avg} (m/s)	Wind power density (W/m²)	Annual energy density (E = kWh/m²)
1	Bach Long Vi	6.1	349.7	3,063.8
2	Truong Sa	5.5	235.0	2,058.3
3	Phu Quy	5.0	185.6	1,625.6
4	Co To	4.3	114.3	1,001.4
5	Quy Nhon	4.1	106.6	935.4
6	Ly Son	3.6	123.1	1,078.4
7	Hon Dau	3.6	75.8	664.4
8	Van Ly	3.5	53.2	466.0
9	Con Co	3.5	82.2	720.2
10	Vung Tau	3.4	41.8	365.9
11	Phan Thiet	3.1	62.2	544.8
12	Cua Ong	3.1	42.3	370.3
13	Hon Ngu	3.0	81.4	712.8
14	Buon Me Thuot	3.0	47.3	414
15	Thai Binh	2.8	33.3	291.8
16	Phu Quoc	2.7	50.3	440.2
17	Bac Lieu	2.7	26	227.6
18	Tam Đao	2.7	60.9	533.5
19	Bai Chay	2.6	26.8	235.1
20	Pleiku	2.6	29.8	260.7

Table 1.2 Wind power development plans by region

No.	Province	Capacity (MW) 2020	2020–2025	2025–2030	Projected by 2030
1	Thai Binh		40	30	70
2	Quang Tri	110			110
3	Ninh Thuan	220			1,429
4	Binh Thuan	700			1,570
5	Dak Lak	110			1,382
6	Ba Ria Vung Tau		34		107
7	Ben Tre	150			1,520
8	Tra Vinh	270			1,608
9	Soc Trang	200			1,470
10	Bac Lieu	401.2			2,507
11	Ca Mau	350			3,607
Total		**2,511**	**74**		**15,380**

for the period up to 2020. These projects are primarily concentrated in 11 provinces, especially in the Central, Central Highlands, and Mekong Delta regions.

By March 15, 2020, in addition to the projects already included in the plans, the Ministry of Industry and Trade had received proposals from the PPCs for a total

Table 1.3 Statistics on the number of projects and total wind power capacity by regions

TT	Region	Number of projects	Capacity (MW)
1	Bac Trung Bo	51	2,918.8
2	Dong Nam Bo	02	602.6
3	Nam Trung Bo	10	4.193
4	Tay Nam Bo	94	25,540
5	Tay Nguyen	91	11,733
Total		**248**	**44,989**

of 248 projects, with an estimated total capacity of about 45,000 MW. These projects are distributed by region and geographical area as shown in Table 1.3.

To determine the wind speed at a height h from wind speed data measured at a reference height, an approximate formula based on the theory of wind speed distribution with height can be used. The commonly used formula is the **logarithmic law** in atmospheric dynamics theory:

$$V = V1\left(\frac{h}{h1}\right)^n \tag{1.1}$$

where:

V is the wind speed to be determined at height h,
V_1 is the wind speed measured at height h_1, and
n is the ground friction coefficient, which depends on the natural terrain conditions.

For example, water surface or flat ground: $n = 0.1$; areas with tall vegetation: $n = 0.2$; and city center areas: $n = 0.4$.

According to the latest survey by the World Bank, Vietnam has significant potential for wind energy development due to its land availability and favorable wind conditions. Specifically, about 31,000 km^2 of land could be utilized for wind energy projects, with 865 km^2 being ideal areas for this development. The survey estimates that these areas have the potential to generate up to 3,572 MW of electricity, with production costs lower than 6 US cents/kWh. This presents a great opportunity for Vietnam to develop clean and sustainable energy, reduce dependence on fossil fuels, and meet the growing demand for electricity.

The detailed results from the survey are presented in Figure 1.1, providing a visual overview of the wind energy potential in specific regions of Vietnam, helping guide investors and policymakers in developing wind energy projects.

On September 19, 2014, in Hanoi, the Electricity Regulatory Authority of Vietnam hosted a workshop on the topic "Developing Regulations for Wind Power Grid Connection and Studying the Integration of Renewable Energy in Vietnam." At the workshop, AWS Truepower from the United States presented the results of a wind energy potential survey in Vietnam using the geographic information system.

4 Clean energy in South-East Asia

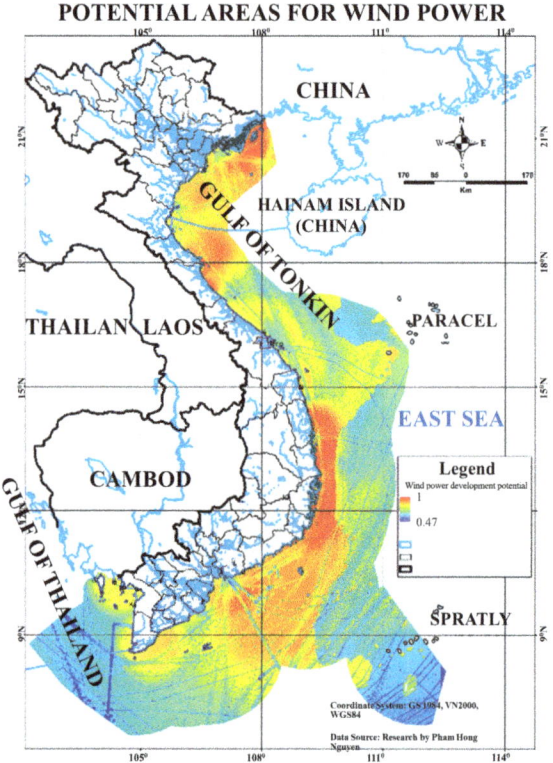

Figure 1.1 Vietnam's wind energy potential as surveyed by the World Bank

The studies screened and identified potential wind energy development sites based on the following factors:

- Wind speed map with a 200-m resolution.
- Potential areas identified based on resource availability, slope, and other natural factors.
- Distance to the existing power grid and transportation routes.
- Sites evaluated based on the cost of electricity generation.
- Energy production potential assessed based on performance or net capacity factor.

1.1.1 Wind energy studies in Vietnam

The research and use of wind energy in Vietnam began in the 1970s with the involvement of various agencies. Since 1984, with the participation of the "State Science and Technology Advancement" program for new and renewable energy, several small-scale wind turbine generators were developed:

- The PD170-6 wind generator, with a power output of 120 W, for charging batteries at the University of Danang.

- The PH500 wind generator, with a power output of 500 W, from the Hanoi University of Science and Technology.
- The 150-W wind generator from the SOLALAB Research Center at the University of Danang.
- Under the "Wind Energy and Applications" training cooperation program, WAT Company supplied the New Energy Research Center at the Hanoi University of Science and Technology with a wind power grid-connected system, which included Airdolphin Mark-Zero wind turbine and generator (1,000 W), battery set, power converter, wind speed and direction measuring devices, power output meters, and data recording and processing software.
- **Some typical wind energy projects in Vietnam:**
  - **Bach Long Vy wind power plant (WPP), Hai Phong city**: The first WPP in Vietnam, installed on the island of Bach Long Vy, with a capacity of 800 kW and a total investment of VND 14 billion.
  - **Phuong Mai 3 WPP, Binh Dinh Province**: Located in the communes of Cat Tien and Cat Chanh, Nhon Hoi Economic Zone, with a total investment of over USD 35.7 million. The plant features 14 wind turbines, totaling 21 MW of capacity, with an annual output of about 55 million kWh. It is developed by Central Vietnam Wind Power Investment and Development Company.
  - **Phu Quy island wind-diesel hybrid power plant Binh Thuan Province**: Comprising three wind turbines with a total capacity of 6 MW, this project has been completed and is operational.
  - **Bac Lieu WPP Bac Lieu Province**: Developed by Cong Ly Construction, Trade and Tourism Co. Ltd, with a total capacity of 99 MW. Located in the coastal area of Bien Dong A Hamlet, Vinh Trach Dong Commune, Bac Lieu Town, Bạc Lieu Province. Phase 1 has been completed with 10 wind turbines, totaling 16 MW, connected to the 110-kV grid.
  - **Tuy Phong WPP, Binh Thuan Province**: Developed by Vietnam Renewable Energy Joint Stock Company, the overall project has a capacity of 120 MW. Phase 1 has been completed with 20 wind turbines, totaling 30 MW, connected to the 110-kV grid. Figure 1.2 shows the entrance and part of the infrastructure of the Tuy Phong Wind Power Plant, one of the first wind energy projects in Vietnam, located in Binh Thuan Province. It illustrates the application of renewable energy in the development of clean electricity in the central region.

These projects represent significant strides made by Vietnam in developing renewable energy, particularly wind power, contributing to energy security and sustainable development.

- The project for building the first wind turbine manufacturing plant in collaboration with the manufacturer Fuhlander AG at Vinh Hao, Binh Thuan Province. The WPP investment projects in Vietnam are presented in Table 1.4.

6 *Clean energy in South-East Asia*

Figure 1.2 Tuy Phong WPP—Binh Thuan

Table 1.4 *Wind power investment projects in Vietnam*

Province	Number of investors	Number of projects	Installed capacity (MW)	Current status				
				IR	IP	TD	UC	IO
Ninh Thuan	9	13	1,068	8	4	1		
Binh Thuan	10	12	1,541	6	4	1		1
Ba Ria Vung Tau	1	1	6		1			
Tien Giang	1	1	100	1				
Ben Tre	2	2	280	2				
Tra Vinh	1	1	93	1				
Soc Trang	4	4	350	4				
Bac Lieu	1	1	99		1			1
Quy Nhon	2	3	156		2	1		
Ca Mau	2	2	300	2				
Dak Lak	1	1	120		1			
Total	**34**	**41**	**4,113**	**24**	**13**	**3**		**2**

1.1.2 *Structural safety of wind turbines and environmental issues*

- **Environmental issues**
  - One of the greatest benefits of wind energy is its fully renewable nature. While the level of technological application may vary between countries and regions, wind remains a natural and ubiquitous phenomenon on Earth. When solar radiation creates air convection currents, these wind currents

generate wind, which can then be harnessed for energy. The ability to generate electricity depends on the speed and frequency of the wind, but as long as sunlight persists, this energy source is nearly limitless.
- Additionally, wind energy has the advantage of not producing waste or harmful emissions like other energy sources such as fossil fuels and nuclear energy. According to the U.S. National Oceanic and Atmospheric Administration, burning fossil fuels for energy generation emits large amounts of carbon dioxide and sulfur gases, which pollute the air and contribute to ozone layer depletion. The World Energy Council also reports that these emissions are the main cause of global climate change and its negative impacts. Transitioning to wind energy helps reduce pollution and supports environmental protection and long-term climate stability.
- Wind turbines were invented to take advantage of wind energy for sustainable development. In fact, many people consider one of the great features of wind energy to be its almost unlimited sustainability. Fossil fuels like coal, oil, and natural gas will eventually be completely extracted from the Earth, while geothermal energy uses accumulated heat from many years of solar radiation. Humans are aware of the importance of maximizing natural resources, but must face the depletion of fuel. In contrast, wind energy will never run out as long as life exists on Earth and the sun continues to shine.
- According to many experts in technical and scientific research reports, wind energy is considered a highly promising energy source capable of supplying energy to the entire planet and meeting the growing energy demand of the world's population. Wind turbines can be installed almost anywhere, from flat lands and the sea to mountainous areas. This makes wind energy a universal energy source, which can be harnessed and used in most countries globally.
- According to the World Energy Council, wind energy currently provides about 1% of the global energy demand. However, in many European countries, especially Denmark, wind energy has met nearly 20% of the country's energy needs. This shows the potential for wind energy development to contribute to the sustainable energy system and help achieve global greenhouse gas emission reduction goals.
- In remote areas without access to the grid, wind turbines can be used to generate and supply electricity independently. With a variety of sizes available, wind turbines are suitable for a wide range of users, from individuals to businesses. Households in small towns and villages can effectively use these wind turbines to meet their energy needs.

- **Disadvantages of wind energy**
  - The main obstacle to using wind turbines for power generation is that wind is intermittent and does not always occur when needed. In other words, sometimes wind cannot be harnessed during peak electricity demand periods.
  - Wind turbines must compete with traditional power sources on a cost basis. Wind farms may or may not be useful depending on the wind energy

of a particular area. While the costs of wind installations have decreased over the past 10 years, wind energy still requires more financial support compared to fossil fuel power generators.
- Impact on animals
  - Birds and bats
    - Birds may die due to collisions with the rotating blades of wind turbines or other structures.
    - Habitat destruction and barrier effects affect their feeding and breeding behaviors.
  - Marine life
    - Wind farms can change the distribution of fish, and nearby wind farms may create artificial reefs, impacting marine biodiversity.
    - Increased water turbidity and additional objects on the seabed (barrier effect) can harm bottom-dwelling animals and plant life and block sunlight from reaching the water.
- Noise issues
  - Noise is one of the major environmental obstacles to the development of the wind energy industry and can cause sleep disturbances and hearing impairment in humans.
  - Exposure to high-frequency noise can lead to headaches, discomfort, fatigue, as well as arterial spasms and weakened immune systems.
  - Many factors influence the propagation and attenuation of noise from wind turbines, such as air temperature, humidity, barriers, reflection, and surface materials.
- Visual impact and landscape
  - The shadows cast by wind turbines can cause undesirable visual effects, or even disturbing flickering when a rotating blade moves across the landscape and houses.
  - The negative visual impact of wind farms on the landscape is another factor that creates negative perceptions of the wind energy industry in people's minds.
  - Factors influencing the intensity of the visual impact of wind turbines include the local landscape and the distance between the viewer and the turbines.
- Local climate change issues
  - Various studies have indicated that wind turbines can affect local weather and regional climate such as temperature and wind speed.
  - The wind speed recovery rate after passing through a wind farm shows a reduced curve.
  - Large wind farms cause cooling effects during the day and warming effects at night due to vertical air mixing near the surface.
  - Through improved turbine blade designs and appropriate spacing and turbine configurations, turbulence caused by rotor motion can be minimized, and the meteorological effects of wind farms can be reduced.

- Structural safety of wind turbines
  o Many structural failures of wind turbines due to harsh environmental loads have been reported and studied in recent years.
  o The collapse of two wind turbine towers in Japan due to Typhoon Maemi in 2003.
  o The collapse of a tower and blade damage of a wind turbine in Taiwan due to Typhoon Jangmi in 2008.
  o The fire of a wind turbine in Binh Thuan, Vietnam in 2020 due to an internal turbine fault.
- Wind turbine towers are tall structures, often flexible and vulnerable to vibrations caused by wind loads.

1.2 Wind energy technology, operating principles, equivalent circuits, models, and wind turbine generator controllers

1.2.1 History of wind turbines development

Wind turbine designs are generally categorized into two main types: horizontal-axis wind turbines (HAWT) and vertical-axis wind turbines (VAWT) as shown in Figure 1.3.

In 1982, the maximum power of wind turbines was only about 50 kW. By 1995, commercial wind turbines had been improved, increasing their power output by more than tenfold, reaching around 750 kW. By 2012, modern wind turbines were being manufactured with a capacity of 7.5 MW and a rotor diameter of 150 m.

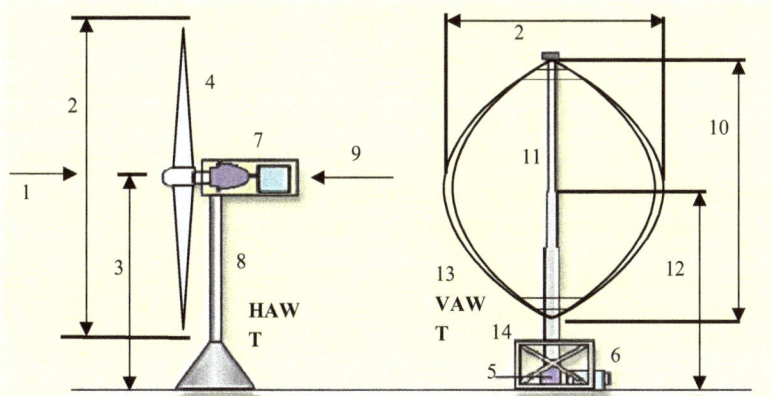

Figure 1.3 Main components of horizontal-axis wind turbine (HAWT) and vertical-axis wind turbine (VAWT). 1. Wind direction of HAWT, 2. Rotor diameter, 3. Turbine height, 4. Rotor blades, 5. Gearbox, 6. Generator, 7. Housing, 8. HAWT tower, 9. Wind direction behind the rotor, 10. Rotor height of VAWT, 11. VAWT tower, 12. Equator height, 13. Rotor blades with fixed pitch angle, 14. Rotor foundation.

10 *Clean energy in South-East Asia*

Notably, by 2024, wind turbine technology had made significant advancements, with a power capacity of up to 15 MW and a rotor diameter of 236 m.

The strong development process of wind turbine power capacity through various stages is illustrated in detail in Figure 1.4.

1.2.2 Types of generators used in wind power systems

Wind energy conversion system model

Based on aerodynamic principles, wind turbines are designed with blades to capture wind energy and convert it into mechanical energy to drive a generator. For large-scale wind turbines (in the MW range), the rotational speed typically varies between 10 and 15 rpm. To convert the low-speed rotational torque into high-speed rotational torque, a gearbox is commonly used. After conversion, the rotational speed is increased to 1,000–1,500 rpm and transmitted to the generator through a drive shaft, as illustrated in Figure 1.5.

In addition to selecting a gearbox, it is possible to choose a generator with multiple pole pairs to meet the speed requirements. However, for optimal connection to the power grid, it is best to use power electronic converters.

Generator manufacturing technology solutions

The generator is responsible for converting mechanical energy from the rotor into electrical energy. In wind energy conversion systems (WECS), both synchronous and asynchronous generators are used. The technological solutions for manufacturing generators in the wind energy sector to meet the requirements of converting wind energy into electrical energy are shown in Figure 1.6. The output power can be either alternating current (AC) or direct current (DC), ensuring compatibility for connection to the power grid.

Some types of wind-powered generators

Table 1.5 introduces the types of generators from leading wind turbine manufacturers around the world, which have been installed and are operational in many countries. The most common types of generators currently include:

- **Induction generators with wound rotor (Type C model)**
- **Synchronous permanent magnet generators (Type D model)**

The Type A and B models have gradually been replaced by the Type C and D models. The control strategy symbols such as VS, FS, VP, and FP in Table 1.5 will be detailed in Chapter 2.

Offshore wind power technology

Offshore wind turbines are typically installed on towers ranging from 60 to 105 m above sea level. These towers allow the wind turbines to capture optimal energy while minimizing the impact of ocean waves. The foundation technology for offshore turbines is designed to suit the specific geographical conditions of each project. Factors such as maximum wind speed, water depth, wave height, current, and soil characteristics all significantly influence the foundation design of the turbine.

Figure 1.4 Development process of wind turbines

12 *Clean energy in South-East Asia*

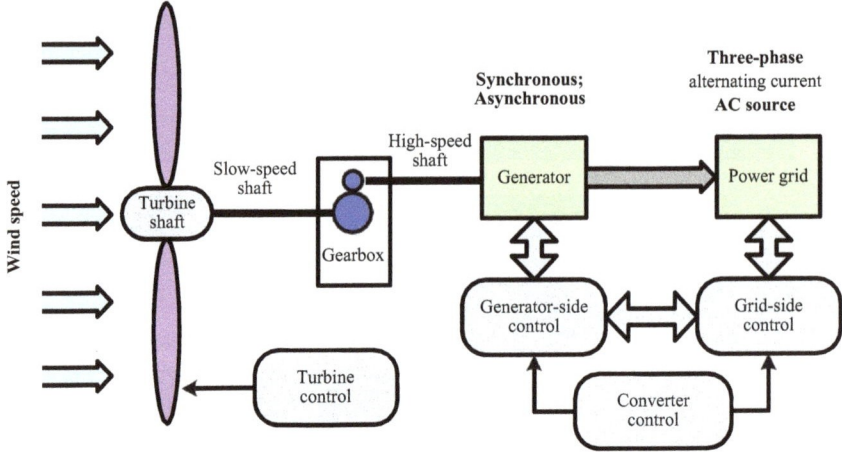

Figure 1.5 Wind energy conversion to electrical energy in a wind power generation system

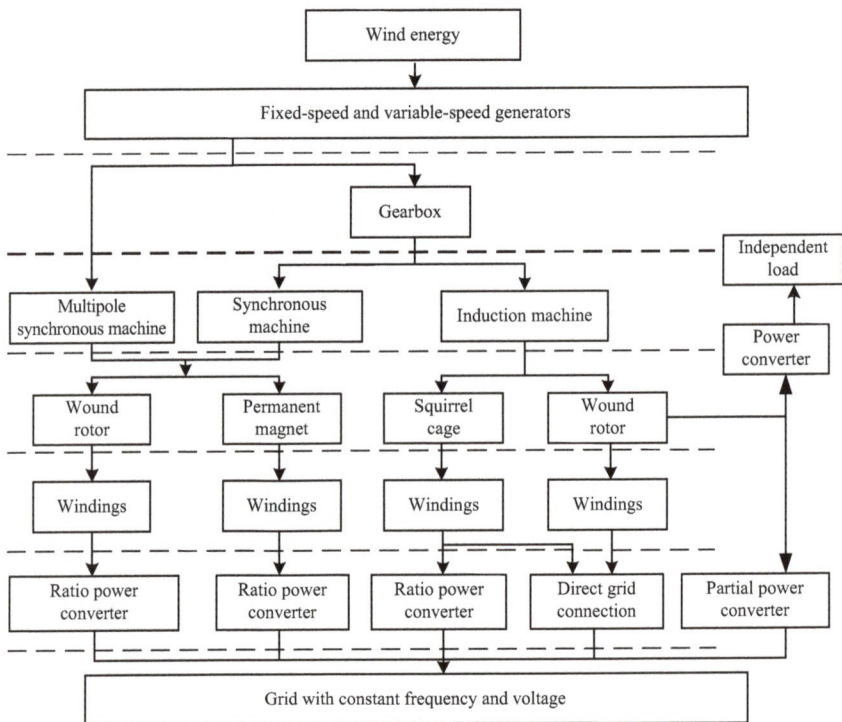

Figure 1.6 Block diagram describing types of wind-powered generators

Table 1.5 Types of wind turbines from leading manufacturers worldwide

Manufacturer (country)	Model	Control strategy		Basic specifications
1. Vestas (Denmark) V80/2.0 MW	Type C	VS–VP	– – – – – – –	Generator voltage (WRIG): 690 V Rotor diameter: 80 m Rotor speed range: 9–19 rpm Generator speed range: 905–1,915 rpm Start-up wind speed: 4 m/s Nominal wind speed: 15 m/s Shutdown wind speed: 25 m/s
2. Enercon (Germany) E112/4.5 MW	Type D	VS–FP	– – – – – –	Generator voltage (WRIG): 440 V Rotor diameter: 114 m Rotor and generator speed range: 8–13 rpm Start-up wind speed: 2.5 m/s Shutdown wind speed: 28 m/s Gearbox-less design
Enercon (Germany) E66/2 MW	Type D	VS–FP	– – – –	Generator voltage (WRIG): 440V Rotor diameter: 66 m Rotor and generator speed range: 10–12 rpm Gearbox-less design
3. Gamesa (Spain) G80/ 1.8 MW	Type B	VS–FP	– – –	Generator voltage (WRIG, Optislip): 690 V Rotor speed range: 15.1–16.1 rpm Generator speed range: 1,818–1,944 rpm
Gamesa (Spain) G128/ 4,5 MW	Type D	VS–FP	– – – –	Generator voltage (WRIG): 690 V Rotor diameter: 128 m Rotor speed: 12 rpm Start-up speed: 3 m/s; rated speed: 13 m/s; shutdown speed: 25 m/s
4. GE Wind (USA) GE104/ 3,2 MW	Type C	VS–VP	– – – –	Stator/rotor generator voltage (WRIG): 3.3 kV/690 V Rotor speed range: 7.5–13.5 rpm Generator speed range: 1,000–1,800 rpm Start-up speed: 3 m/s; rated speed: 12 m/s; shutdown speed: 25 m/s
GE 77/1,5 MW	Type C	VS–VP	– – –	Rotor speed range: 10–20.4 rpm Generator speed range: 1,000–2,000 rpm Start-up speed: 3.5 m/s; rated speed: 13 m/s; shutdown speed: 25 m/s
5. Sinovel (China) S70/ 1.5 MW	Type D	VS–FP	– – – –	Rotor diameter: 70 m Rotor and generator speed range: 9–19 rpm Start-up speed: 3 m/s; rated speed: 11.8 m/s; shutdown speed: 25 m/s Gearbox: None

(Continues)

Table 1.5 (*Continued*)

Manufacturer (country)	Model	Control strategy	Basic specifications
6. Goldwind (China) GW100/2.5 MW	Type D	VS–FP	– Rotor diameter: 100 m – Rotor and generator speed range: 7.6–14.4 rpm – Start-up speed: 3 m/s; rated speed: 12.5 m/s; shutdown speed: 25 m/s – Gearbox: None
7. NEG Micon (Denmark) NM80/2.75 MW	Type C	VS–VP	– Generator stator/rotor voltage (WRIG): 960/690 V – Rotor diameter: 80 m – Rotor speed range: 12–17.5 rpm – Generator speed range: 756–1,103 rpm
8. Nordex (Germany) N80/2.5 MW	Type C	VS–VP	– Generator voltage (WRIG): 660 V – Rotor diameter: 80 m – Rotor speed range: 10.9–19.1 rpm – Generator speed range: 700–1,300 rpm
S77/1,5 MW	Type C	VS–VP	– Generator voltage (WRIG): 690 V – Rotor diameter: 77 m – Rotor speed range: 9.9–17.3 rpm – Generator speed range: 1,000–1,800 rpm
9. Made (Spain) Made AE–90 2 MW	Type D	VS–FP	– Generator voltage (WRSG): 1,000 V – Rotor diameter: 90 m – Rotor and generator speed range: 7.4–14.8 rpm – Generator speed range: 747–1,495 rpm
Made AE–61 1,32 MW	Type A	FS–FP	– Generator voltage (SCIG): 690 V – Two-speed generator: 1,010 and 1,519 rpm – Two-speed rotor: 12.5 and 18.8 rpm
10. Repower (Germany) MM 82/2 MW	Type C	VS–VP	– Generator voltage (WRIG): 690 V – Rotor diameter: 82 m – Rotor speed range: 10–12 rpm – Generator speed range: 900–1,800 rpm
MD77/1,5 MW	Type C	VS–VP	– Generator voltage (WRIG): 690 V – Rotor diameter: 77 m – Rotor speed range: 9.6–17.3 rpm – Generator speed range: 1,000–1,800 rpm

WRIG: wound-rotor induction-generator.

1.3 Typical connection parameters for wind power projects selection of connection circuit parameters

1.3.1 Connection distance

The distance between the electrical collection busbar of the WPP and the connection point to the grid depends on several factors, including the plant's capacity,

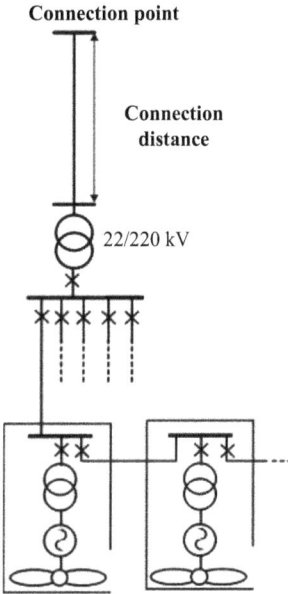

Figure 1.7   Connection diagram corresponding to the power capacity of a wind farm

conductor cross-section, number of circuit lines, the location of the plant, and the voltage drop regulations at the connection point.

Figure 1.7 presents the connection voltage levels corresponding to different power capacities. For wind farms with a capacity of less than 10 MW, to ensure that the voltage drop does not exceed 5%, calculations indicate that the connection distance should not exceed 14 km. For larger wind farms with a capacity greater than 100 MW, connection requirements may vary depending on the design and technical regulations.

1.3.2   Conductor cross-sect

The cross-sectional area of the conductor is an important factor in determining the power capacity of a wind farm. Typically, for wind farms with a capacity of less than 10 MW, the transmission lines connecting the plant to the grid typically use conductors with a cross-sectional area of 185 mm^2. For capacities greater than 10 MW, conductors with a cross-sectional area of 240 mm^2 are used.

In addition to power capacity, the conductor cross-section depends on the connection distance between the wind farm and the grid connection point, as shown in Table 1.3. Selecting the appropriate conductor cross-section ensures efficient power transmission and meets the required voltage drop limits.

1.3.3   Connection power capacity

In Vietnam, most WPPs are connected to the transmission grid system. However, there has not yet been any specific study on the connection power capacity with the

Table 1.6 Relationship of parameters in the wind turbine connection circuit to the power grid

Wind farm capacity (MW)	Maximum allowable	Voltage drop	Connection distance (km)	Voltage level
$P \leq 10$	5%	≤ 14	22 kV	Single power line 185 mm^2
		>14	22 kV	Double power line 185 mm^2
$10 < P \leq 20$		≤ 15	22 kV	Double power line 185 mm^2
		>15	110 kV	Single power line 240 mm^2
$20 < P \leq 100$		≤ 48	110 kV	Single power line 240 mm^2
		>48	110 kV	Double power line 240 mm^2
$100 < P \leq 200$		≤ 48	110 kV	Double power line 240 mm^2
		>48	220 kV	Double power line 400 mm^2

grid. This is because, according to the planning, wind power capacity is not significant compared to the total capacity of the region to which it is connected. Therefore, the connection issue is primarily considered based on the current grid's ability to receive and distribute power, without requiring complex calculations for specific power capacities.

1.3.4 Connection voltage

Since the generation capacity of WPPs is generally limited, wind power is usually connected to medium-voltage and low-voltage electrical networks. However, there are currently no regulations on the maximum voltage level when connecting wind power. Countries that develop wind power typically perform the connection at the distribution or transmission voltage level, depending on the WPP's capacity.

The selected voltage level is based on the protection coordination requirements, rated capacity, and the location of the system connection. In case of medium-voltage grid connection, transformers may play a role in protecting the wind generator. Typically, in countries with concentrated wind power capacity, plants with a capacity of less than 20 MW will be connected to the distribution grid, while plants with capacities greater than this will be connected to the transmission grid, as shown in Table 1.6.

1.4 Calculation of the wind power generation: wind speed

In many regions of Vietnam, wind patterns can fluctuate significantly over time within a very wide range. At times, the wind speed may be very high, but just minutes later, it may drop to almost zero. Furthermore, both the wind speed and the duration of wind from each direction are always changing.

Based on survey and measurement data collected over several years at a specific location, it is possible to construct a probability graph of wind direction

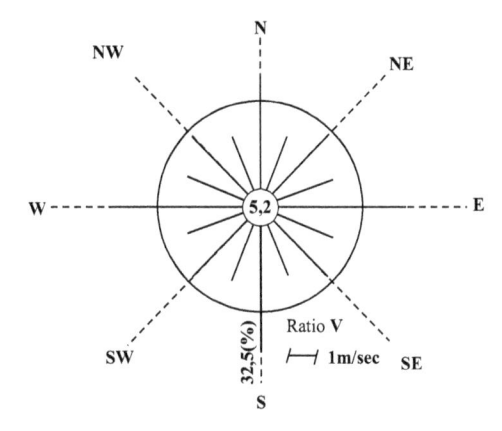

Figure 1.8   Wind flower

changes throughout the year, month, or day, as shown in Figure 1.8. This graph, also known as a "wind flower," is depicted in the form of 16 rays, each corresponding to a wind direction. The length of each ray is proportional to the average wind speed (m/s) and the probability of wind from that direction (%), as measured at the survey location. The data recorded at the center of the circle represents the probability of no wind at the surveyed site.

Wind speed also varies depending on the terrain and the altitude of the measurement point. Some studies suggest using the following relationship:

$$\frac{V}{V_0} = \left(\frac{H}{H_0}\right)^n \quad (1.2)$$

where:

V_0 is the measured wind speed at height H_0,
V is the calculated wind speed at height H, and
n is the terrain coefficient at the surveyed location, typically ranging from 0.1 to 0.4 (lower values are used for areas with strong winds, and higher values for areas with light winds).

In Vietnam, in locations where dedicated wind measurement towers are not available, data from meteorological stations with $h = 10$ mh $= 10$ m are often used. At ground level, terrain significantly affects wind speed distribution, and the higher the measurement height, the less the terrain influence.

Application example

The Tuy Phong WPP in Binh Thuan is connected to the regional distribution grid, according to the simplified diagram shown in Figure 1.9.

When a WPP is in operation, the supply capacity to the consumer load during the survey hour t_j will increase to $(S_i(t_j) + SG(t_j))$ where $SG(t_j)$ is the WPP's

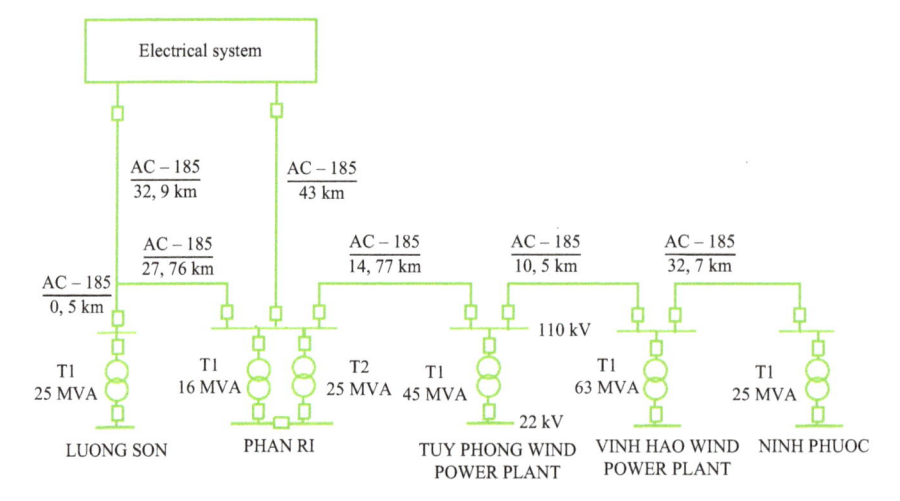

Figure 1.9 Simplified diagram of the 110-kV grid area connected to the Tuy Phong WPP

generation capacity at hour t_j, corresponding to the probability of the supply state $P_i = P_{HTi}*P_{Gi}$, P_{HTi} with P_{Gi} being the probability states of the system and the WPP, respectively.

The power output of the WPP depends on wind conditions, the location of the plant, and various other environmental factors. However, to operate the grid economically and ensure power quality, it is necessary to calculate the amount of power connected to the grid. For example, in Belgium, regulations state that the design capacity of wind power must be lower than the transformer capacity of the WPP and comply with the $(N - 1)$ standard for the connection area. In Italy, the installed wind power capacity must not exceed 65% of the capacity of the connected area, while in Spain, this limit is 50%. The short-circuit capacity of the wind power network should not exceed the capacity of the switching devices. The voltage level is also determined as a standard to set the maximum capacity of the wind turbine generator.

Figure 1.10 presents the power generation profile over a typical day at the Tuy Phong WPP in Binh Thuan Province. The graph shows the variation in power output from the wind turbines under the influence of wind conditions, divided into three main phases:

- **Early day (00:00–07:45)**: Low power generation, reflecting weak wind during this period.
- **Middle (07:45–21:00)**: Power generation increases sharply and peaks, indicating strong and stable winds throughout the day, especially due to the coastal advantage.
- **Late day (21:00–24:00)**: Power generation gradually decreases, consistent with the decline in wind speed as night falls.

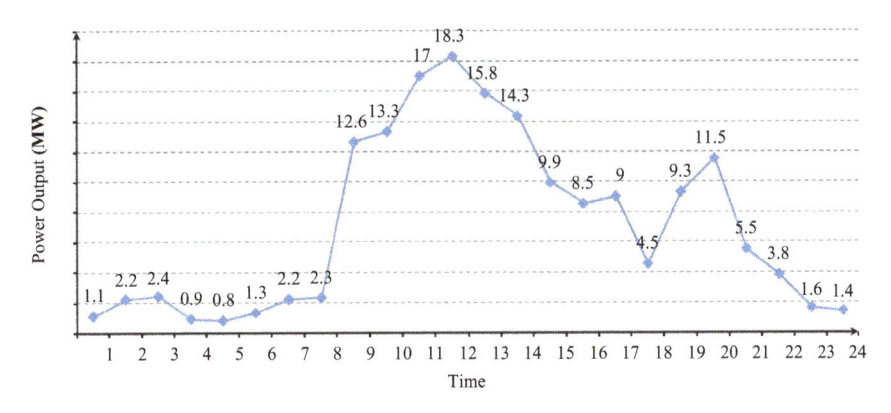

Figure 1.10 Power generation profile of the Tuy Phong WPP – Binh Thuan throughout a day

The graph not only clearly illustrates the natural cyclic nature of wind energy but also emphasizes the critical role of optimizing operations to efficiently harness this renewable energy source.

1.5 Survey of installation and operating costs for wind power projects globally and in Vietnam

1.5.1 Onshore wind power projects

Data from onshore WPPs in Denmark (2013 and 2014) indicate that the average investment cost is approximately 1.4 million USD per MW. In Germany, the average cost in 2012 was higher, around 1.8 million USD per MW, which could be relevant to Vietnam's context, as many areas have low wind conditions suitable for turbines that operate in light winds.

According to IRENA, the total investment cost for onshore WPPs in 2017 was 1.5 million USD per MW, based on extensive data. In the United States, the average cost for onshore wind projects in 2012 was nearly 2 million USD per MW, but it decreased to about 1.7 million USD per MW in 2015. In India and China, costs were lower during the 2013–2014 period, ranging from 1.3 to 1.4 million USD per MW.

A typical example is the Phu Lac Wind Farm in Tuy Phong District, Binh Thuan Province. This plant has a total capacity of 50 MW, with the first phase being 24 MW, which began construction in July 2015 and became commercially operational in September 2016. The wind farm uses 2 MW turbines, 95 m in height, with a blade diameter of 100 m, a startup wind speed of 3 m/s, and a cut-off speed of 22 m/s. The plant has a capacity factor of 19.6% and covers a total land area of 400 ha, with the long-term direct impact area being 9.3 ha (approximately 3,800 m^2 per MW).

The total investment for the project was USD 46.95 million (based on 2016 prices), excluding costs such as administrative, consultancy, project management, site preparation, taxes, and interest during construction. With this investment, the nominal cost reached USD 1.94 million per MW of capacity.

When including all of these costs, the total investment reached USD 51.05 million, corresponding to a cost of USD 2.1 million per MW.

1.5.2 Offshore wind power projects

Costs have historically been a major barrier to the development of offshore wind power. In 2015, the levelized cost of energy (LCOE) for these projects ranged from USD 150 to 200 per MWh, three to four times higher than onshore wind power. However, this situation changed significantly from 2016 to 2017, when competitive bidding in Europe reduced prices below USD 100 per MWh, with Dutch projects leading the way in bidding without subsidies.

In 2017, the bid price for a 30 MW offshore wind farm in the United States dropped to USD 65 per MWh. By September 2019, the United Kingdom set a record with the lowest bid price of GBP 39.65 per MWh (USD 49.6 per MWh), a 30% decrease from the previous bidding round in 2017.

The information above provides an overview of the costs and development status of WPPs globally and in Vietnam, highlighting the potential and challenges of implementing this renewable energy source. Challenges such as a lack of technical expertise, infrastructure limitations, and legal barriers, though significant, are entirely surmountable. In fact, these challenges open up opportunities for the establishment of public-private partnerships, capacity building, and policy reforms. To achieve this goal, a comprehensive approach is needed, including policy reforms, financial incentives, and the promotion of technological applications, with close collaboration between the government, industry stakeholders, and the international community.

If the above proposals are successfully implemented, Vietnam could become a regional leader in renewable energy in Southeast Asia, laying the foundation for sustainable development and economic resilience.

1.6 Optimizing power calculation through nonlinear control and sensor fault monitoring for wind turbines

Using wind energy as a renewable energy source has attracted significant attention, with wind turbines playing a crucial role in converting wind energy into electricity.

This scientific research presents a comprehensive method to optimize power calculation and monitor sensor faults in wind turbines deployed in Bình Định Province, Vietnam. The study aims to improve the efficiency, reliability, and maintainability of wind energy systems by integrating advanced control techniques, stability analysis, and sensor fault detection.

1.6.1 Introduction

The introduction discusses the growing installation of wind energy systems, especially in Bình Định Province, and highlights the challenges of accurately measuring wind speed for real-time control systems in WECS. Existing methods, such as anemometers, gray models, Kalman filters, and genetic algorithms, have been used but have many limitations. An alternative approach is proposed, which involves estimating aerodynamic torque, directly related to wind speed. In control systems, aerodynamic torque is often treated as a disturbance. Therefore, the use of observers to estimate aerodynamic torque instead of direct measurement is proposed.

This research also introduces a new polynomial system that allows modeling nonlinear WECS systems as linear systems, reducing linearization errors and facilitating observer and controller design using the sum of squares (SOS) technique in MATLAB®.

Key contributions of the research

- **New polynomial observer**
  – Simultaneously estimates aerodynamic torque, d-axis, and q-axis currents in the WECS, requiring only generator speed measurement, thus reducing the number of necessary sensors.

- **More flexible observer framework**
  – Eliminates constraints such as the need for first- or higher-order derivatives of aerodynamic torque, making the method more practical for real-world applications.

- **Research structure**
  – The research continues with the system description, LQR controller design, polynomial observer design, stability analysis, simulation results, discussion, and conclusions.

- **Practical benefits**
  – This method helps reduce deployment costs by optimizing the number of sensors, improving accuracy and reliability of the wind energy system, and enhancing practicality for implementation in Vietnam.

1.6.2 Observer-based controller and fault-tolerant control design

Based on the estimated information about reference speed, aerodynamic torque, and system faults, tracking errors and compensators are represented as follows:

- **Tracking error**: This error reflects the discrepancy between the actual and reference values of state variables such as turbine speed, generator speed, or

22   Clean energy in South-East Asia

current. This error is used to improve the accuracy of tracking the desired speed.
- **Compensators**: Compensators are designed to minimize the impact of noise and sensor faults on system performance. This ensures that the system operates stably and efficiently even in the presence of errors or abnormal noise.
- **Practical application**: The observer-based controller not only allows precise tracking of generator speed but also integrates fault-tolerant control (FTC) strategies, ensuring stable operation of the wind turbine system even under less-than-ideal conditions such as sensor faults or measurement noise.

This method is highly beneficial in improving the reliability of wind energy systems, minimizing maintenance time, and ensuring long-term performance, particularly in harsh or rapidly changing environmental conditions.

In this section, we will simulate the WECS with parameters listed in Table 1.7. Figure 1.11 illustrates the structure of the proposed FTC diagram.

Assume that the WECS is affected by faults described by the formulas discussed and shown in Figure 1.12. The wind speed has a waveform as displayed in Figure 1.13. It should be noted that these faults affect both the actuator and the sensor of the WECS.

1.6.3 Observer design for generalized high-level disturbances

In this section, we introduce the key concepts of a variable-speed wind turbine and the design of a generalized high-level disturbance observer (GHODO) to handle cases with rapidly varying disturbances. Although using traditional wind speed measurement devices such as anemometers may seem like a simple solution,

Table 1.7 WECS parameters

Parameter	Symbol	Value	Unit
Rated power	P_{rated}	3.465	
Pole pairs	P	11	
Stator resistance	R_S	0.08	Ω
Stator inductance	L	0.344	H
Magnet flux linkage	Ψ_m	136	V.s/rad
Mechanical inertia	J	2.5×10^6	kg·m^2
Viscous friction coefficient	B_v	0.001	kg.m^2/s
Rotor radius	R	66	m
Air density	ρ	1.205	kg/m^3
Gear ratio	n_{gb}	1	
Cut-in wind speed		4	m/s
Cut-off wind speed		25	m/s
Rated wind speed		12.08	m/s

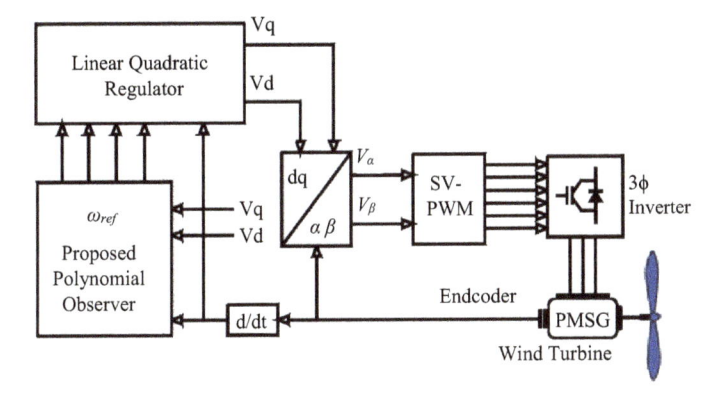

Figure 1.11   Observer-based controller for WECS

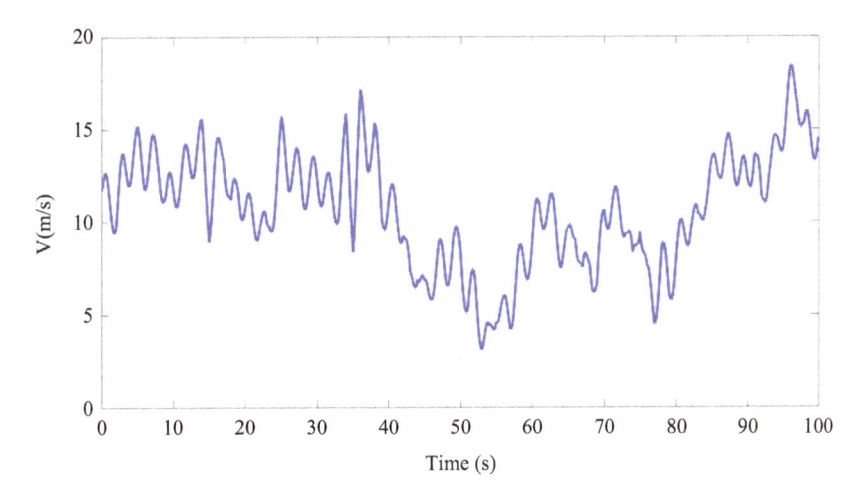

Figure 1.12   Wind speed profile with nominal parameters

employing observers to estimate aerodynamic torque not only provides a cost-effective solution but also ensures high measurement accuracy.

However, using inappropriate observers may result in poor overall system performance and reduced control accuracy. While many studies suggest that rapidly changing disturbances have a minimal impact on the system, complete elimination of errors in steady-state conditions has yet to be observed.

Integrating a high-quality observer with a controller can significantly improve the system's performance. Therefore, using GHODO is considered an appropriate method for estimating aerodynamic torque and disturbances on the d–q axes, rather than relying on measurement devices like anemometers.

Figure 1.14 presents the performance evaluation of speed tracking for aerodynamic torque, electromagnetic torque, and the d-axis current in two scenarios:

24 *Clean energy in South-East Asia*

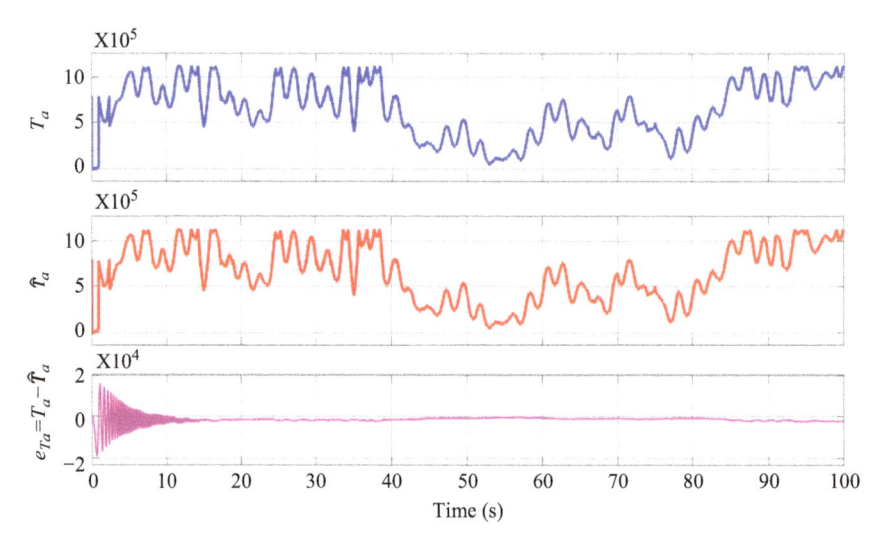

Figure 1.13 Estimation performance of aerodynamic torque with nominal parameters

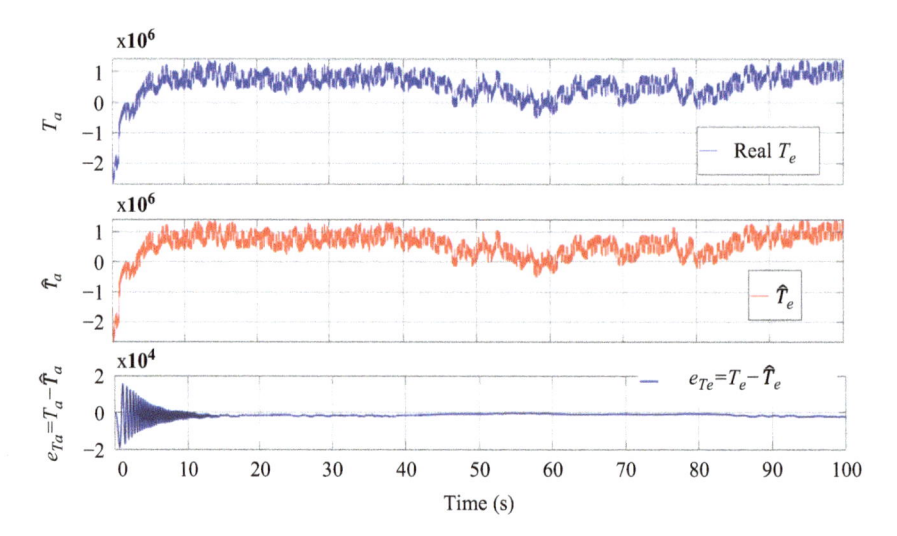

Figure 1.14 Estimation performance of electromagnetic torque with nominal parameters

the standard parameters and a parameter variation with a 40% increase in stator resistance and a 10% decrease in stator inductance. The waveforms display the aerodynamic torque (*Ta*), estimated aerodynamic torque, reference speed, feedback speed, and speed tracking error. Figure 1.14 also illustrates the electromagnetic torque (*Te*), estimated electromagnetic torque (*eTe*), *d*-axis current, estimated *d*-axis current, and their corresponding estimation errors.

Under standard conditions, the estimates are observed accurately, with the steady-state error being below 0.5%. Under parameter variation, the estimation error increases to approximately 10%. Notably, the speed tracking error remains unchanged. Figures 1.15–1.17 illustrate the power coefficient (C_p) and the obtained power (P_a), demonstrating the robust control performance of the pitch angle and the effective maximum power point tracking despite parameter variations. Specifically, under standard conditions, the steady-state estimation error is below 0.5%: the aerodynamic torque estimation error is 0.45%, the electromagnetic torque estimation error is 0.43%, and the *d*-axis current estimation error is 0.00%. However, in the case of parameter variation, the estimation error increases to approximately

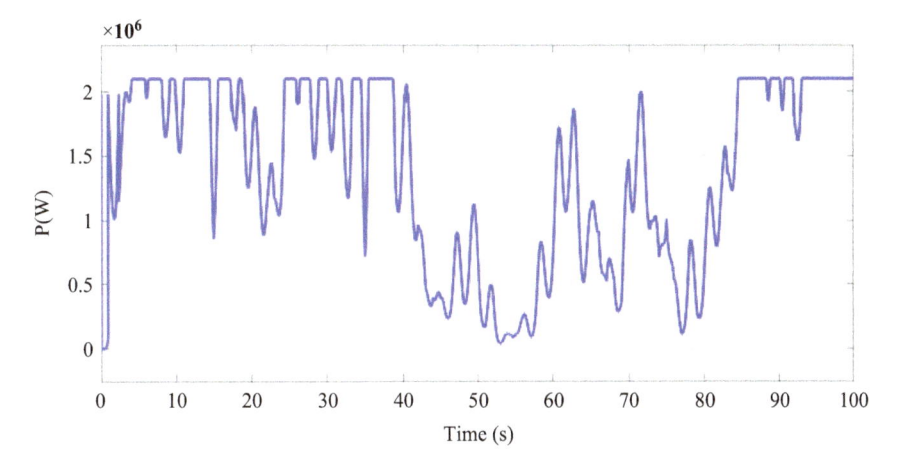

Figure 1.15   Captured power with nominal parameters

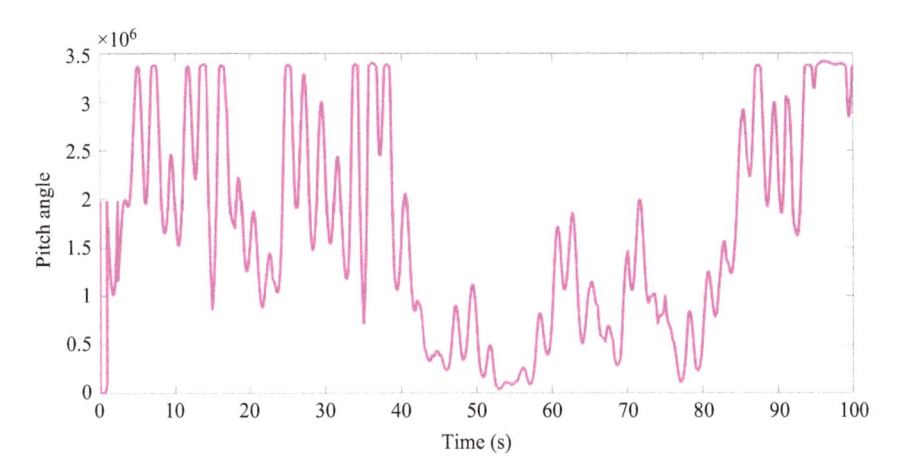

Figure 1.16   Pitch angle with nominal parameters (β)

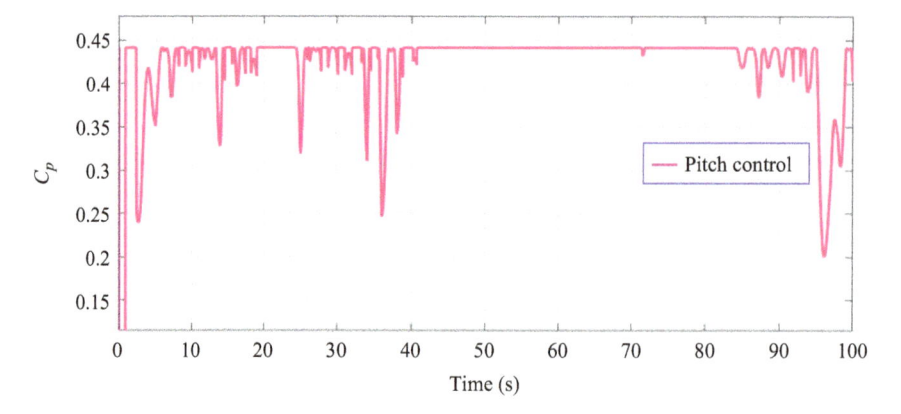

Figure 1.17 Power coefficient with nominal parameters (C_p)

10%: the aerodynamic torque estimation error is 9.5%, the electromagnetic torque estimation error is 9.4%, and the d-axis current estimation error is 10.2%.

1.6.4 Discussion

Discussion 1: The proposed method enables the estimation of unknown states, thereby reducing the number of sensors required in WECS and offering cost-effective benefits. The observer-based approach improves reliability by minimizing measurement errors caused by sensor noise.

Discussion 2: The newly applied polynomial observer in this study removes constraints on aerodynamic torque, presenting a significant difference from traditional methods. This enhancement increases the practicality and applicability of the method in real-world scenarios.

Discussion 3: Modeling errors resulting from linearization and parameter inaccuracies have been acknowledged. While the polynomial framework can address errors due to linearization, parameter errors may still affect the observer's estimation results. However, simulation results demonstrate that these errors do not significantly affect the tracking error in the system.

Conclusion: This study introduces a novel nonlinear observer based on polynomial theory for WECS. The proposed observer can estimate both aerodynamic torque and stator current, eliminating the need for wind speed measurement. This approach allows the control system to rely solely on rotor speed measurements from the generator.

Furthermore, reducing the number and type of sensors needed makes the system more reliable and cost-effective. To achieve optimal speed reference, we developed a well-known Linear Quadratic Regulator (LQR) controller and systematically determined its tuning parameters.

Additionally, conducted a comprehensive stability analysis for the LQR-based observer. Simulation results using MATLAB/Simulink software validated the effectiveness of the proposed observer and control system.

1.7 Chapter summary and new research opportunities

To summarize the main ideas in this chapter:

- The development of renewable energy, especially wind energy, is becoming a global and national trend in Vietnam as a substitute for traditional energy sources. Both domestic and international studies recognize Vietnam's large potential for wind power development. If the right strategies are implemented, wind power could make a significant contribution to the national energy balance in the next 15–20 years. After a prolonged period of research and discussion, Vietnam has launched its first two industrial-scale wind projects, connecting them to the national grid. Researching the operational parameters and the impact of these wind farms on local grids will provide valuable conclusions, laying the foundation for the future growth of wind energy.
- With abundant potential, Vietnam is focusing on wind power as a key part of its future energy strategy. According to Decision 428 on the adjustment of the National Power Development Master Plan for the 2011–2020 period, with a vision toward 2030, approved by the Prime Minister on March 18, 2016, Vietnam's wind power targets are 800 MW by 2020, 2,000 MW by 2025, and about 6,000 MW by 2030. The share of wind power in the total national electricity production is expected to reach 0.8% in 2020, 1% in 2025, and 2.1% in 2030. Currently, the total registered capacity of domestic wind projects has reached 5,700 MW, with four projects already operational, totaling 160 MW.
- Along with wind power development, the government also aims to build the renewable energy equipment industry, including wind power. According to the Renewable Energy Development Strategy by 2030, with a vision toward 2050 (Decision 2068, dated November 25, 2015), the target is to increase the domestic production ratio for renewable energy equipment to around 30% by 2020 and 60% by 2030. By 2050, this industry will not only meet domestic demand but also target exports to regional and global markets.

To efficiently operate a large amount of wind power capacity, it is necessary to establish wind power grid connection standards for the national electricity system. These standards will define voltage levels, permissible voltage deviations, frequency, power quality criteria, and the evaluation of the wind farm's impact on the local grid at the connection point. Chapter 1 provides an overview of global wind power development, wind power potential in Vietnam, and explores issues related to wind power grid connection. This chapter also considers methods for adjusting wind turbine operating modes and the impact of wind farms on the grid, with specific illustrations from the Tuy Phong wind farm in Binh Thuan. These studies play a vital role in establishing grid connection standards for wind farms in Vietnam.

Recommendations

- To sustainably develop wind power in Vietnam, appropriate investment policies, technical standards for grid connection and operation, and coordination

- with other renewable energy sources are needed to build a smart and sustainable energy system.
- The Vietnamese government has made initial subsidies for wind power, but the subsidy level should be increased to attract stronger investments.
- Wind power is a viable option for islands, in combination with solar, wave, and diesel energy, to provide stable electricity, supporting national sovereignty.
- A comprehensive proposal should be made regarding the wind power potential in Vietnam, focusing particularly on offshore WPPs. Furthermore, policies should be developed to encourage investor participation in this field.

The strategic importance for Vietnam is clear: learning from the global wind energy development experience will help Vietnam not only meet domestic energy needs but also promote economic growth, create jobs, and ensure environmental sustainability. Transitioning to wind energy is not only a shift in energy sources but also a strong commitment to a sustainable future, where economic development and environmental protection coexist and grow together.

Further reading

[1] General Department of Meteorology and Hydrology, (2007), "Wind Measurement Data," Hanoi.
[2] Dang Danh Hoang, Nguyen Phung Quang, (2011), "Design of a Passivity-Based Controller for Controlling Doubly-Fed Induction Generators," *Journal of Science & Technology of Technical Universities*, No. 76, Hanoi.
[3] World Bank, (2001), "*Wind Energy Resource Atlas of Southeast Asia*," True Wind Solutions, LLC.
[4] Electricity Regulatory Authority, (2014), Specialized Workshop "Developing Regulations for Wind Power Grid Connection and Renewable Energy Integration in Vietnam," Hanoi.
[5] Nguyen Duy Khiem, (2008), "Research on Wind Power System Control," Master's Thesis in Engineering, Hanoi University of Science and Technology, Hanoi.
[6] Ministry of Industry and Trade, (2012), Specialized Workshop, "Grid Integration and Operation Issues of Wind Power Plants," Hanoi.
[7] Ministry of Industry and Trade, (2010), "Circular on Distribution Electricity Systems," No. 32/2010/TT-BCT, Hanoi.
[8] Ministry of Industry and Trade, (2010), "Circular on Transmission Electricity Systems," No. 12/2010/TT-BCT, Hanoi.
[9] Prof. Dr. La Van Ut, (2009), "Analysis and Stability Control of Power Systems," Science and Technology Publishing House, Hanoi.
[10] Prof. Academician Dr. Tran Dinh Long, (2001), "System Theory," Science and Technology Publishing House, Hanoi.
[11] Ministry of Industry and Trade, Southeast Asia International Wind Energy Association, (2012), "Technical Workshop on Offshore Wind Power Development and Opportunities for Vietnam," Hanoi.

[12] Nguyen Duy Khiem, (2009), "Simulation of the Impact of Wind Power Plants in Binh Dinh Province on the National Grid Using MATLAB-Simulink," *Journal of Science and Technology of Technical Universities*, No. 71, Hanoi.

[13] Nguyen Thi Mai Huong, (2012), "Control Strategies to Enhance the Grid Stability of Wind Power Systems Using Doubly-Fed Induction Generators," Doctoral Dissertation, School of Industrial Technology - Thai Nguyen University, Thai Nguyen.

[14] Tuy Phong Wind Power Plant – Binh Thuan, (2013), "Technical Data Tables, Failures, and Operational Parameters from 2009–2013," Binh Thuan.

[15] Assoc. Prof. Dr. Tran Bach, (2008), "Power Grids & Electrical Systems – Volume 1," Science and Technology Publishing House, Hanoi.

[16] Assoc. Prof. Dr. Tran Bach, (2008), "Power Grids & Electrical Systems – Volume 2," Science and Technology Publishing House, Hanoi.

[17] Decision No. 37/2011/QĐ-TTg, (29/06/2011), "On Support Mechanisms for Wind Power Development Projects in Vietnam," Hanoi.

[18] Vietnam Electricity Group, (2013), "Technical Handbook on Connecting Wind Power to the Vietnamese Grid," Hanoi.

[19] Tran Dinh Long, Nguyen Sy Chuong, Le Van Doanh, Bach Quoc Khanh, Hoang Huu Than, Phung Anh Tuan, Dinh Thanh Viet, (2014), "Reference Book on Power Quality," Hanoi University of Science and Technology Publishing House, Hanoi.

[20] Energy Institute, (2010), "Electricity Planning and Development for the 2011–2020 Period with a Vision to 2030" (Plan 7), Hanoi.

[21] Vietnam's Eighth National Power Development Plan (PDP VIII).

[22] Ngo Dang Luu, Nguyen Hung, Nguyen Anh Tam, and Long D. Nguyen, (2021), "Wireless Power Transfer simulation for EV", Ministry of Culture, Sports and Tourism.

[23] Ngo Dang Luu, Nguyen Hung, and Long D. Nguyen, (2021), "Power flow calculation of the electric power system by MATLAB", Ministry of Culture, Sports and Tourism.

[24] Ngo Dang Luu, Nguyen Hung, Nguyen Anh Tam, and Long D. Nguyen, (2022), "Distance Relay Protection in AC Microgrid", Ministry of Culture, Sports and Tourism.

[25] Ngo Dang Luu, Nguyen Hung, Nguyen Anh Tam, and Long D. Nguyen, (2022), "Analysis of Solar Photovoltaic System Sunlight Shadowing", Ministry of Culture, Sports and Tourism.

[26] Ngo Dang Luu, Nguyen Hung, Le Anh Duc and Long D. Nguyen, (2022), "Estimate Frequency Response at the Command Line", Ministry of Culture, Sports and Tourism.

Chapter 2
Inexhaustible energy sources: the future of solar energy in energy scenarios

2.1 Potential and development prospects of solar energy in Vietnam: solar radiation coefficients in potential localities and some typical solar energy projects in Vietnam

2.1.1 Potential and development prospects of solar energy in Vietnam

2.1.1.1 Solar radiation coefficients in high-potential areas

Vietnam is considered a country with significant solar energy potential, particularly in the Central and Southern regions, where the average solar radiation intensity is approximately 5 kWh/m^2/day (1,825 kWh/m^2/year). In contrast, the Northern region experiences lower solar radiation intensity, estimated at around 4 kWh/m^2/day, due to frequent cloudy and drizzly weather conditions in the winter and spring months.

Solar radiation levels depend on cloud cover and the atmospheric conditions of each locality, leading to significant discrepancies in radiation levels across the country. Generally, the Southern region has higher radiation intensity compared to the North.

Provinces from Thua Thien–Hue northward receive an average of about 1,800–2,100 h of sunshine annually. Meanwhile, provinces from Da Nang southward enjoy 2,000–2,600 h of sunshine per year, with solar radiation intensity about 20% higher than the northern provinces. Therefore, the Central and Southern regions possess abundant solar energy resources that can be efficiently harnessed.

Northwest region: Sunshine is abundant in August, with the longest hours of sunlight occurring in April, May, September, and October. June and July receive the least sunlight due to thick clouds and heavy rain. The highest average daily solar radiation intensity reaches about 5.234 kWh/m^2/day, with an annual average of 3.489 kWh/m^2/day.

Highland areas (above 1,500 m): These areas generally receive less sunlight, with thick clouds and frequent rainfall, especially from June to January. The average solar radiation intensity is low (<3.489 kWh/m^2/day).

Northern and central northern regions:
- In the North, May is the month with the most sunshine.
- In the Central Northern region, the further south you go, the earlier and more abundant the sunlight, particularly in April.
- The highest solar radiation intensity in the North typically occurs in May, while in the Central Northern region, it peaks in April.
- February and March experience the lowest sunshine hours (around 2 h/day), while May sees the highest (6–7 h/day) and continues with high levels from July onwards.

Central region:
- From Quang Tri to Tuy Hoa, the longest sunshine duration occurs in mid-year, around 8–10 h/day.
- From March to September, the average sunshine duration is 5–6 h/day, with average solar radiation exceeding 3.489 kWh/m^2/day, and some days reaching up to 5.815 kWh/m^2/day.

Southern region:
- The southern provinces experience abundant sunshine year-round.
- During January, March, and April, the sunshine lasts from 7:00 AM to 5:00 PM.
- The average solar radiation intensity is greater than 3.489 kWh/m^2/day.

Particularly in Nha Trang, the solar radiation intensity exceeds 5.815 kWh/m^2/day for 8 months of the year.

Based on the solar energy potential map of Vietnam established and published on January 21, 2015, by three research institutes from Spain (CIEMAT, CENER, IDAE), Vietnam is assessed to have high solar radiation potential, with an average radiation level (Global Horizontal Irradiation – GHI) of 4 kWh/m^2/day in the Northern region and 5 kWh/m^2/day in the Central and Southern regions (Figure 2.1).

With its energy potential and the number of sunlight hours throughout the year, as shown in Table 2.1, Vietnam is considered to have great potential for the development of solar energy. Along with support from the government (ministries and sectors) and international organizations, many provinces and cities in Vietnam have successfully implemented the construction of solar power plants (PV) with various capacities to meet the needs of local communities, including remote areas and projects in regions without access to the national grid.

2.1.2 Notable solar energy projects in Vietnam

Dau Tieng Solar Power Plant: This plant is located on the semi-flooded land of Dau Tieng Lake, covering more than 504 ha with an installed capacity of 420 MWp. The total investment is over 9,100 billion VND. The project is located in Tan Chau District, Tay Ninh Province, and is a joint venture between Xuan Cau Co., Ltd (Vietnam) and B. Grimm Power Public Co., Ltd (Thailand). Upon operation, the plant is expected to supply about 688 million kWh of electricity annually to the national grid.

Inexhaustible energy sources 33

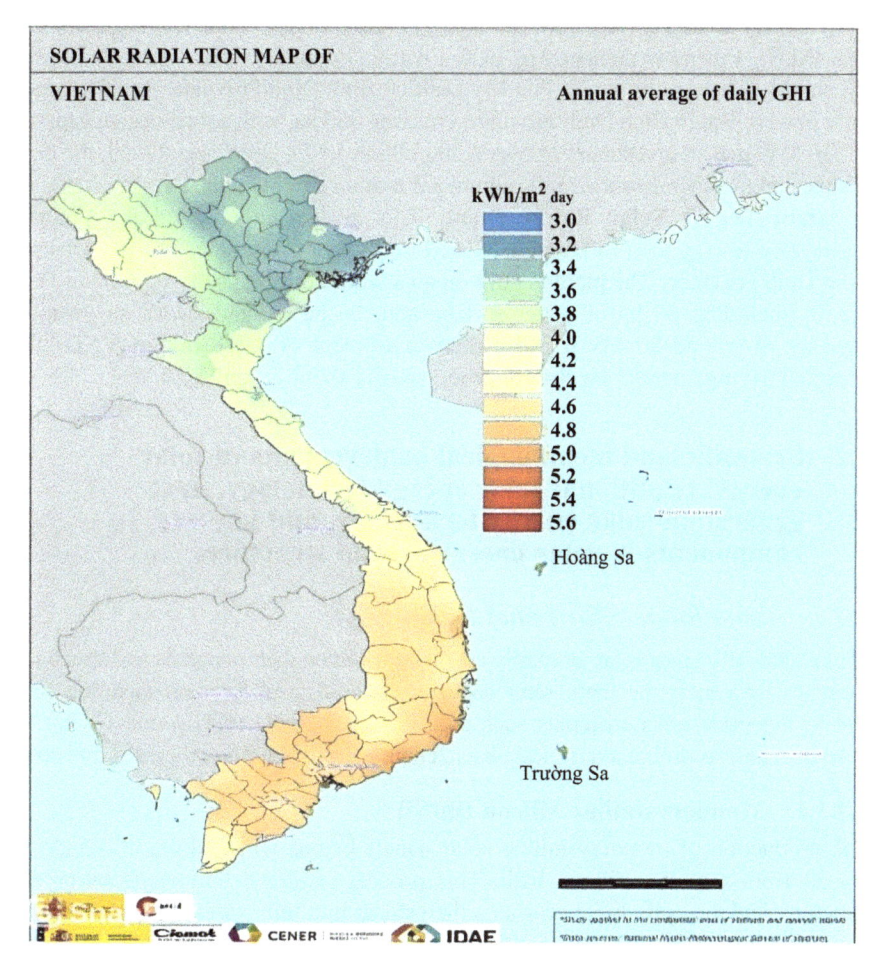

Figure 2.1 Map of average daily solar radiation in Vietnam (Source: Ministry of Industry and Trade)

Table 2.1 Solar radiation levels by region

Region	Annual sunshine hours	Solar radiation intensity (kWh/m^2/day)	Solar energy applications
Northeast	1,600–1,750	3.3–4.1	Average
Northwest	1,750–1,800	4.1–4.9	Average
North Central region	1,700–2,000	4.6–5.2	Good
Central Highlands and South Central Coast	2,000–2,600	4.9–5.7	Very good
Southern region	2,200–2,700	5.3–5.9	Very good
National average Top of Form	1,700–2,500	4.6	Good

34   *Clean energy in South-East Asia*

Phu My Solar Power Plant: Invested by Clean Energy Vision Development Co., Ltd. (NLS), a member company of BCG Energy Group. This plant is located in My An and My Thang communes, Phu My District, Binh Dinh Province. It is the largest solar power plant in Binh Dinh Province, covering 380 ha, with a total design capacity of 330 MW and an investment of over 6,200 billion VND. Once operational, the plant is expected to generate about 520 million kWh of electricity annually.

Dam Tra O Solar Power Plant: This project is invested by Vietnam Renewable Energy Co., Ltd. and is located in My Loi Commune, Phu My District, Binh Dinh Province. The plant is built on an area of approximately 60.6 ha of Dam Tra O (including 60 ha of water surface and 0.6 ha of land), with an installed capacity of around 50 MWp and a total investment of 1,440 billion VND. The expected average annual electricity production is about 78 million kWh.

2.2   Scientific and technological achievements in solar energy: traditional solar energy technology, next-generation solar energy technology, and key components in solar energy system structures

2.2.1   *Traditional solar panel technology*

Photovoltaic (PV) panels are primarily classified based on their materials and structures. Currently, silicon is the main semiconductor material used for manufacturing solar panels, alongside other materials such as cadmium telluride (CdTe) and CIGS/CIS. Each material has different technical characteristics, manufacturing processes, and costs.

2.2.1.1   Monocrystalline silicon (m-Si)

The production of monocrystalline solar panels begins with pulling a cylindrical crystal from a molten silicon bath. This process creates a homogeneous crystal structure, and the cylindrical crystal is then sliced into thin wafers with thicknesses ranging from 120 to 200 μm, commonly referred to as "wafers".

- The cell structure is uniformly arranged in one direction, allowing monocrystalline cells to achieve high efficiency.
- These cells have a consistent black color due to their effective light absorption capabilities (Figure 2.2).

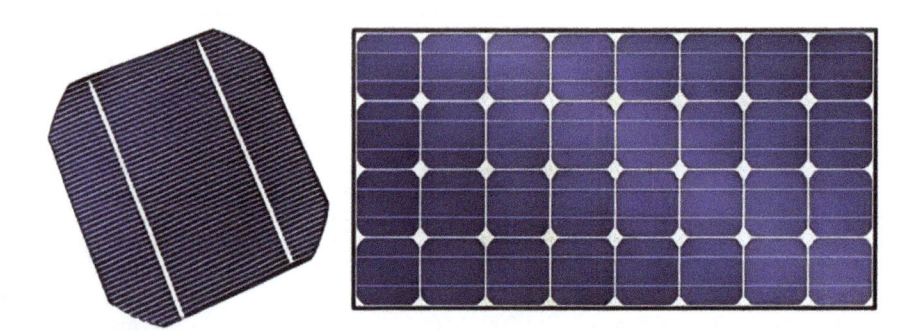

Figure 2.2   Monocrystalline cell and monocrystalline solar panel

2.2.1.2 Monocrystalline cells

Monocrystalline cells typically have an octagonal shape, which results in noticeable gaps at the corners of the solar panel. These cells generally have higher efficiency. Each silicon cell has an output voltage of about 0.5 V_{DC}. Solar panels are designed by connecting these cells in series and parallel configurations. Specifically, monocrystalline or polycrystalline solar panels are commonly sold with configurations of 60 or 72 PV cells, with rated voltages of 30 and 36 V_{DC}, respectively.

Most crystalline solar panels are built with glass faces ranging from 3.2 to 4 mm thick and aluminum frames (ranging from 35 to 50 mm thick) to increase durability. Electrical connections to the junction box are identified by three bypass diodes, and the cables are pre-terminated with quick-connect plugs.

Advantages
- **High efficiency**: Monocrystalline solar panels offer the highest efficiency as they are made from the highest quality silicon.
- **Space-saving**: These panels are space-efficient, requiring less area compared to other types due to their higher output power.
- **Long lifespan**: Monocrystalline panels have the longest lifespan, with most manufacturers offering a 25-year warranty.
- **Performance in low light conditions**: These panels perform better under low-light conditions compared to polycrystalline panels with the same power rating.

Disadvantages
- **Cost**: Monocrystalline solar panels are typically more expensive than panels made from polycrystalline silicon.
- **Sensitivity to shading**: If the panel is partially shaded by objects such as trees, dirt, or snow, the performance of the entire circuit may decrease. In cases of shading issues, it is recommended to use micro-inverters instead of central inverters.
- **Temperature performance**: While monocrystalline panels are more efficient at high temperatures, they have slightly lower thermal tolerance compared to polycrystalline panels. They are better suited for cooler climates.

2.2.1.3 Polycrystalline silicon (p-Si)

In contrast to monocrystalline silicon, polycrystalline PV cells are made from surplus silicon (Figure 2.3).

2.2.1.4 Polycrystalline cells

Polycrystalline PV cells are made from molten silicon, which is then cooled and solidified into blocks, forming crystals with varying purity levels. These silicon blocks are then sliced into smaller pieces. Due to the crystals having different orientations, the surface of polycrystalline cells resembles a patchwork, with some energy loss occurring at the boundaries between crystals. As a result, polycrystalline cells typically have lower efficiency compared to monocrystalline cells.

Figure 2.3   Polycrystalline photovoltaic cell and polycrystalline solar panel

However, because the manufacturing cost of polycrystalline cells is slightly lower, solar panels made from polycrystalline cells are less expensive than monocrystalline panels. This makes polycrystalline panels more popular in large-scale solar energy applications.

Advantages
- **Simpler and cheaper production process**: The process of producing polycrystalline silicon is simpler and more cost-effective, with less silicon waste compared to monocrystalline production.
- **Better performance at high temperatures**: Polycrystalline panels tend to perform better than monocrystalline panels in high-temperature conditions.

Disadvantages
- **Lower efficiency**: Polycrystalline solar panels typically have an efficiency range of 13–17%. Due to lower silicon purity, they are less efficient than monocrystalline panels.
- **Lower space efficiency**: These panels require a larger surface area to generate the same amount of electricity as monocrystalline panels. However, this does not mean that all monocrystalline panels are more efficient than polycrystalline panels.
- **Aesthetic differences**: Monocrystalline and thin-film panels are often considered more aesthetically pleasing due to their uniform appearance, in contrast to the deep blue color and patchwork-like design of polycrystalline silicon.

2.2.1.5   Thin-film solar panels (thin-film PV modules)

Thin-film technology uses semiconductor materials that operate in thin layers, which are deposited onto a substrate or backing layer. The most common substrate material is glass. Thin-film PV cells are typically sandwiched between two layers of glass to form a solar panel. The main semiconductor materials currently used in the market include:

- Amorphous silicon (a-Si)
- Cadmium telluride (CdTe)

- Copper indium selenide (CIS)
- Copper indium (Gallium) di-selenide (CIGS/CIS)

Thin-film solar panels can be classified into two types: rigid and flexible. Rigid thin-film PV cells and panels are produced on the same production line. The primary semiconductors used in this technology are CdTe, a-Si, and CIGS.

Applications
- Rigid thin-film panels are commonly used in large-scale power plants and public infrastructure projects.
- Flexible thin-film panels are optimized for mobile and portable applications. The main manufacturers of thin-film panels include First Solar (using CdTe technology) and Solar Frontier (using CIS technology).

Advantages
- **Simplified mass production**: Thin-film solar panels are easier and cheaper to produce compared to crystalline-based PV cells.
- **Uniform appearance**: They have a uniform appearance, making them more aesthetically pleasing.
- **Flexibility**: Thin-film panels can be made flexible, opening up new applications.
- **Less affected by high temperatures and shading**: Thin-film panels are less impacted by high temperatures and shading, which can affect the performance of other types of solar panels.

Disadvantages
- **Higher cost**: Despite these advantages, thin-film panels are still relatively expensive.

2.2.1.6 Amorphous silicon

Amorphous silicon (a-Si) is the oldest thin-film solar technology. To manufacture an a-Si solar panel, glass is used as the substrate, and a thin layer of silicon is deposited onto it using a method called plasma-enhanced chemical vapor deposition (Figure 2.4).

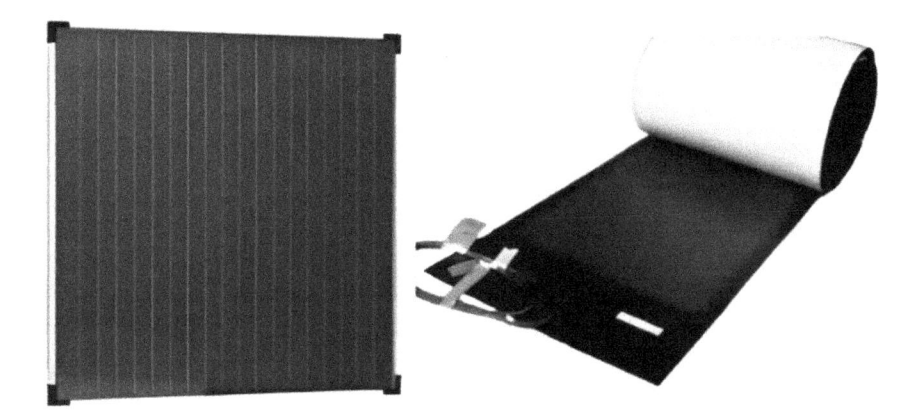

Figure 2.4   Amorphous silicon solar panel (rigid and flexible)

2.2.1.7 Copper indium selenium solar cells

Copper indium selenium (CIS) solar cells use an alloy of copper, indium, and selenium as the semiconductor material. When gallium is added to this alloy, it forms copper indium gallium selenide (CIGS). Although CIGS is the most common variant, both CIS and CIGS are often grouped together under the same thin-film solar cell technology.

Key features
- **High light absorption**: CIS and CIGS materials have a high light absorption coefficient, meaning they can absorb sunlight more efficiently. This allows the production of thinner solar cells compared to other semiconductor materials.
- **Compatibility with flexible substrates**: The thin nature of the material makes it possible to apply these materials onto flexible substrates. However, the best performance is typically achieved when these materials are applied to glass due to the high-temperature deposition process used.

Advantages of a-Si solar cells:
- **Low production cost**: a-Si cells are cheaper to produce than crystalline silicon (c-Si) technology. This is because they require only about 1% of the silicon compared to a typical c-Si cell.
- **Multilayer structure for optimized wavelength absorption**: a-Si cells can be stacked in multiple layers, with each layer optimized for a specific wavelength range. The first layer absorbs the shortest wavelengths, while longer wavelengths pass through it and are absorbed by subsequent layers. This design, known as dual-layer or triple-layer a-Si cells, increases the efficiency by expanding the usable wavelength range of light.

2.2.1.8 Amorphous silicon (a-Si) solar modules

a-Si solar panels can be either rigid or flexible, with the flexibility allowing them to be adapted for mobile applications, curved surfaces, or spaces with unique design requirements.

2.2.1.9 Next-generation solar technology: half-cell technology

Recently, solar panel manufacturers have shifted from traditional full-cell panels to half-cell technology for their premium product lines. This transition provides significant benefits in terms of performance, efficiency, and lifespan of solar panels.

- **Cell size**: Half-cell panels are 3×6 inches, which is half the size of traditional full cells (6×6 inches).
- **Number of cells**: Each panel using half-cell technology contains twice the number of cells compared to traditional full-cell panels of the same size. For example, a typical 2×1-m solar panel would have 72 cells in a full-cell configuration, whereas the half-cell version of the same panel would have 144 cells.

Benefits of half-cell technology
- **Enhanced efficiency**: This technology helps reduce performance losses due to shading or obstructions because each cell is smaller, allowing for more even electrical load distribution.
- **Increased durability**: Half-cell panels are less affected by temperature changes and physical stress, extending the lifespan of the solar panel.
- **Better performance in shaded conditions**: When there is shading or obstruction, the performance of half-cell panels is less impacted compared to full-cell panels.
- **Easier transportation and installation**: With fewer cells and higher panel density, half-cell technology reduces the weight of panels, making them easier to transport and install.

Summary
- Half-cell technology is becoming the new trend in the solar energy industry, improving the performance, efficiency, and durability of solar panels. This change not only increases the energy output but also extends the lifespan of the panels, making them ideal for high-performance applications.

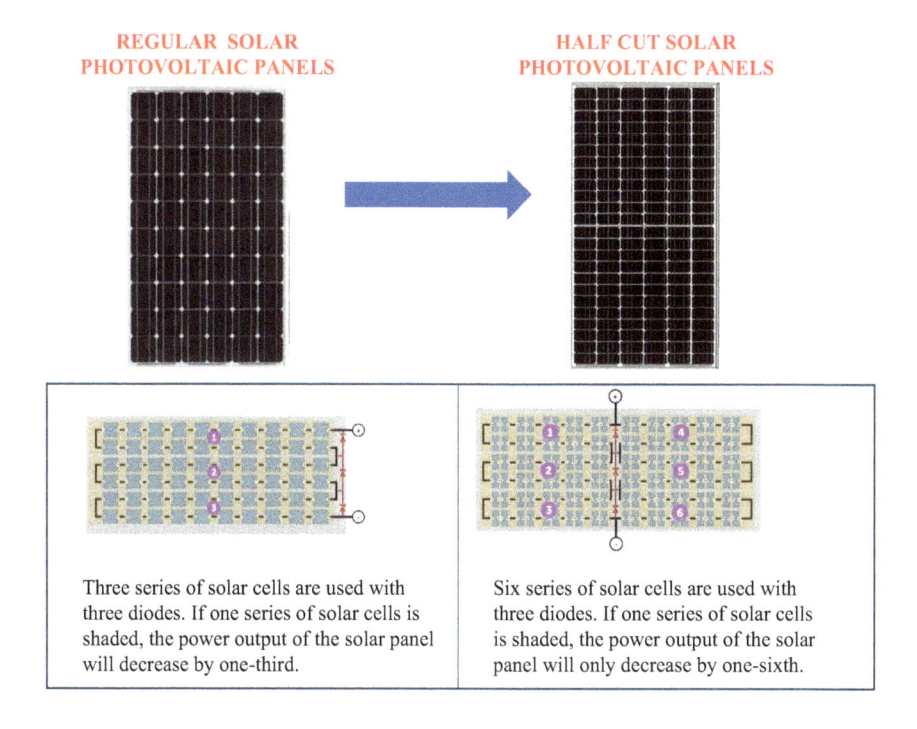

REGULAR SOLAR PHOTOVOLTAIC PANELS → **HALF CUT SOLAR PHOTOVOLTAIC PANELS**

Three series of solar cells are used with three diodes. If one series of solar cells is shaded, the power output of the solar panel will decrease by one-third.	Six series of solar cells are used with three diodes. If one series of solar cells is shaded, the power output of the solar panel will only decrease by one-sixth.

Key advantages of half-cell solar panels
- **Reduced current intensity on busbars**: Half-cell solar panels use PV cells that are split in half, improving energy efficiency. Traditional solar panels with 60 and 72 PV cells will have 120 and 144 half-cells, respectively. When the PV

cells are split, the current flowing through them is halved, which reduces resistive losses and allows the panel to produce more energy. According to the loss formula: $P = I^2 \times R$, when the current (I) is halved, the loss is reduced by a factor of four.

- **Smaller busbar size increases light absorption efficiency**: Busbars are metallic strips on the surface of solar panels that carry current but partially block light, reducing light absorption efficiency. Reducing the busbar size enhances the light absorption efficiency of the PV cells.
- **Optimal performance under shaded conditions**: In traditional full-cell solar panels, the cells are often connected in series. Therefore, if one cell is shaded, the output power of the entire panel can be significantly reduced. Manufacturers have addressed this by dividing the cells into three series and using three diodes to separate the shaded series. As a result, if one cell is shaded, the power of the panel only decreases by one-third. However, in half-cell panels, the cells are divided into six series. If one cell is shaded, the power of the panel only decreases by one-sixth. Therefore, under the same shading conditions, the performance of half-cell panels is only halved compared to full-cell systems.
- **Increased lifespan and durability**: Experiments have shown that smaller PV cells are less affected by mechanical stress, reducing the likelihood of cracking. Half-cell solar panels provide higher and more reliable power output than traditional solar panel technology. As a result, half-cell panels not only improve performance but also enhance the durability of the solar system. When one solar cell is shaded, the energy produced by the non-shaded cells in the same series can flow into the shaded cell, generating heat. If this situation persists for a long period, it could damage the solar panel. For half-cell panels, shorter electrical paths, more busbars, and reduced current intensity in the busbars help minimize the heat generation caused by shaded cells.
- **Dual-glass solar panels**: Dual-glass solar panels use internal circuit connections similar to traditional framed solar panels by connecting 6×10, 6×12, or 6×24 cells in series. However, these cells are encapsulated in tempered glass on both sides. The rear glass is manufactured using a unique drilling technique to ensure reliability during the junction box and solar panel installation process. Compared to traditional panels, dual-glass panels can have or lack a frame and replace the organic backsheet with glass, enhancing the product's lifespan. From this perspective, the structural design of dual-glass solar panels overcomes outdoor issues such as material degradation leading to aging and frequent energy loss, which are common in traditional panels. Furthermore, dual-glass panels can avoid low performance and high damage, which are often considered inherent weaknesses of traditional solar panels.
- **Structure**: Traditional solar panels consist of three main layers: a tempered glass top layer, a middle cell layer, and a Tedlar–PET–Tedlar (TPT) backsheet. For dual-glass solar panels, the TPT backsheet is replaced with another tempered glass layer, similar to the top layer. This replacement increases the stiffness and durability of the panel. The PV cells are placed between the two glass layers, minimizing the risk of weather impact, reducing the likelihood of small cracks,

Inexhaustible energy sources 41

and reducing energy loss over time. The glass used in these panels can withstand hailstones up to 2.5 cm in diameter. The thickness of the glass used by manufacturers is only 2 mm, making it very thin. This design ensures that dual-glass panels weigh the same as traditional panels, with the total glass thickness being only 4 mm (two 2-mm glass layers) (Figure 2.5).

Dual-glass solar panels come in various sizes, each corresponding to a different power range to meet the diverse needs of customers. With two layers of glass, these panels can withstand impacts and reduce the risk of cracking.

The absence of an aluminum frame in dual-glass solar panels makes them ideal for replacing traditional roofing materials. They can be used for roofing glass houses, organic vegetable farms, villa rooftops, carports, and more, offering a unique, environmentally friendly, and aesthetically pleasing solution.

These panels come with a technical warranty of up to 12 years from the manufacturer and a performance warranty of over 83% for 30 years, backed by a reputable international insurance company (Figure 2.6).

Figure 2.5   Schematic structure of a double glass module

Figure 2.6   Actual installation image

2.2.2 Structure and technical specifications of main components in solar power systems

Currently, the primary material used in the production of PV solar panels is silicon, which is divided into three types:

- **Monocrystalline silicon**: This type offers high efficiency and is usually more expensive as it is cut from cylindrical silicon blocks.
- **Polycrystalline silicon**: Made from molten silicon blocks that are cooled and solidified. It is cheaper than monocrystalline silicon but has lower efficiency.
- **Silicon ribbon**: Created from thin layers of molten silicon with a polycrystalline structure. It has the lowest efficiency but is the least expensive because it does not require cutting from silicon blocks.

2.3 Efficient solar power system models: stand-alone and grid-tied solar power systems

2.3.1 Stand-alone solar power systems

Operating principle: Solar energy is absorbed and converted into direct current (DC) by solar panels. This DC electricity is then stored in batteries through a solar charge controller. The DC from the batteries is converted into alternating current (AC) by an inverter, which is used to power loads. The total load capacity must be less than the inverter's output power. The duration for which loads can be powered depends on the solar energy generated and the battery capacity (Figure 2.7).

Applications: Suitable for areas without grid access or where grid electricity costs are high.

Advantages

- **Independence**: Operates independently from the grid.

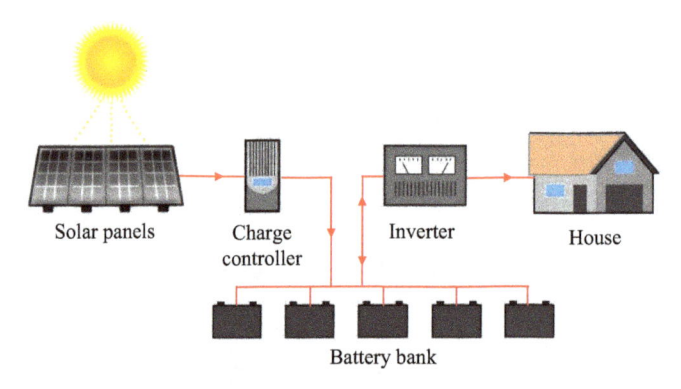

Figure 2.7 Diagram of an off-grid solar power system

Disadvantages

- **Load limitation**: Limited to small loads due to power capacity restrictions.
- **Weather dependency**: Efficiency depends on weather conditions.
- **Maintenance**: Maintenance and repairs can be challenging.
- **Cost**: Higher initial investment cost.

2.3.2 Grid-tied solar power system

Operating principle: In a grid-tied solar power system, the DC generated by solar panels is converted into AC by a grid-tied inverter. This AC electricity is then fed directly into the electrical grid, operating in parallel with the power from the utility company. The system can handle the following scenarios:

- **Solar energy meets or exceeds load demand**: Scenario: If the solar power output is greater than or equal to the load's energy consumption, the excess electricity is fed into the grid and can potentially be sold back to the utility company.
- **Solar energy below load demand**: Scenario: If the solar power output is less than the energy consumption, the load will first use the available solar energy. Any additional electricity needed will be drawn from the grid.
- **Grid power outage**: Scenario: In the event of a grid power outage, the solar power system will also stop operating due to safety requirements from the utility company. This prevents backfeeding electricity into the grid and ensures the safety of utility workers.

Applications: Grid-tied solar power systems are suitable for all facilities that use electricity, as they help reduce the amount of electricity purchased from the utility company.

Advantages:

- **Ease of use and maintenance**: These systems are generally easy to operate and maintain.
- **Automatic operation**: The system operates automatically and is not dependent on the load's electricity consumption.
- **Cost-effective**: The initial installation cost is lower compared to other types of solar power systems.
- **Return on investment**: Provides the best return on investment by reducing electricity bills and potentially selling excess electricity back to the grid.

Disadvantages:

- **Grid dependency**: The system does not operate during a grid outage, even if sunlight is available.
- **Utility rate dependence**: The system's effectiveness depends on the rate at which the utility company compensates for the excess electricity fed into the grid (Figure 2.8).

44　*Clean energy in South-East Asia*

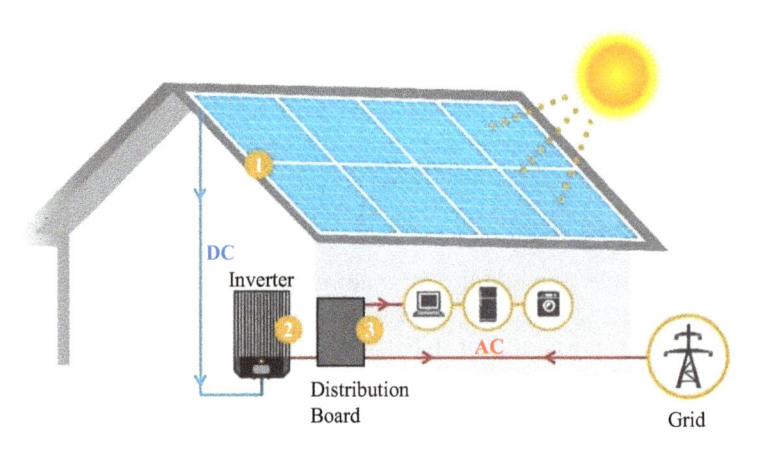

Figure 2.8　Diagram of a grid-tied solar power system

2.3.3　Hybrid grid-tied solar power systems

Operating principle: A hybrid grid-tied solar power system combines the features of both off-grid and grid-tied solar power systems. It operates similarly to a grid-tied system but with the added flexibility of an energy storage system. The operation is as follows:

- **Normal operation**: During normal grid operation, the system functions as a grid-tied system. The energy generated by the solar panels is converted into AC and used to supply power to the load. Any excess electricity is fed into the grid.
- **Grid power outage**: In the event of a grid power outage, the hybrid system continues to operate normally. The electricity generated by the solar panels is first used to power the load. Any excess energy is stored in the battery, or if there is a shortfall, the system will draw power from the battery to supply the load.

Applications: The hybrid system is suitable for all types of electrical facilities. It is particularly effective for households where a stable power supply and energy storage are highly valued.

Advantages

- **Combined benefits**: Addresses the limitations of both off-grid and grid-tied systems.
- **Uninterrupted power supply**: Provides continuous power to critical devices even during a grid outage. The transition to backup mode occurs in less than 20 milliseconds, ensuring no interruption.
- **Energy storage**: Excess energy generated during the day is stored in batteries for use at night or during a power outage, optimizing solar energy usage and preventing energy waste.

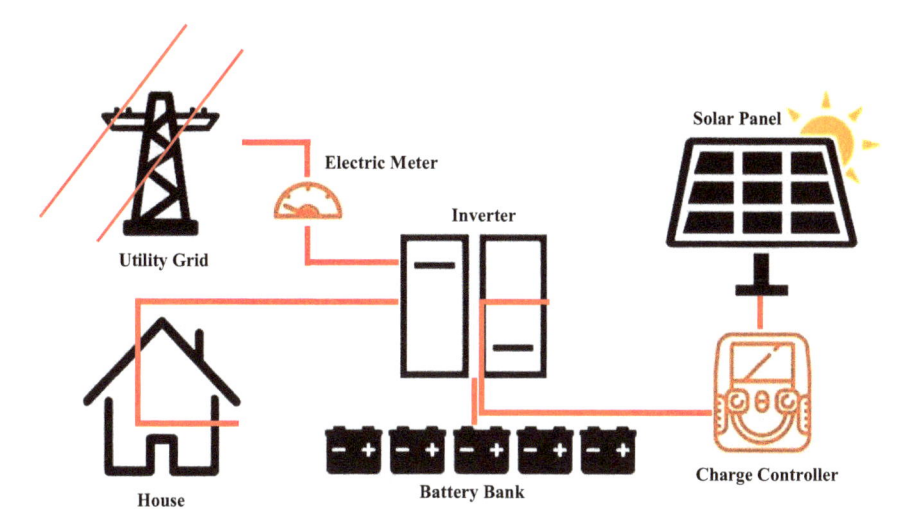

Figure 2.9 Diagram of a hybrid grid-tied solar power system

Disadvantages

- **High cost**: The addition of an energy storage system (batteries) significantly increases the overall cost of the system.
- **Limitations for high loads**: The system may be less efficient for very high power demands due to storage limitations (Figure 2.9).

Description: This diagram will show the configuration of a hybrid solar power system, including:
- **Solar panels**: Convert sunlight into DC electricity.
- **Hybrid inverter**: Converts DC to AC and manages both grid-tied and battery storage functions.
- **Battery**: Stores excess energy generated during the day for use at night or during a power outage.
- **AC loads**: Devices and appliances that use AC electricity.
- **Grid**: Connects to the system, allowing for energy exchange.
- **Net meter**: Measures the flow of electricity into and out of the grid.
- **Transfer switch**: Ensures a smooth transition between the grid power source and battery backup.

2.3.4 Rooftop solar power systems for self-generation and self-consumption: a new direction for connection to the national grid in Vietnam

The rooftop solar power system for self-generation and self-consumption is currently developing rapidly, with the capability to connect to the national grid. This is an effective solution to harness renewable energy sources and meet domestic energy consumption needs. The current regulations specify the implementation process as follows:

2.3.4.1 General regulations

- **Registration or notification**: Organizations or individuals must notify or register with the relevant authorities when implementing the system.
- **Excess power generation option**: The investor can choose whether or not to feed excess electricity into the national grid.
- **Development based on scale**: Installation must be based on the actual scale and electricity consumption needs.

2.3.4.2 Specific regulations by installation capacity

In other cases, the investor must comply with Item b, Clause 1, Article 8 of the relevant Decree.

(a) **Systems under 100 kW**

Organizations, individuals, or households installing rooftop solar power systems with a capacity under 100 kW must notify the relevant authorities using Form No. 04 (Appendix of the Decree).

(b) **Systems from 100 kW to under 1,000 kW:**
- **Notification and design documents**: The investor must send a notification (Form No. 04) along with design documents to the management authorities, such as:
  - Department of Industry and Trade;
  - Building and Fire Prevention and Fighting Authorities (PCCC);
  - Local power utility company.
- Connection conditions:
  - If the system capacity is within the planned and allocated deployment of the local area, the investor is allowed to sell excess electricity to the national grid as per regulations.

2.3.4.3 New highlight: toward a flexible and modern connection solution

A new form of connection is being encouraged in Vietnam: the rooftop solar system combined with smart monitoring and control technology.

- **Integration of modern equipment**: Systems with a capacity of 100 kW or more must be equipped with connection, monitoring, and control devices according to technical standards published by Vietnam Electricity (EVN).
- **IoT technology application**: These solutions allow remote management, optimize system performance, and ensure operational safety.
- **Increased flexibility**: Supports both self-consumption mode and excess power generation into the national grid, depending on the user's actual needs.

2.3.4.4 Proposed expansions

To optimize solar power utilization, the relevant authorities can:

- **Simplify procedures**: Streamline the registration and notification process for systems under 100 kW.

- **Encourage policy support**: Implement financial support mechanisms or tax incentives for households and organizations installing solar power systems.
- **Research on energy storage integration**: Combine energy storage solutions, such as batteries, to enhance system stability and provide electricity during the evening or on cloudy days.

This new form of connection will not only help promote renewable energy but also contribute to building a sustainable energy system, meeting the economic and social development needs of Vietnam in the new phase.

2.4 Solar power with grid-tied rooftop systems: a power source for smart homes, automatic navigation solar power systems, and agricultural irrigation systems

Grid-tied rooftop solar power system – providing electricity for smart homes

Project introduction

The rooftop solar project installed on the rooftop of the Hung Long Joint Stock Company's factory spans approximately 6,738 m^2. To maximize the use of the existing rooftop structure, the investor is building a rooftop solar power system with a capacity of 998.73 kWp.

Project details

- **Location**: Hung Long Solar Power Project, Long Cang Commune, Can Duoc District, Long An Province.
- **System capacity**: 998.73 kWp.

Objective

- To utilize the available space on the factory's rooftop for solar power installation, contributing some of the electricity to the local grid.
- To reduce the impact of temperature from weather on the factory's roof.

Project benefits

- **Energy production**: By installing a large-scale solar power system, the project generates renewable energy, contributing to the local electricity supply and reducing reliance on non-renewable energy sources.
- **Roof temperature reduction**: The solar panels help insulate the building, reducing heat absorption from the sun, which can lower cooling costs and improve overall comfort.
- **Sustainability**: The project supports sustainable energy practices and demonstrates a commitment to environmental protection through the use of renewable resources (Table 2.2).

Connection plan

Based on Document No. 352/ĐLLT-KT dated February 26, 2020, from Ben Luc Power Company, the connection plan for the project requires the investor to build a new substation (TBA) with a capacity of 1,000 kVA. This substation will connect

Table 2.2 Project investment scale

No.	Parameter	Characteristics – quantity
1	Installed capacity	998.73 kWp
2	Number of solar panels (405-Wp capacity from AE Solar, produced using German technology)	2,466 panels
3	Number of 75-kW inverters from SMA, produced using German technology	11 units
4	Transformer station 1,000 kVA-22/0.4-kV and 22-kV connection line	01 system
5	AC, DC electrical wiring system and other connection accessories...	01 system
6	Lightning protection system	01 system

Table 2.3 Technical specifications of the proposed solar panels

	Parameter STC	
	Specification	
Nominal power	P_{MAX} (Wp)	405
Voltage at maximum power PMPP	V_{MP} (V)	41.10
Current at maximum power PMPP	I_{MP} (A)	9.86
Open circuit voltage	V_{OC} (V)	49.1
Short circuit current	I_{SC} (A)	10.37
Module efficiency	%	20.2
Maximum system voltage	V_{SYS} (V_{DC})	1,500
Dimensions	L×W×H (mm)	2,015×996×30 mm
Weight	kg	26.0
	Operating	
Temperature range	°C	−40 to +85
Temperature coefficient of P_{MAX}	P_{MAX} (W)	−0.35%/ °C
Temperature coefficient of V_{OC}	V_{OC} (V)	−0.25%/ °C
Temperature coefficient of I_{SC}	I_{SC} (A)	+0.04%/ °C

to the company's 22-kV medium-voltage grid through a bi-directional meter (provided by the power company).

Solar power system
The project uses high-efficiency monocrystalline silicon solar panels with a lifespan of over 20 years, each panel having an output power of 405 Wp. With a total capacity of 998.73 kWp, the project will use approximately 2,466 solar panels. The panels are connected in series, and these series are then connected in parallel to form arrays. The arrays are also connected in parallel to form blocks. These blocks connect to the DC string junction boxes and then to 75 kW DC/AC inverters to convert the electricity to AC. There will be 11 inverters, each with a capacity of 75 kW (Table 2.3).

Table 2.4 Technical specifications of the inverter station

Technical specifications	SHP 75-10
A. DC input	
Parameter specification maximum input voltage	1.000 V
Startup voltage	600 V
Nominal input voltage	630 V
Voltage range for nominal power	570–800 V
Number of MPPT	1
Maximum number of strings per MPPT	
Maximum input current per MPPT	140 A
Circuit current	210 A
B. AC output	
Parameter specification AC output power	75 kW
Maximum output current	109 A
Nominal AC voltage	400 V
Voltage range	360–530 V
Frequency	50 Hz
Power factor	
Maximum efficiency	98.80%
C. General Specifications	
Parameter specification dimensions ($W \times H \times D$)	570×740×306 mm
Weight	77 kg
Protection rating	IP65
Operating temperature	−25 to 60

Inverter DC/AC

The inverter is a critical component in a solar power project, responsible for converting the DC generated by the solar panels into AC. The number of inverters required depends on the installation capacity and the system design.

For this project, with a total capacity of 998.73 kWp, a 75-kW inverter is recommended. The inverters will be connected to the low-voltage side of a step-up transformer (0.4/22 kV—1,000 kVA) (Table 2.4).

Project layout solution
- **Long Thanh 1 Solar Power Project** has been designed with a carefully planned connection strategy to ensure efficiency and safety. The connection plan includes the construction of a new 1,000 kVA substation, which will be connected to the medium-voltage 22 kV grid via a bi-directional meter, according to the document No. 352/ĐLLT-KT dated February 26, 2020, from Ben Luc Power Company.
- **Solar panel system**
  - **Panel type**: The system utilizes high-efficiency monocrystalline silicon solar panels with a lifespan of over 20 years, with each panel having a power output of 405 Wp.
  - **Total installed capacity**: The project has a total installed capacity of 998.73 kWp, requiring approximately 2,466 solar panels.

- **Panel arrangement**: The panels are installed on the factory roof, primarily arranged on both sides of the roof. Each panel string consists of 18 panels connected in series, with a total of 137 strings.
  - The tilt angle of the solar panels is set at 8.5°, matching the roof's existing angle.
  - The azimuth of the panels is aligned with the roof's orientation to optimize solar energy capture.

- **Solar panel array-inverter layout**
  - **Panel strings**: Each string consists of 18 panels connected in series. The system uses 11 DC combiner boxes, with varying input connections:
    - 8 boxes with 12 inputs each.
    - 1 box with 13 inputs.
    - 2 boxes with 14 inputs each.
  - **Connection cables**:
    - 1C × 4 mm^2 cable from the solar panels to the DC combiner boxes.
    - 2 × 1C × 70 mm^2 cables connecting the DC combiner boxes to the inverters.
    - The inverter output is connected to the AC combiner box via underground copper cables with XLPE insulation, 3 × (4 × (CXV/DSTA–240 mm^2–0.6/1 kV) + 1×(2×(CXV/DSTA – 240 mm^2–0.6/1 kV)), which connects to the terminals of the step-up substation (22/0.4 kV—1,000 kVA).

- **Lightning protection system**
  - **LIVA AX 210 lightning rod**: The system uses the LIVA AX 210 lightning rod to provide direct protection for the entire factory area, with a protection radius of 131 m (Level IV protection).
  - **Lightning discharge system**: The lightning discharge system uses M95 copper cables connected from the lightning rod to the substation grounding network. The grounding is done with galvanized steel poles (Φ12 and 63×63×6 mm), 2.5 m in length, welded together to ensure good grounding.

- **Grounding system**
  - **Insulated copper wire**: The grounding system uses 25 mm^2 insulated copper wire connected to the substation grounding network. The grounding resistance must not exceed 4 Ω, as per current regulations.
  - **Equipment grounding**: All solar panels, inverters, and electrical panels are grounded to ensure safe operation for both personnel and equipment.

- **Surge protection devices**
  - **Low-voltage electrical panel**: The system is equipped with surge protection devices (SPDs) in the low-voltage electrical panel to protect against electrical faults.

- **Inverter protection**: SPDs, designed by the manufacturer, are also installed at the inverter's input and output terminals.
- **Project power generation simulation results**
  - **Average power generation**: The main simulation results show the actual average power output of the plant supplied to the grid, measured hourly in different months. These results are presented in tabular form.
- **Annual power generation chart**
  - The chart illustrates the annual power generation provided, offering an overview of the project's performance and its contribution to the regional grid.
  - This design and layout ensure that the Long Thanh Solar Power Project operates efficiently, sustainably, and contributes significantly to regional electricity supply, while minimizing environmental impacts.
- **Key operating parameters for rooftop grid-connected solar power system**
  - **Grid connection capacity of the rooftop solar power system**: The capacity at which the rooftop solar power system connects to the national grid is an essential parameter. This value represents the maximum power the solar system can supply to the grid. For example, a system may have a nominal connection capacity of 998.73 kWp, indicating the amount of solar power that can be supplied to the grid under optimal conditions.
  - **Power generation of the rooftop solar power system**: The total amount of electricity the rooftop solar system generates over a specific period, usually measured in kilowatt-hours (kWh). This output depends on factors such as solar radiation, the efficiency of the panels and inverters, and environmental conditions. For example, with the given capacity and local sunlight conditions, the system may generate thousands of kilowatts per year.
- Operating efficiency of solar panels, inverters, and the entire solar system:
  - **Solar panel efficiency**: The ratio of the electrical output of a solar panel to the incoming solar energy, typically expressed as a percentage. High-efficiency monocrystalline silicon panels can have efficiencies of up to 20% or more.
  - **Inverter efficiency**: The ability of the inverter to convert DC electricity from the solar panels into AC electricity with minimal loss, typically around 96–98% for high-quality inverters.
  - **Overall system efficiency**: The combined efficiency of all components in the solar power system, including panels, inverters, cables, and other electrical equipment. This is a crucial indicator for assessing the overall performance and power generation of the system.
  - **Solar radiation at the project site**: The intensity of solar radiation received at the project site, typically measured in kilowatts per square meter (kW/m^2). This parameter is vital for estimating the electricity production potential. In areas with high solar radiation, such as those

52 *Clean energy in South-East Asia*

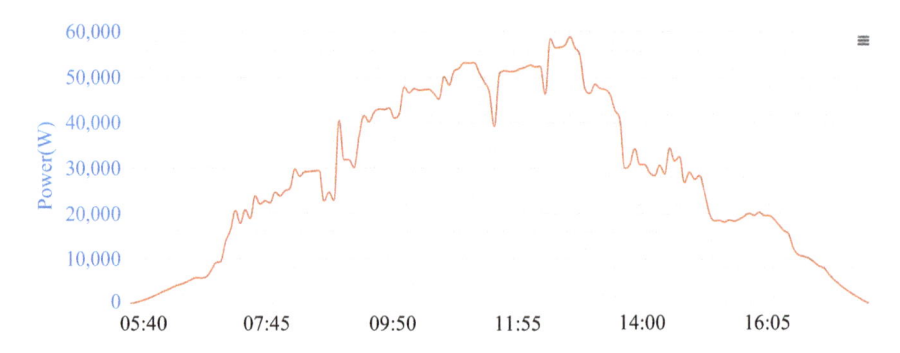

Figure 2.10 Daily power output curve of a rooftop grid-tied solar power system

near the equator or with minimal cloud cover, solar power systems can generate more electricity.
- **Average power generation chart over time**: This chart illustrates the average power output of the solar power system over time, often shown on a daily or monthly basis. It helps to understand the system's performance throughout the seasons and times of day, reflecting the impact of varying solar radiation, shading, and other environmental factors. For instance, power generation may peak at midday and fluctuate with the seasons due to changes in weather patterns.

Summary: These parameters are crucial for assessing the operational capabilities, efficiency, and economic benefits of a rooftop grid-connected solar power system. By analyzing these factors, stakeholders can optimize system design, predict power output, and make informed decisions about investments in solar technology.

Figure 2.10 illustrates the typical daily electricity output curve of a grid-connected rooftop solar power system. This chart shows how the electricity generation of the solar system varies throughout the day. The key features of this graph typically include:

- **Morning increase**: As the sun rises, solar radiation increases, leading to a gradual rise in electricity output. This phase begins when the solar panels start receiving sunlight and lasts until mid-morning.
- **Peak electricity output**: Around midday, when the sun is at its highest point, the solar panels receive maximum radiation. The electricity output peaks during this period, usually from 11:00 AM to 2:00 PM, depending on the location and season.
- **Afternoon decline**: After reaching the peak, electricity output begins to gradually decrease as the sun moves toward the horizon. This decline continues until sunset, at which point electricity production drops to zero.

Factors affecting the curve
- **Weather conditions**: Clouds, rain, and dust can cause fluctuations in electricity output, creating dips in the curve.

- **Seasonal variations**: The time of day and the sun's angle change with the seasons, affecting the shape and timing of the electricity output curve.
- **Shading**: Objects such as trees, buildings, or other obstacles can cast shadows on the solar panels, leading to variations in electricity output throughout the day.

This daily electricity output curve is essential for understanding the performance of the solar power system and for planning energy usage and storage. It helps predict the amount of electricity that can be produced on a specific day, which is crucial for grid management and optimizing solar energy use.

Solar tracking systems
- Solar panels achieve their highest efficiency when sunlight strikes them at a perpendicular angle. To maximize energy output, solar tracking systems are used to adjust the angle of the solar panels throughout the day, ensuring they are always optimally aligned with the sun's position (Figure 2.11).
- The operating principle of this model is to always keep the solar panels facing the sun, ensuring that sunlight strikes the panels perpendicularly. This system can be integrated with a maximum power point tracking (MPPT) controller to track the maximum power point (MPP). This principle serves as the foundation for designing the rotation mechanism of the solar panel system.
  - **Single-axis solar tracking system**: This system uses inclined panel mounts and electric motors to move the panels along a trajectory that closely approximates the sun's position. The axis of rotation can be horizontal, vertical, or inclined. A general diagram of the single-axis solar tracking system shows both the rotation axis (unit vector e) and the energy-collecting plane (unit vector perpendicular to the energy-collecting plane). The angle between these two unit vectors is typically kept fixed in this type of tracking system [1].
  - **Dual-axis solar tracking system**: This system can achieve maximum energy collection capacity because it has complete freedom of movement in two directions, allowing it to track the sun's position anywhere in the sky (Figure 2.12) [1].
  - Currently, there are two types of solar tracking systems: single-axis tracking systems and dual-axis tracking systems. The single-axis tracking

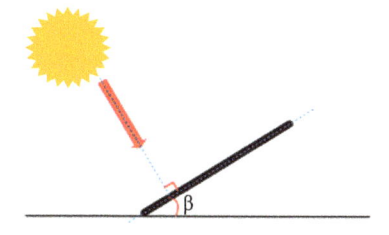

Figure 2.11 Illustration of the sunlight incidence angle on the surface of a solar panel

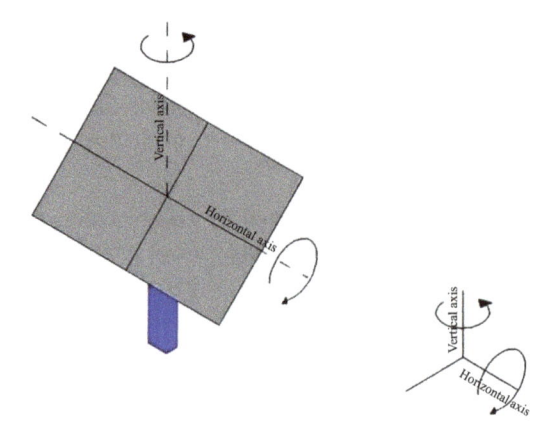

Figure 2.12   Guide of the dual-axis solar tracking system

system adjusts the solar panels along a single axis, aligned from north to south, following the movement of the sun from east to west. Meanwhile, the dual-axis tracking system adjusts the solar panels along two axes, from east to west and from north to south.
- The dual-axis solar tracking system with rotating panels features both a vertical and a horizontal axis of rotation. In this type of dual-axis tracking system, the panels are mounted on a separate frame. The main advantage of this system is the large rotation angle of the frame, up to 360°, allowing it to capture more sunlight and achieve higher efficiency. However, this type of frame lacks high stability, requires initial setup before operation, and involves complex construction, assembly, and operation (Figure 2.13).

Solar-powered automatic water pumping system
- We will explore a project involving a solar-powered automatic water pumping system.
- This system may consist of solar panels, controllers, and water pumps, designed to optimize the use of solar energy for providing water for agricultural or domestic purposes. The solar panels convert solar energy into electricity, supplying power to the pump, which draws water from sources such as wells, ponds, or rivers and delivers it to irrigation systems or water storage tanks. The system can operate autonomously, helping save energy and operational costs while providing a reliable water source for areas without access to the electrical grid (Figure 2.14).

Design of a solar-powered water pumping project
- In the previous chapters, we examined the basic principles of solar panel systems. Building upon these theoretical foundations, we will develop a model for a project designed to use solar energy for water pumping purposes, integrated with MPPT technology.

Inexhaustible energy sources 55

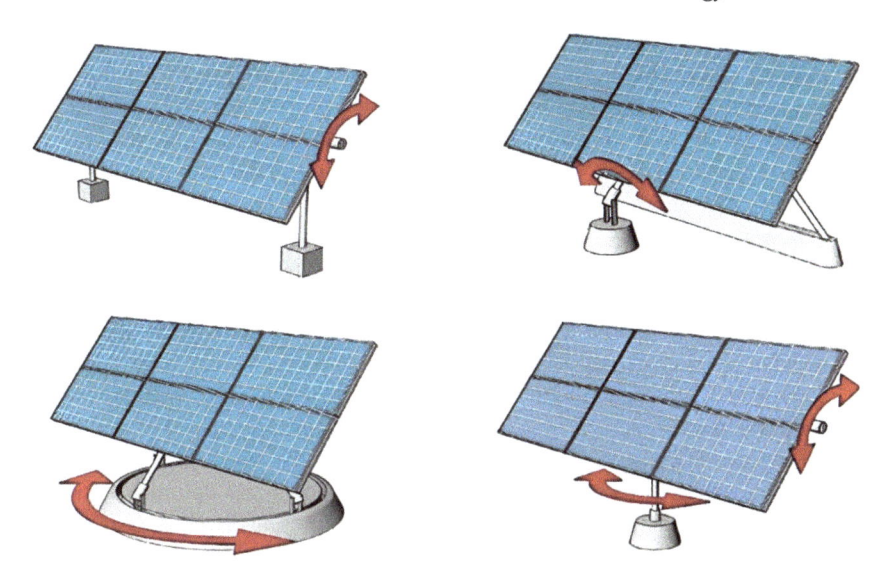

Figure 2.13   Solar panel tracking system

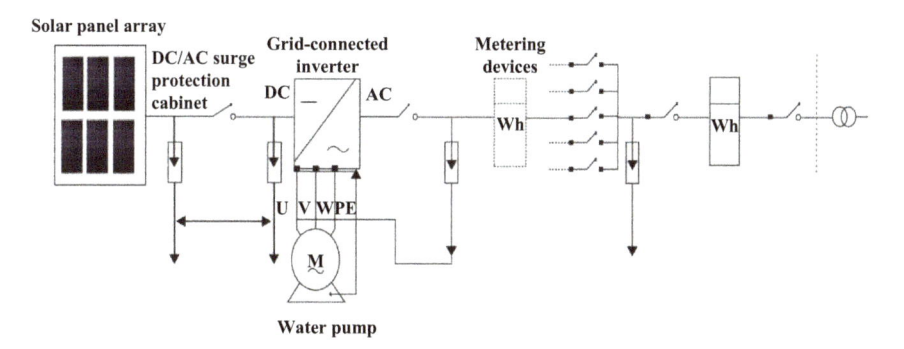

Figure 2.14   Single-line diagram of a grid-connected solar power system

- The system will utilize solar panels to convert solar energy into electrical power, controlling the automatic water pump via the MPPT controller. This will optimize energy use and ensure the maximum pump capacity throughout the operation. The aim of this project is to provide a stable and cost-effective water source for areas using water for domestic purposes, helping to reduce energy costs and protect the environment.

2.5   Summary and future research directions

- In conclusion, Vietnam possesses significant potential for solar energy development due to favorable solar radiation conditions, especially in key regions.

In recent years, several successful solar energy projects have been implemented, demonstrating the country's readiness to expand this field. Advances in science and technology have played a crucial role in transforming the solar energy industry, from traditional technologies to more modern, efficient systems. Key components of solar energy systems have been improved to enhance performance and reliability.

- Efficient solar energy models include both off-grid and grid-connected systems, meeting diverse energy needs. These systems are particularly versatile in providing power for smart homes, supporting autonomous navigation and agricultural irrigation. Roof-mounted grid-connected solar systems are increasingly popular, providing sustainable and cost-effective energy solutions for residential, commercial, and agricultural applications. With strong governmental commitment and the continuous development of technology, Vietnam is poised to continue growing and establishing itself as a leader in the renewable energy sector.
- However, to achieve sustainable development goals and optimize the use of clean energy, Vietnam must continue to promote research and development of next-generation solar technologies, improve energy storage systems, and enhance the efficiency of solar energy applications in daily life and production. There is a need for policies that support scientific research, encourage international collaboration, and facilitate technology transfer to strengthen the renewable energy sector in Vietnam. At the same time, it is essential to enhance the training of high-quality human resources, raise public awareness about the importance of renewable energy, and encourage businesses to invest in clean energy technologies.

Further reading

[1] Juan Reca-Cardeña and Rafael López-Luque, (2018), "Design principles of photovoltaic irrigation systems". In A. Sayigh (ed.), *Advances in Renewable Energies and Power Technologies* (Vol. 1, pp. 217–248). Elsevier.
[2] Vietnam's Eighth National Power Development Plan (PDP VIII).
[3] Environmental Protection Law: Law No. 72/2020/QH14, signed on 17/11/2020.
[4] Law on Water Resources: Law No. 17/2012/QH13, signed on 21/6/2012, effective from 01/1/2013.
[5] Planning Law: Law No. 21/2017/QH14, passed by the 14th National Assembly on 24/11/2017.
[6] Amendment Law No. 28/2018/QH14: Amendments and supplements to several articles of 11 laws related to planning, passed on 15/6/2018.
[7] Amendment Law No. 35/2018/QH14: Amendments and supplements to several articles of 37 laws related to planning, passed on 20/11/2018.
[8] Decree No. 29/2011/NĐ-CP: Issued by the Government on 18/2/2011, regulating strategic environmental assessment, environmental impact assessment, and environmental protection commitments.

[9] Decree No. 23/2006/NĐ-CP: Issued by the Government on 03/3/2006, on the implementation of the Law on Forest Protection and Development.
[10] Decree No. 46/2012/NĐ-CP: Issued by the Prime Minister on 22/5/2012, amending and supplementing several articles of:
[11] Decree No. 35/2003/NĐ-CP (dated 04/4/2003) detailing the implementation of several articles of the Law on Fire Prevention and Fighting.
[12] Decree No. 130/2006/NĐ-CP (dated 08/11/2006) on compulsory fire and explosion insurance policies.
[13] Decree No. 32/2006/NĐ-CP (2006): Application of the Sorensen Coefficient (S) for comparing species composition in the study area with neighboring areas.
[14] Decree No. 18/2015/NĐ-CP (Article 12) and Circular No. 27/2015/TT-BTNMT (Article 7): Provisions on consultation activities during the implementation of the Environmental and Social Impact Assessment (ESIA).
[15] Circular No. 27/2015/TT-BTNMT: Issued on 29/5/2015 by the Ministry of Natural Resources and Environment (MONRE) on Strategic Environmental Assessment, Environmental Impact Assessment, and Environmental Protection Plans (Appendix 2.3, Chapters 1, 3, 4, and 5).
[16] Circular No. 26/2011/TT-BTNMT: Issued on 18/7/2011 by MONRE, detailing certain provisions of Decree No. 29/2011/NĐ-CP on strategic environmental assessment, environmental impact assessment, and environmental protection commitments.
[17] Circular No. 12/2011/TT-BTNMT: Issued on 14/4/2011 by MONRE, on the management of hazardous waste.
[18] Directive No. 08/2006/CT-TTg: Issued by the Government on 08/3/2006, on urgent measures to prevent illegal deforestation, forest burning, and unauthorized logging.
[19] Ngo Dang Luu, Nguyen Hung, Nguyen Anh Tam, and Long D. Nguyen, (2021), "Wireless Power Transfer simulation for EV", Ministry of Culture, Sports and *Tourism*.
[20] Ngo Dang Luu, Nguyen Hung, and Long D. Nguyen, (2021), "Power flow calculation of the electric power system by MATLAB", Ministry of Culture, Sports and Tourism.
[21] Ngo Dang Luu, Nguyen Hung, Nguyen Anh Tam, and Long D. Nguyen, (2022), "Distance Relay Protection in AC Microgrid", Ministry of Culture, Sports and Tourism.
[22] Ngo Dang Luu, Nguyen Hung, Nguyen Anh Tam, and Long D. Nguyen, (2022), "Analysis of Solar Photovoltaic System Sunlight Shadowing", Ministry of Culture, Sports and Tourism.
[23] Ngo Dang Luu, Nguyen Hung, Le Anh Duc and Long D. Nguyen, (2022), "Estimate Frequency Response at the Command Line", Ministry of Culture, Sports and Tourism.

Chapter 3

Technologies and solutions for smart energy systems with the presence of renewable energy

3.1 Introduction to the application of artificial intelligence and data analysis for fault detection and prediction in multi-source energy systems

Artificial intelligence (AI) and data analysis are becoming powerful tools in managing and optimizing modern energy systems. With the rapid development of renewable energy sources such as solar power, wind power, and traditional energy sources, coordinating operations and monitoring systems has become more complex than ever.

In this context, AI and data analysis are applied to:

1. **Fault detection**: Timely identification of faults in equipment or systems (such as transformers, solar panels) using algorithms that analyze historical and real-time data.
2. **Fault prediction**: AI can learn from operational data to forecast potential issues before they occur, reducing downtime and maintenance costs.
3. **Managing multi-source energy**: When integrating multiple energy sources, AI helps optimize energy distribution and usage, ensuring high performance and system stability.
4. **Analysis of influencing factors**: AI also helps analyze environmental factors, load consumption, and external impacts (such as weather and temperature) on system operations.

The application of AI not only improves operational efficiency but also makes multi-source energy systems smarter, reduces risks, and optimizes costs. This is an inevitable trend in building modern and sustainable energy systems.

3.2 Application of artificial intelligence for calculating and estimating the inertia constants of the power system

3.2.1 Overview of power system inertia

Inertia in a power system (PS) refers to its ability to resist frequency changes due to the rotational energy of rotating masses in synchronous turbines of generators.

When disturbances cause an imbalance between power generation and load on the grid, particularly large disturbances like loss of generation or load, inertia helps limit the rate of frequency change. This provides time for other mechanisms, such as primary frequency control, to intervene and stabilize frequency, preventing the system from tripping or becoming unstable.

3.2.1.1 Inertia in traditional power systems

In traditional PS, energy sources such as thermal power, hydroelectric, and gas turbines use large rotating turbines, which significantly contribute to system inertia. As the PS expands, the overall inertia increases, helping the system better withstand frequency disturbances.

3.2.1.2 Challenges from renewable energy

The rapid increase in the penetration of renewable energy sources, such as solar and wind, poses a significant challenge to PS inertia. These sources typically use inverter technology, which does not provide rotational inertia. Furthermore, to prioritize the use of RE, the number of traditional generators connected to the grid has been reduced, decreasing the system's total inertia.

As a result, the frequency of the PS becomes more sensitive to disturbances, and it can change more quickly when faults occur. In severe cases, the frequency can exceed safe limits before primary frequency control systems can respond, leading to the risk of load shedding, generator tripping, or frequency instability.

In summary, inertia plays a fundamental role in maintaining the frequency stability of a PS. However, the structural changes due to high RE penetration require new technological and operational solutions to ensure grid stability in the current context.

3.2.2 Development of renewable energy and its impact on the inertia of the power system in Vietnam

3.2.2.1 Rapid growth of renewable energy

In recent years, thanks to supportive policies, Vietnam has witnessed a boom in renewable energy, particularly solar and wind power. By the end of 2020, the total RE capacity connected to the grid reached approximately 17,540 MW, accounting for 24.6% of the national grid's installed capacity. This includes:

- 8,700 MW from solar power plants,
- 540 MW from wind power,
- 8,300 MW from rooftop solar systems.

RE has become an important source of power, ensuring energy security, especially during off-peak times such as middle day or holidays, when load demand decreases significantly. At these times, RE can account for up to 45-60% of total load demand.

Growth prospects for RE
In the short term, it is expected that by the end of 2021, an additional 3,000 MW of wind power will be added, raising the RE share to 18.6% of the installed capacity. In the long term, renewable energy will continue to grow, occupying an increasing share of the energy mix, while the rate of load growth may not keep pace, influenced by uncertain factors such as the COVID-19 pandemic.

Impact on power system inertia
However, the rapid increase in RE is posing a significant challenge to PS inertia. Renewable energy sources use inverter technology, which lacks rotational inertia, thus unable to provide inertia like traditional power plants. This leads to a trend of decreasing inertia in Vietnam's PS as the RE share increases.

In countries with high RE penetration, such as Germany or Ireland, governments require RE plants to equip synthetic inertia or fast frequency response systems to support system stability. However, in Vietnam, there are no specific regulations on this matter. Current RE plants are unable to provide inertia or support frequency regulation during major disturbances.

Conclusion
The strong growth of RE brings significant benefits for sustainable energy and electricity security, but it also creates considerable operational challenges. The reduction in system inertia requires new solutions, such as the need for synthetic inertia technology or regulatory adjustments to ensure future grid stability.

3.2.2.2 Necessity of real-time calculation and monitoring of power system inertia

Given the challenges posed by the increasing share of renewable energy and the trend of decreasing inertia in the PS, the real-time calculation and monitoring of PS inertia have become a crucial task. This activity not only ensures the safe and stable operation of the system but also supports the optimization of dispatching and operational planning.

Applications of real-time inertia monitoring

- **Real-time and short-term grid dispatch**
  - Inertia data calculated and monitored online will be compared with the critical inertia threshold required for Vietnam's PS.
  - Dispatchers can quickly take appropriate measures to restore system inertia in case of a decline,
  - Inertia data also helps determine the required level of frequency regulation reserves for different operating scenarios, supporting dispatchers in optimizing reserve sources and improving operational efficiency.
  - Additionally, inertia analysis helps retrospectively assess past events, improving future response.
  - Especially during large and extreme disturbances. This helps maintain frequency stability and prevents the risk of sudden load shedding or generator tripping.

- **Support for system operational planning and analysis**
  - The calculated inertia values are used to build frequency response models of the PS, allowing the simulation of future operating modes.
  - Strategies for similar situations.

Conclusion

Real-time calculation and monitoring of PS inertia are strategic steps to enhance the operational capacity of Vietnam's PS in the context of increasing RE penetration. By closely monitoring and proactively responding, Vietnam's PS can ensure stability and safety, while supporting the sustainable development of the energy sector.

3.2.3 Current status

According to international experience, system operators in developed countries such as Europe, North America, and Australia have long implemented real-time inertia calculation and measurement systems for PS. These systems are effective in maintaining the stability of the power grid (PG) when facing large fluctuations.

In Vietnam, however, inertia in the PS has not received adequate attention. Before this research, Vietnam's PS was unable to:

- Accurately calculate the inertia value of the system in real-time.
- Determine the minimum inertia threshold necessary to maintain frequency stability in the event of major disturbances, thus reducing the risk of load shedding.
- Especially during the medium- and short-term operational planning process, particularly during special events like holidays, the national PS dispatch center (A0) often calculates the minimum number of generators needed to meet voltage and power flow distribution criteria to avoid overload. However, this calculation does not include ensuring the minimum inertia. This leads to the risk that the minimum operational configuration may not be reliable or optimized.

To address this situation, initiatives have been researched, developed, and implemented with two main objectives:

- Develop methodologies and tools for calculating and monitoring Vietnam's PS inertia in real-time.
- Identify the minimum inertia threshold to maintain frequency stability during special operating conditions.

3.3 Initiative content

3.3.1 Research on methodology for calculating and measuring PS inertia

3.3.1.1 Theoretical basis

PS inertia refers to the ability to resist frequency changes due to the resistance provided by the rotational energy of rotating masses in turbines and synchronous generators.

The inertia constant H of a single electrical machine is calculated using the following formula:

$$H = \frac{1}{2}\frac{J\omega_{n^2}}{S_n} \tag{3.1}$$

where:

J: Combined moment of inertia of the turbine-generator assembly (kg/m^2);
ω_{n^2}: Angular velocity of the rotor (rad/s); and
S_n: Mechanical power input from the turbine (MVA).

The inertia constant H represents the time required for the kinetic energy stored in the rotating mass of the turbine-generator to supply energy equivalent to its rated power.

3.3.1.2 Practical implications

The real-time calculation of system inertia based on this method not only ensures accuracy but also helps in building frequency response models, thereby improving the operation and planning of the PS in the context of increasing shares of renewable energy.

The inertia constant H of a generator can be determined through manufacturer tests prior to shipment or during the commissioning process. This is an important parameter for evaluating the generator's ability to withstand frequency fluctuations in the PS.

When a synchronous generator is connected to the grid, the dynamic relationship between mechanical power (mechanical torque) and electrical frequency (the rotor's angular speed) is represented by the swing equation. This equation describes the dynamics of the generator during the short period immediately following an imbalance between mechanical and electrical power on the generator's shaft.

The balance equation is written as:

$$\frac{df_i}{dt} = \frac{P_{i,m} - P_{i,e}}{2H_i S_{in}} f_n \tag{3.2}$$

where:

$P_{i,m}$: The mechanical power output of turbine i (MW);
$P_{i,e}$: The electrical power output of generator i (MW);
H_i: The inertia constant of the turbine-generator unit i (s);
f_i: The electrical frequency at the generator bus i (Hz);
f_n: The system frequency in the steady-state (nominal) condition (Hz); and
S_{in}: The rated apparent power of generator i (MVA).

When the influence of the damping factor D is neglected, this equation represents the relationship between the rate of frequency change and the discrepancy between mechanical and electrical power.

64   Clean energy in South-East Asia

Significance
- **Reflecting natural inertia**: The larger the inertia constant H, the slower the rate of frequency change ($d\Delta f/dt$), enabling the PS to better withstand major disturbances and reduce the risk of frequency instability.
- **Frequency response analysis**: This equation is used to simulate and analyze the response of generators in emergency operating conditions, assisting system operators and engineers in formulating appropriate operational strategies.
- **Supporting operational optimization**: By leveraging the inertia constant H of each generator and aggregating system-wide inertia, it is possible to calculate system inertia, ensuring stable operation even when integrating a high proportion of renewable energy sources.

The implementation of real-time inertia calculations based on this theoretical foundation is crucial for Vietnam to enhance the reliability and efficiency of its PS, particularly as the share of renewable energy continues to grow.

By aggregating the swing equations for all synchronous machines across the system, assuming the electrical frequency is uniform at all points and disregarding load damping effects during the initial stages of a power imbalance event, the resulting equation is:

$$\frac{df}{dt} = \frac{\Delta P}{2H_{sys}S_{sys}} \times f_0 \tag{3.3}$$

where $H_{sys}S_{sys}$ represents the total system inertia (unit: MW·s), which is the quantity to be determined.

From the equations above, the total system inertia can be directly calculated as the sum of the inertia contributions from all synchronous machines operating in synchronization with the PS. The formula for calculation is as follows:

$$H_{sys}S_{sys} = \sum H_i S_i I \tag{3.4}$$

Calculation results for minimum inertia requirements
At 12:00 PM during the Lunar New Year holiday, rooftop solar (9,600 MWp) and grid-connected solar (10,000 MWp) could supply up to 80% of the load demand. Even after curtailing 4,000 MW of renewable energy, solar power could still account for 70% of the load. However, solar sources do not provide rotational inertia, leading to a significant drop in total system inertia and impacting frequency stability during generator outages or sudden load/solar generation fluctuations.

3.3.1.3   Results of calculations and recommendations for power system operation

Minimum inertia requirements
The research team calculated the minimum total inertia required to maintain frequency stability under special operational scenarios during holidays. The specific results are as follows:

Technologies and solutions for smart energy systems 65

- **New Year's Day 2021**: Minimum required inertia: **83,610 MWs**.
- **Lunar New Year (Tan Suu) 2021**: Minimum required inertia: **89,610 MWs**.

Operational considerations to meet system inertia requirements

1. **Dispatch of coal-fired power units:**
   - Maintain coal-fired units operating at minimum load but not exceeding 450 MW during off-peak hours (9 AM–3 PM and nighttime).
   - **Objective**: Prevent deep frequency drops during large generator outages.
2. **Prioritize units with strong primary frequency response:**
   Recommended units:
   - Vinh Tan 2 thermal power plant
   - Mong Duong 1 thermal power plant
   - Phu My thermal power plant (Units 1, 21, 4)
   - Vinh Tan 1 thermal power plant
   - Ca Mau thermal power plant
   - Hai Phong thermal power plant

Reason: These units provide fast and accurate frequency support, crucial during major system disturbances.

1. **Response to system inertia falling below required levels:**
   - Dispatch hydropower units to compensate for inertia loss.
   - **Objective**: Enhance system inertia without overloading the grid, ensuring safe and stable operations.

Conclusion

Accurate identification of minimum inertia thresholds has enabled Vietnam's PS to implement appropriate operational strategies during periods with special configurations. Recommendations, such as optimized unit dispatching and prioritization of responsive units, are vital to ensuring stable and secure system operations under all conditions.

3.3.1.4 Observations

On January 1, 2021, at 11:11 AM, Vietnam's PS reached the lowest real-time inertia of **113,419 MWs**, which was still above the minimum required value. This was due to the minimal connection of traditional generators capable of supplying inertia to prioritize renewable energy. This was also the lowest real-time inertia recorded between December 30, 2020 and January 3, 2021.

- Coal-fired plants contributed the most to system inertia, followed by gas turbines and hydropower. The number of online coal-fired units reached its lowest at 8:00 AM on January 1, while gas turbines reached their lowest at 11:00 AM on the same day. These levels were maintained throughout the day. Note that changes in output did not affect the contribution of these units to system inertia.
- Hydropower units showed the most variation in operation throughout the day to meet load profiles.

Table 3.1   Inertia statistics of the power system

Inertia statistics of the electrical system (week from February 8–14, 2021)						
No.	Date	Maximum system inertia	Duration	Minimum system inertia	Duration
1	Feb 8, 2021	171,198.1	Feb 8, 2021 18:24	137,171.8	Feb 8, 2021 11:40
2	Feb 9, 2021	156,219.1	Feb 9, 2021 18:26	124,005.7	Feb 9, 2021 9:42
3	Feb 10, 2021	143,699.1	Feb 10, 2021 19:44	117,303.5	Feb 10, 2021 13:22
4	Feb 11, 2021	141,237.4	Feb 11, 2021 18:52	106,222.3	Feb 11, 2021 12:52
5	Feb 12, 2021	133,997.4	Feb 12, 2021 19:06	99,520.1	Feb 12, 2021 10:34
6	Feb 13, 2021	134,978.3	Feb 13, 2021 18:28	98,548.3	Feb 13, 2021 11:12
7	Feb 14, 2021	135,295.5	Feb 14, 2021 18:26	97,882.7	Feb 14, 2021 12:54

The calculated inertia results indicate that both actual and minimum inertia thresholds were well defined, providing system operators with practical guidance for managing real-time configurations.

3.3.1.5   Operation during lunar new year 2021 (February 8–14, 2021)

A summary table of minimum/maximum inertia levels, timestamps of measurements, and charts illustrating the contributions of different generation sources to system inertia during the Lunar New Year holiday are as shown in Table 3.1.

3.4   Evaluation of harmonic analysis and equipment loss calculation using specialized software and effective selection of power system protection devices

3.4.1   *Study of heat generation and vibration in transformers caused by voltage and current harmonics*

3.4.1.1   Introduction

In addition to the degradation of power quality, which disrupts the operation of the PS and nearby loads and reduces efficiency, harmonics introduce several other issues related to losses, aging, wear, and reduced useful lifespan of electrical equipment in the distribution network. The problem of harmonics is becoming increasingly severe due to the growing number of harmonic-generating sources in the PS, such as non-linear loads, frequency conversion devices, and renewable energy sources. Medium-voltage transformers are highly sensitive to harmonics, widely distributed, and essential in power distribution networks. Furthermore, they are a crucial component in the PS, both technically and economically. The increase in power losses caused by harmonics can lead to overheating, degradation of insulation materials such as paper, fabric, rubber seals, and oil, and may even result

Figure 3.1 Description of the components of a transformer

in transformer explosions, affecting the operational safety of the grid, as well as property, human safety, and the environment.

3.4.1.2 Transformer structure (MBA)

The components in Figure 3.1 include:

1 – transformer tank	9 – pressure relief valve
2 – transformer cover	10 – tap changer
3 – auxiliary oil tank	11 – lifting hook
4 – oil level indicator	12 – transformer base
5 – oil dehumidifier	13 – high-voltage winding
6 – high-voltage bushings	14 – low-voltage winding
7 – low-voltage bushings	15 – steel core
8 – radiator fins	16 – magnetic core clamp

The transformer is composed of a combination of several basic components, including the core, windings, and the transformer tank. Additionally, the transformer tank houses elements such as insulation bushings, oil level indicators, temperature indicators, pressure relief valves, auxiliary oil tanks, tap changers, and oil dehumidifiers, as shown in Figure 3.1. All of these components are interconnected to form a complete transformer.

The increase in eddy current losses due to harmonic currents can lead to excessive losses in the windings, thus causing abnormal temperature rise. Since the total harmonic distortion of current (THDi) is typically much higher than the total harmonic distortion of voltage (THDu), the impact of harmonic currents is generally more severe in most cases. Figure 3.2 illustrates the losses caused by harmonic currents by harmonic order.

The increase in losses and temperature rise of the transformer when harmonic waves in the electrical system are high significantly reduces the efficiency of the transformer. Figure 3.3 illustrates the correlation between transformer efficiency and harmonics in the system.

68   *Clean energy in South-East Asia*

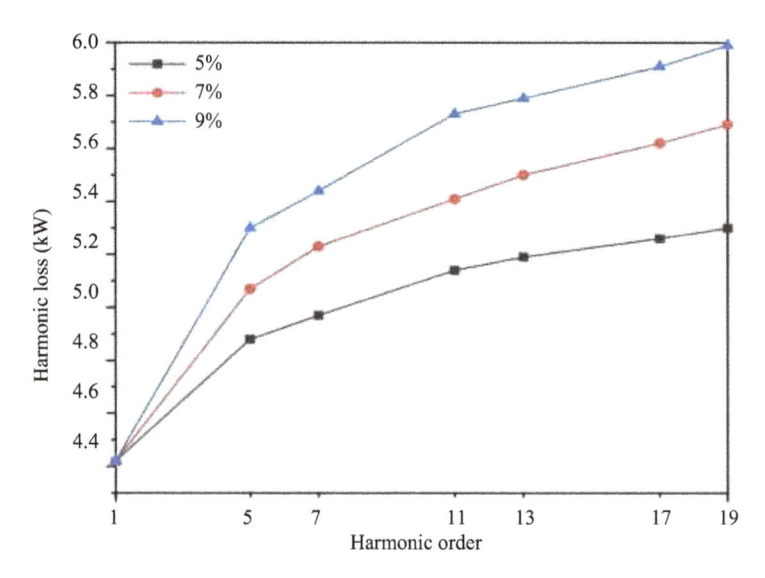

Figure 3.2   Losses according to the harmonic order of the winding when surface effects are not considered

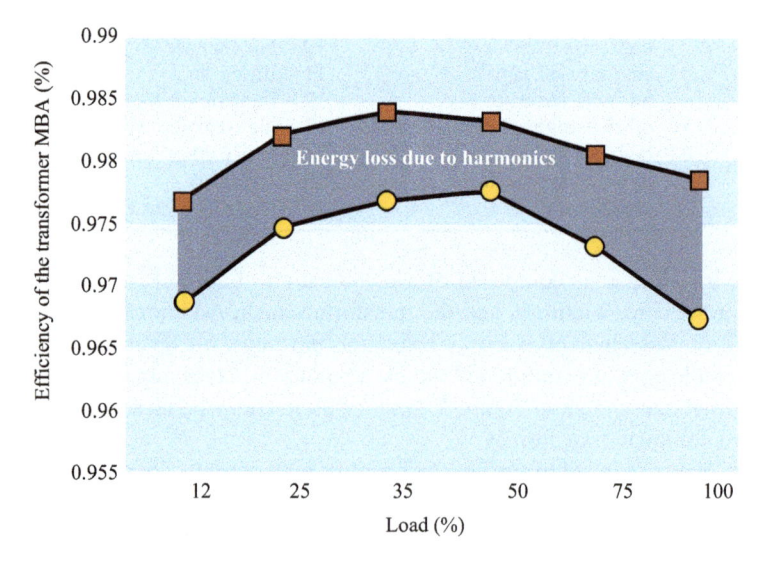

Figure 3.3   Transformer efficiency decrease due to harmonics in the power system

3.4.1.3   Rooftop solar project on three factory roofs of the company
- Total capacity: 2.5 MWp
- Total number of panels: 7,096 panels

Technologies and solutions for smart energy systems 69

- Total number of inverters: 25 inverters
- Voltage harmonics: Highest is the fifth harmonic
- Current harmonics: Low

Location: Provincial Road 830B, Hamlet 3, Long Cang Commune, Can Duoc District, Long An Province.

From Figure 3.4, it can be observed that the total harmonic distortion of the current reaches its highest value at the third harmonic order. However, this value still falls within the permissible limits according to IEEE standards, thus meeting the necessary conditions for operation.

Based on the theoretical research foundations regarding power quality, particularly concerning harmonics; the causes of harmonic generation; the different types of voltage and current harmonics produced by grid-connected photovoltaic (PV) systems and a set of typical nonlinear loads in practice; the impacts of harmonics on electrical equipment, especially transformers; the solutions that have been developed, as well as the methods for performing calculations, analysis, evaluation, and determining the critical thresholds for medium-voltage and low-voltage grids when involving PV systems, it allows us to implement similar projects on PSCAD software (or software with similar functions). This enables the analysis, evaluation, and forecasting of the acceptance capabilities of current and future PV systems, helping grid management and operational units proactively address their related tasks with a focus on efficiency, operational safety, and optimal future capital allocation.

3.4.2 *Protection equipment check and selection at the workshop*
- **Analysis of protection coordination between ACB and MCCB**:
  – In the event of a short circuit at inverter number 5 in Workshop 1, with a short-circuit current of 2,950 A:

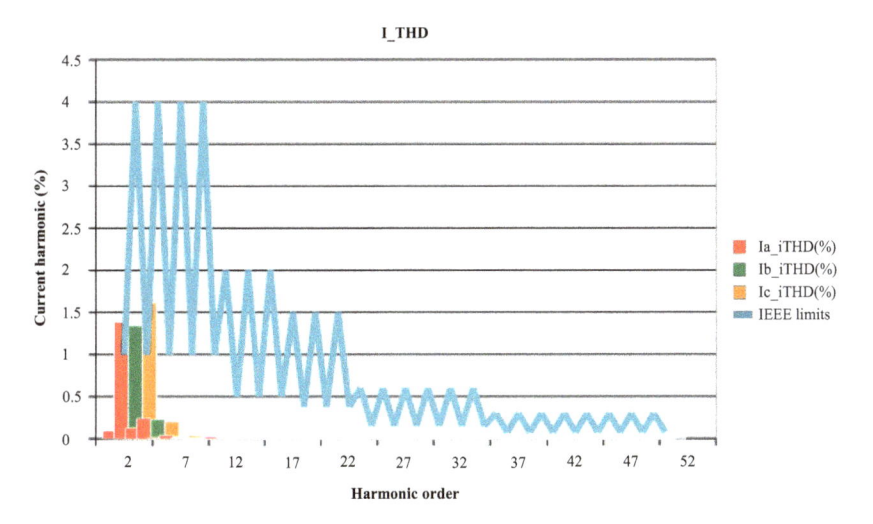

Figure 3.4 Harmonic spectrum of current at high voltage (22 kV) concentrated mainly at harmonic orders 3 and 5

70 Clean energy in South-East Asia

- ○ **MCCB 200A (NE250-SV)**: Has a tripping time from 0.01 to 4.2 s.
- ○ **ACB 2000A (AE2000-SW)**: Has a tripping time from 30 to 45 s.

- **Conclusion**: The MCCB will trip before the ACB in this short-circuit scenario, ensuring effective protection coordination. Therefore, the selection of MCCB and ACB is appropriate.

Based on the specification table, the analysis is as follows:

- **Principle of protection coordination**:
  - MCCB (molded case circuit breaker) is designed to trip quickly in small or localized short circuit faults.
  - ACB (air circuit breaker) has a longer tripping time, ensuring it only operates when the MCCB fails to trip or in the case of large short circuits.

- **Short-circuit current (kA)**:
  - The short-circuit current decreases from inverter 1 to inverter 9, ranging from 7.92 kA (highest) at inverter 1 to 2.40 kA (lowest) at inverter 6.

- **Tripping time**:
  - MCCB: Has a very fast tripping time (0.01–8 s), depending on the short-circuit current at each inverter.
  - ACB: Has a longer tripping time (4–72 s) to ensure that MCCB trips first.

- **Reasonable selection**:
  - In all cases from inverter 1 to inverter 9, the tripping time of the MCCB is always shorter than that of the ACB, ensuring that the MCCB operates before the ACB. This ensures that the protection system functions according to the principle and prevents wide-area power outages.

- **Conclusion**:
  - All combinations of MCCB and ACB are designed and selected appropriately according to the specification table.
  - This coordination ensures the system operates stably and safely during short-circuit events (Table 3.2).

Table 3.2 The protection and coordination characteristics of the inverter

Inverter	Short-circuit current (kA)	MCCB operation time (s)	ACB operation time (s)	Is the selection reasonable?
1	7.92	0.01–0.02	4–6.1	Yes
2	5.67	0.01–0.02	11–16	Yes
3	4.38	0.01–0.02	17–22	Yes
4	3.48	0.01–0.02	25–30	Yes
5	2.92	0.01–4	30–45	Yes
6	2.40	0.01–8	52–72	Yes
7	2.75	0.01–4	36–52	Yes
8	3.91	0.01–0.02	19–24	Yes
9	4.86	0.01–0.02	15–20	Yes

3.5 Some energy storage technologies for renewable energy systems and load forecasting in solar and wind power systems

3.5.1 Energy storage technologies

3.5.1.1 The necessity of energy storage systems due to increasing variability from the integration of large-scale renewable energy sources

- **Balancing production and consumption**: The imbalance between energy production and consumption needs to be effectively managed.
- **Flexibility for uncontrollable sources**: Solar energy (PV) and wind energy (WE) require the use of flexible sources to adjust to their fluctuations.

3.5.1.2 Classification based on application

- **Short-term storage**
  - **Frequency regulation**: Helps maintain grid stability by responding quickly to frequency changes.
  - **Voltage support**: Provides reactive power to stabilize voltage levels in the PG.
  - **Spinning reserve**: Acts as a fast backup to replace lost electricity production in the event of an unexpected outage.

- **Medium-term storage**
  - **Load shifting**: Stores surplus energy during periods of low demand and releases it during high demand periods to balance the load.
  - **Renewable energy integration**: Reduces the fluctuations of renewable energy sources like solar and wind by storing surplus energy and providing it when production is low.

- **Long-term storage**
  - **Seasonal storage**: Manages energy supply over extended periods, such as storing energy during high renewable production seasons for use in low production seasons.
  - **Backup power supply**: Provides reliable electricity in cases of prolonged power outages or emergencies.

- **Applications of energy storage systems in the electrical grid**
  - **Usage time**: The duration for which the energy storage system (ESS) needs to provide or absorb electricity.
  - **Frequency/number of uses**: The number of times the ESS is used.

3.5.1.3 Battery storage technologies

- **Lead-acid batteries**
  - **Characteristics**
    - Very low energy-to-weight ratio.

72 *Clean energy in South-East Asia*

- o Low energy-to-volume ratio.
- o Relatively high power-to-weight ratio.
– **Applications**
  - o Storage in internal combustion engine vehicles (ICE).
  - o Deep-cycle batteries for mobile applications, e.g., storage in micro-grid Bronsbergen.

- **NiCd & NiMH batteries**
  – **Characteristics**
    - o NiCd batteries have high energy density and long lifespan. However, due to cadmium toxicity, NiCd has been banned for consumer use since 2006.
    - o NiMH is a replacement technology with higher energy density, but lower capacity (temperature-sensitive).
  – **Applications**:
    - o Suitable for mobile and portable applications.
    - o This technology remains stable and efficient despite many applications being replaced.

- **Li-Ion batteries**
  – **Characteristics**:
    - o High cell voltage (~3.7 V) – one Li-ion cell can replace three NiMH cells (1.2 V each).
    - o High energy-to-weight ratio.
    - o High cost due to specialized packaging, overcharge protection circuits, and limited lithium resources.
  – **Applications**:
    - o Suitable for mobile applications.
    - o Promising technology for plug-in hybrid electric vehicles and electric vehicles.

- **Comparison of storage technologies**
  – **NiMH (nickel-metal hydride)**: High energy density, efficient but sensitive to temperature.
  – **Lead-acid batteries**: Cheap but heavy.
  – **NiCd (nickel-cadmium)**: High power but polluting.
  – **Li-Ion (lithium-ion)**: Highest energy density, but costly and requires strict safety controls.

- **Overview**: Each storage technology has unique characteristics in terms of energy density, cost, and safety. NiMH and Li-ion are preferred due to their flexibility and high energy density, while NiCd and lead-acid are still used in some applications due to low cost or reliability. However, environmental and safety factors should be carefully considered when choosing a storage technology.

3.5.2 Research on the application of seasonal time series models for monthly electricity production forecasting

This study uses four seasonal time series models for forecasting monthly electricity production in four countries: Vietnam, Thailand, Spain, and South Korea. The models used are Holt–Winters, SARIMA, PROPHET, and N-BEATS, with data from actual electricity production between January 2010 and December 2022. The results were evaluated using mean absolute percentage error (MAPE) and root mean square error (RMSE) metrics, supported by Python in the Google Colab environment. The study showed that for each dataset, different models yielded varied results and suitable models. For Vietnam, the classical Holt–Winters model performed best with a MAPE of 3.6%. Thailand was best suited for SARIMA and N-BEATS with a MAPE of 2.8%. For Spain, Holt–Winters with a MAPE of 4.7% and N-BEATS with a MAPE of 4.6% were most appropriate. South Korea excelled with the deep learning model N-BEATS, achieving a MAPE of 1.7%.

3.5.2.1 Problem statement

Vietnam's Electricity Plan VIII, approved on May 15, 2023, under Decision No. 500/QĐ-TTg, marks a significant step in the country's electricity sector development. It aims to transition to renewable energy and improve the electricity transmission system. The goal is to ensure sufficient electricity for economic development with a GDP growth rate of about 7% per year from 2021 to 2030 and 6.5–7.5% per year from 2031 to 2050. The plan also aims to export electricity, with a target of 5,000–10,000 MW of export capacity by 2030. Priority will be given to the development of renewable energy sources such as hydropower, wind, solar, biomass, as well as new and clean energy such as hydrogen and green ammonia. Particularly, the use of rooftop solar energy for self-production and self-consumption is encouraged.

According to various studies, electricity forecasting is divided into four types corresponding to different time frames:

- **Real-time operation**: Forecasting for each operating cycle (half-hour or hourly) for real-time operation applications, aimed at managing the flexibility and efficiency of the electricity system.
- **Short-term load forecasting (STLF)**: Forecasting for each trading cycle, ranging from a few days to a few weeks. This is important for applications such as the day-ahead market, economic scheduling, load management, and unit commitment planning.
- **Medium-term load forecasting (MTLF)**: Forecasting electricity consumption for weekly or monthly cycles, from a few months to a few years. This supports medium-term operational planning, fuel supply planning, and maintenance scheduling.
- **Long-term load forecasting (LTLF)**: Forecasting total electricity consumption, peak and off-peak capacity for each year, from 5 to 25 years. This helps support power capacity planning, grid planning, investments, and financial planning for the PS and utilities.

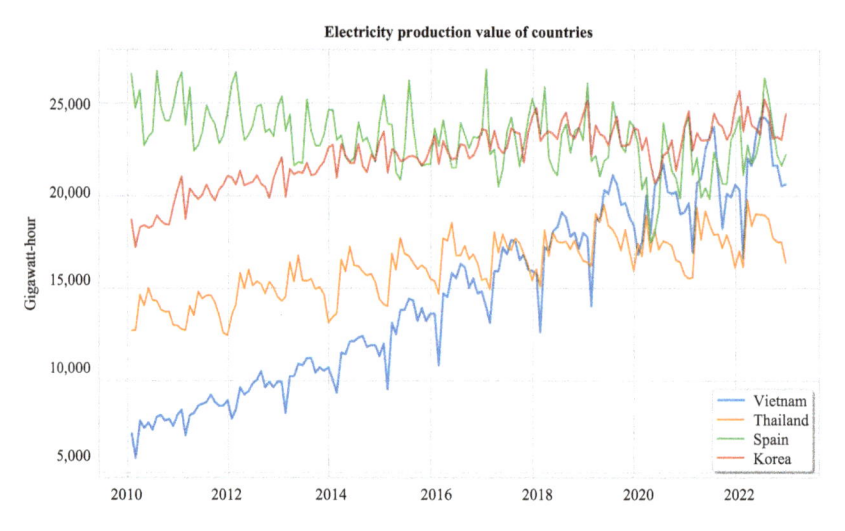

Figure 3.5   Graph of electricity production of different countries

With the distinct seasonal characteristics of Vietnam's electricity market, as illustrated in Figure 3.5, along with the continuous fluctuations in electricity and fuel prices, the frequent changes in customer demand, and the instability in the production of solar, wind, or hydropower energy, significant risks are posed to electricity production companies. Therefore, the research and application of seasonal time series forecasting models are essential, not only to meet domestic consumption needs but also to support economic and social development activities. Specifically, forecasting monthly electricity production plays a crucial role in ensuring a stable electricity supply, enabling electricity producers to manage and distribute energy in a timely and efficient manner.

In recent years, various forecasting methods considering seasonal factors have been developed. These methods are typically divided into two groups: traditional statistical methods and AI-based techniques. The first group includes simpler methods such as exponential smoothing (ETS), linear or logistic regression, and time series-based methods. The AI-based group includes models such as artificial neural networks, support vector machines—linear, recurrent neural networks (RNN), long short-term memory (LSTM), temporal convolutional networks, N-Beats, and so on. In some cases, hybrid methods have been developed by combining several approaches.

The research process shows that medium- and long-term forecasting has received less attention than short-term forecasting. Medium-term forecasting is particularly challenging due to factors such as consumption patterns, political and economic decisions, and energy sector management. MTLF and LTLF require consideration of many complex factors, requiring accurate data and extensive industry knowledge, including reliance on new technologies and integration of energy sources. Meanwhile, STLF provides more detailed data, reflecting short-term changes such as consumption habits and weather. In contrast, LTLF focuses on forecasting long-term energy demand, with less detail, mainly analyzing trends in economics, population, energy policies, and infrastructure development. This shows that the application of different models and methods in MTLF forecasting is still limited. Therefore, using seasonal time series models for forecasting monthly electricity production will contribute to providing more detailed insights into this gap.

This study focuses on analyzing and applying four popular time series forecasting models—Holt–Winters, SARIMA, Prophet, and N-BEATS—to forecast monthly electricity generation in Vietnam and compare it with other countries such as Thailand, Spain, and South Korea, using monthly electricity production data from 2010 to 2022. According to the findings, the Holt–Winters model is the most effective, and the SARIMA model has also proven to be an effective tool for long-term forecasting. Meanwhile, Prophet and N-BEATS, as newer deep learning models, have also shown certain successes.

3.5.2.2 Data and research methodology
(a) **Research data**

In this study, the data used is the monthly electricity production from January 2010 to December 2022 for four countries: Vietnam, Thailand, Spain, and South Korea. The data was collected from the General Statistics Office (GSO), EVN (Electricity of Vietnam), and Trading Economics. The electricity production graphs for these countries are shown in Figure 3.6.

Data processing

First, the data was explored to identify any outliers. The method used for detecting outliers is the interquartile range (IQR). IQR is calculated as the

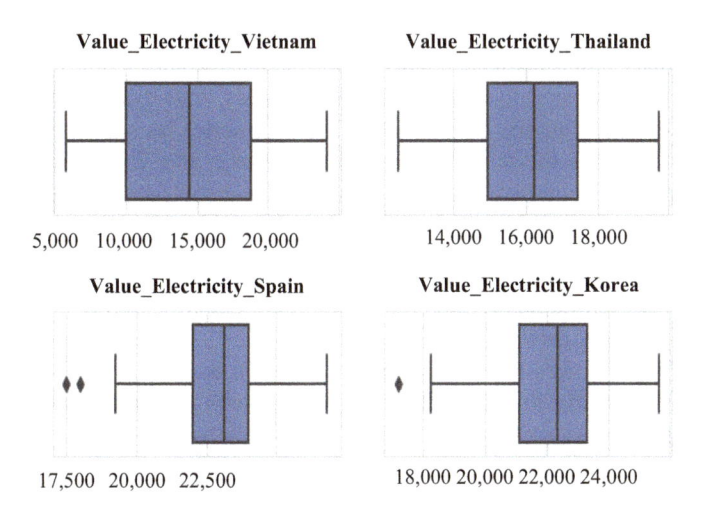

Figure 3.6 Outlier detection chart

76 Clean energy in South-East Asia

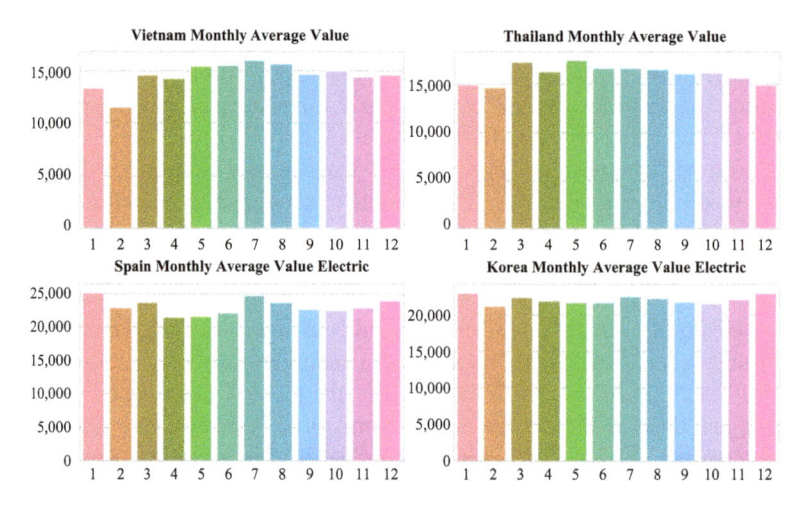

Figure 3.7 Trend and seasonality chart

difference between the third quartile (Q3) and the first quartile (Q1) of the dataset. Q1 represents the value below which 25% of the data falls, and Q3 represents the value below which 75% of the data falls. After removing the outliers, the missing values were filled using the Corresponding Method, where the value from the previous month was used to replace the missing data. Figure 3.6 shows the outlier data, which was identified and presented through a boxplot. Outlier processing will be applied to the Holt–Winters and SARIMA models because they are sensitive to large fluctuations that can distort model estimations. The Prophet model is designed to automatically detect and handle outliers, while the N-BEATS model, being a deep learning model, has the ability to learn from complex data patterns. According to the boxplot, data for Vietnam and Thailand did not contain any outliers, Spain had 2 outliers, and South Korea had 1 outlier.

When decomposing the data from the countries, it is evident that there is a clear presence of trend and seasonality in the time series, as shown in Figure 3.7. Figure 3.7 illustrates the seasonal patterns of Vietnam and Thailand very clearly. Spain shows some seasonal variation, though less distinct, with noticeable fluctuations between months. In contrast, this seasonal pattern is not as pronounced for South Korea.

(b) **Research methodology**
   Holt–Winters model
   The Holt–Winters ETS model, popularized in the 1960s by Winters, is an extension of the simple ETS model, incorporating trend and seasonal components. The model consists of three main components:
   - **Level component**: Represents the average value of the time series.

$$L_t = \alpha \times (Y_t - S_{t-m}) + (1 - \alpha) \times (L_{t-1} + T_{t-1}) \tag{3.5}$$

where:
- L_t is the level estimate at time t.
- Y_t is the actual value of the time series at time t.
- S_{t-m} is the seasonal component of the time series at time $t-m$, where m is the length of the seasonal cycle.
- α is the smoothing coefficient for the level, within the range [0,1].

- **Trend component**: Describes the increasing or decreasing trend of the time series.

$$T_t = \beta \times (L_t - L_{t-1}) + (1 - \beta) \times T_{t-1} \tag{3.6}$$

where:
- T_t is the trend estimate at time t.
- β là is the smoothing coefficient for the trend.

- **Seasonal component**: Describes the repeating seasonal patterns in the time series.

$$S_t = \gamma \times (Y_t - L_t) + (1 - \gamma) \times S_{t-m} \tag{3.7}$$

where:
- S_t is the seasonal estimate at time t.
- γ is the smoothing coefficient for the seasonal component.

The Holt–Winters model has two main variants based on how the trend and seasonality are handled:
- **Additive**: Used when seasonal variations are fixed and do not change over time.
- **Multiplicative**: Suitable when seasonal variations change and are correlated with the level of the time series.

SARIMA model

The SARIMA (seasonal autoregressive integrated moving average) model is an extension of the ARIMA (autoregressive integrated moving average) model, incorporating seasonal components. The SARIMA model is represented by the notation $SARIMA(p, d, q)(P, D, Q)m$, where:
- p: number of lags in the autoregressive (AR) model.
- d: number of differencing operations required to make the time series stationary.
- q: number of lags in the moving average (MA) model.
- P: number of lags in the seasonal autoregressive model.
- D: number of seasonal differencing operations required.
- Q: number of lags in the seasonal moving average model.
- m: The length of the seasonal cycle (periodicity).

SARIMA combines the AR, integrated (I), and MA components, both for the regular and seasonal versions:

- **Autoregressive (AR)**: The autoregressive component describes the dependence of the current value on previous values in the time series:

$$AR : Y_t = \emptyset_1 Y_{t-1} + \emptyset_2 Y_{t-2} + \cdots + \emptyset_p Y_{t-p} + \cdots$$

- **Integrated (I)**: Used to make the time series stationary, typically by taking the difference of the series $I : (1 - B)^d Y_t$; where B is the differencing operator.
- **Moving average (MA)**: The MA model integrates information about past errors: $MA : \theta_1 \varepsilon_{t-1} + \theta_2 \varepsilon_{t-2} + \cdots + \theta_p \varepsilon_{t-p} + \cdots$
- **Seasonal component**: Similar to the above, but applied to data with seasonal cycles: $SAR : \Phi_1 Y_{t-m} + \cdots + \Phi_p Y_{t-pm}; SMA : \Theta_1 \varepsilon_{t-m} + \cdots + \Theta_Q \varepsilon_{t-Qm}$

Prophet model

The Prophet model is based on three main components:

- **Trend component**: This component is flexible in capturing increasing, decreasing, or irregular changes in the time series. The trend can be modeled using various functions, such as linear or logistic functions.
- **Seasonal component**: Prophet can model seasonal factors on daily, weekly, monthly, or yearly cycles. Seasonality is modeled using Fourier series, allowing it to capture complex seasonal patterns flexibly.
- **Holiday and events component**: Prophet allows users to customize holidays and special events, making the model more accurate in forecasting during these specific periods.

Model equation: The Prophet model is represented by the following equation:

$$y(t) = g(t) + s(t) + h(t) + \varepsilon_t \qquad (3.8)$$

in which:
$y(t)$: The forecast value at time t.
$g(t)$: The trend component.
$s(t)$: The seasonal component.
$h(t)$: The holiday and events component.
ε_t: The error of the model.

Parameter adjustment: Prophet provides the ability to fine-tune various parameters to suit the specific characteristics of the data:

- **Adjusting trend**: Parameters such as "changepoint_prior_scale" allow for adjusting the model's sensitivity to changes in trend.
- **Adjusting seasonality**: Parameters like "seasonality_prior_scale" enable adjusting the model's sensitivity to seasonality.
- **Adding holidays and events**: Specific holidays can be added, and their impact on the forecast can be adjusted.

N-BEATS model
N-BEATS (neural basis expansion analysis for time series) is a time series forecasting model that employs deep neural networks. N-BEATS was introduced as a novel approach to time series forecasting. The key difference between N-BEATS and traditional time series models is its avoidance of time-based structures (such as RNN or LSTM). Instead, it focuses on learning a set of basis functions to represent the data.

N-BEATS block
- **Basic structure**: Each block in N-BEATS is a fully connected deep neural network.
  - **Input and output**: The input to each block is past data, $X \in RT$, where T is the length of the observed time series. The output consists of two parts: predictions for the future $(\hat{Y})$ and a backcast vector (*backcast, B*), which is used to reconstruct the input.
  - **Functionality**: The computation performed is $\hat{Y}, B = f(X;\theta)$, where f represents the neural network and θ are the network's parameters.

Basis expansion
N-BEATS uses the concept of basis expansion for forecasting. It learns a set of basis functions (such as polynomials and trigonometric functions) and combines them to generate predictions for the time series. The goal of this component is to represent the prediction $(\hat{Y})$ as a linear combination of these basis functions. The formula for this is: $\backslash \hat{Y} = \sum_{i-1}^{P} V_i g_i(H;\theta)$, with V_i as the optimization coefficients and g_i as the basis functions, learned through the training process.

- **Multi-step forecasting:** The model is designed to perform multi-step regression prediction. This means that it has the ability to predict multiple data points in the future simultaneously.
- **Stacked architecture:** The N-BEATS model involves stacking multiple N-BEATS blocks on top of each other. Each block learns a part of the prediction, contributing to the final forecast, $\hat{Y}_{final}$, is the result of combining the outputs from all blocks: $\hat{Y}_{final} = \sum_{j=1}^{K} \hat{Y}_j$. This method enhances the model's representational capacity and improves prediction accuracy.
- **Trend and seasonal blocks**: There are two main types of blocks: trend blocks (learning long-term trends in the data) and seasonal blocks (capturing recurring cyclical patterns).
- **Deep learning and transfer learning**: The model leverages deep learning techniques and supports transfer learning, meaning it can be trained on one dataset and then applied to another dataset effectively.

3.5.2.3 Results and discussion
Method for evaluating results

In forecasting research, there is always an error between the model's output and the actual values. Typically, this arises from two main sources. The first is sampling error, which can be mitigated through outlier processing. The second is model error, which is limited by the model's generalization capability.

In forecasting studies, prediction accuracy is used to measure the model's fit to the actual values. For selecting evaluation metrics, mean absolute error (MAE), MAPE, and RMSE are commonly used as evaluation functions. However, while MAE can introduce bias in the evaluation of multivariate time series forecasting models, MAPE addresses this limitation. The specific formulas are presented in (3.9) and (3.10).

$$RMSE = \sqrt{\frac{\sum_{i=1}^{n}\left(y(i) - y'(i)^2\right)}{n}} \quad (3.9)$$

$$MAPE = 100 \times \frac{1}{n}\sum_{i=1}^{n}\left|\frac{y(i) - y'(i)}{y_i}\right| \quad (3.10)$$

where $y(i)$ represents the actual value for the month i, $y'(i)$ represents the predicted value for the month i, and n represents the length of the time series.

Holt–Winters model
The model is developed through the following steps:
- Reading and preparing the data.
- Splitting the data using the time series cross-validation (TSCV) method into five parts (*n_splits*=5n_splits=5) for cross-validation. This approach is more optimal than the traditional train/test split method.
- Identifying the best parameters using the GridSearch method for hyperparameters: trend, seasonal, and seasonal_periods.
- Evaluating model performance using MAPE and RMSE.
- Forecasting results for the final split of the data.
- Visualizing the results.

SARIMA model
The SARIMA model is developed through the following steps:
- Reading and preparing the data.
- Splitting the data using the TSCV method into five parts for cross-validation.
- Determining the best parameters using the Auto-ARIMA method to find hyperparameters, including p, d, q for the non-seasonal part, and P, D, Q, m for the seasonal part. Model selection is based on the Akaike information criterion.
- Training the model with the best parameters and evaluating model performance using MAPE and RMSE.
- Forecasting results for the final split of the data.
- Visualizing the results.

PROPHET model
The PROPHET model is developed through the following steps:

- Reading and preparing the data.
- Splitting the data using the TSCV method into five parts for cross-validation.
- Assigning parameters to the model, including fixed parameters such as yearly_seasonality,weekly_seasonality,daily_seasonality,seasonality_mode= "multiplicative",changepoint_prior_scale=0.01,seasonality_prior_scale=10, and the holidays component for the corresponding country. Model selection is based on the best-fit parameter criteria.
- Training the model with the best parameters and evaluating model performance using MAPE and RMSE.
- Forecasting results for the final split of the data.
- Visualizing the results.

N-BEATS model

The N-BEATS model is developed through the following steps:
- Reading and normalizing the data using MinMaxScaler.
- Defining the model with two main components:
  - **NBeatsBlock class**: This is the fundamental building block of the N-BEATS model. Each NBeatsBlock is responsible for forecasting a portion of the time series, processing backcasts and forecasts through trend or seasonality functions depending on the block_type.
  - **NBeatsNet class**: This creates the overall N-BEATS model by combining multiple NBeatsBlocks.
- Splitting the data with a prediction window of 12, creating train/validation datasets.
- Initializing the model with specific parameters such as input_size, theta_size, layer_sizes, EPOCHS, and horizon. The Adam optimizer is selected, with early-stopping implemented to avoid overfitting.
- Training the model with the best parameters and evaluating its performance using MAPE and RMSE.
- Forecasting results for the last 24 points of the data.
- Visualizing the results.

Discussion of results

The results of the four models across four countries are evaluated using two performance metrics: MAPE and RMSE. The objective of forecasting is to minimize errors. Depending on the type of forecast and whether the output is system-level or customer-specific, short-term load forecasts typically require a MAPE error below 5%. Medium- to long-term forecasts have higher allowable errors, often targeting below 10%.

As such, the forecasts indicate that all four models are suitable for conducting monthly electricity demand forecasts for all four countries.

Vietnam

- **Holt–Winters**: With a MAPE of 3.6% and RMSE of 918.2 GWh, this model demonstrates the best performance.

82 *Clean energy in South-East Asia*

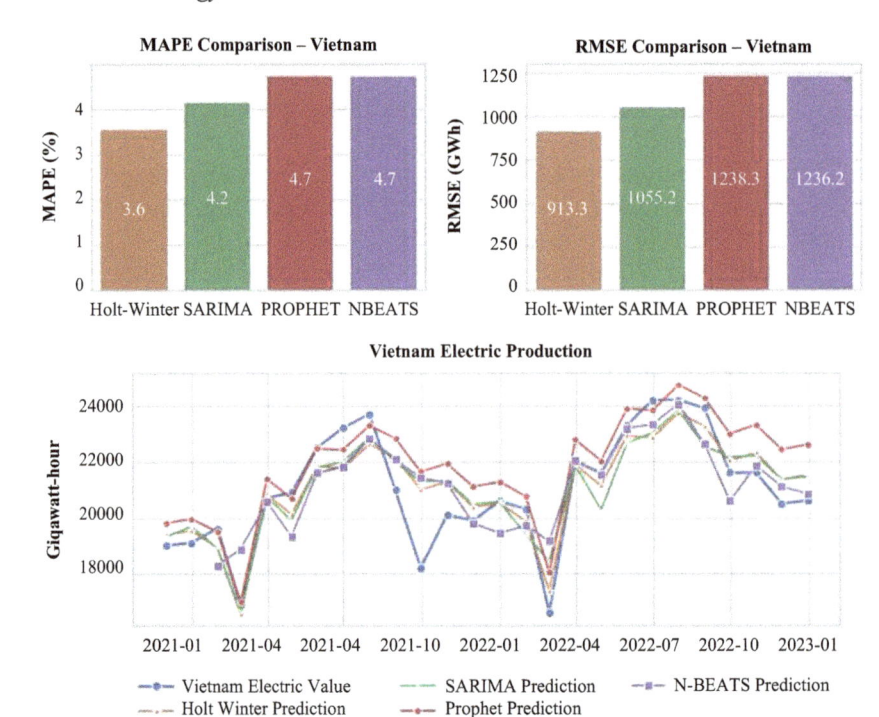

Figure 3.8 Vietnam electricity production prediction by different models

- **SARIMA**: With a MAPE of 4.2% and RMSE of 1055.2 GWh, SARIMA is less effective than Holt–Winters but still performs very well.
- **PROPHET**: With a MAPE of 4.7% and RMSE of 1238.3 GWh, this model is suitable for the data from Vietnam.
- **N-BEATS**: With a MAPE of 4.7% and RMSE of 1236.2 GWh, this model performs similarly to PROPHET (Figure 3.8).

Thailand
- **Holt–Winters**: With a MAPE of 3.4% and RMSE of 738.0 GWh, this model demonstrates good performance with Thailand's data.
- **SARIMA**: The model performs excellently with a MAPE of 2.8% and the lowest RMSE of 600.2 GWh.
- **PROPHET**: With a MAPE of 3.2% and RMSE of 662.7 GWh, it shows very good performance.
- **N-BEATS**: With a MAPE of 2.8% and RMSE of 690.7 GWh, this model performs similarly to SARIMA (Figure 3.9).

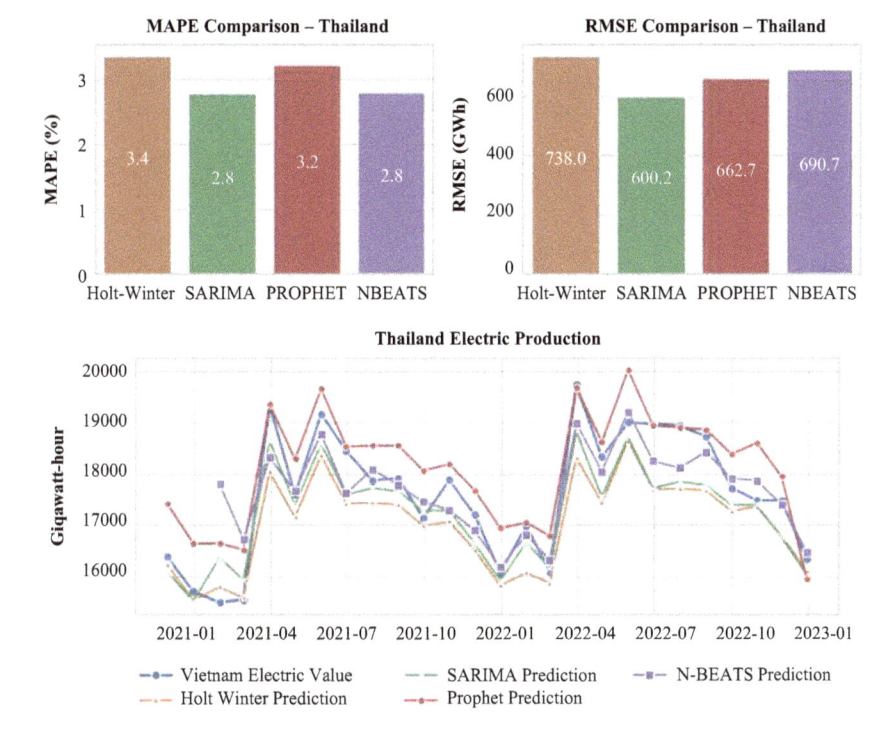

Figure 3.9 Thailand electricity production prediction by different models

Spain

- **Holt–Winters**: With a MAPE of 4.7% and RMSE of 1328.0 GWh, this model performs very well, ranking just behind N-BEATS.
- **SARIMA**: This model has the lowest performance among the four, with a MAPE of 8.5% and RMSE of 2396.1 GWh.
- **PROPHET**: With a MAPE of 5.1% and RMSE of 1507.0 GWh, it shows fairly good performance.
- **N-BEATS**: This model delivers the best performance and is most suitable for the data, with a MAPE of 4.6% and RMSE of 1299.4 GWh (Figure 3.10).

South Korea

- **Holt–Winters**: With a MAPE of 5.9% and RMSE of 1524.7 GWh, this model is quite effective in South Korea.
- **SARIMA**: This model performs poorly with a MAPE of 8.9% and RMSE of 2264.4 GWh.
- **PROPHET**: It delivers excellent performance with a MAPE of 3.3% and RMSE of 948.8 GWh.

84   *Clean energy in South-East Asia*

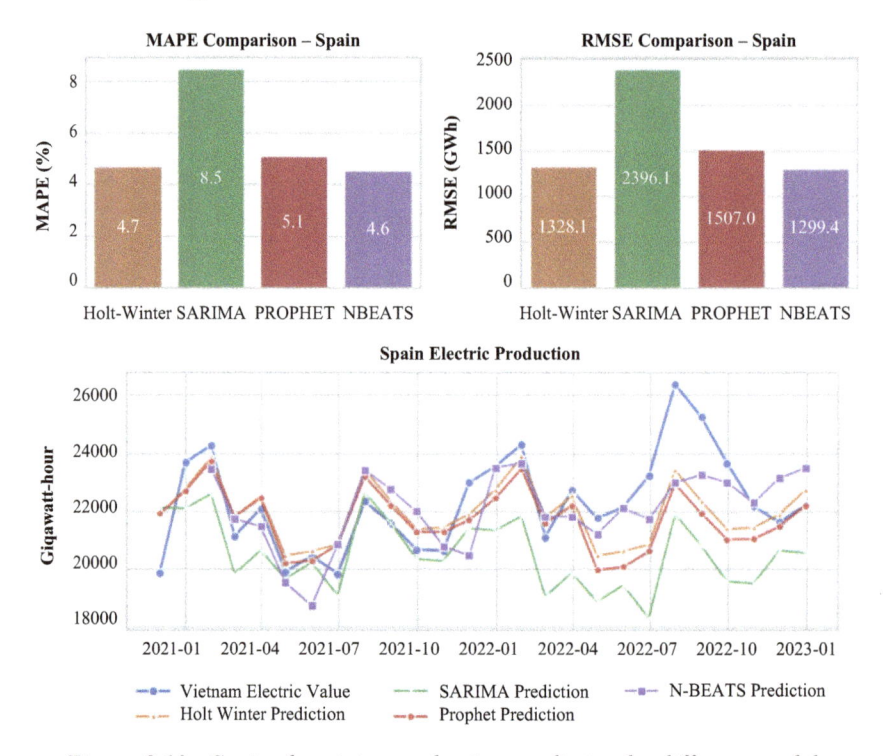

Figure 3.10   Spain electricity production prediction by different models

- **N-BEATS**: Achieves the best performance with the lowest MAPE of 1.7% and the lowest RMSE of 458.0 GWh (Figure 3.11).

Overall comparison:

- **Performance by country**: Each country has a model that works best for it. For instance, Vietnam suits traditional statistical models, while deep learning models like N-BEATS perform excellently in South Korea. SARIMA works particularly well in Thailand. This shows the diversity of the data and the need for model customization based on each specific case.
- **Model diversity**: No model excels in every situation. Each model has its strengths and weaknesses, depending on the characteristics of the data.
- **Model selection**: The choice of model must consider the specific characteristics of the data. For example, the N-BEATS model may be more suitable for data with unclear seasonality, as seen in South Korea, while Holt–Winters and SARIMA models may perform better with data exhibiting stable trends and cyclical patterns, as in Vietnam and Thailand.

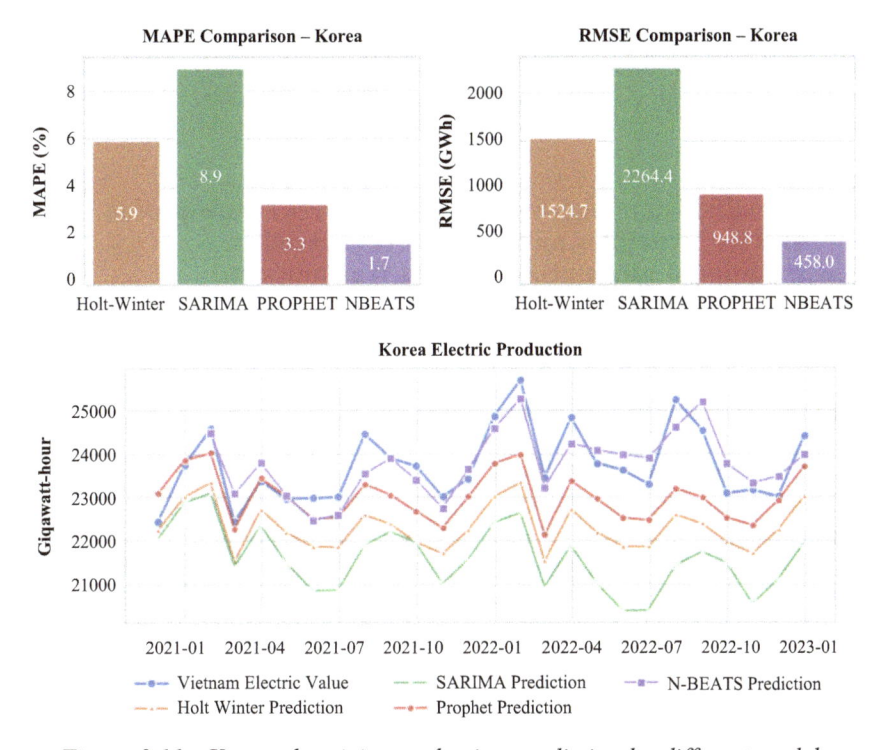

Figure 3.11 Korea electricity production prediction by different models

Conclusion

The research results indicate that when comparing the performance outcomes of each model with the four forecasting methods across different datasets, the results vary. This detailed analysis, with a sufficient number of models and countries, provides reliable insights into building forecasting methods. This analysis could be useful for improving time series forecasting methods for both short-term and long-term objectives, while offering significant support for operational units and electricity trading companies to plan activities in the competitive electricity market.

The effectiveness of the models is publicly disclosed, showcasing specific performance results that can be applied in real operations. Therefore, this study represents a valuable contribution to both the research community and the energy sector. From the research outcomes, it can be concluded that traditional statistical models can still perform well and remain competitive with modern machine learning models, although their results may not be as strong.

Compared to other studies, while the combination of models to achieve better results has not yet been realized, due to the use of various datasets and model-building approaches, it is not yet possible to provide a definitive conclusion and comparison. A common data source would be needed to make the most accurate comparison. Future work will focus on using a unified algorithm to select the most suitable model for forecasting monthly electricity production across different datasets, integrating external factors, improving data quality, and enhancing execution speed.

3.6 Introduction to high-voltage direct current transmission and its principles and benefits compared to traditional alternating current transmission

3.6.1 Recent achievements in transmission technology

Since the 1980s, the development of science and technology has led to remarkable advancements in power transmission technology.

3.6.1.1 Superconducting technology

High-temperature superconducting (HTS) technology is rapidly advancing. Conductors using heat-resistant superconducting materials can carry currents 2–3 times greater than conventional conductors. Superconducting materials are currently used in cables with voltages up to 138kV. Superconducting cables (HTS-cables) have reached their third generation, developed by the American superconductor corporation (AMSC, USA). The longest superconducting cable line to date is 600 m, with a voltage of 138 kV and a capacity of 574 MVA, connecting the Holbrook station to the Long Island Power Authority's system (USA), expected to be operational in 2008.

Overhead lines using composite aluminum core conductors can replace conventional steel-reinforced aluminum conductors, providing twice the transmission capacity, which is ideal for upgrading power transmission systems in large cities and areas with limited transmission corridors.

3.6.1.2 Trend toward smaller power systems

In North America, the total capacity of the Eastern Interconnection system is 600,000 MW, while the Western Interconnection system has a capacity of 130,000 MW. When one side experiences a grid failure, it can propagate to the other side. There is a trend toward dividing large systems into smaller PS, making operations more manageable. These smaller systems are interconnected by high-voltage direct current (HVDC) lines or back-to-back converter stations. For the USA, the cost of this approach is estimated at 8–10 billion USD (according to a study by the Northeast Energy Cooperation Council), which, compared to the estimated 6 billion USD loss from the 2003 blackout, makes this project highly relevant, especially as the development of power electronics technology is reducing the cost of HVDC and flexible AC transmission systems (FACTS).

Currently, Swiss company ABB has successfully developed small-scale HVDC systems (HVDC Light) with capacities of several tens of MW at acceptable costs. HVDC Light uses transistor gate isolated (IGBT) technology, which is much more affordable than traditional thyristors. IGBT technology is also used in converter stations as voltage source converters, reducing issues like voltage fluctuations, harmonic distortion, and reactive power compensation in AC systems. HVDC Light technology has been implemented in the USA (40 km submarine cable—330 MW connecting Connecticut to Long Island), Australia (180 km—200 MW connecting Murray Link to the south), and the US–Mexico interconnection (back-to-back station 36 MW).

3.6.1.3 Variable frequency transformers

GE Energy (Atlanta, USA) has developed the variable frequency transformer (VFT), which can continuously change frequency and voltage phase angle. Along with HVDC, VFTs are used to interconnect two asynchronous PS. Currently, VFTs are used to connect asynchronous systems in Québec (Canada) and Laredo (Texas).

In addition to interconnecting asynchronous systems, VFT technology is being developed to transmit power between synchronized systems. In this case, the VFT acts as a phase angle regulator, first used in the Pennsylvania to New York City interconnection with a capacity of 300 MW. Three additional projects with a total capacity of 900 MW are under development to connect Linden (New Jersey) to New York, expected to be operational in 2009.

3.6.1.4 ETO thyristor applications in FACTS and HVDC

The Sandia National Lab (USA) has successfully developed the ETO thyristor (emitter turnoff thyristor), which has fast response times (5 kHz), can handle large currents (4 kA), and high voltages (6 kV), while being much cheaper than traditional thyristors. This type of thyristor is ideal for controlling FACTS systems and HVDC converters. ETO thyristors are currently being developed for use in static compensators (STATCOMs) and for grid shock absorbers.

3.6.1.5 Fault current limiters

As PS grow in scale, fault currents increase, leading to the need for device upgrades. One alternative to upgrading devices is to use fault current limiters (FCLs), which operate by combining low-inductance coils with HTS cables. FCLs are being tested in the USA and Japan.

US manufacturers of superconducting materials are also developing high-resistance stabilizer cables. These cables have the feature that under normal conditions, current flows through the HTS superconducting material, but when a fault occurs, the high-resistance layer activates and isolates the fault. Once the fault is cleared, the superconducting layer resumes normal operation.

3.6.1.6 Efforts in ultra-high-voltage direct current and alternating current transmission

Currently, AC transmission technology up to 800 kV has been mastered with 25 years of experience, and there are no technical barriers. The ±600 kV DC transmission technology has also been mastered (with over 20 years of experience), but for voltages above ±600 kV, further development is required, mainly for testing equipment in converter stations. For power lines, design parameters for both AC

88 *Clean energy in South-East Asia*

and DC transmission have been fully established. According to statistics, if pursued rigorously, a super high-voltage DC system (±600 kV) can be designed in 3 years, while an EHVAC system can be designed in 1 year.

3.6.2 Key technical requirements for high-voltage direct current transmission

Basic components

An essential component of the HVDC power converter is the valve. If the valve is made from one or more series-connected power diodes, it is called an uncontrolled valve. If it is constructed from a series of thyristors, it is called a controlled valve.

Valve symbols according to IEC (International Electrotechnical Commission) as shown in the diagram below:

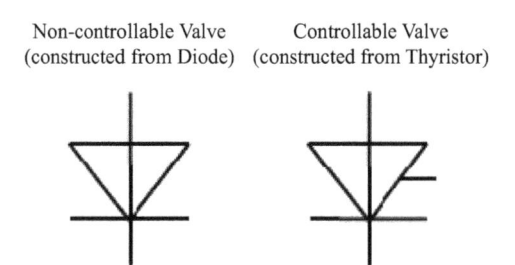

Non-controllable Valve   Controllable Valve
(constructed from Diode) (constructed from Thyristor)

3.7 Potential integration of HVDC into existing power grid systems and real-world HVDC transmission projects

This part shows the applications in specific cases in Vietnam.

3.7.1 Assumptions used in the calculations

The source and grid data used to calculate the system operation mode come from the following sources:

- The "National Power Development Plan for the period 2006–2015, with a view to 2025," prepared by the Institute of Energy in 2006 and approved by the Prime Minister through Decision No. 110/2007/QD-TTg on July 18, 2007.
- The "Research Project on the Possibility of Energy Cooperation between Vietnam and China," prepared by the Institute of Energy in 2004.

3.7.2 Technical assumptions

- Each 500 kV single-circuit overhead line with a cross-section of 4×AC 330 mm^2 can carry a maximum load of 1,500 MW, while a 500 kV three-phase 6×480 mm^2 line can carry 2,000 MW.

- Each 500 kV AC substation can carry a maximum load of 2,000 MW (equivalent to 3 × 900 MVA transformers).
- Each 500 kV DC single-circuit line has a maximum load capacity of 1,500 MW, while a ±500 kV double-circuit line can carry up to 3,000 MW.
- Each 765 kV single-circuit overhead line can carry an average load of 2,500 MW (natural capacity), with fault loading up to 3,750 MW (1.5 times the natural capacity).
- Reactive power compensation devices (SVC) and transverse compensation reactors need to be installed to ensure the proper operation of ultra-high voltage transmission lines.
- The 500 kV transmission network that supplies power to the Southeastern region must deliver at 500 kV voltage.
- The project stops at calculating power flow in the PS, determining system losses and losses on transmission lines, without considering PS stability factors.

3.7.3 Economic assumptions

- Investment costs for substations and transmission lines are calculated based on unit investment costs from similar projects in Vietnam and worldwide.
- Investment costs for AC-DC and DC-AC converter stations are based on unit investment costs from similar projects worldwide.
- The investment cost for double-circuit lines is 1.6 times that of single-circuit lines.
- The investment cost for DC lines is 0.8 times the investment cost of AC lines with the same voltage and circuit count.
- T_{max} = 6,000 h for thermal and nuclear power plants, T_{max} = 5,000 h for the system linked to Chinese electricity imports.
- Loss electricity price = 4.5 US cents.
- Discount rate: 10%.
- Annual maintenance and operation cost = 2% of capital investment.
- Construction time: 5 years.
- Calculation period: 30 years.

3.7.3.1 Investment cost for power transmission lines

The investment cost for 500 kV transmission lines in recent years (Table 3.3):

The average investment cost for 500 kV double-circuit transmission lines is approximately 481,832 USD. With a 20% contingency, the calculated investment cost for 500 kV double-circuit transmission lines is 0.6 million USD per circuit.

According to data from KOPEC in 2006, the investment cost for constructing transmission lines is as shown in Table 3.4.

It is observed that when the transmission voltage doubles, the investment cost approximately doubles. According to statistics on 500 kV transmission lines in Vietnam, construction costs typically account for over 60% of the total cost. Compared to construction costs in South Korea, Vietnam's construction costs are lower, resulting in a lower overall cost for building transmission lines.

Table 3.3 Investment cost per power transmission line

No.	Project name	Number of circuits	Cross-sectional area	Length (km)	Total estimate (USD)	Investment rate (USD/km)
1	Phu Lam–Long An	Single power line	4×ACSR795MCM	62.0	19,323,972	311,677
2	Song May–Tan Dinh	Double power line	4×ACSR330	41.0	18,363,397	447,888
3	Phu My–Song May	Double power line	4×ACSR330	65.9	33,989,647	515,776
	Average	Double power line				481,832
	Approximate Investment Rate (+20%)	Double power line				**578,198**

Table 3.4 Investment cost for power transmission line construction

No.	Transmission line item	Number of circuits	Cross section	Unit price (Million USD)
1	765 kV transmission line	Double circuit	6 × 480 mm^2	3.5
2	345 kV transmission line	Double circuit	4 × 480 mm^2	1.8
3	154 kV transmission line	Double circuit	2 × 480 mm^2	0.7

The investment cost for 765 kV transmission lines in Vietnam is approximately two times the cost of 500 kV lines. Therefore, the investment cost for a 765 kV double-circuit line is 1.2 million USD/km.

The investment cost for a ±500 kV DC double-circuit line is 0.8 × 0.6 = 0.48 million USD/km, while for a single circuit, it is 0.3 million USD/km.

3.7.3.2 Investment cost for AC substations

The investment cost for substations is mainly for the equipment, based on the average global investment cost.

For a 500 kV substation with three transformers, capable of handling 2,000 MW, the investment cost is 50 million USD per substation (with three transformers totaling 24 million USD, reactive power compensation equipment costing 10 million USD, and other terminal equipment amounting to 16 million USD).

The average investment cost for a 500 kV substation is 25,000 USD/MW.

For a GIS 765/345/22 kV substation with 3 × 2,000 MVA transformers in South Korea, the investment cost is 200 million USD per substation. The calculated investment cost for a 765/500 kV substation is 30,000 USD/MW (which is 20% higher than the cost for a 500/220 kV substation).

3.7.3.3 Investment cost for AC–DC, DC–AC conversion stations

The investment cost is based on the average global cost.

According to a 1997 study by Oak Ridge National Laboratory, USA, which investigated turnkey project costs for four HVDC projects from three different manufacturers, the results were as shown in Table 3.5.

Breakdown of costs for each project (cost adjusted to 1996 prices) (Table 3.6):

It is observed that the larger the capacity of the converter station, the lower the investment cost per unit. For capacities ranging from 1,000 to 3,000 MW per station, the project suggests using an investment cost of 100 USD/kW per station for the calculations.

Table 3.5 Transmission system parameters (DC and AC)

System	DC voltage	Power (MW)	AC voltage
1	±250 kV	500	230 kV
2	±350 kV	1,000	345 kV
3	±500 kV	3,000	500 kV
4	back-to-back	200	230 kV

Table 3.6 Cost breakdown of transmission system components by power capacity and voltage

No.	Item	Back-to-back 200 MW	±250 kV 500 MW	±350 kV 1,000 MW	±500 kV 3,000 MW
1	Valve system	19.0%	21.0%	21.3%	21.7%
2	Transformer substation	22.6%	21.3%	21.7%	22.0%
3	DC distribution yard	3.0%	6.0%	6.0%	6.0%
4	AC distribution yard	10.7%	9.7%	9.7%	9.3%
5	Control, protection, and communication	8.7%	8.0%	8.0%	7.7%
6	Construction cost	13.0%	13.7%	13.7%	13.7%
7	Auxiliary costs	2.0%	2.3%	2.3%	2.3%
8	Management costs	21.0%	18.0%	17.3%	17.3%
	Total cost (million USD)	43.3	145.0	213.7	451.7
	Cost $/kW/substation	217	145	107	75

3.7.3.4 Investment cost for SVC compensation devices

The global average cost for a 300 MVAr SVC is 30 million USD.

3.8 Calculation of current cost adjustments when input factors change

3.8.1 Power system simulation in PSS/E

The PSS/E program, copyrighted by Power Technologies, Inc.®, is an integrated computer-based PS simulation program designed for calculating and researching the operation and planning of electrical systems. The system analysis calculations include:

- Power flow calculations and system analysis
- Analysis of symmetrical and asymmetrical faults
- System equivalencing
- Electromechanical transient simulation

The input data structure for PSS/E is organized into records, where the simulation parameters for all equipment and system structures are specified. The use of the PSS/E program has been implemented at the National Dispatch Center (A0) and the Institute of Energy for system operation and planning calculations. Currently, the highest voltage in the Vietnamese PS is 500 kV AC. The project will not present the simulation steps for each element in the system but will focus more on the parameters of the expected ultra-high voltage AC and DC transmission lines that will be calculated in the study. These parameters will be used for power flow calculations and system losses.

For the 765 kV overhead transmission line, the phase conductor consists of $6 \times ACSR480$ mm², with a phase spacing of 400 mm (according to South Korean standards) (Table 3.7).

Main parameters of the 500 kV DC system
- AC voltage: 500 kV
- Phase conductor: $4 \times ACSR480$ mm²
- Maximum transmission capacity for each 500 kV DC line: 1,500 MW

Table 3.7 Technical specifications of 765 kV transmission system

Parameter	765 kV
Nominal voltage (kV)	765
Permissible voltage (kV)	800
Resistance ($\times 10^{-6}$ pu/km)	1.951
Reactance ($\times 10^{-5}$ pu/km)	4.475
Charging (capacitance) ($\times 10^{-2}$ pu/km)	2.4
Natural power (MW)	2,315
S_{base} = 100 MVA	
U_{base} = 765 kV	

- Line unit resistance: R_{dc} = 0.06351 Ω/km
- Utilizes a 12-pulse rectifier/inverter scheme for each DC line (2 blocks of 6 pulses)
- Transformer for voltage conversion: 500/220 kV with load-tap-changer
- Maximum α angle = 300°, minimum α angle = 80°, maximum γ angle = 300°, and minimum γ angle = 80°

Planned transmission distances
- Transmission over 270 km from South Central Vietnam (Nuclear Power Plant 1) to Ho Chi Minh City with power levels ranging from 2,500 MW to 12,000 MW.
- Transmission over 450 km from HongHe, China to Hiep Hoa, Vietnam with power levels ranging from 1,500 MW to 4,500 MW.

3.8.2 Transmission distance of 270 km

According to the national power development plan and the planning of thermal power plants nationwide, the electricity source in the South Central region (around Nuclear Power Plant 1 – Ninh Thuan) by 2025 will be as shown in Table 3.8.

If the 2015–2025 period is separated, the sources available are as follows:
The locations of the thermal power plants are as shown in Table 3.9.

Table 3.8 List of power plants and operating capacity

No.	Power plant name	Capacity (MW)	Operational year	Notes
1	Southern Pumped Storage Power Plant	1,200	2019–2020	
2	Regional plants (Da Nhim, Dai Ninh, Ham Thuan, Da My)	937	đang V/h	
3	Cam Ranh I–#1,2	1,200	2015–2016	
4	Cam Ranh II–#1,2	1,200	2020	
5	Vinh Tan I–#1	600	2011	CSG
6	Vinh Tan I–#2	600	2012	CSG
7	Vinh Tan II–#1	600	2013	EVN
8	Vinh Tan II–#2	600	2014	EVN
9	Vinh Tan III–#1	1,000	2015	EVN
10	Vinh Tan III–#2	1,000	2016	EVN
11	Power Plant 1&2–#1	2,000	2020	EVN
12	Power Plant 1&2–#2,3,4	6,000	2021–2025	EVN
	Total	16,937		

Table 3.9 List of power plants and operating capacity (updated)

No.	Power plant name	Capacity (MW)	Operational year	Notes
1	Southern Pumped Storage Power Plant	1,200	2019–2020	
3	Cam Ranh I–#1,2	1,200	2015–2016	
4	Cam Ranh II–#1,2	1,200	2020	
9	Vinh Tan III–#1	1,000	2015	EVN
10	Vinh Tan III–#2	1,000	2016	EVN
11	Power Plant 1&2–#1	2,000	2020	EVN
12	Power Plant 1&2–#2,3,4	6,000	2021–2025	EVN
	Total	13,600		

With a distance of 270 km and a transmission capacity above 12,000 MW, we will consider the following transmission voltage levels: alternating current (AC) 500 kV, 765 kV, and direct current (DC) ±500 kV (Figure 3.12).

The simulation of transmission options through a simplified PS in PSS/E consists of a power generation source with capacities ranging from 2,500 MW to 12,000 MW, transmitted to corresponding load buses with the assumption that the load station capacity is as large as possible and the number of transmission lines is minimized. From this, the total transmission loss can be determined.

The process of simulating the simplified PS is represented in the diagrams below, where the scope for determining investment capital and current costs includes the items within the enclosing rectangle:

With the transmission power range from 2,500 MW to 12,000 MW and the assumptions as outlined above, the transmission options are summarized in Table 3.10.

The results of the power loss calculation are as shown in Table 3.11.

Comment: The transmission option using 765 kV AC has significantly lower losses compared to the other two options (only about 40% of the losses in the AC 500 kV option). The power loss for the 500 kV DC transmission is only about 80% of the losses in the AC 500 kV option (Figure 3.13).

Determining investment costs: For each transmission configuration, there are different levels of investment, with the investment cost per unit as previously mentioned. The total investment for the transformer station is approximately double the investment for the load transformer station, as it must account for the source transformer. The results are as shown in Table 3.12.

The graph representing the correlation between investment cost and transmission capacity is as shown in Figure 3.14.

Comment: At a distance of 270 km, the option using DC transmission lines always has the highest investment cost (more than twice that of the AC 500 kV option). The investment cost for the AC 765 kV system is higher than the AC 500 kV option by 20–50%, with the investment cost difference decreasing as the transmission capacity increases.

3.8.3 Calculation of present value

For a calculation period of 30 years, with investment data for the construction of the transmission system, operational costs of 2% of the investment, a discount rate of 10% per year, power loss, T_{max} = 6,000 h (thermal power plants may have a higher T_{max}), and a loss electricity price of 4.5 US cents, we have the following summary table (Table 3.13):

The graph representing the relationship between the present value cost and transmission capacity is as shown in Figure 3.15.

Comment: For transmission capacities between 2,500 and 5,000 MW, the AC 500 kV option has a lower present value cost than the AC 765 kV option by 15–30%. When the transmission capacity increases from 6,000 to 12,000 MW, the

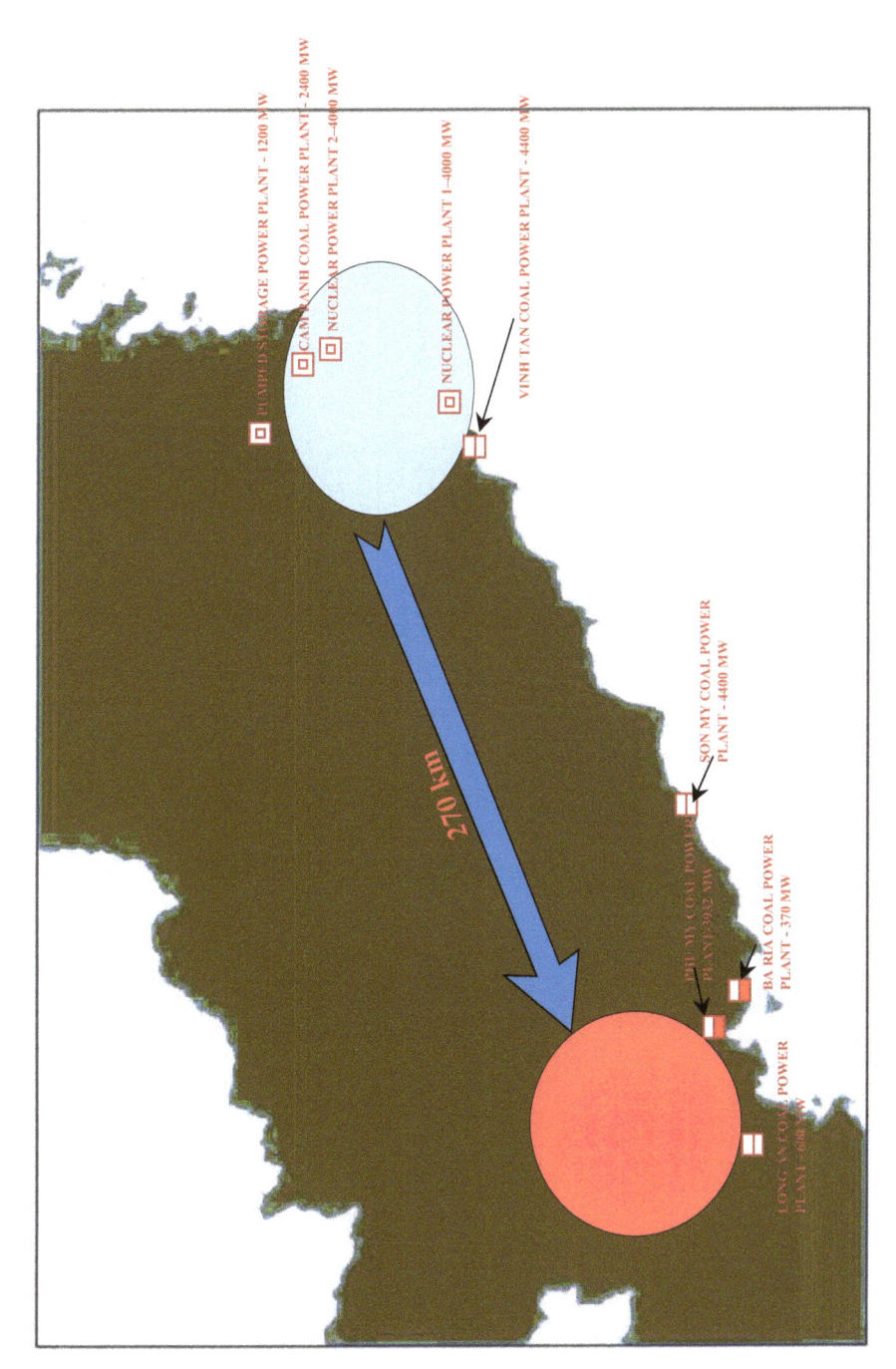

Figure 3.12 Map of power plant locations in the south central region

Table 3.10 Transmission voltage level and power line configuration

No.	Power level (MW)	AC 500 kV Transmission voltage level	AC 500 kV Substation for load	AC 765 kV Transmission line	AC 765 kV Substation for load	DC 500 kV Transmission line	DC 500 kV Converter station, load
1	2,500	Three power lines (1 double, 1 single)	2 substations 500 kV	1 power line (1 single)	1 substation 765 kV, 2 substations 500 kV	2 power lines (1 double)	2 converter stations, 2,500 kV stations
2	3,000	Three power lines (1 double, 1 single)	2 substations 500 kV	2 power lines (1 double)	1 substation 765 kV, 2 substations 500 kV	2 power lines (1 double)	2 converter stations, 500 kV stations
3	4,000	Four power lines (two double, 1 single)	2 substations 500 kV	2 power lines (1 double)	1 substation 765 kV, 2 substations 500 kV	3 power lines (1 double, 1 single)	2 converter stations, 2,500 kV stations
4	5,000	5 power lines (2 double, single)	3 substations 500 kV	3 power lines (1 double, 1 single)	1 substation 765 kV, 3 substations 500 kV	4 power lines (2 double)	2 converter stations, 3,500 kV stations
5	6,000	6 power lines (3 double)	3 substations 500 kV	3 power lines (1 double, 1 single)	1 substation 765 kV, 3 substations 500 kV	4 power lines (2 double)	2 converter stations, 3,500 kV stations
6	8,000	8 power lines (4 double)	4 tram 500 kV	4 power lines (2 double)	1 substation 765 kV, 4 substations 500 kV	6 power lines (3 double)	2 converter stations, 4,500 kV stations
7	10,000	10 power lines (5 double)	5 substations 500 kV	5 power lines (2 double, 1 single)	1 substation 765 kV, 5 substations 500 kV	7 power lines (3 double, 1 single)	2 converter stations, 5,500 kV stations
8	12,000	12 power lines (6 double)	6 substations 500 kV	6 power lines (3 double)	1 substation 765 kV, 6 substations 500 kV	8 power lines (4 double)	2 converter stations, 6,500 kV stations

Technologies and solutions for smart energy systems 97

Table 3.11 Power loss and transmission loss by voltage class

Voltage class	Power loss in MW							
	2,500	3,000	4,000	5,000	6,000	8,000	10,000	12,000
	Transmission loss (MW)							
AC 500 kV	57.7	81.1	110.8	140.5	170.2	229.8	289.5	349.2
AC 765 kV	20.9	28.4	46.5	50.9	69.7	93.0	117.4	143.0
DC 500 kV	49.5	68.4	83.7	106.5	136.8	174.9	220.5	273.6
	Loss correlation							
AC 500 kV	100%	100%	100%	100%	100%	100%	100%	100%
AC 765 kV	36%	35%	42%	36%	41%	40%	41%	41%
DC 500 kV	86%	84%	76%	76%	80%	76%	76%	78%

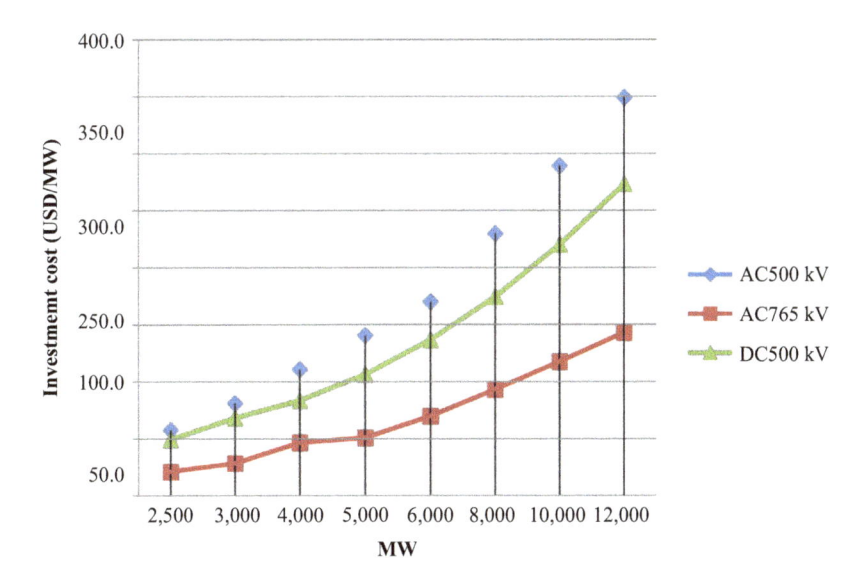

Figure 3.13 Graph representing losses based on transmission capacity when L = 270 km

present value cost of the AC 765 kV and AC 500 kV options become comparable. The DC ± 500 kV transmission option has a present value cost 1.8 times higher than the corresponding AC transmission option, due to the very high construction cost of the conversion stations which cannot be offset by the lower costs of the transmission lines and power losses.

Consideration of how present value costs change with investment costs:

The investment cost for a 765/500 kV substation decreases to 25,000 USD/MW (the same as for a 500/220 kV substation). When the power is between 4,000 and

Table 3.12 Investment capital by voltage level

Voltage level	Investment capital (million USD)							
	2,500	3,000	4,000	5,000	6,000	8,000	10,000	12,000
Investment capital								
AC 500 kV	388.25	413.25	524.00	675.25	786.00	1048.00	1310.00	1572.00
AC 765 kV	596.50	633.00	736.00	981.50	1036.50	1280.00	1640.50	1920.00
DC 500 kV	754.60	879.60	1210.60	1509.20	1759.20	2388.80	2969.80	3518.40
Investment capital comparison with AC 500 kV								
AC 500 kV	100%	100%	100%	100%	100%	100%	100%	100%
AC 765 kV	154%	153%	140%	145%	132%	122%	125%	122%
DC 500 kV	194%	213%	231%	224%	224%	228%	227%	224%

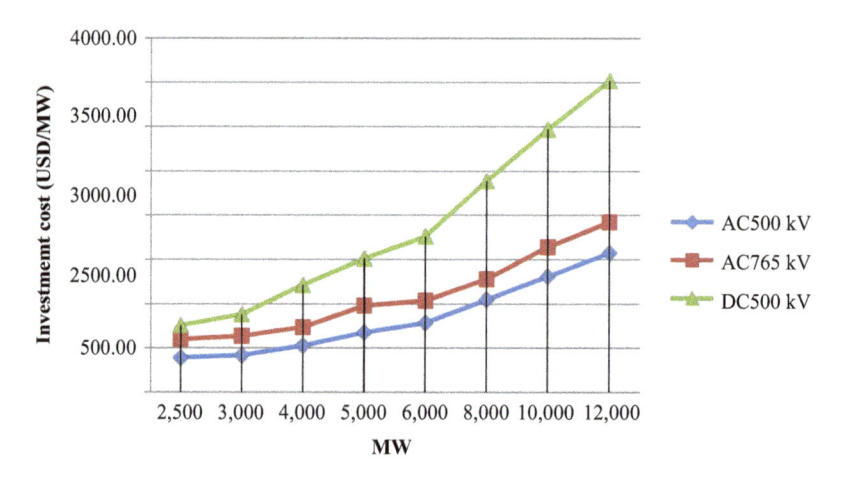

Figure 3.14 Graph representing investment costs based on transmission capacity when L = 270 km.

6,000 MW, the present value cost of both AC 765 kV and AC 500 kV options is similar. For power greater than or equal to 8,000 MW, the AC 765 kV option becomes more advantageous than AC 500 kV (Figure 3.16).

Investment cost for the 765/500 kV substation is reduced to 20,000 USD/MW: When the power is between 3,000 and 5,000 MW, the present value cost of both AC 765 kV and AC 500 kV are similar. For power greater than or equal to 6,000 MW, the AC 765 kV option becomes more advantageous.

Investment cost for the AC–DC, DC–AC conversion station is reduced to 75 USD/kW: The present value cost of the DC ± 500 kV transmission option remains 50–60% higher than the AC 500 kV option (Figure 3.17).

Table 3.13 Current cost by voltage level

Voltage level	Current cost (million USD)							
	2,500	3,000	4,000	5,000	6,000	8,000	10,000	12,000
	Current cost							
AC 500 kV	411.20	467.90	605.54	776.82	914.45	1223.66	1533.04	1842.41
AC 765 kV	527.70	569.55	682.92	893.65	968.23	1206.32	1543.32	1814.86
DC 500 kV	702.98	835.87	1134.38	1417.49	1671.74	2253.36	2806.11	3343.47
	Correlation compared to AC 500 kV							
AC 500 kV	100%	100%	100%	100%	100%	100%	100%	100%
AC 765 kV	128%	122%	113%	115%	106%	99%	101%	99%
DC 500 kV	171%	179%	187%	182%	183%	184%	183%	181%

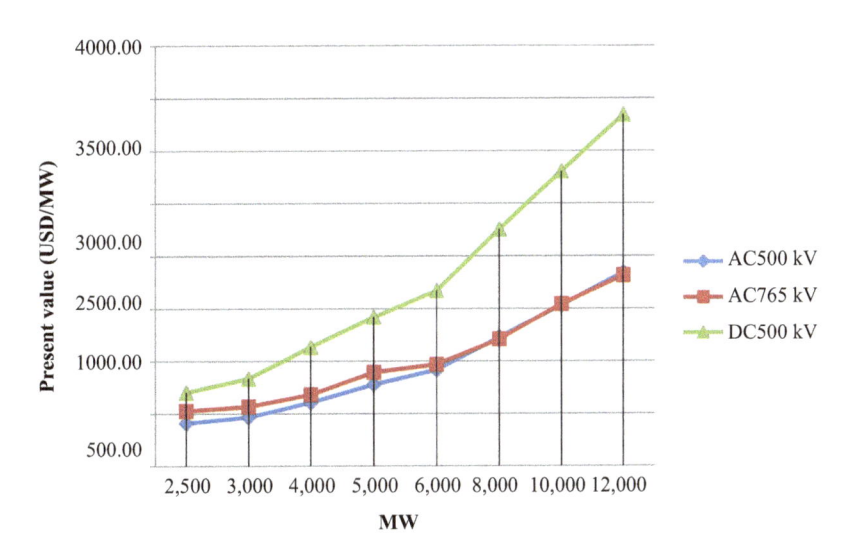

Figure 3.15 Graph representing the present value of costs based on transmission capacity when L = 270 km

3.8.4 Transmission distance of 450 km

According to the "Study on the Potential for Energy Cooperation between Vietnam and China" conducted by the Institute of Energy in December 2004, the 2011–2020 phase, and the "500 kV and 220 kV Grid Connection Plan between Vietnam and China by 2025" also conducted by the Institute of Energy in December 2006, Yunnan Province (China) could supply 3,000 MW to Vietnam. The potential transmission distance is from the HongHe substation to the 500 kV Sóc Sơn substation (now the 500 kV Hiệp Hòa substation), with a distance of 450 km.

100 *Clean energy in South-East Asia*

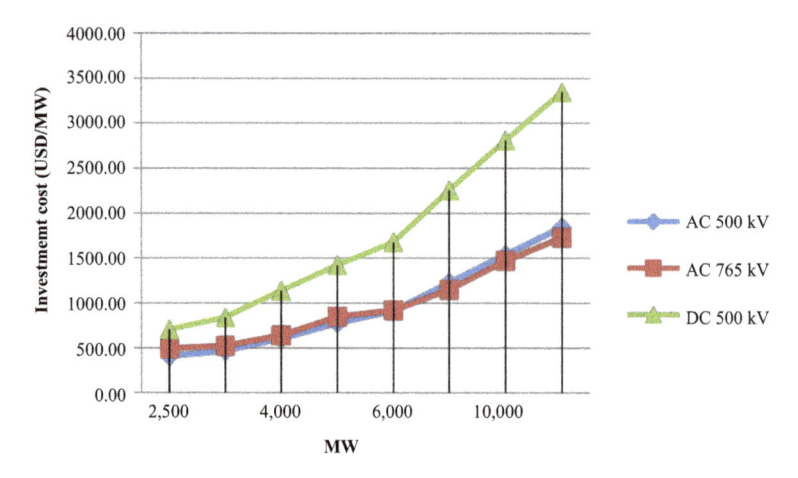

Figure 3.16 Investment cost for the 765/500 kV transformer station decreases by 25,000 USD/MW

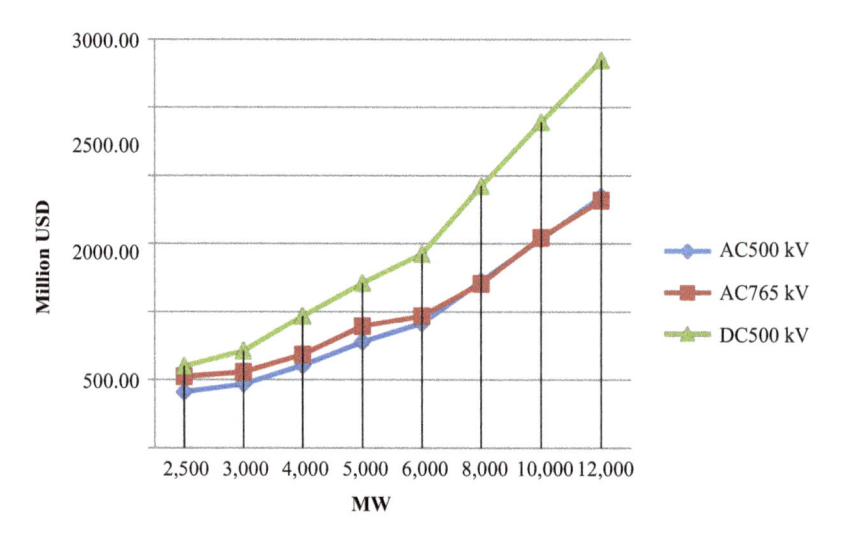

Figure 3.17 The investment cost for the AC–DC and DC–AC converter stations decreases by 75 USD/kW

Yunnan has a huge hydropower potential (about 100,000 MW). According to the Southern power grid company (YEPG) plan, the power capacity to be supplied outside the province by 2020 is as follows:

- 2005: 1,600 MW—for Guangdong
- 2006: 2,800 MW—for Guangdong

- 2007–2010: 7,800 MW—for Guangdong (4,800 MW already agreed between Yunnan and Guangdong)
- Potential to supply 3,000 MW for Guangxi and 3,000 MW for Vietnam
- 2011–2020: 23,800 MW—for Guangdong, Guangxi
- 3,000 MW—for Thailand (under an agreement)
- 15,000–20,000 MW—for the Central China region
- Potential to supply 3,000 MW for Vietnam

3.9 Potential DC transmission projects

Transmission of electricity from the south central coast to the southeast region

According to the National Power Development Plan for Phase VI, by 2020, the power load from the Nuclear Power Plant 1 in the South Central Coast will be 3,555 MW through the 500 kV grid—5 transmission lines, from sources such as nuclear power, pumped storage hydropower, and Binh Thuan thermal power (Vinh Tan). By 2025, the power load to the South Central Coast will be 7,976 MW through 10,500 kV transmission lines. The 500 kV double circuit lines in the calculation have a maximum transmission capacity of about 1,600 MW. Based on the transmission line specifications, these are 4×330 mm^2 phase conductors, a common size for 500 kV lines in Vietnam's PS.

According to the calculations in the study above, for a transmission power of 8,000 MW in 2025, with a distance of 270 km, we can consider two transmission options: AC 500 kV and AC 765 kV, as they have similar present value costs. The AC 765 kV transmission option has about 20% higher investment costs but much lower transmission losses (only about 40% of AC 500 kV). The AC 765 kV option also has the advantage of fewer circuits, meaning narrower transmission corridors. The 765 kV AC transmission technology is well-established and widely used in countries like Canada, the USA, India, and South Korea. The transmission distance of 270 km is not too long, so the issues of reactive power compensation and stability are less complicated compared to longer lines (over 1,000 km). Given the phased investment process over multiple years, a combined 765 kV AC and 500 kV AC transmission system would be flexible and economical, and should be considered on a case-by-case basis.

The DC $\pm$ 500 kV transmission option is not effective in this area due to the short transmission distance, which does not fully leverage the low cost advantages of the transmission lines, nor does it transmit reactive power, resulting in smaller losses than AC 500 kV.

The DC $\pm$ 800 kV transmission option is not considered in the proposal, as no country currently uses it and there are still several issues that need further study, especially regarding the equipment at the two ends of the converter station.

The HVDC transmission system has a 60-year development history, with over 20 years of operational experience with the $\pm$600 kV systems. With advancements in power electronics technology, the cost of HVDC systems has been decreasing. HVDC systems have been widely deployed around the world and have become an essential part of national PS in countries like the USA, Brazil, and China.

In Vietnam, HVDC current transmission with ±500 kV has very high application potential when purchasing electricity from China at a capacity of 1,500 MW over a distance of 450 km. Although the present value cost of the DC system is about 40% higher than that of AC systems, DC lines ensure that the two countries' PS can operate independently.

For transmitting large amounts of power from the South Central region to the Ho Chi Minh City area, over a distance of 270 km, using a DC 500 kV transmission line is not economically viable compared to the AC 500 kV and AC 765 kV transmission systems. This is because the construction costs for the converter stations at both ends of the line are too high, resulting in an investment increase of 80–90% compared to the AC 500 kV option. Furthermore, the advantages of the DC system, such as lower line investment costs, lower losses, and the absence of reactive power compensation issues, are not fully realized in this case.

Recommendations

Currently, there are no standard documents or regulatory frameworks for HVDC transmission lines in Vietnam. The most recent "Electrical Equipment Standards" issued by the Ministry of Industry (now the Ministry of Industry and Trade) in 2006 does not include regulations for HVDC systems. If the plan to purchase electricity from China through the HVDC 500 kV system becomes a reality, there will be a need for a regulatory framework for ultra-high-voltage direct current, which will serve as the basis for the development, management, and operation of the system.

Similarly, a new regulatory framework for 765 kV voltage levels should also be developed, as the feasibility of using 765 kV transmission for the Nuclear Power Plant 1, 2, and pumped storage plants, as well as thermal power plants in Ninh Thuan and Binh Thuan, is very promising.

3.10 Impact of renewable energy sources on the power quality of the grid and discussion on reducing energy losses and power losses, improving voltage quality and reliability

3.10.1 Building a model to calculate power supply reliability and assess the impact of wind power plants on the distribution grid

To evaluate the reliability of power supply and analyze the impact of wind power plants (WPPs) on the distribution grid, it is essential to construct a model to calculate the expected power deficit for local load groups near the connection area with the WPP.

Research objective: This study introduces a method for building a model to calculate the system's state probability when connected to wind power sources and simultaneously estimates the expected power deficit for consumers.

Research methodology: The method presented in the study is illustrated through a real case study of the Tuy Phong Wind Power Plant in Binh Thuan. This model helps not only determine the potential for stable power supply but also assess the impact of wind power on the stability and efficiency of the local distribution grid.

Significance of the study: The development and application of this model provide an effective tool for managers and engineers to:

- Forecast and minimize the risk of power shortages in areas near WPPs.
- Optimize the integration of renewable energy sources into the existing PG.
- Enhance the reliability and stability of the distribution grid, ensuring continuous and efficient power supply to consumers.

By applying this calculation model, Vietnam can advance in integrating renewable energy sources while ensuring the stability and reliability of the national PS.

3.10.2 Analyzing the impact of wind power plants on the power grid and statistical analysis of wind turbine failures at the Tuy Phong Wind Power Plant—Binh Thuan

3.10.2.1 Power exchange characteristics between WPP and the power grid

To analyze the impact of the WPP on the PG, it is necessary to build characteristics (charts) for power exchange between the WPP and the PG through the connection elements, with data being collected for daily, monthly, and yearly periods.

WPPs are power sources that are entirely dependent on wind speed, resulting in fluctuating and unstable power output. Therefore, to evaluate the impact of this energy source on power quality and supply reliability, surveying, collecting, and analyzing the failure parameters of wind turbines is essential.

3.10.2.2 Failure parameters of wind turbines at Tuy Phong Wind Power Plant—Binh Thuan

a) **Failure classification**

Operational experience with wind turbines shows that the primary failure types fall into four main categories:

- **Mechanical system**: Includes components such as the gearbox, rotating shaft, and other mechanical parts.
- **Electrical system**: Includes the generator, wiring, and other equipment related to power transmission.
- **Control system**: Devices that control the operation and monitoring of the wind turbine.
- **Other failures**: Includes issues outside the above categories, such as environmental impacts or specific design flaws.

b) **Failure data statistics**

Failure data for the wind turbines at the Tuy Phong Wind Power Plant in Binh Thuan was collected from August 21, 2009 to April 15, 2013 (nearly 4 years). This

104 *Clean energy in South-East Asia*

statistical data provides detailed information on the occurrences and frequency of failures, which are presented in Table 3.1.

3.10.2.3 Practical operation experience

Vietnam currently lacks extensive experience in operating and collecting data from large-scale WPPs. However, the Tuy Phong Wind Power Plant in Binh Thuan, with nearly 5 years of commercial operation, has provided valuable data, albeit limited, for estimating some characteristic failure parameters of wind turbines under the climate and operational conditions in Vietnam.

3.10.2.4 Significance of the study

Based on real data from the Tuy Phong Wind Power Plant, this study contributes to:

- Assessing the adaptability of wind turbines to the climatic and environmental conditions in Vietnam.
- Building a database to improve the operational quality and maintenance of future WPPs.
- Proposing solutions to minimize adverse impacts on power quality and enhance the reliability of the grid when integrating wind power sources.

These analyses and statistics will serve as a crucial foundation for the development and efficient operation of WPPs in the context of Vietnam's energy transition (Table 3.14).

For the Tuy Phong Wind Power Plant, due to the high failure rate in the initial years of operation, it is necessary to consider the state of the plant with 2 turbines out of service. The probability of the plant's state with 0, 1, or 2 turbines out of service is presented in Table 3.15.

Connecting the WPP to the distribution grid increases the reliability of power supply for local consumers, reducing the expected energy shortfall for these consumers.

Table 3.14 Types of wind turbine failures

Types of wind turbine failures	T	T_{sc}	ω	q
1. Mechanical system failures	1,817.817	78.788	3.762	0.0338
2. Control system failures	3,111.417			
3. Electrical system failures	508.2			
4. Other types of failures	300.1			

T – Total downtime for repairs during the survey period (20 turbines) (hours)
T_{SC} – Average repair time per wind turbine per year per failure (hours)
ω – Average failure frequency (failures per year)
q – Probability of failure.

Table 3.15 Calculation results of wind power source state probability

State	Number of faulty units	State probability (*Pi*)
1	0	0.502736784
2	1	0.351738839
3	2	0.116894371

The expected energy shortfall is determined based on the regional load curve, the probability of supply availability of the system, and the power generation curve of the WPP, taking into account the failures of wind turbine units.

The method is demonstrated for the Tuy Phong—Binh Thuan Wind Power Plant case.

Further reading

[1] Vietnam's Eighth National Power Development Plan (PDP VIII).
[2] Environmental Protection Law: Law No. 72/2020/QH14, signed on 17/11/2020.
[3] Law on Water Resources: Law No. 17/2012/QH13, signed on 21/6/2012, effective from 01/1/2013.
[4] Planning Law: Law No. 21/2017/QH14, passed by the 14th National Assembly on 24/11/2017.
[5] Amendment Law No. 28/2018/QH14: Amendments and supplements to several articles of 11 laws related to planning, passed on 15/6/2018.
[6] Amendment Law No. 35/2018/QH14: Amendments and supplements to several articles of 37 laws related to planning, passed on 20/11/2018.
[7] Decree No. 29/2011/NĐ-CP: Issued by the Government on 18/2/2011, regulating strategic environmental assessment, environmental impact assessment, and environmental protection commitments.
[8] Decree No. 23/2006/NĐ-CP: Issued by the Government on 03/3/2006, on the implementation of the Law on Forest Protection and Development.
[9] Decree No. 46/2012/NĐ-CP: Issued by the Prime Minister on 22/5/2012, amending and supplementing several articles of:
[10] Decree No. 35/2003/NĐ-CP (dated 04/4/2003) detailing the implementation of several articles of the Law on Fire Prevention and Fighting.
[11] Decree No. 130/2006/NĐ-CP (dated 08/11/2006) on compulsory fire and explosion insurance policies.
[12] Decree No. 32/2006/NĐ-CP (2006): Application of the Sorensen Coefficient (S) for comparing species composition in the study area with neighboring areas.
[13] Decree No. 18/2015/NĐ-CP (Article 12) and Circular No. 27/2015/TT-BTNMT (Article 7): Provisions on consultation activities during the implementation of the Environmental and Social Impact Assessment (ESIA).

[14] Circular No. 27/2015/TT-BTNMT: Issued on 29/5/2015 by the Ministry of Natural Resources and Environment (MONRE) on Strategic Environmental Assessment, Environmental Impact Assessment, and Environmental Protection Plans (Appendix 2.3, Chapters 1, 3, 4, and 5).

[15] Circular No. 26/2011/TT-BTNMT: Issued on 18/7/2011 by MONRE, detailing certain provisions of Decree No. 29/2011/NĐ-CP on strategic environmental assessment, environmental impact assessment, and environmental protection commitments.

[16] Circular No. 12/2011/TT-BTNMT: Issued on 14/4/2011 by MONRE, on the management of hazardous waste.

[17] Directive No. 08/2006/CT-TTg: Issued by the Government on 08/3/2006, on urgent measures to prevent illegal deforestation, forest burning, and unauthorized logging.

[18] Ngo Dang Luu, Nguyen Hung, Nguyen Anh Tam, and Long D. Nguyen, (2021), "Wireless Power Transfer simulation for EV", Ministry of Culture, Sports and Tourism.

[19] Ngo Dang Luu, Nguyen Hung, and Long D. Nguyen, (2021), "Power flow calculation of the electric power system by MATLAB", Ministry of Culture, Sports and Tourism.

[20] Ngo Dang Luu, Nguyen Hung, Nguyen Anh Tam, and Long D. Nguyen, (2022), "Distance Relay Protection in AC Microgrid", Ministry of Culture, Sports and Tourism.

[21] Ngo Dang Luu, Nguyen Hung, Nguyen Anh Tam, and Long D. Nguyen, (2022), "Analysis of Solar Photovoltaic System Sunlight Shadowing", Ministry of Culture, Sports and Tourism.

[22] Ngo Dang Luu, Nguyen Hung, Le Anh Duc and Long D. Nguyen, (2022), "Estimate Frequency Response at the Command Line", Ministry of Culture, Sports and Tourism.

[23] KOPEC (Korea Power Engineering Company), (2006), *Investment Cost Data for Transmission Line Construction*.

[24] Oak Ridge National Laboratory, (1997), *Turnkey Construction Cost Study*, United States.

[25] Ministry of Trade, Industry and Energy, Republic of Korea, *South Korean Technical Standards for Transmission Line Construction*, various years.

[26] Ministry of Industry and Trade, *Vietnam National Power Development Plan VIII (PDP8)*, Hanoi, Vietnam.

Chapter 4
Study and application of smart grid

4.1 Introduction to smart grid and its importance in improving performance and network management

In recent years, with the strong development of the economy, the pace of industrialization, and the rapid growth of the tourism and service sectors in Da Nang City, the demand for electricity has significantly increased. Along with this growth, there is an increasing demand for high-quality power supply from customers. The research team needs to develop plans for the development of the power grid that align with the load demand, while also implementing reasonable operational measures to improve power quality and reduce electricity supply interruptions.

The distribution grid in Da Nang city currently has nearly 67 points capable of performing 22 kV closed-loop and operations. However, at these points, operators still need to manually isolate and transfer power in the event of a fault, resulting in significant time delays during fault resolution.

Isolating faults, sectioning faults, and transferring power to minimize the customer outage time are still primarily based on the experience of dispatchers and operations staff during shifts. However, with the development of intelligent electronic devices and communication infrastructure, the distribution automation system (DAS) is now being implemented to quickly detect faults, perform isolation, and reconfigure the grid. These are essential components of the smart grid, which significantly assist the electricity sector in improving operational efficiency and service quality (Figure 4.1).

Implementation of a combined model using Load Break Switches (LBS) and Reclosers, with strategic placement and spacing of these devices based on load distribution across each segment, to optimize the Distribution Automation System (DAS).

Based on the existing hardware and software infrastructure of the control center, the DAS data linkage will be implemented through the supervisory control and data acquisition (SCADA) system, utilizing common communication protocols such as IEC 60870-5-104, OPC, and other compatible protocols.

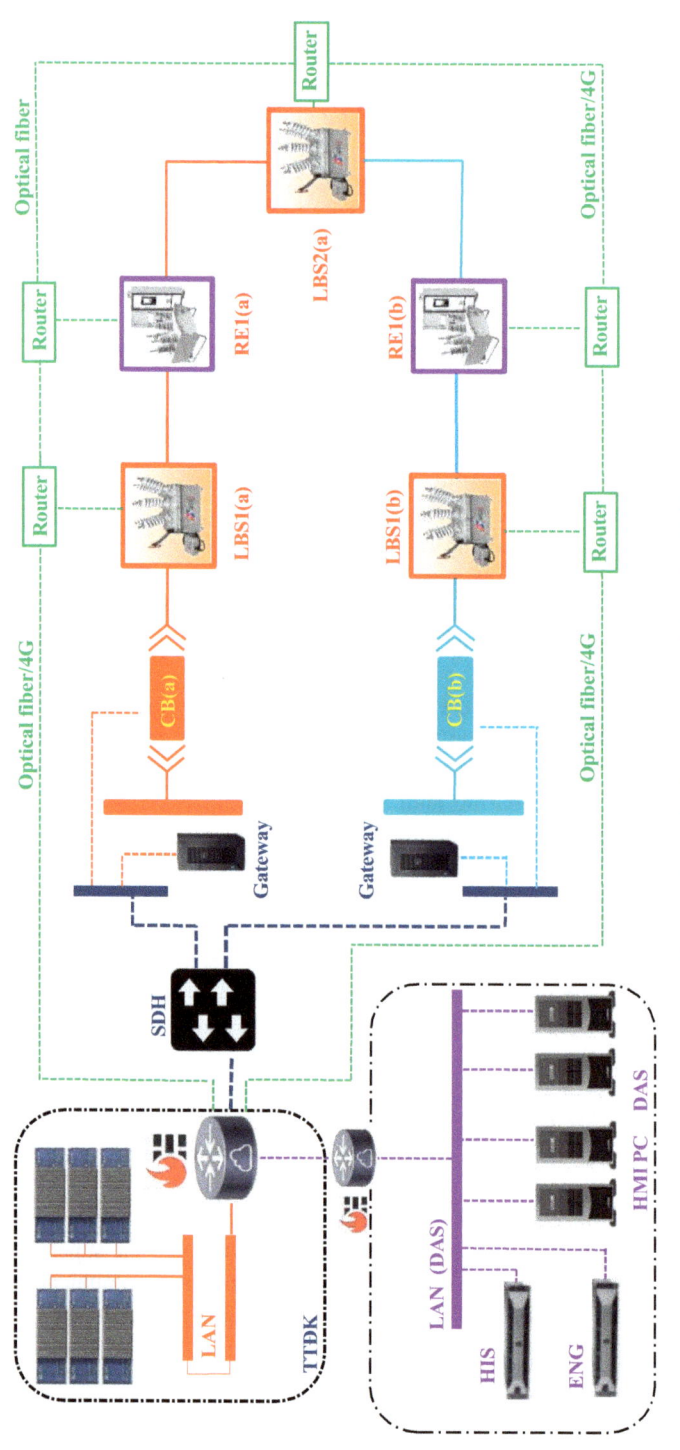

Figure 4.1 Architecture model of the DAS

4.1.1 Required input data for DAS application

1. **At circuit breaker relays on line exit points**
   - Device status signal.
   - Protection group.
   - Trip and pickup signals.
   - Fault indication signal.
   - Load power.
   - Control command.
   - Local/Remote mode.

2. **At reclosers on the grid**
   - Device status signal.
   - Protection group.
   - Trip and pickup signals.
   - Fault indication signal.
   - Control command.
   - Communication status (Good/Fail).

3. **At LBS**
   - Device status signal.
   - Fault indication signal.
   - Control command.
   - Communication status (Good/Fail).

4.1.2 Software requirements

1. **Operating system**
   - Windows Server 2019 or higher: For the server.
   - Windows 10 or higher: For operator machines.

2. **Data linking software**
   - Software with no time usage limitation, ensuring data linkage between the existing SCADA/distribution management system (DMS) and the DAS.

3. **DAS software**
   - FLISR (fault location, isolation, and service restoration) function: Identifying fault zones, isolating, and restoring unaffected areas.
   - System development and fault simulation function: Supporting extended configuration, design, testing, and simulation of fault scenarios before practical implementation, without affecting the running FLISR function.
   - Protection setting management function: Adjusting protection parameters based on grid status.

4. **Historical data storage software**
   - Storing historical data and generating operational reports for the DAS.
5. **SCADA software (if needed)**
   - If the contractor proposes a new SCADA system, the software must meet the technical requirements as specified.
6. **Other supplementary software**
   - Ensuring stable, secure operation of the DAS, protecting data security and meeting related technical requirements.

4.2 Technology and advances in smart grid

To meet Vietnam Electricity Central Power Corporation's (EVNCPC) objectives, the modernization process of the power grid has been deployed in several stages:

4.2.1 Level 1: Mini-SCADA

Mini-SCADA is a combination of SCADA and operation diagrams for each feeder line (station → feeder → devices), aimed at supporting operators in fault handling and isolation, transfer operations within 15 min.

4.2.2 Level 2: DAS

Based on the limitations of Mini-SCADA and the goal of reducing isolation and transfer time to under 5 min, the DAS model has been implemented. DAS integrates SCADA functions with FLISR applications. This system uses fault signals (TRIP, FI) from SCADA to locate faults and load signals and power data prior to the fault to calculate suitable transfer plans.

Key steps in the DAS operation include:

- **Fault location**: Instead of relying on operators, the FLISR fault location function performs this task using device status and TRIP signals provided by SCADA.
- **Transfer method**: After identifying the fault location, FLISR uses historical data from 5 min before the fault along with current line status and current signals from SCADA to determine power and sources, proposing an optimal transfer plan.
- **Transfer operation–semi auto**: FLISR can operate in two modes: Semi-Auto (program proposes a transfer plan, but the operator checks and gives the final command) or Auto (program automatically proposes and executes the plan).

This approach not only optimizes fault resolution time but also improves the operational efficiency and automation of the power grid.

4.2.3 Level 3: DMS model

After DAS reaches the fully automated stage, the grid achieves a high level of reliability in power supply. However, the DAS model mainly focuses on reducing outage time due to faults ("isolation and transfer time"). To reduce planned outage time and improve power supply quality, an automation system that meets multiple goals, such as planning, loss calculation, grid reconfiguration, and ensuring power quality (over-voltage, under-voltage), is needed. In this context, the DMS model is deemed more appropriate.

4.2.4 Definition and function of DMS

DMS is an automated software system that supports the management, monitoring, and control of the distribution grid optimally. The system provides comprehensive tools for grid management, including:

- **BLA**: Load distribution calculation.
- **DPF**: Power flow calculation.
- **FL**: Fault location identification (similar to DAS).
- **FISR**: Fault isolation and service restoration (similar to DAS).
- **AFR**: Grid reconfiguration to reduce losses and prevent overloads.
- **LVM**: Load and voltage/var management.
- **PLOS**: Planned outage scheduling.

4.2.5 Operating principle

The functions within DMS operate based on power flow calculations, where the accuracy of input parameters is crucial. To successfully implement DMS, data integration from multiple systems, including geographic information system, customer management information system, pavement management information system, outage management system, and others, is required. Ensuring data quality and communication exchange time between systems is a major challenge in the implementation process.

DMS not only helps reduce outage time but also supports optimizing grid operation, meeting increasingly higher demands for reliability and power supply quality.

4.2.6 Application of distribution grid automation in enhancing power supply reliability at EVNCPC

The automation of distribution grids to improve power supply reliability is not a new concept globally. However, selecting solutions suitable for the current grid conditions and economic conditions in Vietnam remains a significant challenge, requiring responsibility and innovation from electrical engineers. EVNCPC has progressively researched, learned, and derived valuable lessons while demonstrating strong determination in deploying grid automation. These efforts not only prove the effectiveness of applying scientific and technological advancements to grid operation in Central Vietnam but also contribute to enhancing the stability of power supply, aiming to maximize customer satisfaction.

The distribution grid is a crucial part of the power system with the goal of delivering energy reliably, efficiently, and safely to customers. The rapid growth of loads, along with the strict implementation of distribution grid regulations, compels us to introduce utilities and solutions to improve and enhance reliability. Distribution grid automation is seen as a powerful tool for improving reliability indices, contributing to the construction of a smart grid with high automation capabilities, thus promoting overall economic growth.

FDIR (fault detection, isolation, and restoration) is a function of the automation system designed to automatically detect faults, isolate the fault area from the rest of the system, and restore supply to unaffected areas, minimizing the outage zone. The FDIR function operates in real time and is displayed intuitively on the screen.

Table 4.1 presents the reliability indices of the Da Nang distribution grid in 2019 and the reliability targets set for 2021. The indices analyzed include MAIFI (momentary average interruption frequency index), SAIDI (system average interruption duration index), and SAIFI (system average interruption frequency index), categorized by two causes: faults and maintenance.

- **MAIFI**: In 2019, the momentary interruptions were 0.08 times due to faults and 0.037 times due to maintenance. The target for 2021 is to reduce this to 0.02 times due to faults and 0.03 times due to maintenance.
- **SAIDI**: The total average outage duration in 2019 was 222 min due to faults and 940 min due to maintenance. The 2021 target aims for a significant reduction to 40 min due to faults and 790 min due to maintenance.
- **SAIFI**: The average number of outages in 2019 was 4.62 times due to faults and 4.33 times due to maintenance. The target for 2021 is to reduce this to 0.80 times due to faults and 3.90 times due to maintenance.

Overall, the 2021 targets reflect a significant goal to improve the reliability of the distribution grid, with a particular focus on minimizing faults and optimizing maintenance to enhance the quality of electricity supply.

Table 4.1 Reliability indexes achieved in 2019 and reliability targets for 2021 for the Da Nang distribution network

Index	Year 2019		Year 2021	
	Fault	Operation	Fault	Operation
MAIFI (time)	0.08	0.037	0.02	0.03
SAIDI (min)	222	940	40	790
SAIFI (time)	4.62	4.33	0.80	3.90

4.3 Application of artificial intelligence and automation in smart grids on cloud computing platforms

4.3.1 Project effectiveness analysis
4.3.1.1 Reduction of outage duration

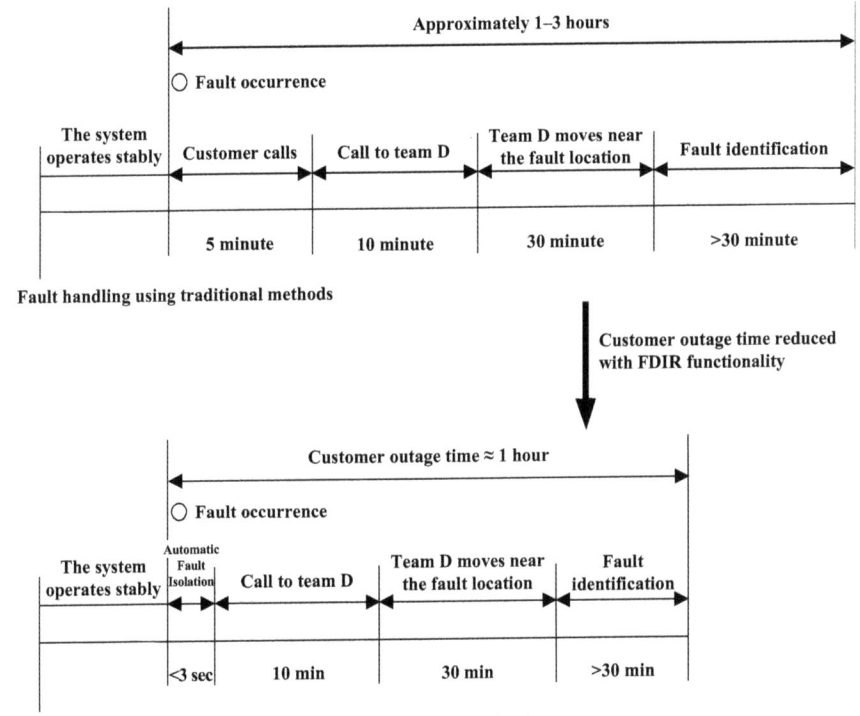

Fault handling using traditional methods

Fault handling with FDIR functionality of the DAS system

4.3.1.2 Improving the reliability index set

Table 4.2 presents an analysis of the improvement in reliability indexes—specifically the System Average Interruption Duration Index (SAIDI)—for a simple

Table 4.2 Analysis of improving reliability indexes with an example of a simple power distribution network

Case	Method	SAIDI	Improvement (%)
1	No recloser	8.8	–
2	Breaker with recloser function	3.3	62.5
4	Feeder with 1 recloser	2.6	70.5
5	Ring circuit with manual switching	2.3	73.9
6	Ring circuit with 3 reclosers	2.1	76.1
7	Ring circuit with 5 reclosers	1.7	80.7

power distribution network under various configurations. The table compares different methods ranging from no recloser to increasingly sophisticated ring circuit setups with multiple reclosers. As shown, the addition of reclosers and the implementation of ring circuits significantly reduce SAIDI values, thereby enhancing system reliability. The percentage improvement in SAIDI is also provided to highlight the effectiveness of each method.

4.3.1.3 Structure of DAS with FDIR function

There are three common structural configurations applied by the FDIR system, each with its own advantages and disadvantages:

Conventional coordination FDIR structure: For the sectionalizer, it is necessary to have a voltage transformer (VT) installed on the main power source side (the source currently supplying electricity). No wireless connection, 3G, or fiber optic cables are required. However, the system does not include a current monitoring system for the load circuit used for power transfer. This structure has limited flexibility and scalability (Figure 4.2).

Distributed FDIR structure (coordination among circuit breakers only): For sectionalizing circuit breakers, VT are not required. However, a wireless connection, 3G, or fiber optic network is necessary for the circuit breakers to coordinate with each other. This structure does not include a load current monitoring system for the circuit used for power transfer. Its flexibility and scalability are rated as moderate (Figures 4.3 and 4.4).

Centralized FDIR structure (server-based): VTs are not required for sectionalizing circuit breakers. However, a wireless connection, 3G, or fiber optic network is essential for the circuit breakers to send data to the central server. This structure includes a load current monitoring system, offering high flexibility and scalability. However, the system may face issues if there is a communication failure between the circuit breakers and the server (Table 4.3).

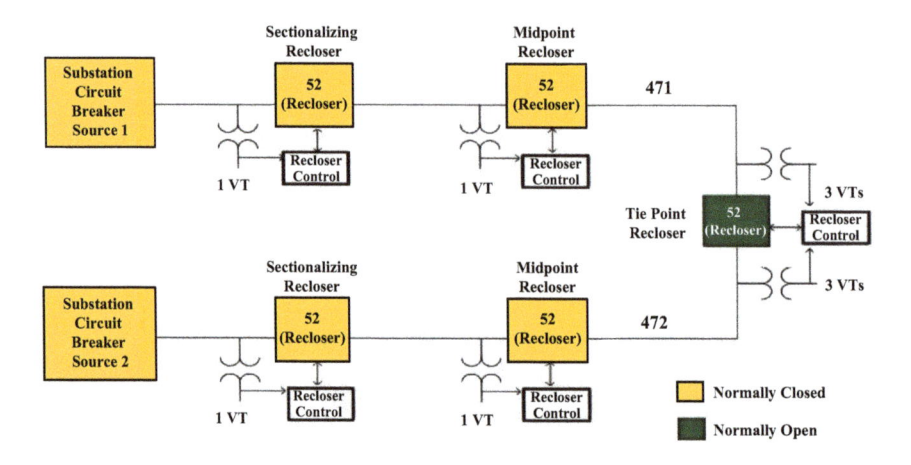

Figure 4.2 Basic structure of a conventional FDIR system

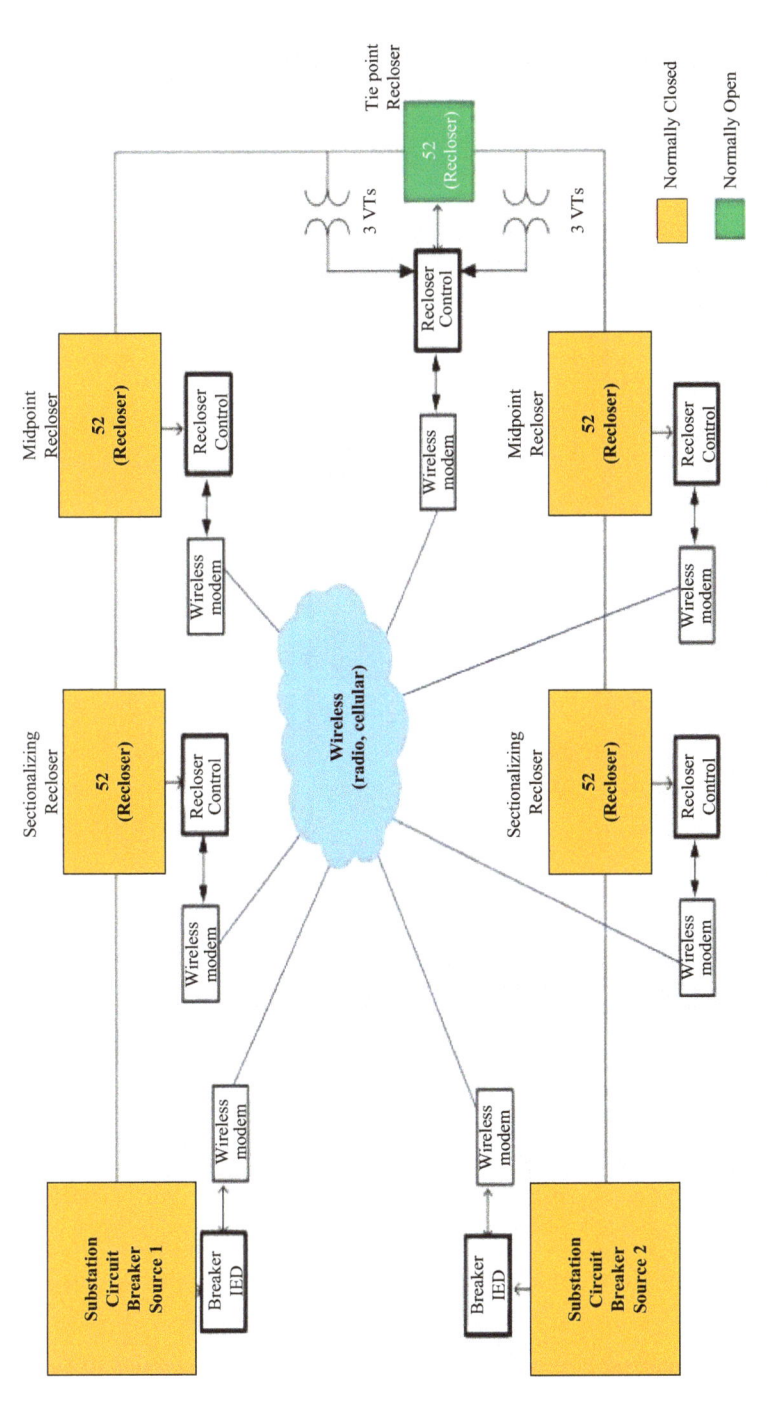

Figure 4.3 Basic structure of a distributed FDIR system

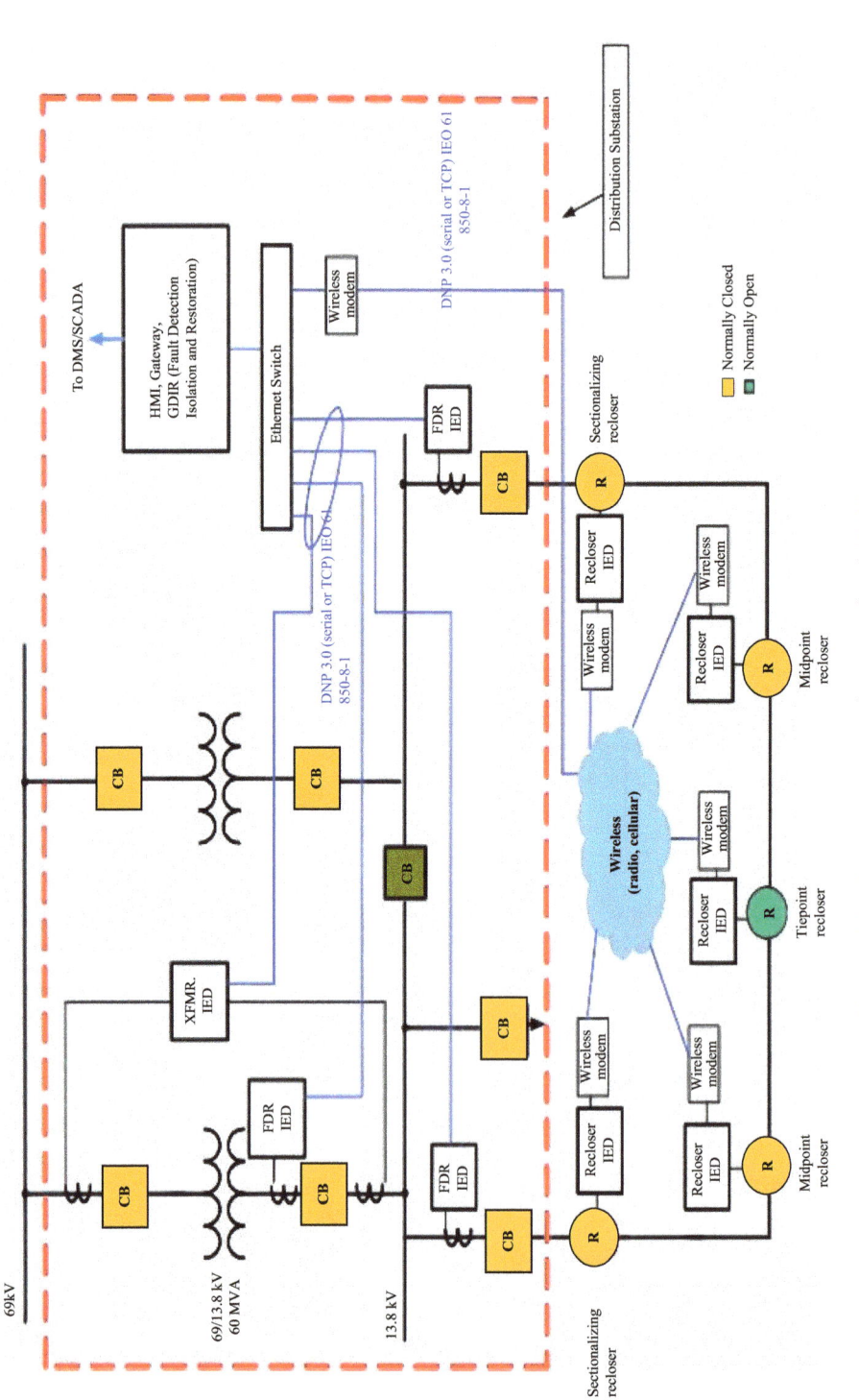

Figure 4.4 Basic structure of a centralized FDIR system

Table 4.3   Comparison of differences between FDIR solutions

FDIR solutions	VT for sectionalizing recloser and communication	Communication connection (fiber optic, 3G)	Scalability	Load current checking for other reclosers	Flexibility
Conventional FDIR	X	–	Low	–	Low
Distributed FDIR	–	X	Medium	–	Medium
Centralized FDIR	–	X	High	X	High

Table 4.4   Current status of reclosers on the feeder pairs

471E13–472E13	471 Hoa Hiep	to	DCL48.1-4 Nguyen Luong Bang
	471 Le Van Hien	to	These two feeder lines are connected via
	471 Non Nuoc	to	DCL 153-4 Dong Tra

- **Selection of DAS structure for Da Nang city power grid**
- **Proposed solution**

After analyzing the architecture of the DAS, we propose adopting the DAS solution with FDIR functionality based on a centralized structure within the DMS/SCADA system.

- **Basis for solution selection**
  - **Operational experience with DAS**
    This part has referred to the experience in designing and operating DAS at Tan Thuan Power Company and Thu Thiem Power Company, both under the Ho Chi Minh City Power Corporation.
  - **Hardware configuration of Da Nang city power grid**
    o Da Nang's power grid has completed the implementation of the SCADA system, with reclosers (automatic circuit breakers) controllable via central software.
    o The hardware configuration of the reclosers and control cabinets installed on the current grid meets the requirements for centralized FDIR functionality (Table 4.4).
    o The 3G communication network has been fully developed, supporting remote operation and control.
- **Technical specifications of proposed feeders for the pilot project**: (Details of feeder specifications to be included based on the project's scope.)

118   *Clean energy in South-East Asia*

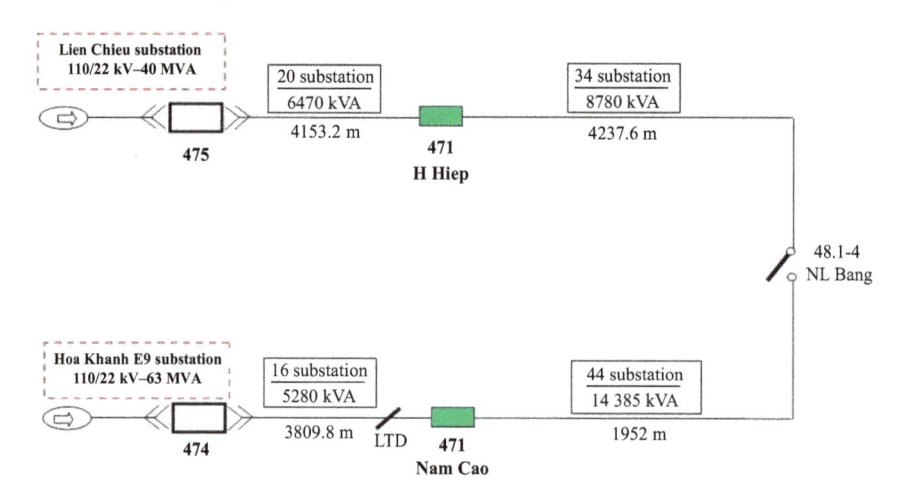

Figure 4.5   Feeder pair 474E9 and 475Elc

- **Feeder pair 474E9–475Elc**: These are two medium-voltage overhead lines with an open-loop connection (operating in an open state) via DCL48.1-4 Nguyen Luong Bang (LDS switch) located at pole C-48 Figure 4.5).
- **Load characteristics of the two feeders**
  – The primary load of the two feeders is residential.
- **Current status of feeder pair 474E9 and 475Elc**
  – Refer to Appendix 1 for detailed information.
- **Equipment on the feeders**
  – On Feeder 475Elc, there is a **recloser 471Hoa Hiep**, and on Feeder 474E9, there is a **recloser 471Nam Cao**.
  – Both Reclosers are manufactured by **Cooper**, model **NOVA27i**, and are controlled using **F6 cabinets**.
- **Peak load current of the two feeders in 2019**
  – (Details of the peak load current to be provided based on data analysis.)

Feeder line	Maximum load current in 2019 (A)
474E9	378A (10/6)
475Elc	334A (25/5)

4.3.1.4   Load transfer and grid upgrades for the DAS project

To ensure the ability to transfer loads and fully utilize the FDIR functionality of the DAS, load transfers and upgrades for the two feeder pairs must be implemented.

This process involves:

- Adjusting the configuration of the feeders to optimize load distribution.
- Upgrading the infrastructure to meet the technical requirements for FDIR operations.
- Ensuring that the network design supports seamless FDIR.

These improvements aim to enhance the reliability and operational efficiency of the distribution network (Tables 4.5 and 4.6).

Table 4.5 Conductor parameters and maximum current for feeder lines

No.	Wire type/size	Allowable current (A)	Standard
1	M(3 × 240)	387	Manufacturer
2	M(3 × 185)	335	Manufacturer
3	M(3 × 150)	296	Manufacturer
4	M120	485	Electrical Equipment Standards
5	3M(1 × 185)	391	Manufacturer
6	AV150	440	Manufacturer
7	AV185	504	Manufacturer
8	AV240	593	Manufacturer
9	3AV(1 × 300)	402	Manufacturer
10	2A(3 × 240)	2 × 304	Manufacturer
11	3A(1 × 300)	402	Manufacturer
12	A(3 × 240)	304	Manufacturer
13	AC185	510	Electrical Equipment Standards
14	ACKII300	585	Electrical Equipment Standards

Table 4.6 Load current at circuit breakers and reclosers in the pilot project

Related line	Circuit breaker/recloser	Maximum load current (A)	Notes	Data source
471E13–472E13	471E13	167	Circuit Breaker at the Starting Point	AMIS Software
	471 Le Van Hien	80	Recloser for Segmentation	Management Unit
	472E13	181	Circuit Breaker at the Starting Point	AMIS Software
	471 Non Nuoc	88	Recloser for Segmentation	Management Unit
474E9–475Elc	474E9	378	Circuit Breaker at the Starting Point	AMIS Software
	471 Nam Cao	288	Recloser for Segmentation	Management Unit
	474E10	217	Circuit Breaker at the Starting Point	AMIS Software
	472HVThai	131	Recloser for Segmentation	Management Unit
	475Elc	334	Circuit Breaker at the Starting Point	AMIS Software
	471 HHiep	101	Recloser for Segmentation	Management Unit

Implementation of the DAS project with FDIR functionality at Da Nang power company

The implementation of the DAS project with FDIR functionality at Da Nang power company in 2016 was essential and achieved many critical objectives set by the company. The project not only provided practical benefits to the company but also contributed to the transformation of Da Nang's power grid into a smart grid with advanced technical and automation capabilities.

4.3.1.5 Recommendations

To ensure the project's success, we recommend close collaboration and coordination among departments to guarantee that the project meets its objectives and timeline. Additionally, for achieving long-term goals, we propose the following:

- **Selective tripping adjustments**: To minimize non-selective over-tripping of circuit breakers, work with A3 to adjust tripping times in compliance with Circular 39/2015/TT-BCT.
- **Automation development roadmap**: Develop a roadmap for expanding automation with FDIR functionality across Da Nang's entire distribution grid by 2020.
- **Feeder upgrades**: Upgrade feeders to ensure adequate load transfer capability. Specifically, the current load on conductors should utilize only 50–60% of their maximum load capacity.
- **Feeder 474E10 upgrades**: To ensure the load transfer capability of feeder 474E10 when transferring loads from feeder 474E9, upgrade sections of feeder 474E10 that currently fail to meet technical standards.

These measures are critical for maximizing the effectiveness of the DAS project and enhancing the reliability and resilience of the power grid.

Feeder line	Conductor – length (m) not meeting standard	Location	Proposed conductor replacement-conductor – length (m)
474E10	M(3×150) – 110	Column 118 to 119	M (3×300) – 110

4.3.2 Application of artificial intelligence and automation in smart grids on cloud platforms

4.3.2.1 Role of transmission lines

Transmission lines are a critical component of the power system, ensuring the connection between power generation sources and loads. With the rapid expansion of the power system, both the number and length of transmission lines have significantly increased, leading to various operational challenges.

4.3.2.2 Common faults in transmission lines

During operation, transmission lines can encounter faults such as:

- Lightning strikes, short circuits, conductor breakage, and ground faults.
- Equipment or user-related issues.
- Overloading and aging of equipment.

Protective relays are used to isolate faulty components from the system, ensuring safety and maintaining normal operations. However, accurately locating the fault position remains challenging due to significant errors in measurements provided by protective relays.

4.3.2.3 Research needs and objectives

This study focuses on researching fault location methods for transmission lines, particularly:

- Utilizing impedance measurement data from one end of the line.
- Applying artificial neural networks (ANN) to improve fault location accuracy.

4.3.2.4 Research methods and techniques

- **Impedance measurement**
  Fault location techniques based on impedance measurement yield good results for two-phase faults but exhibit significant errors for single-phase-to-ground faults.
- **Application of artificial neural networks**
  – Conducted studies using 740 simulated experimental data points for training, testing, and validating the ANN.
  – The system was designed for a 100 km transmission line with a voltage level of 115 kV and a frequency of 50 Hz.

4.3.2.5 Fault location procedure using ANN

- Simulate and operate the transmission line to collect data.
- Normalize the collected data to ensure effective analysis.
- Extract fault features from parameters such as phase currents I_a, I_b, I_c.
- Use ANN to determine the fault location.
- Evaluate the effectiveness of the fault location method.

4.3.2.6 Types of faults studied

The study focuses on ten types of short-circuit faults, including:

- Single-phase-to-ground faults: AG, BG, CG.
- Two-phase faults: AB, AC, BC.
- Two-phase-to-ground faults: ABG, ACG, BCG.

The fault location method using ANN and impedance data demonstrates significant potential for improving accuracy, especially in complex faults. The findings from this research form the basis for developing more advanced solutions to enhance reliability and efficiency.

Figure 4.6 is the block diagram for the structure of the ANN used in the research on fault location in power transmission lines.

4.3.2.7 ANN structure for fault location

- In the study, ten neural networks were developed to handle ten different types of faults on transmission lines.

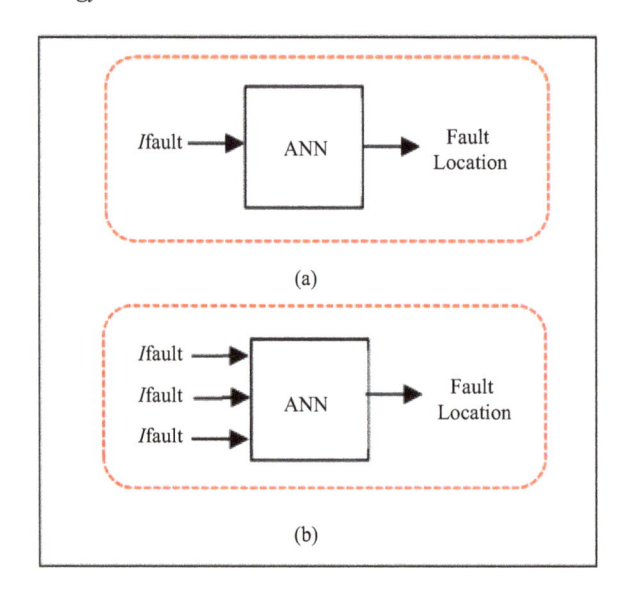

Figure 4.6 Block diagram for the structure of an artificial neural network (ANN). (a) ANN structure for faults occurring on a single line. (b) ANN structure applied for faults on all three phases of the current signal.

- Each neural network consists of three layers:
  - **Input layer**: Receives characteristic data from the current signal.
  - **Hidden layer**: Processes data using the sigmoid activation function.
  - **Output layer**: Outputs results using a linear activation function.

4.3.2.8 Multilayer perceptron technique

- The connection between layers is made through weights and biases to optimize predictions.
- Training algorithm: Levenberg–Marquardt backpropagation.

4.3.2.9 Performance evaluation

- The mean squared error (MSE) is used to measure the model's accuracy.
- Ideal result: MSE equals 0.
- Simulations show that when using information from all three phases, ANN performs better with lower MSE due to richer and more complete information.

4.3.2.10 Key data features

The parameters are prioritized for fault location as follows:

- Standard deviation.
- Output wave energy.
- Mean value.
- Minimum-maximum deviation.
- Signal amplitude.

4.3.2.11 Fault location challenges on lines with series compensation

Lines with series compensation, protected by metal oxide varistors (MOV), produce nonlinear characteristics, making it difficult for conventional fault location algorithms:

- **MOV characteristics**:
    - In normal operation, MOV has high resistance.
    - When the capacitor voltage exceeds a threshold, MOV switches to low resistance, allowing current to flow.
    - MOV's operating time depends on the fault current amplitude, resulting in nonlinear impedance.

- **Effective fault location solution**:
    - A fault location algorithm is proposed, based on:
        ○ Data from a transmission line source.
        ○ Power factor angle.
        ○ Time constant during the fault event.
    - This technique allows fault location with minimal error, regardless of MOV operation conditions.

- **Takagi's improved method for series compensation lines**:
    - Combines voltage and current phase angles at relay positions with the fault signal's time constant.
    - Distinguishes between high and low resistance faults, improving location accuracy.

The study provides modern and optimized solutions for fault location on transmission lines, especially in systems with series compensation. The use of ANN and improved algorithms enhances performance and reliability, meeting the operational requirements of modern power grids.

4.3.3 Fault location process and issues in power transmission lines in Vietnam

4.3.3.1 Current fault location process in Vietnamese power transmission

- **Companies steps**
    - **When a fault occurs**
        ○ The protection relays at the substation automatically disconnect the power on the line and attempt to restore it.
        ○ Distance protection relays or differential relays provide fault distance (measured in kilometers) from the measurement location.
    - **Fault processing**
        ○ Classify the fault as transient (quick recovery, successful restoration) or sustained (requires direct intervention).

- o Line management teams identify the fault location, repair damage, and suggest solutions to prevent recurrence.
- **Searching for the fault location**
  - o Based on the distance reported by the relay, personnel are divided into two groups to search from both sides.
  - o At each pole location within the fault-reported area, a visual inspection is conducted to detect damage or abnormalities.
  - o Upon identifying the exact fault location, photos are taken, and a report is made.

4.3.3.2 Issues in the current fault location process

- **Location error from protection relays**
  - The error can be several kilometers in some cases.
  - For example, on a 300 km long transmission line, an error of $\pm 1\%$ could result in a 6 km search range, equivalent to about 20 poles to inspect.

- **Difficult terrain**
  - Transmission lines often pass through mountainous areas, forests, or places far from settlements, making it difficult and time-consuming to access and identify faults.
  - The time to identify the fault location can take up to one day, with 4–6 personnel required.

4.3.3.3 Technical causes of fault location error

- **Measurement errors**
  - Fault location algorithms rely on voltage and current data measured from the transmission line ends.
  - Measurement errors or signal collection discrepancies can cause misalignment in fault location results.
  - Sources of measurement errors include:
    - o Signal noise.
    - o Calibration errors of measuring devices.
    - o Environmental factors such as temperature and humidity.

- **Inaccurate line modeling**
  - The line model used in location algorithms may not fully reflect the physical characteristics of the actual transmission line.
  - This discrepancy can occur due to differences in:
    - o Impedance parameters.
    - o Line structure (e.g., series or shunt compensation).
    - o Operating conditions (load, temperature, weather).

- The current fault location process in Vietnam still largely depends on information from protection relays and manual inspection, facing many difficulties due to location errors and terrain conditions. Technical causes like measurement errors and inaccurate modeling significantly affect location efficiency.

- Improving technology, using smarter location algorithms (like ANN, big data-based optimization algorithms), and enhancing measurement devices are important development directions to improve efficiency in the future.

4.3.3.4 Current process
- **When a fault occurs**
  - The protection relay system automatically disconnects and attempts to reconnect the power.
  - Provides fault distance information from the relay measurement point.

- **Fault processing**
  - Line management teams are sent to the approximate location based on reports.
  - Actual checks are conducted on each pole to determine the exact fault location.
  - The search time depends on the terrain, usually ranging from several hours to one day.

4.3.3.5 Existing issues
- **Error in protection relay reports**
  - A deviation of $\pm 1\%$ can lead to a search range of up to 6 km.

- **Terrain difficulties**
  - Transmission lines pass through hilly areas, far from settlements, increasing search time and costs.

4.3.3.6 Technical causes of errors
- **Measurement device errors**
  - VTs have an error of $\pm 6\%$, and current transformers (CT) have an error of $\pm 10\%$.
  - CT saturation can distort current measurements.

- **Asynchronous signals**
  - Asynchronous measurements between the transmission line ends can cause phase shifts, affecting fault location.

- **Line modeling**
  - Using inaccurate centralized parameter models for long and complex lines.

4.3.3.7 Fault location in integrated renewable energy systems
- **Challenges with solar power sources**
  - The involvement of changes the fault current characteristics.
  - Limitations of impedance and wave propagation techniques when applied to integrated systems.

- **Proposed solution**
  - Use optimization algorithms, such as an improved cuckoo search algorithm.
  - Integrate signal synchronization technologies like GPS or optical cables.

126 *Clean energy in South-East Asia*

4.3.3.8 New mathematical model

- Develop a system consisting of
  - Traditional power sources: thermal, hydro (S_1).
  - Transmission lines (1–2).
  - Distributed power sources: solar and wind (S_2).
  - Load connected to busbar 2.
- Key components:
  - Photovoltaic (PV) panels.
  - DC/AC, AC/DC/AC converters.
- Applications:
  - Improve the accuracy of fault location for short circuits.
  - Ensure operational efficiency of integrated distributed power systems.

4.3.3.9 Future research directions

- Develop optimized algorithms with faster and more accurate signal processing capabilities.
- Improve measurement technology and signal synchronization.
- Integrate more detailed line modeling, suitable for renewable energy systems (Figure 4.7).

4.3.4 *Some protection solutions for PV sources*

Protecting the electrical system is crucial for safeguarding equipment and minimizing load losses. Protection relays must be coordinated efficiently to function accurately. When integrating PV sources into the distribution grid, several protection issues arise, such as power fluctuations, protection dead zones, and asynchronous coordination.

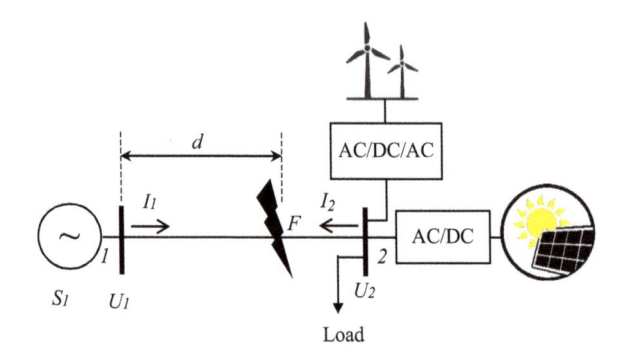

Figure 4.7 Transmission line model with a short-circuit fault in an electric power system considering solar energy source, load

4.3.4.1 Protection strategies for PV sources

- **Optimizing directional overcurrent settings**
  - Use fault current limiters to reduce fault current, enhancing system protection efficiency.

- **Comprehensive genetic algorithm method**
  - Optimize overcurrent relay (OCR) coordination to improve system protection performance.

- **Relay performance testing in high-PV penetration networks**
  - Analyze fault conditions and irradiance fluctuations using the particle swarm optimization (PSO) algorithm to optimize network connectivity.
  - A two-stage inverter model allows for different protection measures for grid-connected PV sources.

- **Differential protection and setting issues**
  - Adjust settings based on input parameters such as irradiation, wind speed, and ambient temperature. However, this method requires modern communication equipment, leading to higher costs.

- **Disconnection solutions according to IEEE 1547 standards**
  - Disconnect PV systems from the grid during faults to reduce fault currents without modifying relay settings or adding new protection equipment. This method is economically feasible and is already implemented in the US and Canada.

4.3.4.2 Power quality and islanding issues

- **Fast protection for PV systems**
  - PV systems must be quickly protected from faults or grid fluctuations. Traditional protection systems typically take 40 ms to 2 s to detect and respond to faults.
  - Islanding is a critical issue as it can cause significant damage if the grid is reconnected suddenly.

- **Islanding detection**
  - Parameters such as voltage, frequency, and harmonic waves are monitored to detect islanding. The rate of change of frequency technique is widely used but can be sensitive to grid disturbances, leading to unwanted fault activation.

4.3.4.3 Practices in Vietnam

In the incident on May 13, 2021, at the 500 kV Tan Dinh substation, the electrical system was receiving significant power from PV sources. When the fault occurred, the simultaneous disconnection of multiple plants caused significant power fluctuations and exacerbated the fault. Some plants activated fault ride through behavior, which helped stabilize the system without significant impact.

4.3.5 Distance protection (F21) for grid-connected PV sources

4.3.5.1 Distance protection (F21)

Overcurrent protection is the simplest and most widely used protection method. However, in complex grid configurations, it may not meet the requirements for selectivity and rapid response time. To address this, more complex protection methods are necessary, and distance protection offers an effective solution.

Distance protection operates by measuring current and voltage values at the location of the relay and calculating the total impedance of the system. If the measured impedance at the time of the fault is less than the preset impedance value in the relay, the relay activates and initiates protection.

The basic principle of distance protection involves using impedance to detect and handle faults. However, relay manufacturers often integrate various algorithms and approaches to optimize operational performance, ensuring greater flexibility and accuracy under different operating conditions.

This method is particularly useful for complex electrical grids and grid-connected PV sources due to its quick and accurate response to faults, reducing risks and impacts on the overall system.

The total impedance measured by the relay in normal operation is determined by the following formula:

$$Z_R^{bt} = \frac{U_R^{bt}}{I_R^{bt}} = \frac{I_R^{bt} \times (Z_D + Z_{pt})}{I_R^{bt}} = (Z_D + Z_{pt}) > Z_D \tag{4.1}$$

with:

Z_R: The total impedance measured by the relay
Z_D: The impedance of the protected line, where $Z_D = R_D + jX_D$
Z_{pt}: The equivalent impedance of the loads connected after the transmission line.

It can be observed that the total impedance measured by the relay in normal operation is always greater than the impedance of the transmission line.

When a fault occurs at a point on the transmission line:

$$Z_R^{sc} = \frac{U_R^{sc}}{I_R^{sc}} = \frac{I_R^{sc} \times Z_{sc}}{I_R^{sc}} = Z_{sc} \leq Z_D \tag{4.2}$$

4.3.5.2 Differential protection for PV sources

Differential protection (F87) operates based on the principle of comparing the total current flowing into and out of the protected object. If this total current is non-zero, the protection system is activated to address the fault.

This principle is illustrated in the diagram in Figure 4.8 on the following page. The CTs are configured with the appropriate polarity to ensure that, under normal operating conditions or when a fault occurs outside the protected zone, the current flowing through the relay follows the schematic in Figure 4.9.

The current through the protective relay is the difference between the secondary currents from the CTs. In other words, this current represents the discrepancy between the input and output of the protected area or equipment. This is why this protection method is called "differential protection," as it relies on comparing and detecting the current discrepancy.

The differential protection method ensures high accuracy and rapid response, making it especially effective for systems with grid-connected PV sources.

$$I_{role} = I_{1tc} - I_{2tc} \tag{4.3}$$

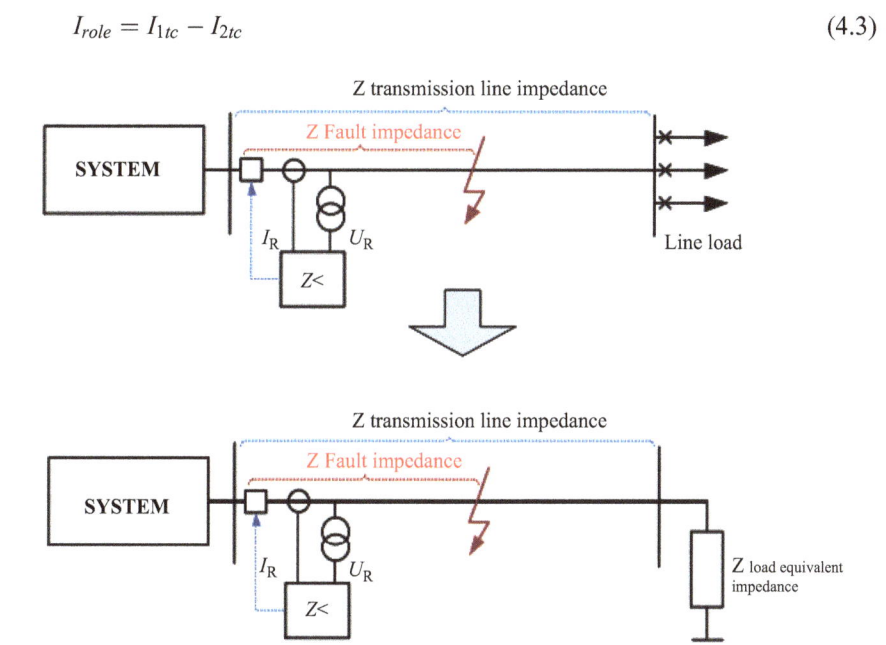

Figure 4.8 Basic configuration of distance protection

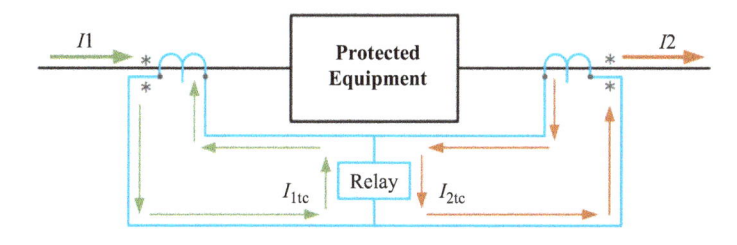

Figure 4.9 Schematic diagram of current differential protection

130   *Clean energy in South-East Asia*

Under normal conditions, the incoming current (I_1) is equal to the outgoing current (I_2), so theoretically, the current flowing through the relay is zero. However, due to the differences in the characteristics of the CTs, a small differential current still flows through the relay:

$$I_{role} = I_{1tc} - I_{2tc} = I_{so\_lech} \tag{4.4}$$

In the event of a fault outside the protected area that shown in Figure 4.10, the incoming and outgoing currents remain equal. However, due to the high value of the fault current, the errors in the CTs also increase, resulting in a larger differential current.

In the case of an internal fault that shown in Figure 4.11, the current on one side reverses direction. The differential current flowing through the relay is equal to the sum of the secondary currents from the CTs and has a very high value, causing the relay to trip.

In practical operation, differential protection is often used as the primary protection for power transformers due to its fast response, high sensitivity, and protection range limited by the positions of the CTs. Additionally, differential protection is also used as the primary protection for transmission lines (87L), combined with a communication channel system (F85–trip relay). The system receives signals from the CTs at both ends of the transmission line, and through the F85 trip relay, information is sent to protect the entire transmission line within the protection zone.

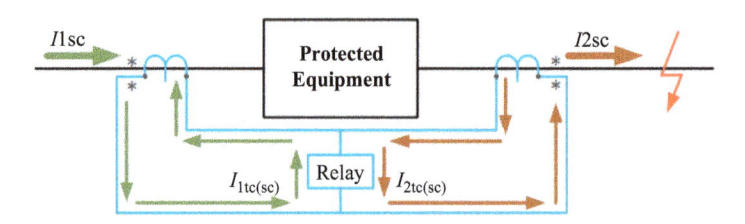

Figure 4.10   Differential protection with an external fault

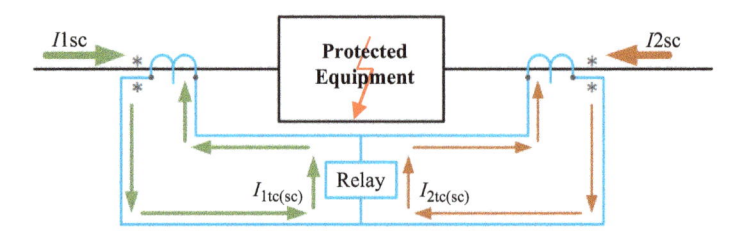

Figure 4.11   Differential protection with in-zone fault

4.3.5.3 Changes with grid-connected PV systems

Traditional power distribution systems use protection devices such as OCR, reclosers, fuses, and sectionalizing devices. Among these, OCR is commonly used due to its time–current characteristic (TCC) curve and its ability to coordinate effectively with other devices to form a protection system.

For example, Figure 4.11 illustrates a system with two OCRs (R1 and R2) serving as primary and backup protection. The inverse TCC curve of the relay is adjusted to detect faults at the end of the system (busbar C). Protection settings are based on the time dial setting (TDS) and the fault current threshold (I_p), according to IEC or IEEE standards. A lower TDS reduces fault clearing time but reduces flexibility when dealing with smaller fault currents that shown in Figure 4.12.

When a PV source is connected to the grid, the current direction and magnitude change due to the increased fault current at the busbar. This affects the coordination between protection relays, requiring evaluation and re-setting of the system.

Furthermore, the assumption that current only flows in one direction in the traditional system no longer holds true. PV sources can cause voltage flicker, incorrect relay operations, unsuccessful recloser actions, or unintended islanding. For example, when a PV branch is connected to bus B that shown in Figure 4.11, the fault current at F1, detected by R1, will increase, while the current seen by R2 may remain unchanged or decrease, leading to a loss of coordination.

Moreover, the severity of the fault depends on the operating mode of the PV system (grid-connected or islanded) and the amount of power that the PV system feeds into the grid. Integrating PV sources requires a complete reassessment of the protection configuration to ensure safety and efficiency.

4.3.5.4 Typical faults in PV systems

A PV power generation system, like other electrical systems, is highly sensitive to faults. Faults in PV systems are divided into two main stages: the DC stage and the AC stage as shown in Figure 4.12.

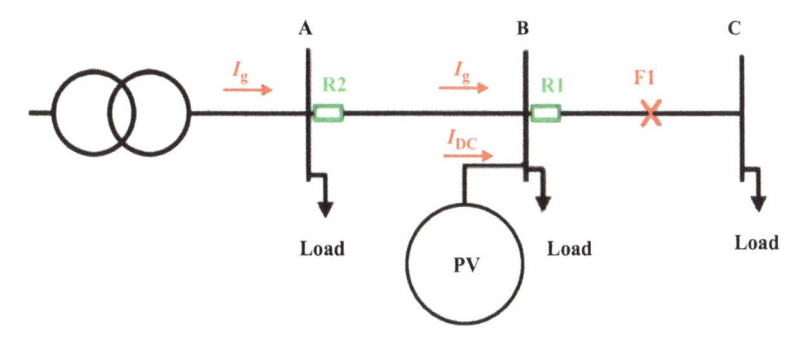

Figure 4.12 Single-line diagram of grid-connected PV source branch

132 *Clean energy in South-East Asia*

(a) **DC stage**
- The voltage and current of the PV modules are limited, depending on factors such as radiation, temperature, humidity, and wind speed.
- With the presence of maximum power point tracking (MPPT), PV panels typically operate near the short-circuit current (I_{SC}) and around 80% of the open-circuit voltage (V_{OC}).
- Fault currents on the DC side are typically low in magnitude, making detection and fault differentiation challenging.

(b) **AC stage**
- Faults on the distribution (AC) side of the PV system are easier to detect and isolate thanks to standard protection schemes.
- The PV diagram in Figure 4.13 illustrates common faults at each stage of the power conversion process. The causes of faults may vary, requiring:
  – Accurate fault type identification.
  – Assessment of the necessity for fault detection.
  – Analysis of the challenges involved in protecting the system.

Therefore, a detailed analysis of the different faults is provided in this section to gain a better understanding. In summary, this section offers insights into the various faults that occur in PV systems, their impact on electrical characteristics, the protection challenges they pose, and the necessity of fault detection.

In each stage and operation of the PV system, faults are categorized into different types of effects and severity levels.

For PV arrays, typical faults include:

- Phase-to-ground (L-G) faults,
- Phase-to-phase (L-L) faults,
- Open-circuit faults,
- PV module bypass diode faults and shading,
- Arc faults (Figure 4.14).

4.3.5.5 Faults in PV systems

In practice, faults on the AC side (at the transformer or after the transformer) are generally more severe and common compared to those on the DC side (Figure 4.15).

(a) **DC stage**
   This is the core stage of the PV system, but the fault current is often too small to be clearly detected. Common faults include:
   - Faults in the converter switch.
   - MPPT faults.
   - Backup battery faults.

(b) **AC stage**
   Faults in the AC stage typically occur when the power converter malfunctions, including:
   - Uncontrolled islanding operation.
   - Imbalances or fluctuations in the grid.

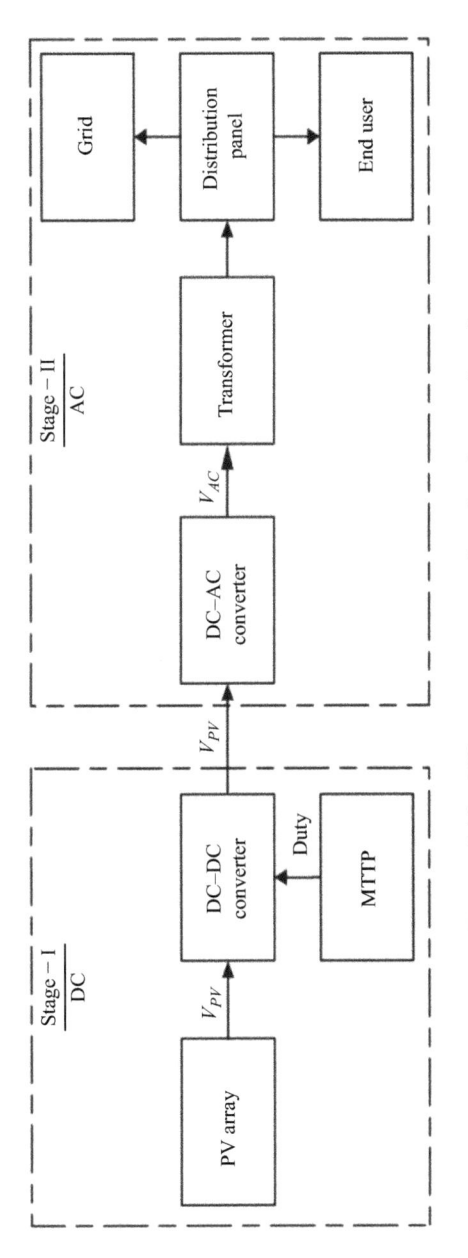

Figure 4.13 PV system connected to the electrical grid

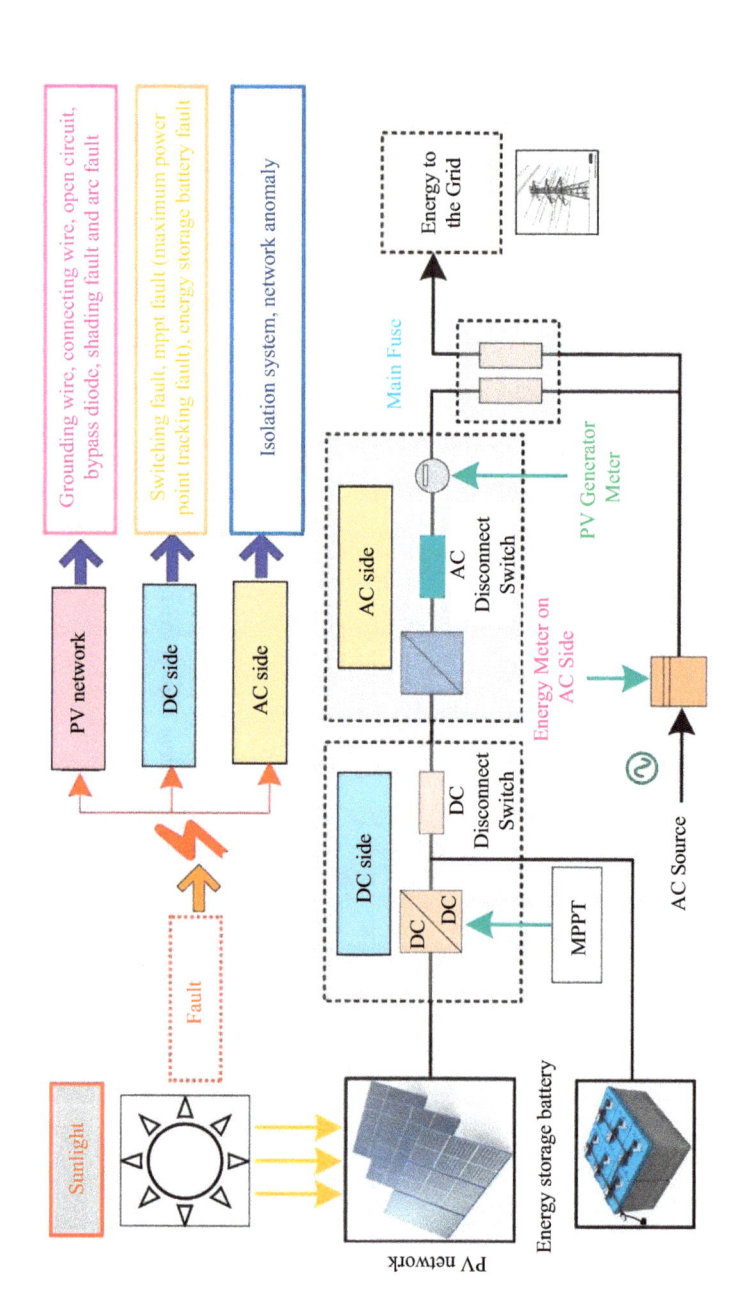

Figure 4.14 Schematic diagram showing common faults in a PV system

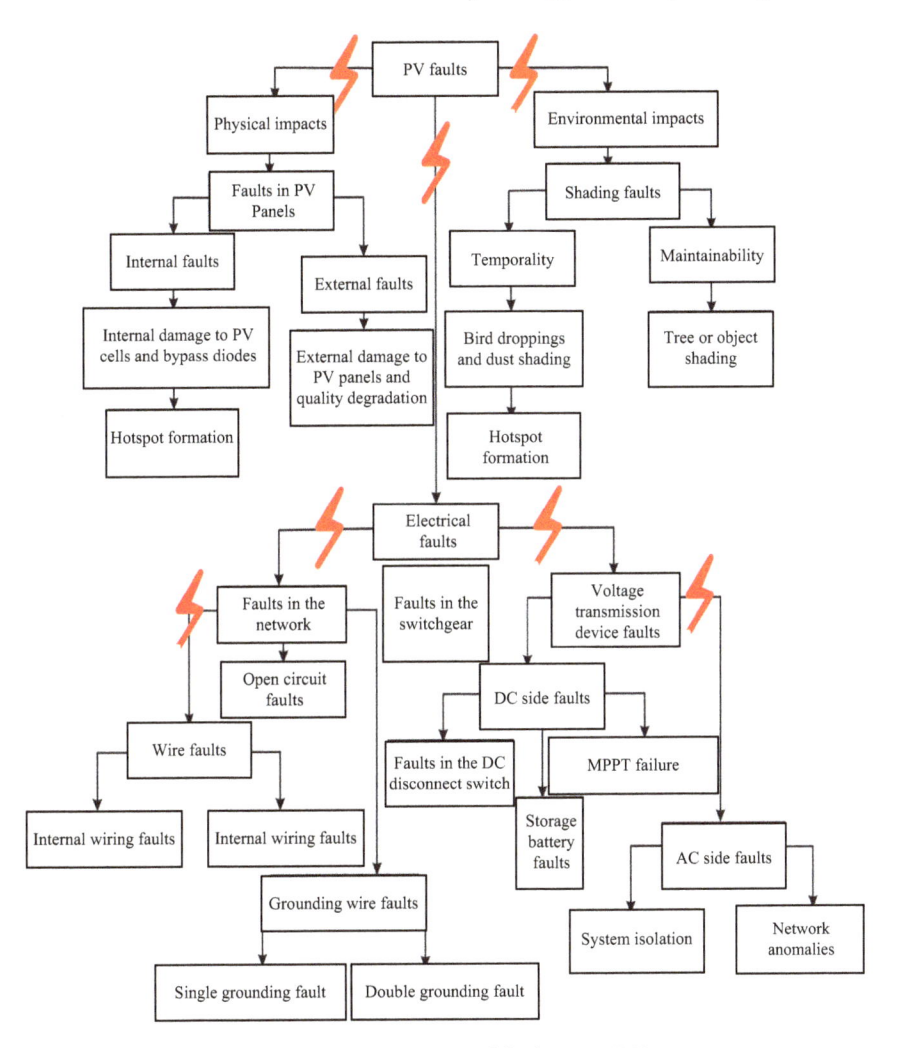

Figure 4.15   Classification of faults in a PV system

4.3.5.6 Solutions for rapid detection of short circuit current in PV systems

PV systems provide lower fault currents to the grid due to the characteristics of the PV panels and inverter operation. The short circuit current from PV inverters typically fluctuates as follows:

- 1.2 times the rated current (inverters >1 MW),
- 1.5 times (500 kW),
- 2–3 times (small inverters).

Although each system contributes a small amount, the total fault current from multiple PV systems can exceed acceptable levels, potentially damaging switching equipment. This requires costly equipment upgrades, leading to high rejection rates for PV system connections in regions such as Ontario, Canada.

In Vietnam, inverters are required to support the grid (by activating low-voltage ride-through (LVRT)) to maintain system stability. Rapid fault detection is critical, with two main solutions:

- **PV disconnection**: Disconnecting before the fault current exceeds the rated value, minimizing the impact on the grid.
- **Switching to STATCOM**: The PV inverter operates as a dynamic reactive power compensator, supporting the grid in case of faults.
- Supporting technologies include:
- **Modeling PV plants**: Predicting short circuit current contributions.
- **Traditional relays**: Detecting faults via over-voltage/over-current signals.
- **Wavelet CWT**: Analyzing transient current and voltage to detect faults within half a cycle.
- **Fault current limiters**: Quickly limiting fault current, extending equipment lifespan.

Detection techniques based on the rate of change of current (di/dt) and fault current values enable quick responses, protecting the system, supporting the grid, and enhancing overall stability (Figure 4.16).

The research system consists of a 27.6 kV distribution line, 25 km in length, connected to the transmission network through a substation with two 47 MVA transformers. The source impedance is combined with the transformer impedance in the PSCAD/EMTDC model. The total load of 15 MVA is divided into three groups of three-phase static loads and a large load at the end of the line. A 7.5 MW solar power farm is connected near the end of the line.

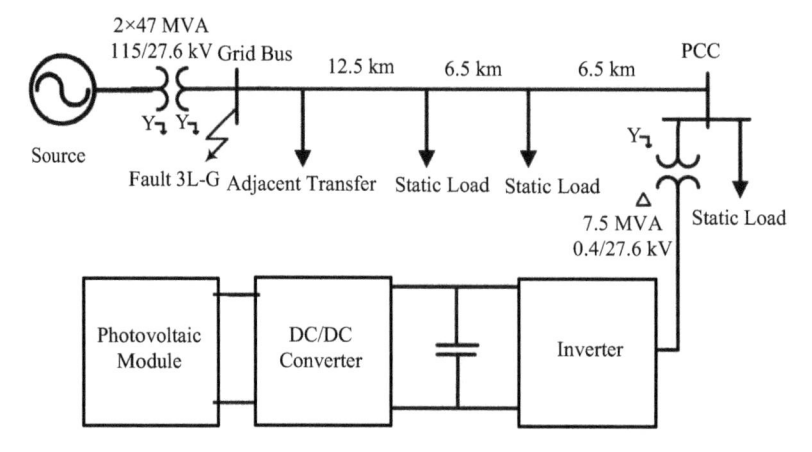

Figure 4.16 Research system diagram

The PV system includes PV modules, a step-up converter, an inverter, an AC filter, an MPPT controller, and an inverter controller. The PV modules are modeled as a DC current source, while the converter and inverter adjust the DC voltage and reactive power to zero. A fault detection module measures current and voltage at the inverter connection, using a low-pass filter, a slope detector, and a magnitude detector to identify faults.

The fault triggering signal is coordinated through AND/OR gates, ensuring activation only when the monitored value exceeds the threshold. The "PVIso" fault signal can stop the operation of the IGBT, isolate the PV system from the grid within a few hundred microseconds, or support the grid through LVRT or PV-STATCOM mode, depending on grid standard requirements (Figure 4.17).

The protection of PV systems aims to minimize the impact of electrical faults on the PV panels, converters, transformers, and the power grid. Protection requirements include:

- Earth fault detection in PV panels.
- Initiation of alarms in the event of an earth fault.
- Disconnection of faulty PV panels.
- Fault current disconnection.

Overcurrent protection is only effective in the case of reverse current but is insufficient to protect against discharge or earth faults. The NEC 690 standard requires earth fault monitoring in PV modules. The direct current from the PV system can saturate transformers and equipment, causing protection disruption. IEEE and IEC standards limit the direct current entering the grid to 0.5–1% of the rated current.

Existing protection functions
- **Earthing for the PV system**: For safety protection, earth faults must be detected and isolated. The PV system may be either grounded or ungrounded, depending on the standards and installation location (Figure 4.18).

4.3.5.7 Introduction to the GE F650 relay

The GE F650 relay is a new generation of digital and multifunctional devices that easily integrate with automation systems and monitoring and control systems for substations.

The F650 is a digital device containing a central processing unit (CPU) that processes various types of input and output signals. It can communicate via a local area network with the operating interface, enabling device programming.

The CPU module contains software that provides protection elements in the form of logic algorithms, as well as programmable logic gates, timers, and decision-making control features.

The input elements accept various types of analog or digital signals from field devices. The F650 isolates and converts these signals into logic signals used for the relay. The output elements convert and isolate the logic signals generated by the relay into digital signals that can be used to control field equipment (Figure 4.19).

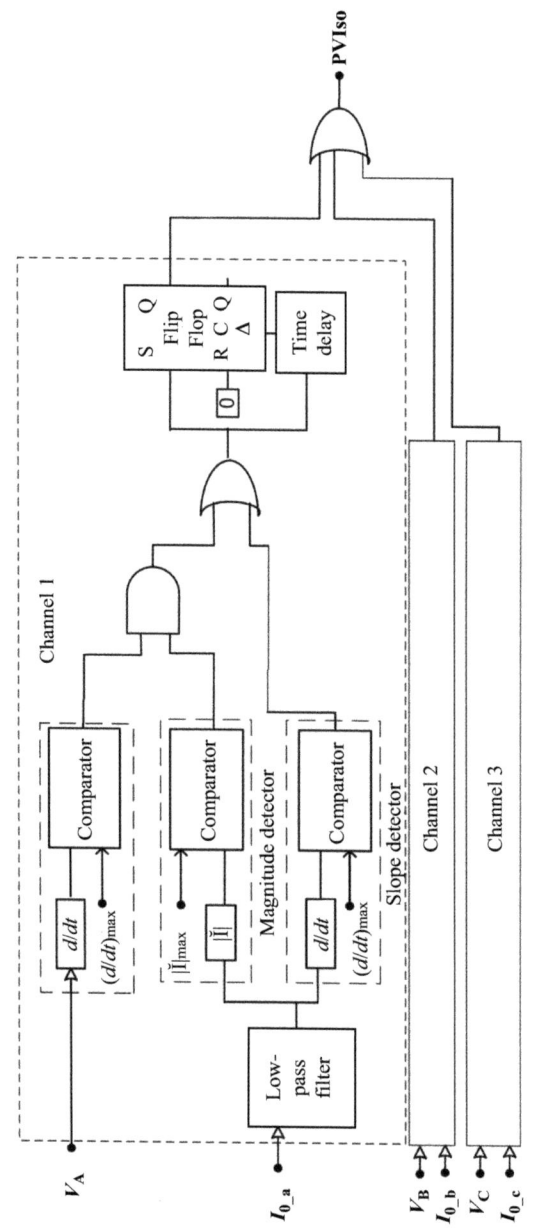

Figure 4.17 Fault detection module

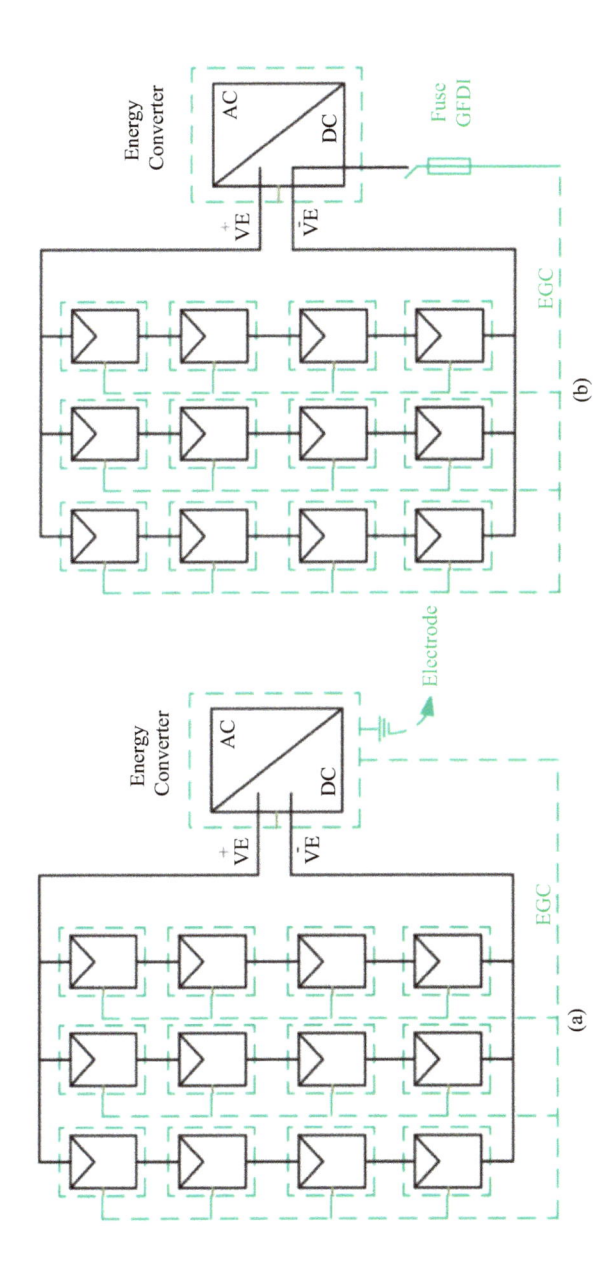

Figure 4.18 Grounded PV system (a), and ungrounded PV system (b)

Figure 4.19 Schematic diagram of the F650 relay

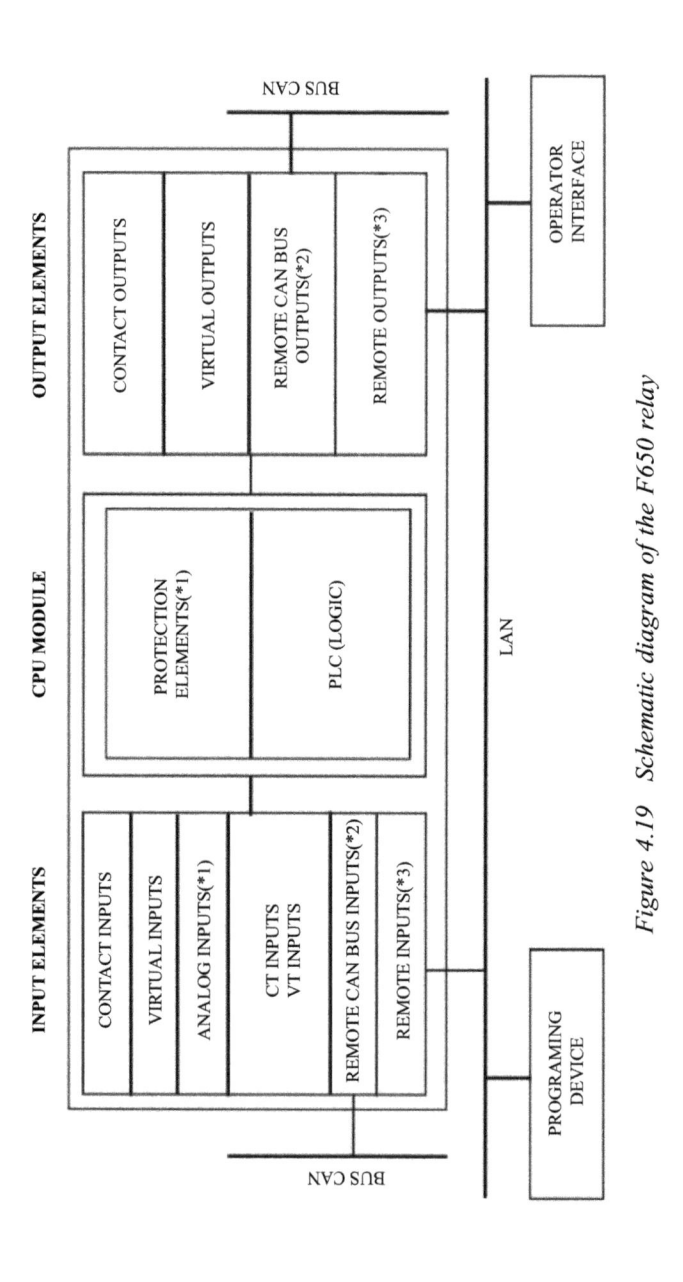

Input/output contacts
- **Digital signals**: Used to transmit control data.
- **CT and VT inputs**: Signals from CT and VT, monitoring the system's power supply.
- **Remote bus CAN I/O**: Links to physical input/output contacts via the CAN Bus, connecting to the 650 device.
- **PLC**: Programmable logic controller, enabling configuration and execution of logic circuits.
- **Protection elements**: Protection relays (overcurrent, overvoltage, etc.).
- **Remote I/O**: Shares status information between remote devices via IEC 61850 GSSE and GOOSE.
- **Analog inputs**: Signals from sensors.

The actual 22 kV distribution network integrated with DG in Ho Chi Minh City, Vietnam, has been simulated using ETAP/RSCAD software to assess the reliability of the new short-circuit analysis method, calculate power flow, and validate the effectiveness of the new protection coordination method for OCR, as shown in Figure 4.20. It is important to note that this study uses a real distribution network with DG/MG 22 kV sources instead of standard IEEE distribution systems. The use of real-time load data from this distribution network is appropriate for determining the minimum-maximum reliability intervals for load power at each load bus, load current, bus voltage, and fault current contributed by the grid and DG units.

4.3.6 Application of STATCOM for dynamic voltage stabilization in Power Systems

The use of wind energy globally is increasing significantly, with forecasts indicating that in the United States, wind energy will contribute 20% of electricity capacity by 2030. However, as the proportion of wind energy integrated into the grid increases, it poses challenges to system stability due to the intermittent and unpredictable nature of this energy source.

To address these challenges, devices from the flexible alternating current transmission system (FACTS) have been proposed for integration into the grid to enhance system stability. Among these FACTS devices, the static synchronous compensator (STATCOM) is one of the most effective solutions. STATCOM not only maintains voltage stability at the connection point but also improves the dynamic stability of the power system.

For example, STATCOM has demonstrated its ability to significantly improve the transient amplitude and LVRT capability of wind farms through both analysis and simulation. Similarly, the integration of STATCOM at the point of common coupling has been proposed to stabilize voltage and protect wind farms when connected to a weak grid.

In another study, a new control strategy combining a permanent magnet generator with STATCOM was introduced to compensate for reactive power, thereby improving transient voltage stability. Additionally, STATCOM has been proposed

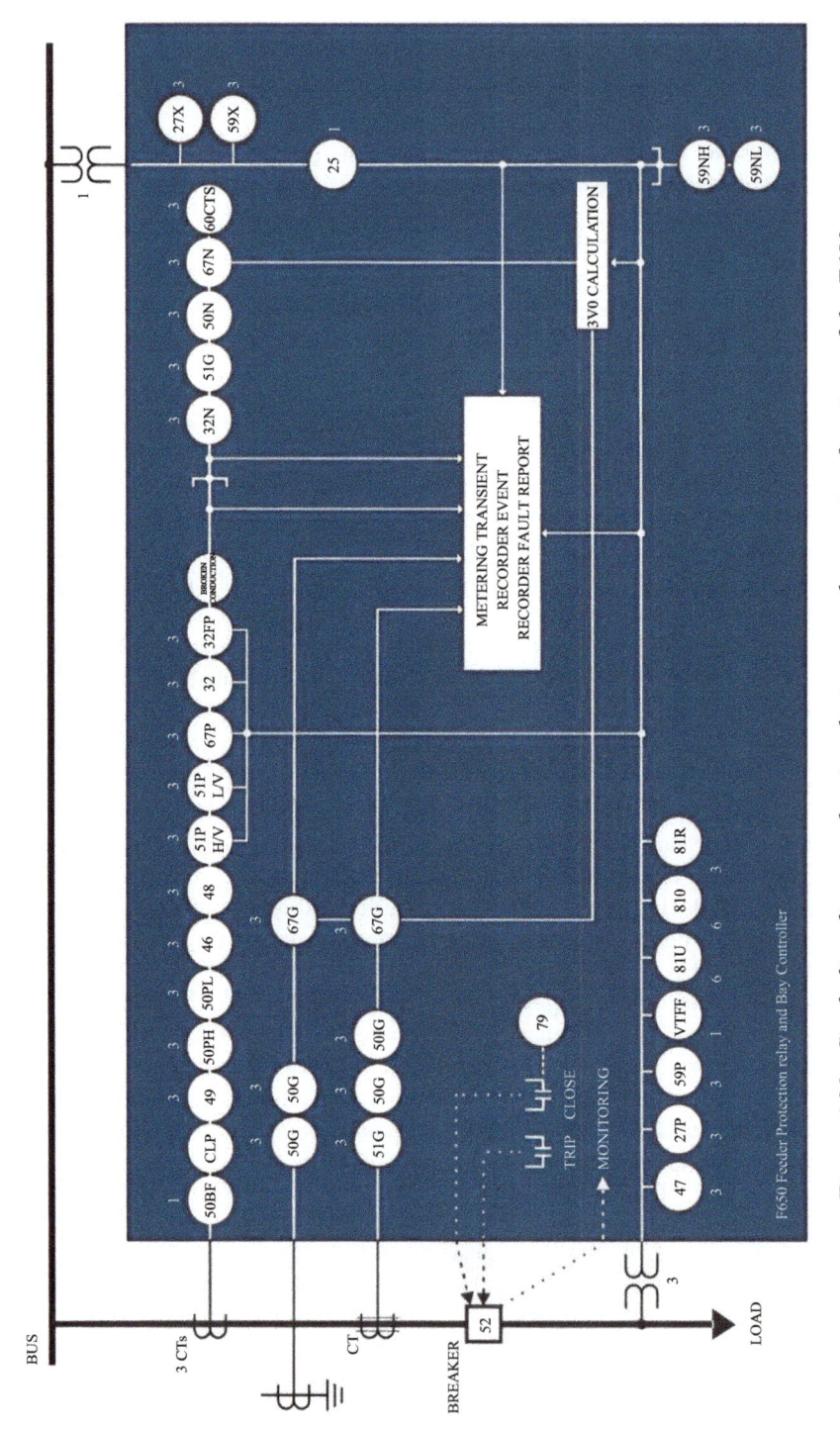

Figure 4.20 Single-line diagram showing the integrated protection functions of the F650

for use in systems with high reactive power loads, where excessive consumption of reactive power from the grid can adversely affect the connected loads.

The application of fuzzy controllers in STATCOM systems has also been studied to enhance transient stability in interconnected power systems. However, studies have shown issues such as large oscillations and long stabilization times in the system's responses. Conversely, a method has been introduced where the proportional–integral (P–I) controller parameters are continuously updated to adjust the voltage, thus improving the voltage profile in multi-machine systems under dynamic disturbances.

Moreover, STATCOM has been integrated with power system stabilizers in multi-machine systems connected to PV sources to minimize transient instability, as studied in some recent works.

4.3.6.1 Objectives of research

This study focuses on the application of STATCOM in power systems utilizing renewable energy sources, emphasizing its potential to improve voltage stability and enhance overall grid operation performance under dynamic conditions.

A STATCOM system has been employed to improve power stability by integrating a fuzzy logic controller into a two-area power system with four interconnected generators. In another study, a combination of PI and fuzzy logic (FL) controllers was proposed to enhance the dynamic performance and stability of a permanent magnet synchronous motor speed controller.

In distribution networks, a distributed STATCOM (D-STATCOM) system has been proposed, and various fuzzy control structures combined with PI control have been applied to control and maintain the total harmonic distortion of the grid current within IEEE standards.

Neuro-fuzzy systems (combining neural networks with FL) have emerged as a new hybrid intelligent system, combining the key features of neural networks and FL systems. Both FL systems and neural networks alone are insufficient to solve problems that require both linguistic and numerical knowledge simultaneously. Adaptive neuro-fuzzy inference system (ANFIS) combines the transparent linguistic reasoning of FL with the learning ability of ANN to perform intelligent self-learning, leading to widespread applications.

There are two basic training methods for ANFIS systems: hybrid learning algorithms and backpropagation algorithms. In this study, the backpropagation algorithm is applied to train the ANFIS. System identification and modeling play a crucial role in analyzing and designing physical systems. Measuring the input and output values of the system to determine the mathematical model relationship is a common method in linear systems. However, system identification in nonlinear systems is much more complex due to the changes in their mathematical characteristics. The universal approximation property of neural networks makes them suitable for modeling unknown nonlinear systems. Narendra and Parthasarathy integrated neural networks into adaptive systems to identify and control nonlinear systems. ANN have been widely used for identifying and controlling unknown nonlinear physical systems.

4.3.6.2 Design of the online ANFIS intelligent controller
(a) **Online ANFIS intelligent controller**

ANFIS differs from conventional FL systems due to its adaptive parameters, meaning that both the premise parameters and the consequent parameters can be adjusted. The performance of the ANFIS system depends on its internal parameters, including the membership function, the number of membership functions, training data, as well as the amount of data and the number of training cycles. These factors need to be carefully adjusted to achieve optimal performance.

The proposed online ANFIS intelligent controller system is illustrated in Figure 4.21. In this design, five membership functions are applied to both inputs: error (e) and the rate of change of error (de).

(b) **ANFIS structure**

A typical structure of the ANFIS system used is illustrated in Figure 4.21. In this structure, circular nodes represent fixed nodes, while square nodes represent adaptive nodes.

To simplify, we consider two inputs x_1, x_2, and one output y_m. Among many fuzzy inference system models, the Sugeno fuzzy model is the most widely used due to its high interpretability, computational efficiency, and integrated optimization and adaptation techniques.

For each model, a general set of fuzzy rules with two "if-then" fuzzy rules can be represented as follows:

Rule i : if x_1 is A_i, va x_2 is B_i, then $y_m = d_{2i}x_1 + d_{1i}x_2 + d_{0i}$

where:

- A_i và B_i are fuzzy sets in the antecedent part.
- $z = f(x_1, x_2)$ is a crisp function in the consequent part.

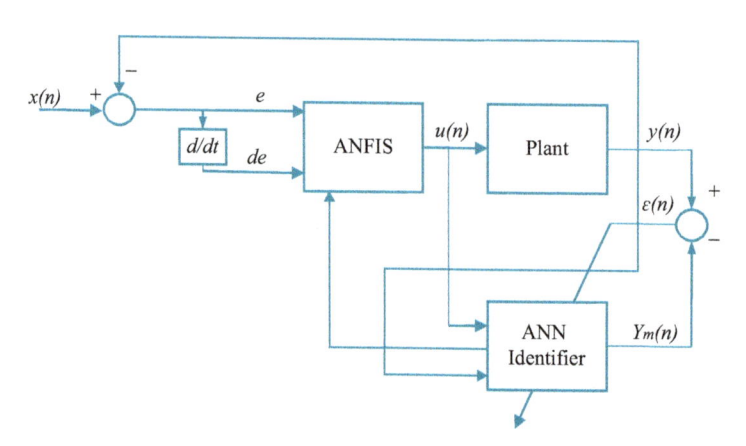

Figure 4.21 Proposed ANFIS-online smart controller

- d_{2i}, d_{1i}, d_{0i} là are the parameters to be updated in the rule.

This ANFIS structure combines the linguistic interpretability of FL with the learning capability of ANN, resulting in high efficiency in various practical applications (Figure 4.22).

- In these studies, each ANFIS consists of five layers as follows:

  Layer 1

  In this layer, the fuzzification process of the inputs takes place. Mathematically, this function can be expressed as follows:

  $$O_{ij}^{(1)} = \mu_j(I_{ij}^{(1)}) \tag{4.5}$$

  where $O_{ij}^{(1)}$ is the output of the node in Layer 1 corresponding to the jth linguistic term of the ith input variable. i represents the number of input variables, and j represents the number of linguistic terms for each input variable.

  Layer 2

  The total number of rules in this layer is 25. Each node's output represents the degree of activation of a rule, expressed as:

  $$O_k^{(2)} = w_k \prod_{i=1}^{q} O_{ij}^{(1)} \tag{4.6}$$

  where k is the number of rules.

  Layer 3

  The output of the kth node in this layer is the activation strength of each rule divided by the total activation value of all the fuzzy rules. This results in the normalization of the activation value for each fuzzy rule. This operation can be expressed as:

  $$O_k^{(3)} = \overline{w}_k = \frac{O_k^{(2)}}{\sum_{m=1}^{y} O_m^{(2)}} \tag{4.7}$$

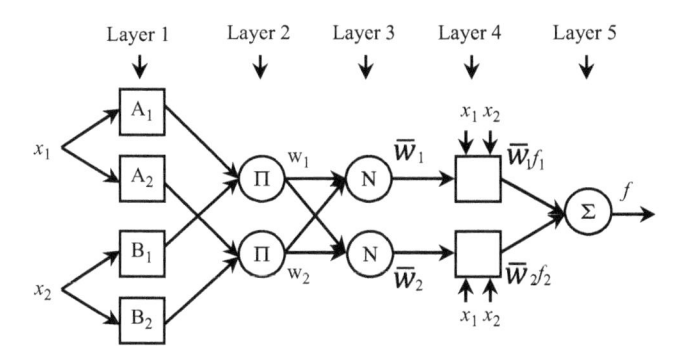

Figure 4.22 Structure of ANFIS

Layer 4
Each node k in this layer is associated with a set of adjustable parameters $d_{1k}, d_{2k}, \ldots, d_{N_{input}k}, d_{yk}, d_0$, and performs a linear function:

$$\begin{aligned} O_k^{(4)} &= \overline{w}_k f_k \\ O_k^{(4)} &= (d_{1k}I_1^{(1)} + d_{2k}I_2^{(1)} + \ldots + d_{N_{input}k}I_1^{(1)} + d_{0k}) \end{aligned} \quad (4.8)$$

Layer 5
The single node in this layer calculates the overall output by aggregating all input signals, as expressed by the following equation:

$$O_k^{(5)} = \sum_{k=1}^{y} O_k^{(4)} = \sum_{k=1}^{y} \overline{w}_k f_k = \frac{\sum_{k=1}^{y} w_k f_k}{\sum_{k=1}^{y} w_k} \quad (4.9)$$

(c) **ANN identification**
A multi-layer perceptron (MLP) network is used to represent the dynamics of the system. The structure of the MLP network is illustrated in Figure 4.23. The proposed network has six inputs, one hidden layer with nine neurons using the tanh activation function, and an output layer with one neuron exhibiting a linear node characteristic. The overall structure of the system with the ANN identifier is shown in Figure 4.23. The output of the ANN identifier is given by:

$$\Delta \widehat{P}_s(n+1) = f(\Delta P_s(n), \Delta P_s(n-1), \Delta P_s(n-2), u(n), u(k-1), u(k-1)) \quad (4.10)$$

where $\Delta P_s(n)$ is the power deviation at the STATCOM bus, and $u(k)$ is the control signal, both at time step n. The output of the ANN identifier is

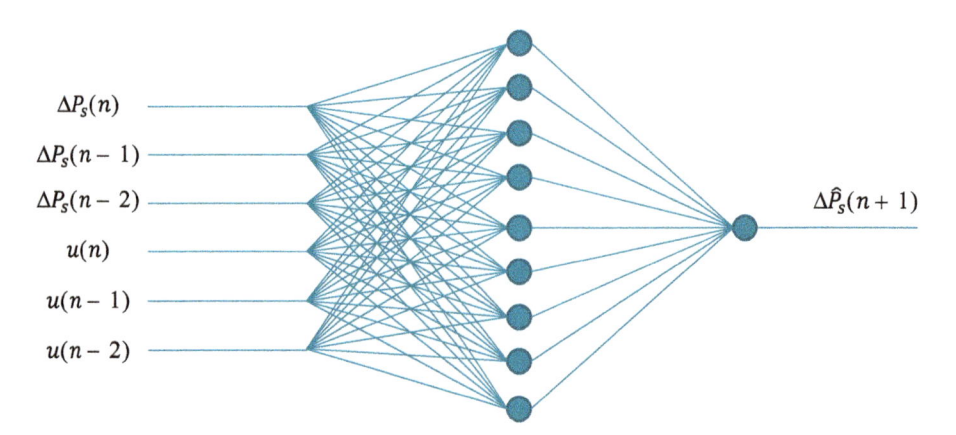

Figure 4.23 ANN identifier

$\Delta \widehat{P}_s(n+1)$ predicted at time step $(n + 1)$. The inputs to the ANN identifier are scaled before being applied to the network, so that their values lie within the range of $[-1, +1]$. To correctly infer the system's output at time step $(n + 1)$, the identifier is trained in advance to make the model's estimated output, $\Delta \widehat{P}_s(n+1)$, closely follow the actual system output, $\Delta P_s(n)$, by minimizing the cost function as follows:

$$F(n) = \frac{1}{2}(\varepsilon(n))^2 = \frac{1}{2}\left(\Delta P_s(n) - \Delta \widehat{P}_s(n)\right)^2 \quad (4.11)$$

The weights of the identifier are updated online using the gradient descent method as follows:

$$W(n) = W(n-1) - \eta \nabla_W J_i(n) \quad (4.12)$$

where $W(n)$ is the weight matrix at time n, η is the learning rate of the network, and $\nabla_W J_i(n)$ is the gradient of $J_i(n)$ with respect to the weight matrix $W(n)$. The gradient is calculated using the following formula:

$$\nabla_W J_i(n) = -\left[\Delta P_s(n) - \Delta \widehat{P}_s(n)\right] \frac{\partial \widehat{P}_s(n)}{\partial W(n)} \quad (4.13)$$

(d) **Online training of ANFIS**

The task of the learning algorithm for this structure is to adjust all the tunable parameters such as the Gaussian membership function variables and the values of the ANFIS rules, including $\{\sigma_{ij}, c_{ij}\}$ và $\{d_{2i}, d_{1i}, d_{0i}\}$. The adjustment is made so that the output of ANFIS matches the training data. This study proposes a neuro-fuzzy control strategy based on ANN recognition. Its structure is presented in Figure 4.21. The ANFIS parameters are updated online using the output of the ANN recognition, as will be introduced in the next section.

The error is defined as follows:

$$\varepsilon = x(n) - y_m(n) \quad (4.14)$$

The performance index used to evaluate the controller's ability is defined as:

$$E(n) = \frac{1}{2}(\varepsilon(n))^2 = \frac{1}{2}(x(n) - y_m(n))^2 \quad (4.15)$$

For both inputs, the authors use the well-known bell-shaped membership function, which is defined as follows:

$$\mu_j(I_{ij}^{(1)}) = e^{-\frac{1}{2}\left(\frac{x_i - c_{ij}}{\sigma_{ij}}\right)^2} \quad (4.16)$$

where the triplet of parameters $\{\sigma_{ij}, c_{ij}\}$ referred to as the premise parameters or nonlinear parameters, and they adjust the shape and position of the membership function. These parameters are adjusted during the training phase using the backpropagation algorithm. In this study, $i = (1, 2)$ và $j = (1, 2, \ldots, 5)$.

To calculate $\Delta\sigma = \sigma(n) - \sigma(n-1)$ và $\Delta c = c(n) - c(n-1)$, the authors use the following equations:

$$\sigma_{ij}(n+1) = \sigma_{ij}(n) + \eta\left(-\frac{\partial E_{(n)}}{\partial \sigma_{ij}}\right) \tag{4.17}$$

$$c_{ij}(n+1) = c_{ij}(n) + \eta\left(-\frac{\partial E_{(n)}}{\partial c_{ij}}\right) \tag{4.18}$$

To update the parameters of the function node $\{d_{2i}, d_{1i}, d_{0i}\}$, the authors use the following equations:

$$d_{2i}(n+1) = d_{2i}(n) + \eta\left(-\frac{\partial E_{(n)}}{\partial d_{2i}}\right) \tag{4.19}$$

$$d_{1i}(n+1) = d_{1i}(n) + \eta\left(-\frac{\partial E_{(n)}}{\partial d_{1i}}\right) \tag{4.20}$$

$$d_{0i}(n+1) = d_{0i}(n) + \eta\left(-\frac{\partial E_{(n)}}{\partial d_{0i}}\right) \tag{4.21}$$

In this case, the number of neurons, as shown in Figure 4.22, in the adaptive neuro-fuzzy controller is 5, 10, 20, 20, and 5 for layers 1, 2, 3, 4, and 5, respectively. This means there are ten center points in the membership functions (five for each input) and five output parameters that need to be updated online. A network with these parameters is considered relatively complex and computationally expensive, especially for real-time applications.

(e) STATCOM model

STATCOM is designed to regulate the voltage at its terminals by compensating for reactive power into or out of the electrical system. When the system voltage is low, STATCOM will inject reactive power into the system; when the voltage is high, it will absorb reactive power. Additionally, STATCOM can be designed to function as an active filter to absorb harmonics in the system. To analyze STATCOM, a mathematical model is used. In this model, the output voltage is separated into two components represented along the d and q axes as follows.

The voltages along the dq-axis of the STATCOM, as studied in Figure 4.24, can be expressed by the following equations:

$$v_{dsta} = V_{dcsta}\, km_{sta} sin(\theta_{bus} + \alpha_{sta}) \tag{4.22}$$

$$v_{qsta} = V_{dcsta}\, km_{sta} cos(\theta_{bus} + \alpha_{sta}) \tag{4.23}$$

where θ_{bus} is the phase angle of the common AC voltage of the bus, V_{dcsta} is the DC voltage in pu of the DC capacitor C_m; v_{dsta} and v_{qsta} are the voltages along the dq pu

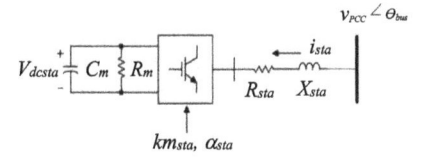

Figure 4.24 STATCOM model

at the STATCOM outputs, respectively; km_{sta} is the modulation index; α_{sta} is the phase angle of the STATCOM.

The DC voltage–current relationship of the DC capacitor C_m can be described by the following equation:

$$(C_m)(\dot{V}_{dcsta}) = w_b[I_{dcsta} - (V_{dcsta}/R_m)] \quad (4.24)$$

where the DC current can be calculated as

$$I_{dcsta} = i_{qsta}km\cos(\theta_{bus} + a) + i_{dsta}km\sin(\theta_{bus} + a) \quad (4.25)$$

where R_m is the equivalent resistance in pu used to calculate the power losses of the STATCOM, and i_{qsta}, i_{dsta} are the currents along the q and d axes in pu flowing into the STATCOM outputs.

4.3.6.3 Simulation results

This section presents the simulation results of the intelligent ANFIS-Online controller for STATCOM in improving the voltage transient stability during a short circuit fault. The scenarios investigated are for the power grid of the Saigon High-Tech park in Ho Chi Minh City. These simulation results, showing voltage responses in the time domain, were conducted using MATLAB® software to assess the effectiveness of the designed controller for STATCOM.

The power grid of the Saigon High-Tech park in Ho Chi Minh City is shown in the single-line diagram in Figure 4.25. The grid supplying the 110 kV Intel substation is fed from three different lines: Line 1 is from 110 kV Cat Lai (171)–Thu Duc East–Intel (171 and 177); Line 2 is from Cat Lai substation (172)–Thu Duc East–Tang Nhon Phu–Intel (172 and 174); and Line 3 is from Thu Duc 220 kV substation (178)–Thu Duc North–Intel (176 and 177). The power grid in the area supplying the Intel substation has many complex branching points, increasing the risk of faults. The area is mainly supplied by two 220 kV substations: Thu Duc and Cat Lai. In this power supply diagram, the circuit breakers at the 110 kV Thu Duc, Intel, Thu Duc North, Tang Nhon Phu, and Thu Duc East substations are always closed to create a loop, ensuring continuous power supply to the substations in the area and enhancing reliability in the event of a fault in any transmission line element.

For the Saigon Hi-Tech park power grid, a STATCOM unit with a capacity of ±15MVAR will be installed at node 7, the Intel substation, due to the numerous transmission lines connected to other loads such as Thu Duc North, Tang Nhon

150 *Clean energy in South-East Asia*

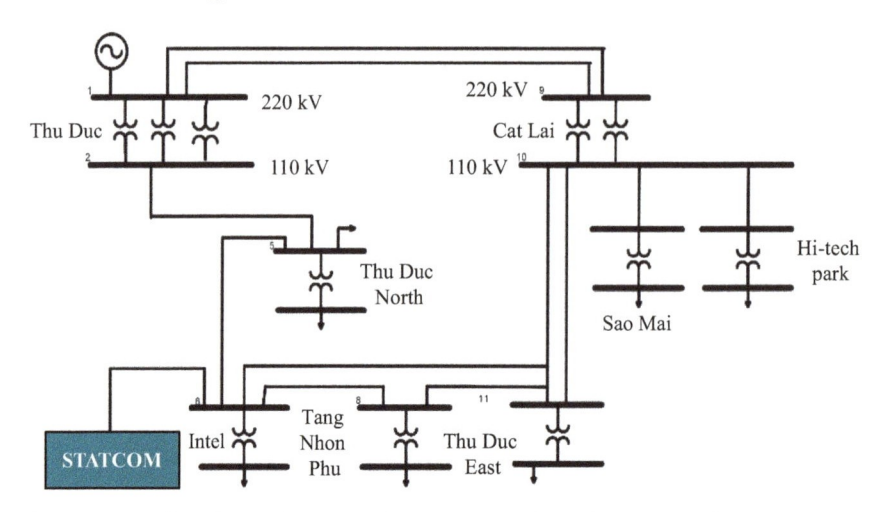

Figure 4.25 Simulation diagram of the power grid in the Saigon Hi-Tech park, Ho Chi Minh city

Phu, and Thu Duc East. According to the technical management report of the Ho Chi Minh City High Voltage Grid Company for the period from 2013 to 2017, a total of 178 incidents occurred on the Ho Chi Minh City grid, affecting various loads, of which 12% were three-phase short circuit faults. Therefore, the author focuses on simulating and analyzing three-phase short circuit faults on the High-Tech Park power grid in Ho Chi Minh City to evaluate the impact on voltage levels at the 110 kV substation in this area. The power grid simulation diagram supplying the High-Tech Park is presented in Figure 4.25. The specific fault scenarios studied include: (i) a short circuit fault on the 110 kV Thu Duc–Thu Duc North–Intel line, and (ii) a short circuit fault on the 110 kV Cat Lai–High-Tech Park line. The results present the voltage sag under these fault scenarios, with the fault occurring at 0.3 s and lasting for 100 ms.

- **Scenario 1: Use of STATCOM controller during fault on the Thu Duc–Thu Duc North–Intel line**
  The power supply method for the 110 kV Intel substation is from the 110 kV ring circuit, starting from the 110 kV busbar at the Thu Duc substation, passing through the 110 kV Thu Duc North, Intel, Tang Nhon Phu, Thu Duc East, Cat Lai 110 kV substations, and ending with the 220 kV Cat Lai–220 kV Thu Duc line. In the event of a fault on the Thu Duc–Thu Duc North–Intel line (section between Thu Duc Bac and Intel), the circuit breakers at the Intel and Thu Duc North substations will operate and isolate the faulty section, but the power supply to the Intel substation is still maintained through the 110 kV Thu Duc–Cat Lai–Thu Duc East–Intel and the 110 kV Thu Duc–Cat Lai–Thu Duc East–Tang Nhon Phu–Intel lines. Therefore, when investigating the voltage at Intel, it was observed that the voltage dropped to 0pu during the fault duration, but

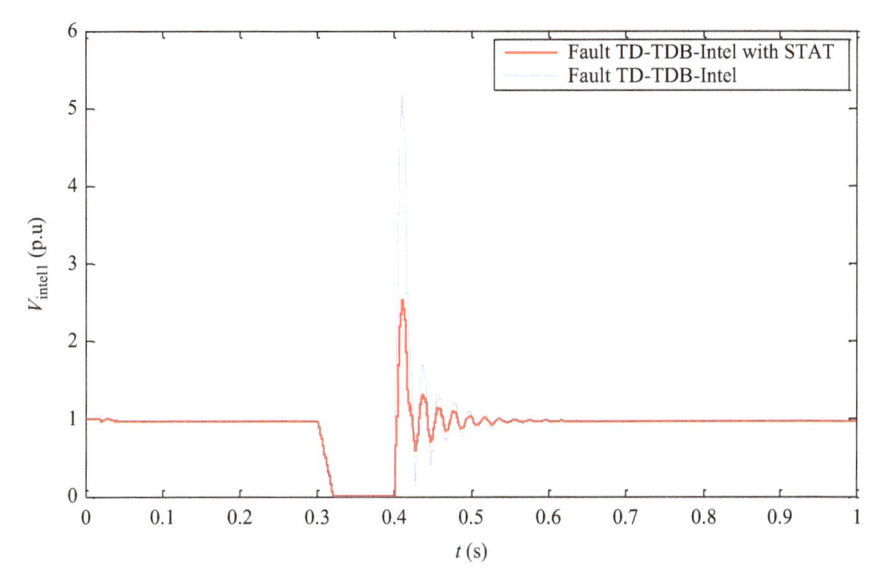

Figure 4.26 Voltage survey results at the intel station with and without the use of STATCOM device

after the faulted section was isolated, the voltage recovered, generating oscillations, while the load at the Intel substation was not isolated (i.e., no power loss). The simulation results for this scenario are detailed as follows:

- **Simulation results in cases with or without the STATCOM unit**

As mentioned earlier, to improve voltage fluctuations at the Intel load node, the research results propose the installation of a 15 MVAr STATCOM unit at the Intel substation. The dynamic response of the voltage at the Intel load node, for the three-phase short circuit fault scenario on the Thu Duc–Thu Duc North–Intel line, is presented in Figure 4.12. As shown in Figure 4.26, the voltage at the Intel substation shows a significant improvement in voltage deviation with the installation of the STATCOM device.

- **Simulation results of the STATCOM controller using ANFIS-online method**

The simulation results in Figure 4.27 show that the ANFIS-Online training method provides better performance than traditional ANFIS. Specifically, at the Intel load node, the time for the voltage amplitude to reach the permissible threshold ($\leq 5\%$) for ANFIS-Online is 0.433 s; the maximum transient voltage value for ANFIS-Online is 1.027 pu; and the voltage deviation for ANFIS-Online is only 5%.

When comparing the time for the voltage to stabilize after the fault is cleared, the time for the voltage to remain within the permissible range of 5% is shorter when using the ANFIS-Online controller compared to ANFIS. Specifically, with the ANFIS-Online controller, the voltage stabilization times for the Intel, Thu Duc East, and High-Tech park nodes are 0.433 s, 0.432 s, and 0.433 s, respectively.

- **Scenario 2: Using STATCOM controller during fault on the Cat Lai–High-Tech park line**

    For the simulation of a fault on the 110 kV Cat Lai–High-Tech Park line, the author simulates this as a transient fault and assumes that it is a self-clearing fault, in order to study the voltage fluctuations at the High-Tech Park substation after the fault. The voltage responses from the simulation results are as follows:

- **Simulation results with and without STATCOM controller**

The dynamic response of the voltage at the Intel, Thu Duc East, and High-Tech Park load nodes for the three-phase short circuit fault scenario on the Cat Lai–High-Tech Park line is presented in Figure 4.27. As shown in Figure 4.27, the voltage at the Intel, Thu Duc East, and High-Tech Park substations shows a significant improvement in voltage deviation when the STATCOM device is installed. Once again, this result demonstrates the effectiveness of using STATCOM in improving voltage quality as desired.

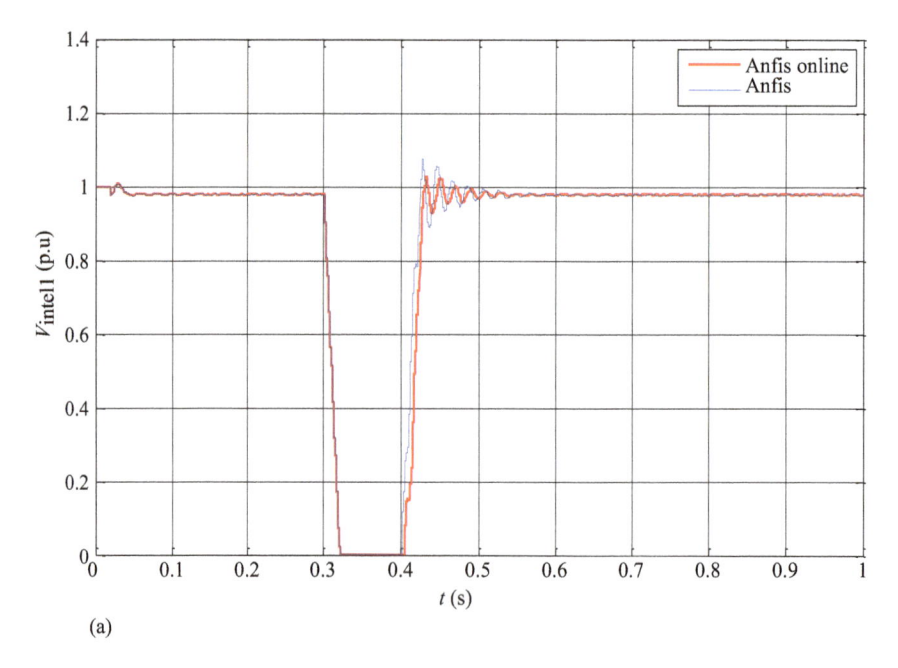

Figure 4.27 Survey results of voltage at intel, Thu Duc East, and High-Tech nodes with and without the use of STATCOM equipment. (a) Transient characteristics of voltage at the intel node (V_{Intel}) during a fault. (b) Transient characteristics of voltage at the Thu Duc East bus ($V_{ThuDucEast}$). (c) Transient characteristics of voltage at the High-Tech bus ($V_{High-tech}$).

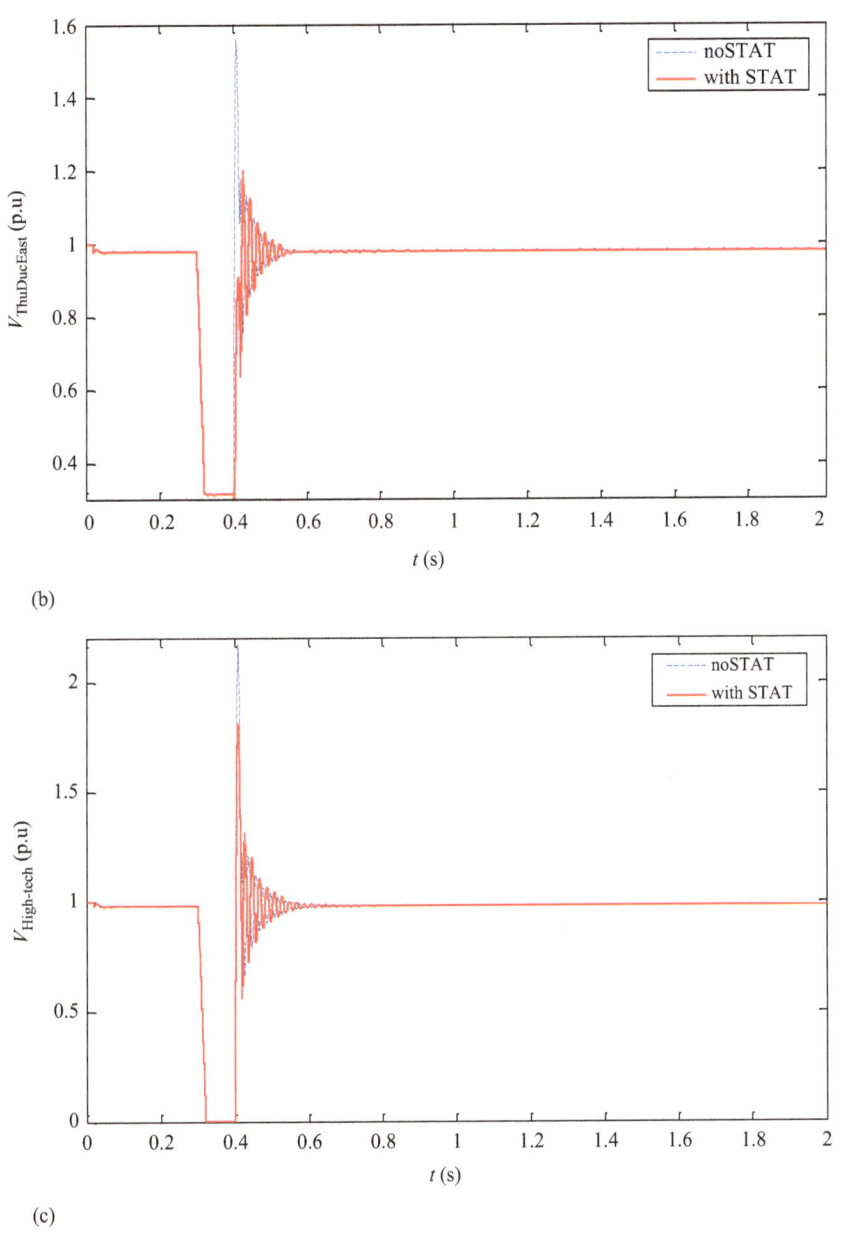

Figure 4.27 (Continued)

- **Simulation results for improved STATCOM controller using the ANFIS-online method**

The author employs the ANFIS-Online controller to regulate the STATCOM device, aiming to enhance voltage quality in the power grid. The performance of the ANFIS-Online controller is evaluated and compared with other controllers, including PID, Fuzzy, ANFIS, ANFIS-PSO, and ANFIS-GA, as reported in previous studies. The simulation results, shown in Figure 4.28 demonstrate that the ANFIS-Online controller provides the best response. Specifically, at the Intel load node, the time for the voltage amplitude to reach the allowable threshold ($\leq 5\%$) for ANFIS-Online is 0.457 s, and the maximum transient voltage for ANFIS-Online is 1.025 pu. The voltage deviation for ANFIS-Online is only 4%. It is noted that the voltage deviation for the ANFIS-Online controller is 4%, which is better than the 14% deviation of the standard ANFIS controller.

- **Dynamic characteristics of voltage at Thu Duc East node ($V_{ThuDucEast}$)**

In the case of comparing the time it takes for the voltage to reach a new steady-state after fault isolation, with the voltage remaining within the allowable range of 5%, the ANFIS-Online controller exhibits a faster stabilization time. Specifically, when using the ANFIS-Online controller, the voltage at the Intel, Thu Duc East, and High-Tech nodes reaches the steady-state values after 0.457 s, 0.457 s, and 0.488 s, respectively.

Furthermore, Table 4.7 highlights the percentage improvement in voltage amplitude achieved by the ANFIS and ANFIS-Online controllers for STATCOM.

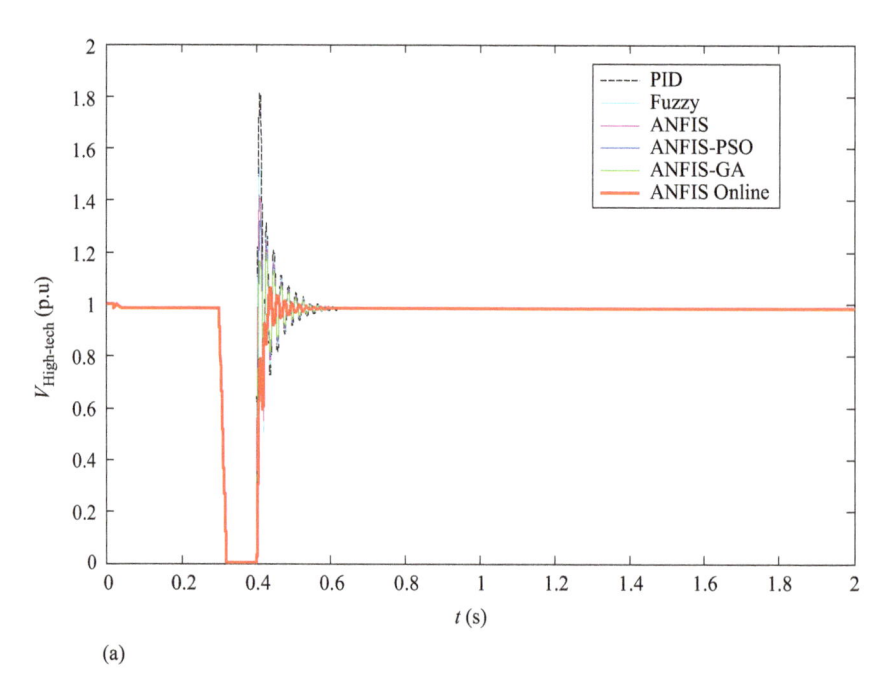

(a)

Figure 4.28 Survey results of voltage at intel load bus, Thu Duc East, and High-tech bus in the case of using the ANFIS-online controller. (a) Transient characteristics of voltage at the High-Tech bus ($V_{High-tech}$). (b) Transient characteristics of voltage at the intel bus (V_{Intel}) during a fault.

(b)

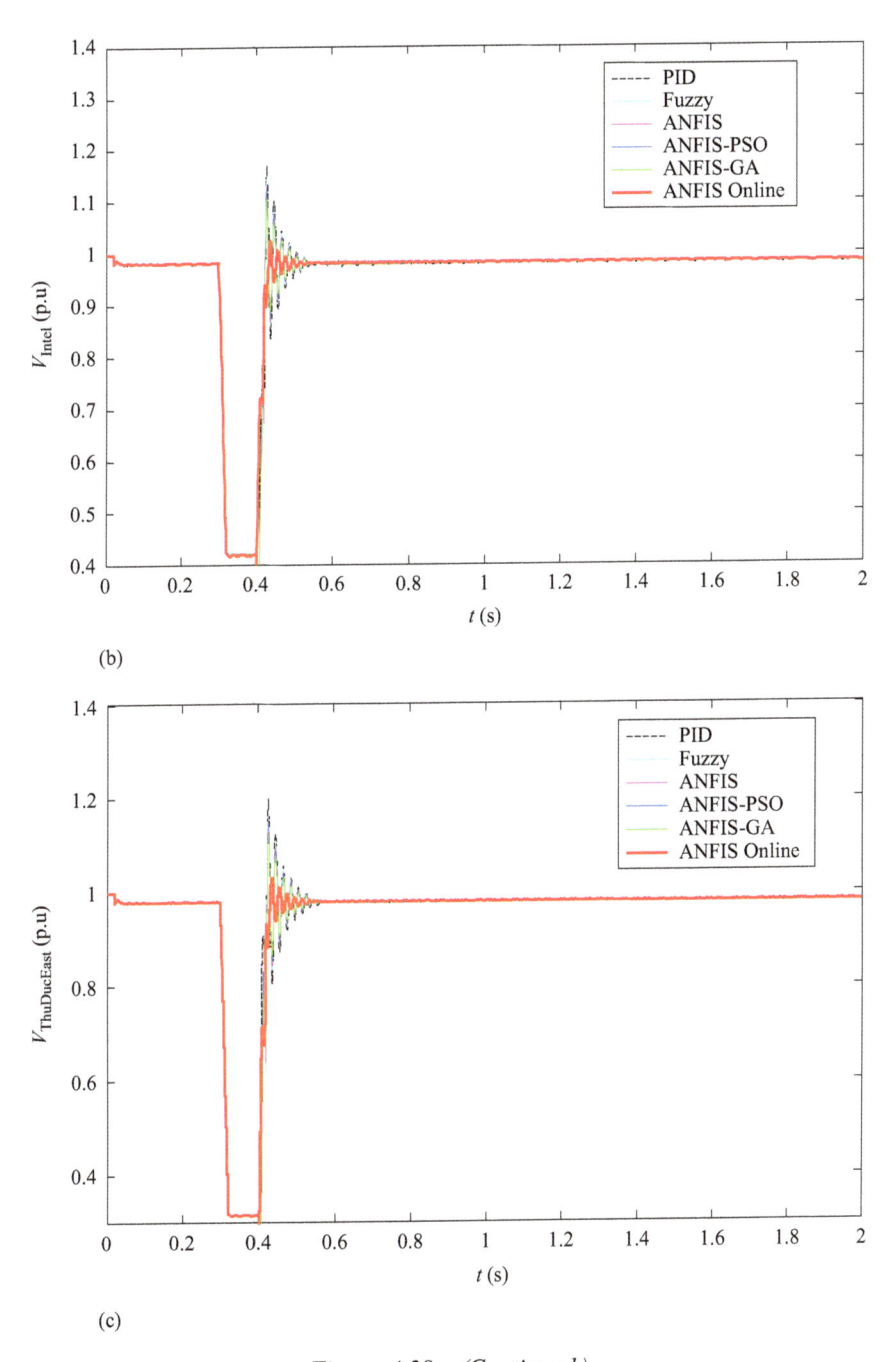

(c)

Figure 4.28 (Continued)

Table 4.7 Comparison table of ANFIS and ANFIS-online controllers for STATCOM devices on the actual power grid

Kinematic parameters	Node	Controller	
		ANFIS	ANFIS Online
Time to reach voltage amplitude within allowable threshold (±5%) (s)	Intel	0.487	0.457
	Thu Duc East	0.489	0.457
	High-Tech nodes	0.527	0.488
	Tang Nhon Phu	0.487	0.457
	Thu Duc North	0.447	0.437
	Cat Lai	0.527	0.477
Maximum overvoltage value (pu)	Intel	1.123	1.025
	Thu Duc East	1.154	1.031
	High-Tech nodes	1.417	1.059
	Tang Nhon Phu	1.141	1.026
	Thu Duc North	1.050	1.007
	Cat Lai	1.293	1.056
Voltage deviation (%)	Intel	14%	4%
	Thu Duc East	17%	5%
	High-Tech nodes	44%	8%
	Tang Nhon Phu	16%	4%
	Thu Duc North	7%	2%
	Cat Lai	31%	7%

In terms of voltage deviation reduction, the ANFIS-Online controller consistently delivers a superior voltage response compared to the ANFIS controller in the actual power grid.

4.3.6.4 Conclusion

This section has detailed the simulation results corresponding to different fault scenarios occurring on the power grid of the Saigon High-Tech park in Ho Chi Minh City. The scenarios investigated include three-phase short-circuit faults on the Thu Duc–Thu Duc North–Intel transmission line and the Cat Lai–High-Tech transmission line, both with and without STATCOM, and with different STATCOM control algorithms such as ANFIS and ANFIS-Online, aimed at improving voltage quality during transient conditions. From the analysis and evaluation of the simulation results, it is evident that the power grid of the Saigon High-Tech park in Ho Chi Minh City requires the use of STATCOM, and the implementation of the ANFIS-Online controller for STATCOM is a feasible solution for the future.

4.4 Challenges and opportunities in implementing smart grids

- **Smart grid: opportunities and challenges**

4.4.1 Introduction to smart grids

A smart grid combines advanced technology, automation systems, and real-time data to improve the efficiency, reliability, and sustainability of the power system. Within this framework, the advanced metering infrastructure (AMI) plays a crucial role by providing detailed information to both suppliers and consumers, promoting energy efficiency, and minimizing environmental impact (Figure 4.29).

- Development journey and technology integration for smart grids

  The diagram outlines the roadmap for the development and application of modern technological solutions aimed at enhancing reliability, integrating distributed generation sources, promoting green energy, ensuring cybersecurity, and improving customer experience.

  – **Training and capacity building**
    o In-depth training programs on standards such as IEC61850, blockchain, and artificial intelligence to support grid operation and development.
    o Development of a professional engineering team proficient in advanced technologies for system management and operation.

  – **Intelligent monitoring and control**
    o Deployment of SCADA, DMS, and advanced technologies at substations, integrating remote monitoring systems.
    o Piloting and expanding applications such as DERMS, VPP, and AMI to optimize power system operations.

  – **Data analytics and integration of renewable energy**
    o Big data analytics to optimize grid operation.
    o Promoting the integration of distributed energy sources like solar, wind, and energy storage systems.

  – **Cybersecurity and customer experience**
    o Application of ISO27001 and IEC62443 standards to protect systems and data.
    o Enhancing customer experience through blockchain solutions and transparent data transmission protocols.

This development journey underscores the importance of integrating technology with sustainability goals, while ensuring safety and meeting the growing demands of customers.

4.4.2 Key challenges in implementation

- **High initial investment costs**
  – Issue: Smart grids require significant infrastructure upgrades, including sensors, smart metering technologies, energy management systems, and telecommunications.
  – Limitation: This is a major barrier for countries or regions with limited budgets.

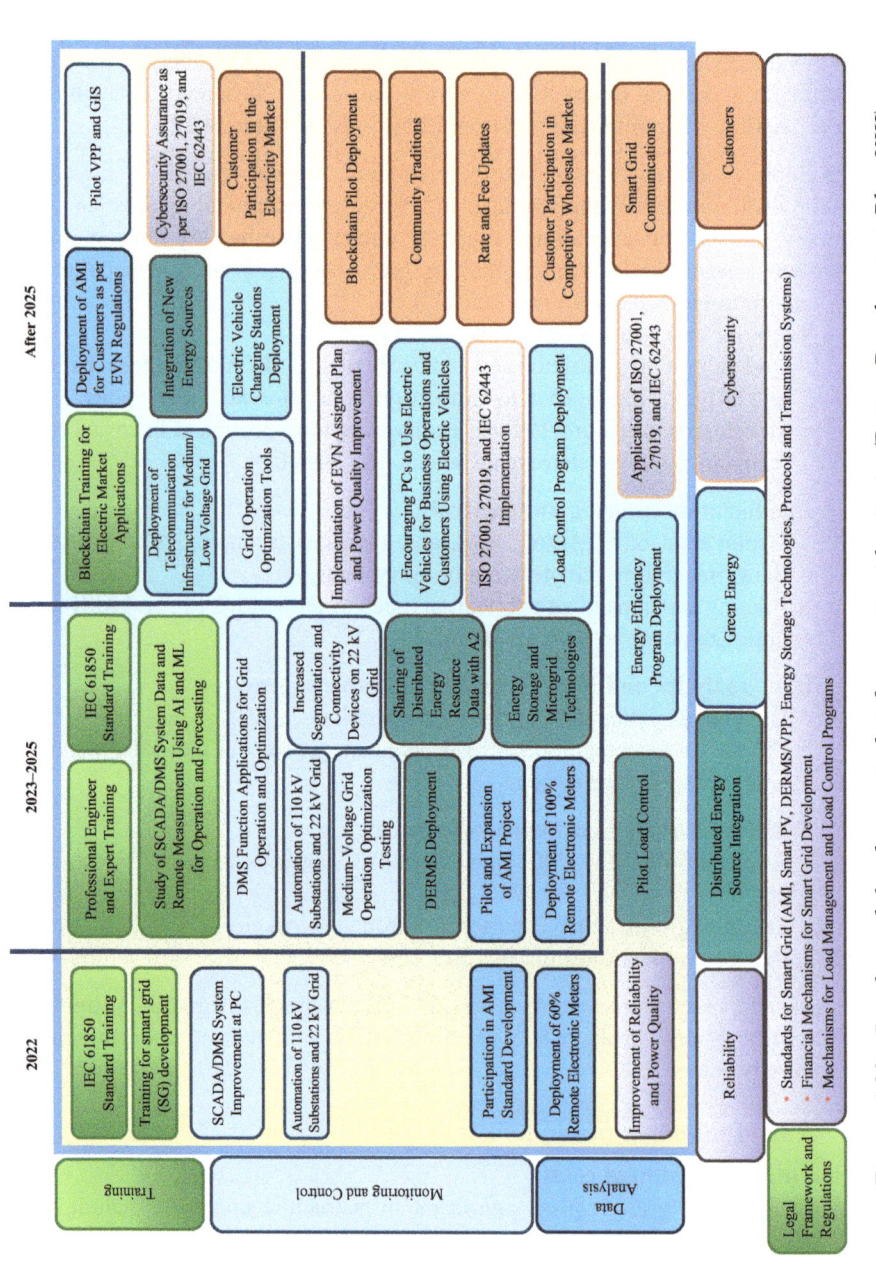

Figure 4.29   Results and development plan for smart grid criteria (Power Development Plan VIII)

- **Cybersecurity and network security**
  - Risk: Dependence on the Internet and automation increases the risk of cyberattacks, affecting user data and system stability.
  - Requirement: Strict cybersecurity measures need to be implemented.
- **Compatibility and integration**
  - Challenge: The system needs to synchronize devices and technologies from multiple suppliers.
  - Solution: Establish common standards and enhance collaboration among stakeholders.
- **Workforce training**
  - Requirement: A skilled workforce must be trained to manage complex systems, which demands time and significant investment.
- **Telecommunications infrastructure limitations**
  - Issue: In rural areas, telecommunications networks may not be robust enough to support real-time data transmission.
- **Legal and policy challenges**
  - Barrier: Inconsistent regulations and policies on privacy and energy could delay deployment progress.

4.4.3 Benefits and opportunities

- **Real-time monitoring and control**
  - Optimizes resource utilization and minimizes environmental damage.
- **High reliability**
  - Systems are self-healing with backup provisions to protect the grid from outages.
- **Demand response programs**
  - Encourages customer participation in adjusting consumption to maintain load balance during peak hours.
- **Renewable energy integration**
  - Facilitates the coordination of energy sources like solar and wind, moving toward a more sustainable grid system.
- **Conclusion and future directions**

Despite the numerous challenges, smart grids present significant opportunities to transform current energy approaches. Overcoming these barriers requires collaboration among governments, businesses, consumers, and technological innovation.

Future directions will focus on:

- Enhancing data security and management capabilities.
- Improving telecommunications infrastructure, especially in rural areas.

- Expanding research and development of automation solutions and renewable energy integration.
- Establishing consistent legal frameworks and financial support for pilot projects.

Smart grids are not just a technological solution but a commitment to creating a future where energy, technology, and the environment develop in harmony.

Further reading

[1] Vietnam's Eighth National Power Development Plan (PDP VIII).
[2] Environmental Protection Law: Law No. 72/2020/QH14, signed on 17/11/2020.
[3] Law on Water Resources: Law No. 17/2012/QH13, signed on 21/6/2012, effective from 01/1/2013.
[4] Planning Law: Law No. 21/2017/QH14, passed by the 14th National Assembly on 24/11/2017.
[5] Amendment Law No. 28/2018/QH14: Amendments and supplements to several articles of 11 laws related to planning, passed on 15/6/2018.
[6] Amendment Law No. 35/2018/QH14: Amendments and supplements to several articles of 37 laws related to planning, passed on 20/11/2018.
[7] Decree No. 29/2011/NĐ-CP: Issued by the Government on 18/2/2011, regulating strategic environmental assessment, environmental impact assessment, and environmental protection commitments.
[8] Decree No. 23/2006/NĐ-CP: Issued by the Government on 03/3/2006, on the implementation of the Law on Forest Protection and Development.
[9] Decree No. 46/2012/NĐ-CP: Issued by the Prime Minister on 22/5/2012, amending and supplementing several articles of:
[10] Decree No. 35/2003/NĐ-CP (dated 04/4/2003) detailing the implementation of several articles of the Law on Fire Prevention and Fighting.
[11] Decree No. 130/2006/NĐ-CP (dated 08/11/2006) on compulsory fire and explosion insurance policies.
[12] Decree No. 32/2006/NĐ-CP (2006): Application of the Sorensen Coefficient (S) for comparing species composition in the study area with neighboring areas.
[13] Decree No. 18/2015/NĐ-CP (Article 12) and Circular No. 27/2015/TT-BTNMT (Article 7): Provisions on consultation activities during the implementation of the Environmental and Social Impact Assessment (ESIA).
[14] Circular No. 27/2015/TT-BTNMT: Issued on 29/5/2015 by the Ministry of Natural Resources and Environment (MONRE) on Strategic Environmental Assessment, Environmental Impact Assessment, and Environmental Protection Plans (Appendix 2.3, Chapters 1, 3, 4, and 5).
[15] Circular No. 26/2011/TT-BTNMT: Issued on 18/7/2011 by MONRE, detailing certain provisions of Decree No. 29/2011/NĐ-CP on strategic

environmental assessment, environmental impact assessment, and environmental protection commitments.

[16] Circular No. 12/2011/TT-BTNMT: Issued on 14/4/2011 by MONRE, on the management of hazardous waste.
[17] Directive No. 08/2006/CT-TTg: Issued by the Government on 08/3/2006, on urgent measures to prevent illegal deforestation, forest burning, and unauthorized logging.
[18] Ngo Dang Luu, Nguyen Hung, Nguyen Anh Tam, and Long D. Nguyen, (2021), "Wireless Power Transfer simulation for EV", [Copyright] Ministry of Culture, Sports and Tourism.
[19] Ngo Dang Luu, Nguyen Hung, and Long D. Nguyen, (2021), "Power flow calculation of the electric power system by MATLAB", Ministry of Culture, Sports and Tourism.
[20] Ngo Dang Luu, Nguyen Hung, Nguyen Anh Tam, and Long D. Nguyen, (2022), "Distance Relay Protection in AC Microgrid", Ministry of Culture, Sports and Tourism.
[21] Ngo Dang Luu, Nguyen Hung, Nguyen Anh Tam, and Long D. Nguyen, (2022), "Analysis of Solar Photovoltaic System Sunlight Shadowing", Ministry of Culture, Sports and Tourism.
[22] Ngo Dang Luu, Nguyen Hung, Le Anh Duc and Long D. Nguyen, (2022), "Estimate Frequency Response at the Command Line", Ministry of Culture, Sports and Tourism.
[23] Ministry of Industry and Trade of Vietnam (2023). *Power Development Plan VIII*. Hanoi: MOIT.

Chapter 5
Study and application of microgrid systems

5.1 Definition and objectives of microgrid systems

A microgrid (MG) typically uses distributed energy sources such as wind turbines (WTs) and solar photovoltaic (PV) modules. When multiple distributed generation sources with different characteristics are used in a MG, managing these sources becomes an important issue. The generation power of solar modules and WTs in a MG continuously fluctuates due to solar radiation and wind speed. Due to the inherent instability and unpredictability of renewable energy sources, energy storage systems (ESS) are often employed in MGs. To control the distributed energy sources and storage devices, maintain the supply–demand balance within the MG, and provide sustainable and stable power to the loads, an energy management system (EMS) is applied.

5.2 Key components of a microgrid system

A MG is composed of key components that work synchronously to ensure sustainable, efficient, and reliable energy supply. The following are the main components of a MG:

- **Distributed energy resources (DERs)**: These are small-scale power generation sources integrated directly into the MG, including:
  - **Renewable energy**
    - **Solar PV**: Converts solar radiation into electricity. This is a clean and widely used energy source in MGs.
    - **Wind turbines**: Uses the kinetic energy of wind to generate electricity, typically integrated in areas with stable wind speeds.
  - **Traditional energy sources**: Diesel generators, powered by fossil fuels, serve as backup sources.
  - **Biomass energy systems**: Converts biological fuel into electricity.
- **Energy storage systems**: Energy storage devices play a crucial role in maintaining the stability of the MG.
  - **Battery storage**: Stores excess energy from renewable sources for use when needed.

- **Mechanical energy storage system**: This includes systems such as flywheels or pumped hydro storage.
  - **Supercapacitors**: These temporarily store energy for applications that require rapid response.
- **Load systems**:
    MGs supply power to various types of loads:
    - **Critical loads**: Such as hospitals, data centers, or telecommunications systems, which require a stable and uninterrupted power supply.
    - **Non-critical loads**: Such as street lighting or household appliances, which can be deprioritized in case of energy shortages.
- **Power electronics converters**:
    These convert energy from generation sources and storage systems to meet load requirements:
    - **Inverter**: Converts direct current (DC) from solar panels or storage batteries into alternating current (AC).
    - **Rectifier**: Converts AC to DC for systems requiring DC.
    - **Regulator**: Ensures voltage and frequency stability within the MG.
- **Energy management system**:
    This central system coordinates and optimizes MG operations:
    - **Monitoring and surveillance**: Tracks electricity generation, storage levels, and load demand in real time.
    - **Energy source coordination**: Automatically balances supply and demand between energy sources and loads.
    - **Cost optimization**: Manages energy use efficiently to reduce operational costs.
- **Protection system**:
    Ensures the safety of the MG and its components:
    - **Protective relays**: Detect faults such as short circuits or overloads and disconnect affected components.
    - **Anti-islanding**: Detects and prevents the MG from operating out of sync with the main grid during grid outages.
- **Communication infrastructure**:
    Communication systems play a crucial role in MG management:
    - **Component communication**: Ensures coordination between EMS, DERs, ESS, and loads.
    - **IoT technology**: Connects and manages devices through sensor networks and remote controls.
- **Main grid interface**:
    MGs can operate in the following modes:
    - **Grid-connected mode**: Exchanges energy with the main grid when needed.
    - **Island mode**: Operates autonomously when disconnected from the main grid.

The main components of a MG include not only energy generation and storage sources but also modern management, protection, and communication systems to ensure efficient, stable, and sustainable operation.

5.3 Application of microgrid technology in renewable energy sources and cloud-based systems

As fossil fuels deplete and environmental pollution from fossil fuel combustion worsens, renewable energy has gained significant attention as a future energy choice. Renewable sources such as wind, solar, geothermal, ocean wave, and tidal energy play a major role in improving the human living environment and mitigating environmental impact. Renewable energy systems are also increasingly deployed in remote and isolated areas where national grid expansion is economically unfeasible.

Most isolated regions around the world are powered by diesel-powered stations; however, more isolated areas are now being supplied by hybrid renewable energy systems, including wind power combined with diesel. Many hybrid power systems in isolated regions already incorporate wind energy.

Vietnam, with significant wind energy potential, has a great opportunity for wind power development, especially as it is considered the largest wind energy resource in Southeast Asia, with an estimated potential capacity of 513,360 MW—200 times greater than the capacity of Son La Hydropower.

Facing electricity shortages and climate change challenges, the Vietnamese government has prioritized developing "green electricity" from renewable sources to ensure energy security and environmental protection. The government recently identified wind energy as a key focus area for the future.

Government policies and incentives for the wind energy sector are outlined in clear legal frameworks, such as Decision No. 1208/2011/QD-TTg, which highlights a priority for renewable energy sources in electricity production. It aims to increase the share of electricity from renewable sources, from 3.5% in 2010 to 6% by 2030. Specifically, wind power's share is expected to grow from a modest 31 MW to approximately 1,000 MW by 2020 and 6,200 MW by 2030.

The government's commitment to renewable energy, especially wind energy, is evident with policies such as Decision No. 37/2011/QD-TTg, implemented in 2011. Vietnam has over 4,000 islands, many of which are far from the mainland, making grid connection economically infeasible. Diesel power stations currently serve these islands but at a high cost, consuming large amounts of diesel and contributing to environmental pollution. Thus, the development of hybrid wind–diesel power systems in remote areas is crucial and aligns with global trends.

5.3.1 Global research on wind–diesel hybrid power systems

Wind energy has been harnessed for centuries, originally used for sailing ships and balloons, and later for mechanical power via windmills. The idea of using wind energy for electricity generation emerged after the invention of electricity and generators. Initially, windmills only converted wind kinetic energy into mechanical power, but later, generators were added to produce electricity. With advances in fluid mechanics, turbine designs were optimized for better efficiency.

In the 1970s, following oil crises, global efforts to develop alternative electricity generation sources, including WTs, accelerated. The development of wind energy systems started in Germany and spread internationally. As part of this trend, hybrid wind–diesel systems were deployed in many isolated regions.

Numerous laboratories worldwide focus on hybrid wind–diesel systems, including National Renewable Energy Laboratory (NREL) (1996) and RERL–Umass (1989) in the USA, CRES in Greece (1995), DEWI in Germany (1992), RAL in the UK (1991), EFI in Norway (1989), IREQ (1986) and AWTS (1985) in Canada, and RISØ in Denmark (1984).

5.3.1.1 Research on optimal hybrid wind–diesel power systems

Several studies focus on optimizing hybrid wind–diesel systems by calculating the number and type of turbines for connected regions. However, these studies do not consider isolated grids. Some studies calculate the appropriate number of WTs but only for a single turbine type, such as 600 kW or 1,500 kW models. Other studies only calculate the combination of one WT with a diesel generator. These models focus on the project's lifecycle rather than operational calculations, which may not be suitable for designing new wind power stations for isolated islands in Vietnam.

5.3.1.2 Research on stable operation of hybrid wind–diesel systems

Various studies examine optimizing the operation of hybrid wind–diesel power systems, particularly focusing on auxiliary equipment.

5.3.1.3 Study of solutions to increase wind power penetration rate

Based on economic-technical indicators (such as equipment cost, efficiency, simplicity, lifespan, payback time, reliability, environmental impact, operating limits, storage capacity, compatibility with wind–diesel systems, and fuel consumption reduction), studies have assessed the performance index of most energy storage technologies. From these comparisons, the pressure accumulator for energy storage was applied in a hybrid wind–diesel power system supplying electricity to a residential area. In the case of a peak load of 851 kW, fuel consumption was reduced by 27%; for a 5 kW load, fuel consumption was reduced by 98%.

5.3.1.4 Study on improving the power quality of hybrid wind–diesel power systems

Using superconducting coils as magnetic storage in a 650 kW isolated grid provides relatively good frequency and power quality. Currently, the integration of an electromagnetic clutch (EMC) into WTs has only been proposed in studies.

However, these studies aim to confirm that WTs with synchronous generators directly connected to the grid and integrated with EMC provide power quality equivalent to that of current variable-speed WTs.

5.3.2 Research status in Vietnam

The research and exploitation of wind energy in Vietnam began in the 1970s with the involvement of many agencies. Since 1984, under the state-level "Scientific and Technological Advancement Program" for new and renewable energy, several wind power systems with capacities ranging from 120 to 500 W have been built, charging batteries at the University of Danang and Hanoi University of Science and Technology, with the aim of lighting and powering radios.

Most studies in Vietnam focus on controlling WTs using doubly fed induction generators (DFIG), with research on control methods for DFIG based on: control algorithms ensuring decoupling between torque and power factor; nonlinear algorithms based on backstepping techniques; passive Euler-Lagrange and Hamiltonian methods; and grid-following control. In addition, there are studies to ensure the power quality of DFIGs, including voltage stabilization, eliminating steady-state errors based on nonlinear algorithms, and adaptive current control based on backstepping techniques. Furthermore, in order for WTs to support asymmetric grid faults and symmetrical grid failures, studies have analyzed and proposed passive methods for DFIG. These studies also propose integrating active filters into the grid-side controller to improve power quality when the grid is stable and under nonlinear loads. Power quality has been partially improved.

Vietnam's wind and climate conditions have specific characteristics, so suitable WT designs are required. Therefore, research has designed and manufactured WTs with capacities ranging from 10 to 30 kW. However, operational results have not fully aligned with Vietnam's climate conditions.

5.3.2.1 Study on selecting the optimal wind–diesel hybrid power system

The research suggests installing WTs using squirrel cage induction generators (SCIG) or permanent magnet synchronous generators (PMSG) in isolated grids. However, the study only asserts that SCIG and PMSG are better than DFIG, without simulation or experimental validation. The study also does not specify the number and capacity of WTs to be installed.

5.3.2.2 Study on stable operation and enhancing wind power penetration in hybrid wind–diesel power systems

Research has detailed the features of auxiliary equipment, analyzing and proposing the use of low-load diesel generators to increase wind power penetration to 70% P_t while ensuring frequency stability. This study also emphasizes the need for policies that encourage load growth to make the use of additional auxiliary equipment economically viable for investors.

The study simulated operating conditions aimed at maximizing wind energy use on MATLAB® for a hybrid wind–diesel power system without auxiliary equipment in an isolated grid. However, this study did not focus on reactive power distribution for wind and diesel power, nor did it account for the operating limits of each generator or show the appropriate wind power penetration rate.

Research objectives and tasks

In a hybrid wind–diesel power system in an isolated grid, two important factors need attention: power quality and wind power penetration. However, these two criteria are often inversely related as wind power penetration increases. Therefore, the goal of the research is to propose an optimal operating method to maximize wind power penetration while ensuring power quality.

During operation, the hybrid wind–diesel power system must meet technical constraints, with two of the most important being:

1. Stable operation: Ensuring the system operates safely and reliably.
2. Power quality: Evaluated through simulation, experimental, and real-time data.

Currently, system stability can be analyzed and assessed using mathematical modeling methods. Therefore, this research will focus on determining the operating region of the system based on stability limits and then assessing power quality under specific conditions.

From the analysis of the current research on hybrid wind–diesel power systems in isolated grids, some issues remain unresolved. Based on the observations and objectives outlined, this research sets out the following specific tasks:

1. Study the stable operation conditions of a hybrid wind–diesel power system in an isolated grid without auxiliary equipment.
2. Propose solutions to enhance wind power utilization while ensuring power quality and system stability.
3. Identify the type and number of suitable WTs to be added to the existing diesel station in isolated areas, aiming to optimize wind power utilization.

In general, the hybrid wind–diesel power system in an isolated grid should be generalized. The capacity of wind power connected to isolated grids with small capacities should be evaluated. It is essential to align wind and diesel power appropriately.

From the survey above, it is clear that the stability of the hybrid wind–diesel power system is a prerequisite for operation. The study on the operating characteristics of this system in an isolated grid identifies factors significantly affecting system stability, proposing ways to improve and enhance system stability.

The study suggests that the current operational mode of hybrid wind–diesel systems is inadequate. The dissertation proposes an algorithm and optimal operating solution to maximize wind energy utilization, thereby saving diesel fuel and protecting the environment.

Surveys also indicate that WTs installed in isolated areas need additional auxiliary equipment to enhance power quality and wind power penetration.

Therefore, specialized WTs for isolated areas are required in the future. The dissertation proposes integrating EMC into WTs to improve power quality and increase wind power penetration.

Surveys of installed wind–diesel hybrid power generation systems have shown that the selection of the number and capacity of WTs has not been optimal. This dissertation proposes an optimal calculation program for the economic efficiency of wind–diesel hybrid power generation systems during operation. It identifies the types of WTs and the corresponding numbers suitable for new installations in isolated regions.

5.3.3 Problem statement

In the past, diesel generators were the main source of electricity in isolated areas; however, they were expensive and caused environmental pollution. Today, a solution to address these limitations is to combine existing diesel generators with newly installed wind power systems. This hybrid system provides the benefit of continuous electricity supply to islands or isolated communities while saving fuel. Generally, the installed capacity of diesel generators is sufficient to meet the peak electricity demand, but in practice, they are used only when wind power generation is insufficient to meet load demands.

Hybrid wind–diesel power systems in isolated grids have been used in many parts of the world, but there is not yet a standardized system for widespread implementation. In Vietnam, this type of system has only been applied in recent years, so there are still many operational challenges. Therefore, there is a need for research and analysis to clarify the technical characteristics of this system. This serves as a foundation for conducting in-depth studies aimed at finding solutions for efficient operation of this system and developing prototype systems for wider application in the future.

This chapter provides an overview of wind power in isolated areas and focuses on analyzing the technical features of the wind–diesel hybrid power generation system in isolated grids.

5.3.4 Standalone wind power generation system

Standalone wind power systems are typically used in isolated areas. These systems usually consist of small WTs and battery chargers to supply electricity to household devices, as shown in Figure 5.1.

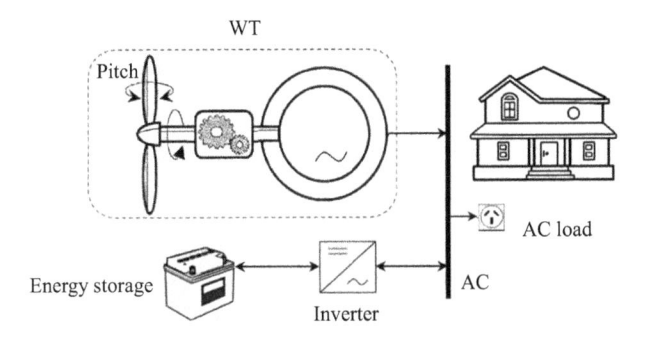

Figure 5.1 Independent wind power generation system

This type of system always requires storage devices (such as batteries) for start-up and storage. In cases where the generator is directly connected to the load, the storage devices help ensure system stability.

These systems typically have small capacities. However, if the system has a larger capacity, it requires excellent coordination of auxiliary equipment, such as storage devices for start-up, charging when excess power is generated, and supplying energy to the load when the power source is insufficient. Additionally, a dummy load is used to balance the power when the generation capacity exceeds the consumption, and the generation system has not yet adjusted. A good control system is also essential. Such systems tend to have a higher cost compared to other electricity generation methods.

5.3.5 Wind–diesel hybrid power system

The wind–diesel hybrid power system is suitable for isolated areas where extending the national grid would be very costly. Currently, the cost of installing a large solar power system is considerable. Furthermore, storing electricity for extended periods in batteries leads to high maintenance and operational costs for the solar system. This factor makes the wind–diesel hybrid power system a more appropriate choice for such areas.

The wind–diesel hybrid power system operates as a small, isolated grid. A typical wind–diesel hybrid system, as shown in Table 5.1, consists of a wind power station, a diesel power station, a storage system, a dummy load, all interconnected via a distribution grid, and transmitting electricity to the end users.

In the report from NREL, it is stated that for a wind–diesel hybrid power system to operate with a high penetration of wind power, a complex control system combined with suitable technical solutions is required. According to this report, many projects of this type have been successfully implemented, though not all have met expectations.

5.3.6 Cloud computing

Cloud computing refers to the use of computer technology that harnesses the power of multiple network-connected processors to perform complex database tasks without requiring a dedicated computer for each user. This technique forms the backbone of current communication networks for IoT applications. The deployment of cloud computing can be explained by the emergence of new technologies on the Internet, with vast amounts of data needing to be processed and organized for users. This leads to the provision of enhanced performance and processing capabilities, thanks to economies of scale at large data centers.

5.3.6.1 Cloud technology

Several key technologies of cloud computing are explored as follows:

- **Applications**: These are components stored on remote servers and run in real time for clients. Users access the application through a user interface (UI), where they use the features of cloud computing.

Table 5.1   Comparison of isolated grid-connected wind power projects

No.		Configuration	Grid connection principle	Auxiliary system	Evaluation
I	Theoretical	Wind power + diesel + energy storage system + dummy load	A wind power generation system should be connected to the grid indirectly through a converter	Energy storage system and dummy load	The system is stable and can achieve high penetration levels (instantaneous penetration level from 100% to 400%, average penetration level from 50% to 150%).
II	Practical				
1	Phu Quy	Wind power (3×2 MW) + diesel (6×0.5 MW)	DFIG, direct grid connection	None	Operation is not yet fully stable, wind power penetration is still low, and the three wind turbines have not been fully utilized.
2	Bach Long Vi	Wind power (800 kW) + diesel generators (2×414 kVA)	SCIG, indirect grid connection	none	Operating the wind–diesel hybrid system is very complex. Operational, maintenance, and repair activities cannot be ensured, leading to system downtime.

- **Clients**: A UI platform, such as the Mozilla Firefox web browser, Google Chrome, or Microsoft Internet Explorer.
- **Infrastructure**: This includes the computer hardware and the buildings that house the hardware. The server environment uses virtualization technology, meaning the internal server framework is independent of the specific number of machines. Thus, it operates entirely on software, harnessing the processing power of many machines to offer more processing benefits for each client.
- **Platform**: The method of deploying applications through Platform as a Service. This involves a web application framework in the form of open-source code with web design programming.
- **Services**: Refers to the user experience of using cloud computing. Numerous online services support users, requiring vast database storage and processing power to perform tasks.

- **Storage**: One of the major features of cloud computing, though it can be costly. Cloud providers must ensure the safety of customer data through service agreements and clear interactions with clients.
- **Processing power**: Significantly expanded with cloud computing. Companies can use this capacity to experiment with new markets and applications on the web. Cloud service providers offer processing power to Internet-connected users. Servers need to be used efficiently to avoid wasting valuable processing resources.

5.3.6.2 Cloud management

Cloud computing is more flexible in resource use, such as the server leasing model. Specifically, clients can decide the amount of storage space and processing power required, as well as when to update applications across the entire enterprise in real time. The scalability of cloud computing allows users to quickly transition from small to large processes while ensuring the processing of large amounts of data at a given time without the need for large servers. However, there are still many challenges with cloud computing in IoT applications, as any innovation comes with its own set of hurdles. Below are some potential challenges requiring new solutions:

- **Reliability**: To receive resources from cloud centers, users need to prove their credibility, such as through an electronic utility model. This energy source needs to be reliable for cloud computing to provide a certain level of service.
- **Security**: Protecting data is always a top priority in cloud IT due to the large volumes of data stored in cloud centers. With various technologies and applications, accessing and communicating over the Internet can be risky, and thin devices may be tampered with or attacked if not well-managed.
- **Service support**: Providers need to assist clients with new solutions, differing from traditional IT solutions. Ensuring stable service levels, such as 24/7 support, is a good example. Reliable connections and data security are vital for improving service quality.
- **Processing capability**: IT tasks organized in the cloud must be operated efficiently to maximize benefits for both the cloud center and the client. Resource management aims to achieve high performance and ensure service quality.

5.3.6.3 Fog computing

The volume and variety of data in IoT are rapidly increasing and exploding. As a result, the time taken to process data in the cloud has significantly increased. To address this issue, a new technology for IoT applications and services, known as fog computing, was developed based on the cloud computing model, supporting large-scale systems, such as a large number of devices/nodes. Fog computing offers many benefits, such as low latency, wide geographic distribution, high mobility, location awareness, a strong role of wireless access, the prominence of real-time applications, and heterogeneity.

5.4 Specific applications and examples of microgrid power grid systems deployment

5.4.1 Wind–diesel hybrid power system in Phu Quy

In Vietnam, several power generation systems combining wind energy with diesel power have been implemented. Among them, the wind–diesel hybrid power system on Phu Quy Island is the only one currently in operation and stands out as the most significant (being the largest and most efficient). Therefore, this system is chosen as the main subject for calculations, analysis, and comparisons in this study (Figure 5.2).

The diesel power station has been operational since 1999, consisting of six Cummins VTA-28 generators with a nominal capacity of 500 kW (625 kVA), a minimum capacity of 165 kW, and a terminal voltage of 0.4 kV, with a power factor of 0.8. The 0.4 kV system is connected to the 22 kV grid through three parallel 1,600 kVA transformers, $22 \pm 2 \times 2.5/0.4$ kV. The 22 kV power grid spans approximately 21.5 km and includes two feeder lines, 471 and 472.

In 2011, the wind power station was commissioned in conjunction with the diesel power station. The wind power station consists of three V80-2MW WTs using DFIG technology from Vestas Group. Each turbine has a nominal power output of 2.0 MW and a nominal voltage of 690 V, connected to the 22 kV grid through a 2.1 MVA dry transformer, 22/0.69 kV, YN/yn–0. The minimum power output of each WT is 500 kW when wind speeds range from 7.2 to 17.8 m/s, and between 550 and 800 kW when wind speeds range from 17.8 to 25 m/s.

5.4.2 Control of wind–diesel hybrid power system

In isolated grid systems with wind–diesel hybrid power generation, control is typically conducted at three different levels as shown in Figure 5.3. This study also includes millisecond-level control. The specific functions of each level are as follows:

- **Level I control**: This is the direct control at each power source, structured as shown in Figure 5.3. The slope characteristic control function is used to share power and adjust voltage and frequency parameters instantaneously when these parameters change. For WTs, this involves both Level I and Level II control.

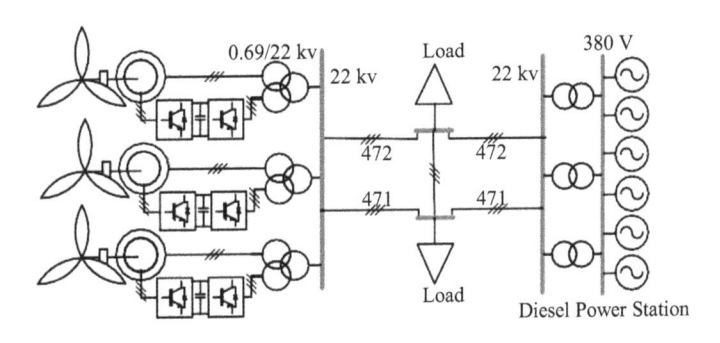

Figure 5.2 Hybrid wind–diesel power generation system on Phu Quy island

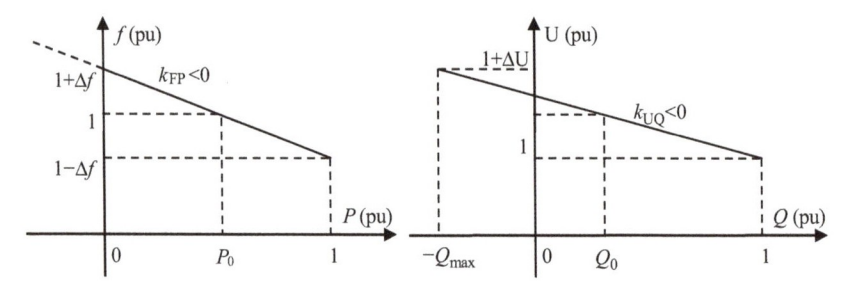

Figure 5.3 Characteristic slope used to control power generation sources

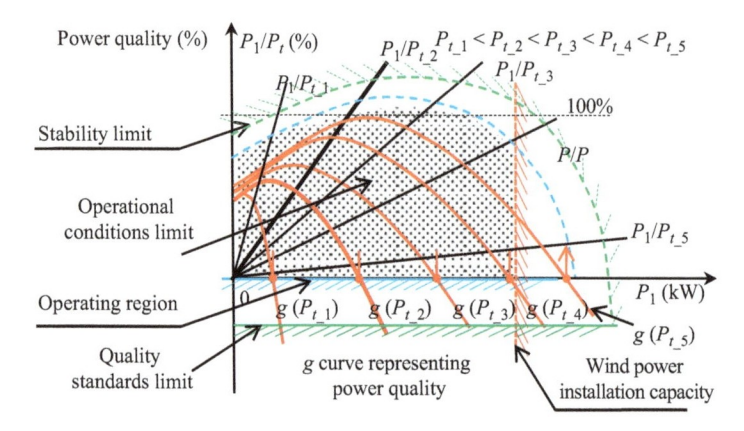

Figure 5.4 Qualitative diagram illustrating the relationship between power quality and wind power penetration under different conditions

5.4.3 Operation of the wind–diesel hybrid power system

In the wind–diesel hybrid power generation system with an isolated grid, the two indicators of power quality and the level of wind power penetration are inversely related in areas with high wind power penetration as shown in Figure 5.4. This phenomenon occurs because when the power output of the wind power station dominates, power fluctuations are significant when wind speeds change. However, since the number of diesel generators in operation is limited, their ability to compensate for these fluctuations is very poor. As a result, the system's state parameters experience strong oscillations, leading to poor power quality. Therefore, depending on the load requirements, priority should be given to improving the relevant performance indicator

The key parameters for power quality that this thesis focuses on are voltage and frequency. According to the Electrical Distribution System Regulations issued by the Ministry of Industry and Trade (32/2010/TT–BCT dated 30 July 2010), the following are specified:

- **Technical standard for frequency**: The nominal frequency in the national electrical system is 50 Hz. Under normal conditions, the system frequency can fluctuate within a range of ±0.2 Hz from the nominal frequency. In cases

where the electrical system is not stable, the frequency may fluctuate within a range of ±0.5 Hz from the nominal frequency.
- **Technical standard for voltage**: Under normal operating conditions, the allowable operating voltage at the connection point can fluctuate around the nominal voltage as follows:
  ○ At the customer connection point: ±5%;
  ○ At the power plant connection point: +10% and −5%.
- During single fault conditions or in the process of restoring stable operation after a fault, the allowable voltage fluctuation at the customer connection point directly affected by the fault can range from +5% to −10% of the nominal voltage.
- In severe fault conditions of the electrical transmission system or during fault recovery, the allowable voltage fluctuation can range from ±10% of the nominal voltage.

It is known that wind speed changes unpredictably, and the load from consumers also fluctuates irregularly. Additionally, when wind power is connected to an isolated grid, the sudden changes in wind power generation degrade power quality and make the system more unstable. Thus, maximizing the power output from natural wind energy while ensuring system stability and good power quality is a very challenging task. Therefore, effective solutions are needed to adapt to these random fluctuations and address the associated challenges.

The goal of this chapter is to develop a mathematical model for solutions aimed at maximizing wind energy utilization while ensuring the stable operation of the wind–diesel hybrid power system. The operation of such a system needs to consider constraints related to: technical limits of equipment, technical characteristics of the energy sources, and the system's ability to operate stably.

Proposal for using wind turbines integrated with an electromagnetic clutch to achieve maximum wind power penetration

Through the research results on the operating modes of the wind power generation system (using DFIG) combined with diesel, it has been found that without auxiliary equipment, the diesel generators cannot be avoided. In reality, there are times when wind energy exceeds the power demand of the load, but this is an unstable energy source and thus cannot operate the wind power station independently. Moreover, most WTs are not suitable for isolated grids (as presented in the literature). Therefore, there have been proposals to add auxiliary equipment. Additionally, many similar proposals have been made worldwide. As a result, most wind–diesel hybrid power systems globally use auxiliary equipment to increase wind power penetration.

Based on Table 5.2, the role of some auxiliary equipment in a wind power system connected to an isolated grid is as follows:

First, it can store energy (in the form of mechanical energy, electrical energy, etc.) when there is excess wind energy and then release it back into the system in a

Table 5.2 Economic and technical characteristics of solutions using auxiliary equipment

No.	Comparison auxiliary equipment	Active power balancing (P)	Reactive power generation (Q)	Frequency stability	Voltage stability	Increased penetration	Main applications	Applicable projects	Reference price
1	Low-load diesel	Generate P from 5% rated power	Generate Q	Good	Good	Increase by 20–25% when replacing conventional diesel	Increase penetration and stability	Rottnest Island, Hopetoun, Bremer Bay, Denham	1.200 USD/kW
2	Dummy load	Power consumption (P)	None	Relatively good	None	None	Increase stability		70.000–100.000 USD/200–500 kW
3	Flywheel	Compensate instantaneous P	None	Very good	None	None	Increase stability	Ross Island	
4	Supercapacitor	Compensate active power (P)	Compensate reactive power (Q)	Good	Good	None	Increase stability		300.000 Euro/1 MW
5	Battery	Balance P	Reactive power (Q)	Good	Good	Increases penetration when storage time is long	Increases penetration and enhances stability	Island	
6	Fuel cell	Balance P	Reactive power generation	Good	Good	Increases penetration when storage time is long	Increases penetration and enhances stability		
7	Pumped hydro storage	Balance P	Generate Q	Good	Good	Increases penetration with longer storage time	Increases penetration and enhances stability		

(Continues)

8	Capacitor bank	None	Generate reactive power	None	Relatively good	None	Power Compensation		50.000 USD/ 600 kVA
9	Load simulator + capacitor bank	Consumes active power (P) for balancing P	Generates reactive power	Good	Relatively good	None	Reactive power compensation and stability improvement		
10	Load simulator + battery	Balances active power (P)	Generates reactive power	Very good	Good	Increases penetration when storage time is long	Increases penetration and enhances stability	Wales Alaska	
11	Load simulator + supercapacitor	Balances active power	Provides dynamic reactive power	Very good	Good	None	Enhances stability		
12	Load simulator + low-load diesel	Balances active power	Generates reactive power	Very good	Good	Increases penetration by 20–25% when replacing conventional diesel	Enhances penetration and stability		893.000 Euro/800 kW diesel + 1.000 kWe tái già
13	Flywheel + low-load diesel	Balances active power	Generates reactive power	Excellent	Very good	Increases penetration	Enhances penetration and stability	Coral Bay	
14	Supercapacitor + low-load diesel	Balances active power	Generates reactive power	Very good	Very good	Increases penetration	Enhances penetration and stability		

short period when wind speed decreases, helping to reduce power fluctuations from the generation source and frequency oscillations.

Second, it can "cut back" the system's generation capacity when there is a sudden drop in load. This is typically achieved using dummy loads.

Third, it can meet the demand for a sudden increase in load. Storage devices are usually used to support in this case.

Fourth, it can generate reactive power (Q) as required by the load. Controlled compensation devices are used to perform this function. If the power source in the surveyed area is a wind–diesel hybrid power system, the role of generating reactive power for the system is taken on by the diesel generator.

Therefore, a solution is needed to use a single simple device integrated into the WT, which can still improve power quality while enhancing the wind power penetration rate and avoiding the waste of this natural energy source. This proposal aims to design a type of WT specifically for isolated grids in the future.

5.4.4 Results from steady-state mode survey

5.4.4.1 Operating characteristics in steady-state mode

The survey calculations were applied to the hybrid power system on Phu Quy Island with parameters as shown in Table 5.3.

Table 5.3 System parameters

Parameter	Value	Unit	Parameter	Value	Unit
Wind generators, 3 units			**Diesel generators, 6 units**		
Rated power	$P_{wN} = 2000$	kW	Rated power	$P_{dsN} = 500$	kW
Rated voltage	$U_{wN} = 690$	V	Rated voltage	$U_{dsN} = 380$	V
Stator reactance XwsX_{ws}	$X_{ws} = 0.01882$	Ω	Converted rotor reactance	$X_{dsd} = 3.76$	pu
Converted rotor reactance	$X'_{wr} = 0.02578$	Ω	Quadrature-axis reactance	$X_{sdq} = 1.74$	pu
			Transient reactance	$X'_{dsd} = 0.178$	pu
Mutual reactance	$X_{wm} = 0.762$	Ω	Power factor	$\cos\varphi_w = 0.8$	
Per-unit transient reactance	$X'_{wd} = 0.185$	pu	Inertia constant including the machine	$T_{Jds} = 1.02$	s
Power factor	$\cos\varphi_w = 0.98$				
Inertia time constant including the turbine.	$T_{Jw} = 8.27$	s			
Transformers at the wind power station, with 3 units.			**Transformers at the diesel power station, with 3 units.**		
Rated power	$S_{b1N} = 2{,}100$ kVA		Rated power	$S_{b2N} = 1{,}600$ kVA	
Rated voltage	$U_{b1N} = 0.69/22$ kV		Rated voltage	$U_{b2N} = 0.4/22$ kV	
Short circuit voltage	$U_{b1ng} = 5.3\%$		Short circuit voltage	$U_{b2ng} = 4.6\%$	
Transformer ratio	$k_1 = 22{,}000/690$		Transformer ratio	$k_2 = 22{,}000/380$	
Power line from wind power station to load:			**Power line from diesel power station to load:**		
Line resistance	$r_{01} = 0.3688$ Ω/km		Line resistance	$r_{02} = 0.3688$ Ω/km	
Line reactance	$x_{01} = 0.380$ Ω/km		Line reactance	$x_{02} = 0.380$ Ω/km	
Line length	$l_1 \approx 6$ km		Line length	$l_2 \approx 6$ km	
Load:					
Load voltage	$U = U_3 = 22$ kV		Maximum power	$P_{tmax} \approx 2{,}000$ kW	
Power factor	$\cos\varphi_t \approx 0.87$		Minimum power	$P_{tmin} \approx 1{,}100$ kW	

5.4.5 Analysis and proposed solutions to enhance stability

Through the survey, the thesis identifies the parameters that strongly influence the stability of the system and proceeds to analyze their effects as presented in Table 5.4.

Table 5.4 Analysis of technical solutions

Parameters	Large value	Small value
Short-circuit cutting time (t_c)	A large cutting time negatively affects the transient stability of the system but is advantageous for setting relay protection parameters for selective tripping based on fault location.	A fast cutting time helps protect equipment and is beneficial for transient stability, but it is difficult to set relay protection parameters for selective tripping based on fault location.
Auxiliary equipment Installing a flywheel to increase the moment of inertia (J) for the diesel generator Installing additional reactance Installing additional controlled capacitors	**With auxiliary equipment:** **Installing a flywheel** Installing a flywheel benefits the transient stability of the system. However, it negatively impacts the diesel generator's ability to generate peak load power. This capability to generate peak load power is essential in a wind–diesel power system, as wind turbines have high inertia and cannot perform this function. **Installing additional electrical reactance** Adding electrical reactance helps reduce the short-circuit current. The reactor must be disconnected simultaneously when the short circuit is cleared (or when a line is disconnected for maintenance) to ensure stability when operating a single line. However, installing additional reactance negatively affects rotor angle stability under small disturbances, reduces the transmission capacity of the line, and can lead to instability if the reactor is not disconnected when operating a single line. **Installing additional compensation capacitors** Adding compensation capacitors helps improve rotor angle stability under small disturbances and also enhances the system's transient stability. Additionally, it reduces the reactive power generation at the diesel power station.	**Without auxiliary equipment:** A small value of J negatively impacts the system's transient stability. If J is too small, there is a risk of instability during a short circuit due to the delayed response of the circuit breaker. The value of the electrical reactance of the 22 kV grid is usually small, so the short-circuit current on the line is typically very large, which can damage transmission equipment and generators. If the power factor ($\cos\varphi$) of the load is low, it will cause significant voltage drop along the transmission line, resulting in transmission losses. In practice, diesel power stations often need to generate large amounts of reactive power, which leads to substantial transmission losses

180   Clean energy in South-East Asia

From this comparison, it is evident that a flywheel should not be used because the diesel engine here has sufficient moment of inertia to extend the cut-off time to 173 ms. This indicates that opting for a quick cut-off method before this time is more economically and technically beneficial. Additionally, installing capacitors to reduce energy losses and minimize voltage drops during transmission is advisable. Installing resistors should be avoided, as they could destabilize the system.

5.4.6   Application of a general control algorithm for the wind–diesel hybrid power generation system on Phu Quy Island

The calculations were applied to the wind–diesel hybrid power generation system on Phu Quy Island. This system is currently operational, making it convenient for analysis and comparison.

The operating mode applied until June 29, 2014, was 16 h/day (from 7:30 AM to 11:30 PM). Since June 30, 2014, the system has been operating 24 h/day. The typical load is in the range of 1–2 MW (Table 5.5).

5.4.7   Comparison with reality and other solutions

In practice, the operation of the wind–diesel hybrid power generation system on Phu Quy Island typically follows a 50–50% wind–diesel generation ratio. Specifically, in cases where the load power is low ($P_{wmin} \leq P_t \leq 1{,}100$ kW) and wind speeds exceed 7.2 m/s, the WT will switch to a fixed power generation mode: $P = P_{wmin} + 50$ (kW), with the remaining load compensated by the diesel generators. In this case, the wind–diesel generation ratio can reach 70%–30%. With this operating approach, the maximum wind energy penetration typically ranges from 50% to 60% of P_t (on May 23, 2012, $P1/P_t \leq 50.366\%$, and on July 13, 2014, $P_1/P_t \leq 54.4\%$).

Table 5.5   Operating conditions at Phu Quy

Condition	Note
Power generation limit	
$165 \text{ kW} \leq P_{ds} \leq 420 \text{ kW}$ $S_{ds} \leq 625 \text{ kVA}$ $P_{wmin} \leq P_w \leq 2{,}000 \text{ kW}$	P_{dp}—Power of the diesel generator P_w—Power of the WT. min—Minimum.
$P_{wmin} = 500 \text{ kW}$	When the wind speed is between 7.2 m/s and 17.8 m/s.
$P_{wmin} = 500\text{–}800 \text{ kW}$	When the wind speed is between 17.8 m/s and 25 m/s.
Set the rotational reserve power for the power system.	
$P_{dp} = P_{dp1} + P_{dp2}$ $P_{dp2min} \leq P_{dp}$	P_{dp}, P_{dp1}, P_{dp2}—These represent the rotational reserve power for the entire system, the wind power station, and the diesel power station, respectively.
$P_{dp2min} = 150 \text{ kW}$	When the wind speed is greater than 7.2 m/s.
$P_{dp2min} = 250 \text{ kW}$	When the wind speed is less than 7.2 m/s.

The wind–diesel hybrid power generation system has been applied to islands and remote areas in Vietnam, with the aim of providing stable, continuous, and cost-effective electricity. However, some wind–diesel hybrid projects in Vietnam have not delivered the expected economic benefits. For example, a project on Bach Long Vi Island, invested in 2004, was shut down, and the wind power station on Phu Quy Island, commissioned in 2011, is still not operating efficiently. In practice, the wind–diesel generation ratio on Phu Quy Island is typically 50–50%.

Globally, many wind–diesel hybrid systems also have low wind energy penetration ratios. Notable examples include:

- Sal Island in Cape Verde: 22%
- Mindelo Island in Cape Verde: 17%
- Dachen Island in China: 26%
- Denham Island in Australia: 23%

To address the issues above, various studies have been conducted. However, as discussed in section "2.1. Global Research Status" (under "Introduction"), the solutions provided in these studies are not suitable for the calculation and design of new wind power stations on islands in Vietnam.

Therefore, a planning calculation method is needed to determine the appropriate wind power station that can maximize energy generation while minimizing investment costs. This chapter aims to provide recommendations for selecting a wind power station suitable for installation alongside existing diesel stations on islands in particular and for isolated areas in general.

To address this, the assumption in this chapter is that the island already has an electrical grid and a diesel power station. The task is to calculate the number and rated capacity of each WT for the new wind power station, ensuring compatibility with the existing load and diesel power station. The results of this study can be applied similarly to other islands to reduce investment costs and maximize benefits for the project.

The calculations in this chapter are carried out for all types of WTs, including those currently available on the market and those with integrated EMC.

5.4.8 Operational constraints

The operational calculation process must satisfy the conditions outlined in Table 5.6.

The central theme of this thesis is aimed at improving the wind power penetration of the wind–diesel hybrid power system in isolated grids. Specifically, it starts with an overview of the wind–diesel hybrid system, theoretical insights, and proposed solutions to enhance wind power penetration considering operational constraints. It also builds models to evaluate the system's stability and assess the effectiveness of the proposed solutions (Chapter 4). Drawing from the lessons learned from existing systems, Chapter 5 suggests a calculation method to determine the appropriate wind power station in alignment with the existing diesel power

182   *Clean energy in South-East Asia*

Table 5.6   Operating conditions

Condition	Note
Power balance (condition 1)	Same as condition 1 in Table 5.6
Power generation limit (condition 2)	Same as condition 2 in Table 5.6
Set rotating reserve for the power system (condition 3)	
Reserve for the case of sudden load increase	
$P_{dpmin} \leq P_{dp}$ $P_{dpmin} = \max(P_{bapt\_i})$	P_{dp}—rotating reserve power of the wind–diesel hybrid power generation system P_{bapt_i}—power of the 22/0.4 kV substation for load group max()—function to select the maximum value
Rotating reserve for a generator failure	
$P_{N-1} \geq P_t$	P_{N-1}—total power of remaining generators after the failure of one generator
Stability condition (condition 4)	Same as condition 4 in Table 5.6

station in isolated areas to maximize operational economic benefits while minimizing investment costs.

- **Achievements of the study**

5.4.8.1   Overview and analysis of the wind–diesel hybrid power system

- Based on previous studies, this research has synthesized an overview of the wind–diesel hybrid power system in isolated grids.
- It analyzes the technical characteristics of the WT using DFIG and synchronous generators in the diesel power station.
- It evaluates the control structure of the wind–diesel hybrid power system.

5.4.8.2   Modeling the wind–diesel hybrid power system

- Building on previous studies, this research develops mathematical models for various operational modes.
- Proposes a control algorithm for the system without auxiliary equipment in isolated grids, including:
  – Power distribution calculation for active and reactive power.
  – Determining the number of generators required for optimal wind power utilization.
  – Proposing a WT structure specifically designed for isolated grids, along with an effective operating method for EMC-integrated turbines.

5.4.8.3   Analysis of stable operating conditions

- Identifying factors affecting static stability: transmission reactance, and the reactive power capability of generators.

- Evaluating factors impacting transient stability: the circuit breaker clearing time.
- Recommendations for the hybrid power system at Phu Quoc Island:
  - Total transmission reactance must be less than 0.518 pu.
  - The reactive power of the wind station must align with the turbine's specifications (V80–2 MW with $\cos\varphi = 0.98$).
  - Circuit breaker clearing time: <173 ms (diesel station) and <500 ms (wind station).
  - Installing additional capacitor banks to increase system stability, reduce power loss, and minimize voltage drop.

5.4.8.4 Application of solutions to the Phu Quoc system

- The control algorithm achieves an average wind power penetration of 80% PtP_tPt, with a maximum of 89.159% PtP_tPt.
- Simulation of EMC-integrated turbines shows the potential to reach 100% PtP_tPt during strong winds or low load periods, ensuring power quality and reducing diesel fuel consumption during peak hours or when wind speeds are low.

5.4.8.5 Proposed algorithm and program to identify an appropriate wind station for isolated grids

- Identifying the power and number of turbines for different types of turbines.
- Recommending wind stations for isolated areas similar to Phu Quoc Island:
  - DFIG turbines ≤ 1 MW: wind power penetration rate of 84.59% AtA_tAt.
  - SG or PMSG turbines: install four turbines of 1 MW each, penetration rate of 87.6% AtA_tAt.
  - EMC-integrated turbines: install three turbines of 1.5 MW, achieving a maximum of A1max = 116.5% AtA_tAt.
- Currently, it is advisable to use D-type turbines with SG or PMSG.

5.4.8.6 Recommendations and future research directions

- It is recommended that regulatory authorities create conditions for implementing new solutions for wind power integration.
- Continued research should focus on increasing wind power penetration and maximizing diesel fuel savings.

5.5 Development of a microgrid model for Phu Quoc Island

5.5.1 Overview of the electrical grid model for Phu Quoc Island

Phu Quoc, Vietnam's first island city, covers an area of 573 km^2 and has a population of nearly 180,000 people. The city includes nine administrative units, including two towns: An Thoi, Duong Dong, and seven communes: Cua Duong, Tho Chau, Ham Ninh, Duong To, Bai Thom, Ganh Dau, and Cua Can.

On February 6, 2014, in Duong To Commune, Phu Quoc District, Kien Giang Province, the Vietnam Electricity (EVN) inaugurated the 110 kV undersea cable project from Ha Tien to Phu Quoc. This key project in the power sector holds significant meaning for the development of Phu Quoc Island.

The project has a total investment of 2,336 billion VND, including the following main components:

- On the mainland: 0.3 km of 110 kV double-circuit transmission line from the 110 kV Ha Tien Substation to the shore point at Ha Tien, and 7.6 km of 110 kV transmission line from the shore point at Phu Quoc to the 110 kV Phu Quoc Substation.
- 110 kV Phu Quoc Substation: Designed to include two transformers with a capacity of 40 MVA each, with one unit installed in the first phase.
- Submarine Cable Section: The submarine cable is 56 km long, stretching from the shore point in HaTien to the shore point in Phu Quoc. It uses a single-core, three-phase underground cable with a total length of 57.33 km, a cross-sectional area of 3×630 mm^2, and a maximum transmission capacity of 131 MVA.

The primary goal of this project is to ensure stable electricity supply from the national power grid to the Phu Quoc archipelago. This project supports the economic and social development needs, enhances competitiveness, and positions Phu Quoc as an important economic zone. Furthermore, it contributes to promoting high-quality tourism on a national, regional, and international level, while also improving and protecting the environment of the island district.

Based on the specific conditions of Phu Quoc, the authors propose the implementation of a MG system on the island to enhance the stability and sustainability of the electricity supply. This system includes:

1. **Wind energy source**: Utilizes the area's natural wind potential to harness renewable energy, reduce reliance on traditional power sources, and minimize carbon emissions.
2. **Energy storage system**: Ensures the storage of surplus energy from the wind source to be used during peak demand periods or when wind resources are insufficient.
3. **Connection to the national power grid**: The system will be integrated with the national grid to maintain a stable electricity supply while minimizing the risk of power disruptions due to weather conditions or technical issues.

The implementation of the MG system will not only ensure a continuous power supply for Phu Quoc's economic and social activities but will also promote a sustainable, environmentally friendly development model. This aligns with the vision of making Phu Quoc an economic special zone and a high-quality international tourist destination.

5.5.2 Wind turbine model

Next, the authors investigate the impact of wind speed on the power output of WT. Wind speed varies with altitude, and the wind speed at the turbine hub height (V_{hub})

is calculated using:

$$V_{hub}(t) = V_{ref(t)} \cdot \left(\frac{H_{hub}}{H_{ref}}\right) \tag{5.1}$$

in which:

H_{hub}, H_{ref}, and α represent the hub height, reference height, and ground roughness coefficient of the WT, respectively.

The power of the WT (P_{WT}) is calculated using:

$$P_{WT}(t) = \begin{cases} 0, V_{hub}(V_{cut-in}, \quad V_{hub} > V_{cut-out}) \\ V_{hub}^3(t) \cdot \left(\dfrac{P_r}{V_r^3 - V_{cut-in}^3}\right) - P_r \cdot \left(\dfrac{V_{cut-in}^3}{V_r^3 - V_{cut-in}^3}\right) \end{cases} \tag{5.2}$$

$V_{cut-in} \leq V_{hub} < V_r$

$P_r, V_r \leq V_{hub} \leq V_{cut-out}$

where V_{cut-in}, $V_{cut-out}$ and V_r are the cut-in wind speed, cut-out wind speed, and reference wind speed of the WT, respectively. P_r is the rated power of the WT. The energy produced by the WT (E_{WT}) over a time period t is calculated as:

$$E_{WT}(t) = N_{WT} \cdot P_{WT}(t) \cdot \Delta t \tag{5.3}$$

where N_{WT} is the time interval (taken as 1 h). The average wind speed is approximately 6.31 m/s over the course of a year. The highest wind speed occurs in December and January, reaching about 7.4 m/s, while the lowest speed is observed in August, at only around 5.03 m/s. In the grid model, the wind speed data is based on actual values recorded over a day and night period in the study area for simulation purposes.

5.5.3 Power balance

Each driver operates a vehicle on a daily basis, covering a random distance across the island. Therefore, the driving distance needs to be modeled using a random distribution. The charging or discharging power of the electric vehicle (EV) must satisfy the constraints as follows:

$$0 \leq P_t^{EV,ch} \leq P_{cap} \times C_{EEV} \quad \forall t \in [T^a, T^d] \tag{5.4}$$

where $P_t^{EV,ch}$ là is the charging power into the EV at time **t**; P_{cap} is the power capacity of the charging station; CE_{EV} is the charging power of the station, and T^a, T^d are the random arrival and departure times of the EV.

The proposed MG system on the island includes the following key components:

- **Wind energy source**: WTs provide renewable energy.
- **Diesel generator**: Serves as a backup power source when renewable sources are insufficient.

- **National grid power supply**: Provides stable energy through the 110 kV power line.
- **Energy storage system**: Includes batteries or other storage devices, which help to regulate power and optimize the use of renewable energy.

The island's load consists of three main categories:

1. Residential load: The electricity demand from households.
2. Industrial load: The electricity demand from production facilities and industrial activities.
3. Electric vehicle charging station: The demand for charging EV batteries on the island.

Power balance:

To ensure stable operation, the power between the components in the system must be balanced over each time period. The power balance equation is described as follows:

$$P_t^{TW} + P_t^{Grid} + P_t^{DG} + P_t^{ESS} = P_t^{EV,ch} + P_t^L \quad \forall t \in [T^a, T^d] \tag{5.5}$$

where:

P_t^{TW}: is the power from the WT at time t.
P_t^{Grid}: is the power taken from the grid at time t.
P_t^{DG}: is the power from the diesel generator at time t.
P_t^{ESS}: is the power from the ESS at time t.
P_t^L: is the power consumption of the load at time t.

In this case, the diesel generator will step in to supply critical loads, minimizing the demand from the main grid and utilizing surplus wind energy to ensure continuous power supply and improve system stability:

$$P_t^{EV,ch} + P_t^L - P_t^{TW} - P_t^{ESS} - P_t^{DG} = P_t^{Grid} \quad \forall t \in [T^a, T^d] \tag{5.6}$$

The economic cost optimization function (C_{NPC}) is expressed in the following equation:

$$C_{NPC} = \frac{C_{annual,\,Total}}{CRF\,(i,\,R_{project})} \tag{5.7}$$

where $C_{Annual,Total}$ is the annual cost; $R_{project}$ represents the project lifespan, and i, is the annual interest rate, with CRF being the capital recovery factor.

5.5.4 Building the power grid model in MATLAB

In this model, the team assumes that the renewable energy sources have the parameters listed in Table 5.7 as follows:

Figure 5.5 illustrates the island power grid simulated in MATLAB, with the load in this section being the EV charging station, which participates in the grid during different time periods.

Table 5.7 Equipment parameters on the grid

System name	Quantity	Value	Unit
Wind turbine	2	1	MW
Storage battery	1	0.5	MVA
Diesel generator	1	1	MW
Fast charging station	20	60	kW
Regular charging station	15	30	kW

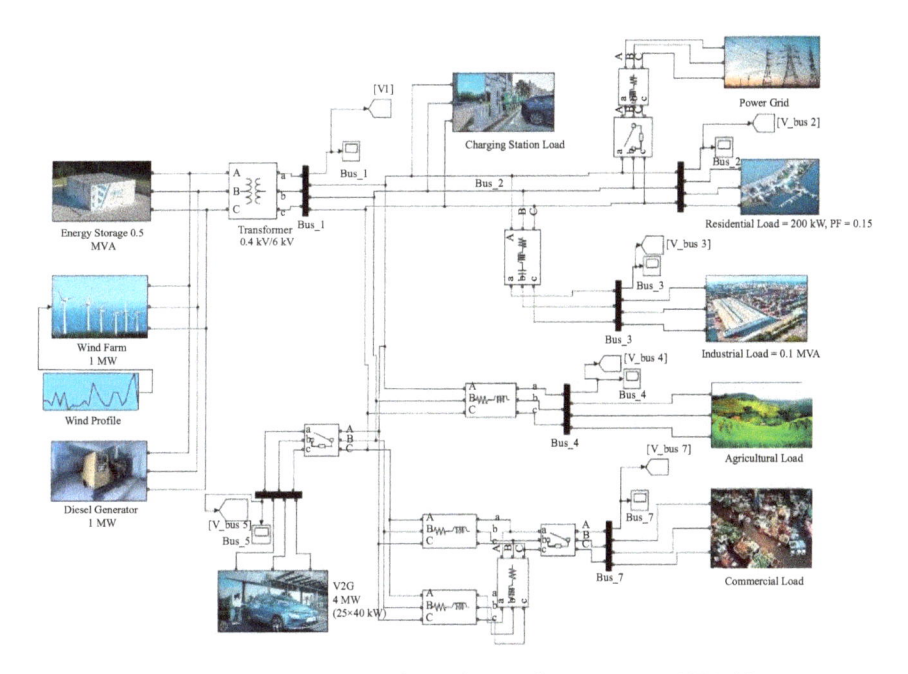

Figure 5.5 Power grid simulation diagram on MATLAB

5.6 Smart energy management algorithm

5.6.1 Building the MATLAB-PLC communication module

To apply the actual power grid model combined with the monitoring and energy management model, the author team developed a communication module between MATLAB and the SIEMENS PLC via the KEPServerEX software. The PLC acts as the OPC client, while MATLAB serves as the OPC server. Real-time data is collected through the PLC and input into MATLAB, where the algorithm generates control signals from MATLAB and SCADA, which are then sent back to the PLC to control the switching devices.

5.6.2 Energy management system algorithm

- The EMS algorithm is as follows:
  - Step 1: The program collects wind speed data, load power, WT power, etc., via the PLC's analog input.
  - Step 2: The algorithm calculates the power of sources P_{WT}, P_{Load}, P_{EV}, P_{ESS} to compare the sources and the load.
  - Step 3: If the WT or ESS has power, the algorithm sends a control signal allowing the renewable source to participate in the grid and charge the storage system.
  - Step 4: The control signal switches the circuit breakers through KEPServerEX software, which connects to the PLC to control the actual devices.
  - Steps 5 and 6: The algorithm recalculates the grid's loading capacity and the load to generate the next scenario.

  In the case of EVs, Algorithm 1 will trigger a subroutine—Algorithm 2 to handle the charging requests for the EVs separately.
- Battery storage system (ESS) management:
  - The ESS will charge under conditions of low load demand, and the battery's state of charge (SOC) will remain within the range: 20% < SOC < 80%.
- Power supply to the grid:
  - Once the ESS is fully charged or exceeds the required load level, energy will be discharged and fed into the grid.
  - The grid-connected energy will be used to supply the loads on the island, including residential, industrial, and the EV charging station.

Alert and forecasting features.
- The algorithm integrates the following capabilities:
  - **Alert**: Detecting faults or abnormal conditions in the renewable energy system or consumption load.
  - **Forecast**: Providing information on expected power from renewable energy sources such as wind or solar at specific times.
  - The forecasting feature helps the system generate optimal scenarios for performance optimization and ensures energy balance.

This algorithm not only supports the efficient operation of the MG system but also enhances its sustainability and flexibility in energy usage scenarios.

Algorithm 1 Start ($i = t = 1$);

Collect P_{WT}, Wind speed, P_{Load}, P_{WT}, P_{ESS}, P_{DG}
% Step 1: Compare time $i = i+1$ (1:8760 h) if $I > 8760$ end.
% Step 2: Calculate P_{WT}, P_{Load}, P_{EV}, P_{ESS}, % balance power using formulas (5), (6).
% Step 3: Check P_{wind}?; $P_{Energy} > 85\%$ or $P_{Energy} < 15\%$.
% Step 4: Control ACB3, VCB1, ACB1, ACB2.

% Step 5: Calculate the loading capacity of P_{EV}, P_{Load} and display warnings on screen.
% Step 6: Calculate (7) CNPC, run Algorithm 2.
% Step 7: Save the scenario and end.

5.6.3 Electric vehicle charging algorithm

In this section, the author team proposes an algorithm for charging EVs at charging stations. Currently, there are two typical types of charging stations: fast charging and regular charging, which are used at charging posts and residential homes. Due to the random nature of EV movement, the charging process is divided into different priority levels.

In this algorithm, the case when the WT has power is considered. Pe is the power when the WT is insufficient to supply the load. Px is the power difference between the WT and the load. According to the priority order of vehicles at the charging station and the time of day, the charging requirements for EVs will vary.

When the WT's power is insufficient, the MG will draw additional power from the national grid. Conversely, if there is surplus power, it will be fed back into the grid through ACB3 in the synchronization cabinet. Figure 5.6 shows the detailed steps of Algorithm 2.

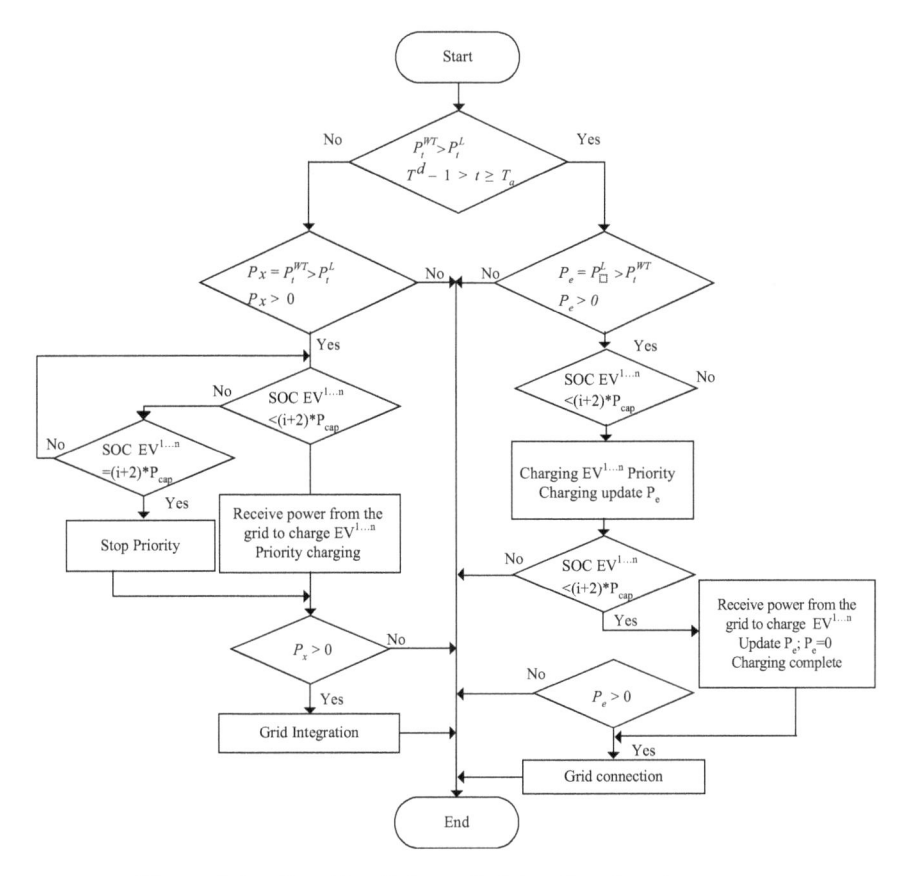

Figure 5.6 Algorithm 2 for optimizing the charging process

5.7 Simulation results

After designing the MG system, the author team proceeded with the system simulation in two parts: using Algorithm 1 to control the devices in the laboratory and running Algorithm 2 to check the charging capabilities at the charging stations. The hardware and software used in the simulation system include a wind energy model, a battery storage system, and charging station load objects in MATLAB, in combination with a diesel generator and actual switching devices in the laboratory.

5.7.1 Energy management system test

First, the author team used the control hardware to manage the system. The control of opening and closing ACB1 and ACB2 is displayed on the screen, and the hardware system results are shown in Table 5.8.

5.7.2 Electric vehicle charging scenario simulation

Next, to test the coordination capabilities of the EV charging stations, the team conducted simulations with a random number of EVs at the charging stations through two scenarios:

- **Scenario 1**: Without using Algorithm 2, the number of charging EVs was gradually increased, and the voltage, frequency, and power of the grid were examined.

  Figure 5.7 illustrates the power balance graph without considering the impact of EVs. During the peak period from 18:00 to 19:30, the primary energy source is from the grid, accounting for about 78% of the total power generation of the grid. Meanwhile, the ESS power and wind energy power contribute 10% and 12.5%, respectively. The period from 20:00 to 22:30 corresponds to when the WT produces the highest power, approximately 0.9 MW, which accounts for 65% of the total grid power. Additionally, this is a time when the power consumption of the island's residents decreases, allowing for the charging of ESS.

- **Scenario 2**: Using Algorithm 2 to monitor the number of EVs charging and the operation of the charging stations, the results are shown in Figure 5.8.

Table 5.8 Variable connection states in MATLAB—PLC

Name	Address	Status	Check
O_ACB1_close/Trip	%Q136.0 %Q136.1	Good	Good
O_ACB2_close/Trip	%Q136.2 %Q136.3	Good	Good
O_ACB3_close/Trip	%Q136.4 %Q136.5	Good	Good
O_ACB4_close/Trip	%M1.0 %M2.0, %M3.0	Good	Good

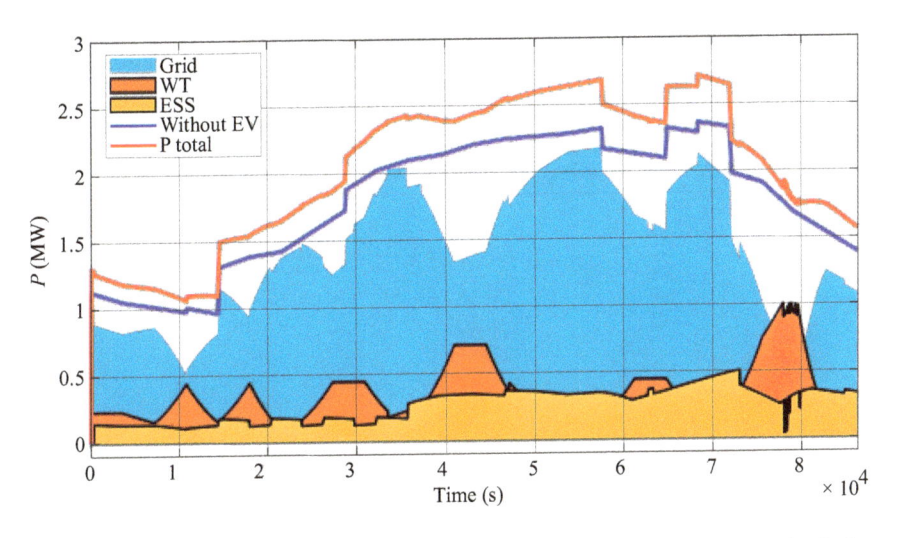

Study and application of microgrid systems 191

Figure 5.7 Power balance graph without considering EV (electric vehicles)

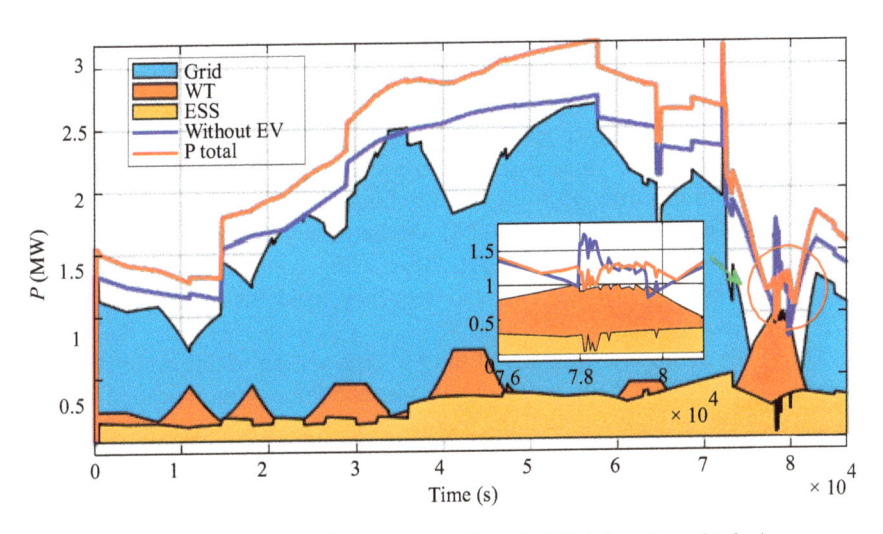

Figure 5.8 Power balance graph with EV (electric vehicles)

The power consumption of the load increases as the number of EVs charging rises. Between 3:00 PM and 5:00 PM, the load consumption is at its highest, reaching over 3.2 MW. From 9:00 PM onward, the residents tend to use fewer devices, allowing the opportunity for EV charging during this period. On the other hand, the power output from the WT is at its peak, reaching 95%, so the

Table 5.9 Proposed results for the estimated number of electric vehicles to be charged

Percentage (%)	Number of test vehicles	Number of limited vehicles	Charging stations	Power (MW)
10	15	12	2	0.6
30	40	34	4	1.9
50	68	53	6	2.8
75	80	68	10	3.1
90	124	105	15	3.9
120	>200	–	–	–

surplus power is utilized for charging. Additionally, to assess the load capacity at various nodes on the MG, the authors increased the number of EVs charging and activated the charging stations to test the algorithm, with the results presented in Table 5.9.

Based on the results from Table 5.9, the 30 experimental trials consistently showed that the number of EVs and charging stations within the specified limits met the requirements. In the final case, where the number of EVs was not limited during peak hours, the system immediately issued a warning and disconnected the non-priority charging stations from the grid. The authors then tested the voltage and frequency at several nodes connected to the charging stations to assess the power quality. The frequency at some nodes with EV charging stations between 5:00 AM and 6:30 AM dropped to 49.4 Hz, with a frequency deviation of about 1.2%. Similarly, between 6:00 PM and 7:00 PM, the frequency dropped to approximately 49.48 Hz and rose to about 50.45 Hz. After using Algorithm 2, the frequency deviations did not violate the voltage and frequency standards according to Circular 25/2016/TT-BCT.

This study simulated the EMS of the MG grid on the island using laboratory hardware and a model built on MATLAB. In addition, communication between Siemens PLC, KEPServerEX, and MATLAB was integrated into the simulation system to study the characteristics of the island's grid. In practice, the ability to intervene and test the electrical system on the island is highly complex and not feasible for real testing, so the research model developed serves as a method for testing scenarios and algorithms with some real equipment. Furthermore, the report examined the impact of charging stations on the grid and the use of surplus wind energy to charge EVs, based on optimization and control algorithms, and proposed the number of EVs at critical nodes during peak hours. Ultimately, this research model supports the development of smart grids and remote monitoring and control systems for future power grids.

5.8 Study on the impact of wind power plants on the reliability of the distribution grid

5.8.1 Application of reliability indices calculation for the independent grid of Ly Son Island—Quang Ngai with the integration of wind power

5.8.1.1 Wind power potential in Ly Son

- Ly Son, also known as Cu Lao Re, has been a district of Quang Ngai Province since 1992. Located in the East Sea, Ly Son plays a vital role as the frontline island of the nation, situated near the Paracel and Spratly Islands.
- The district covers an area of approximately 9.97 km^2 but has a population of over 20,460 people. Ly Son consists of three main islands:
  - **Big island** (Ly Son or Cu Lao Re)
  - **Small island** (Cu Lao Bo Bai), located to the north of Big Island
  - **Hon Mu Cu**, situated to the east of Big Island
- The district is divided into three communes:
  - **An Vinh** (the district's administrative center, located on Big Island)
  - **An Hai** (on Big Island)
  - **An Binh** (on Small Island)
  - **Climate characteristics**
- Ly Son lies within the tropical monsoon climate zone, characterized by hot and humid conditions with an off-season rainy period from September to February of the following year. The island's notable climate features include:
  - **Abundant sunshine**: Ly Son is one of the sunniest areas in Vietnam's coastal islands, with an average annual total of 2430.3 sunshine hours. This provides favorable conditions for the development of solar power stations to meet the island's energy needs.
  - **Irregular rainy season**: Rain is primarily concentrated in the rainy season, which accounts for approximately 71% of the total annual precipitation, with an annual rainfall of about 2,260 mm.
  - **Extended dry season**: From March to August, the weather is dry and hot due to the influence of the southwest monsoon winds.
  - **Humidity**: The average relative humidity is around 85%, typical of tropical coastal regions.
  - With its unique geographical position, along with diverse natural and climatic conditions, Ly Son not only holds significant potential for developing renewable energy but also plays a strategic role in protecting Vietnam's sovereignty over its maritime territories (Figure 5.9).

5.8.1.2 Power supply system on Ly Son island
(a) **Power supply sources**
- As of 2005, the power supply system for Ly Son included two diesel power stations:

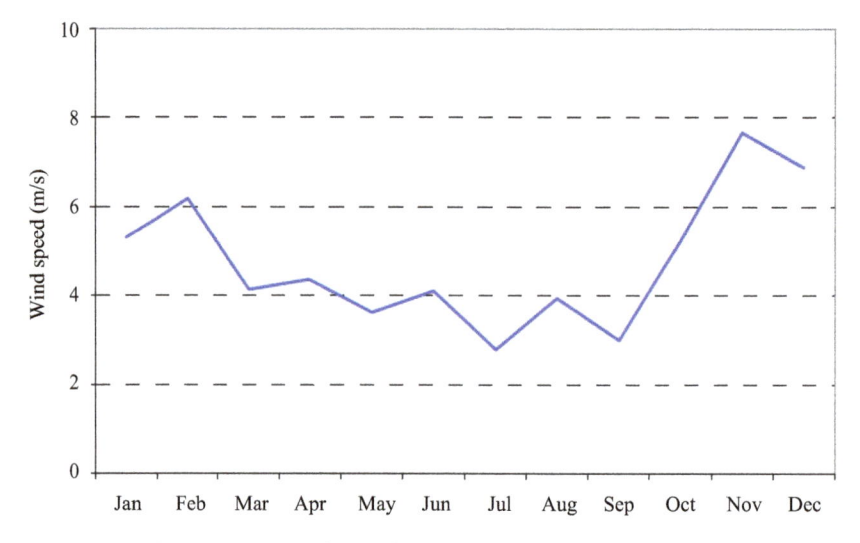

Figure 5.9 Average wind speed measured at a height of 12 m in 2023

- **Power station 1 (five generators)**
    - 1 × 304 kW—Wilson
    - 2 × 366 kW—IVECO
    - 2 × 200 kW—EGM These generators were installed in Ly Hai in 1999. Currently, all the generators are very old and have undergone multiple refurbishments and maintenance. This power station ceased operations when Power Station 2 was commissioned at the end of 2005.
- **Power station 2 (two generators)**
    - 2 × 780 kW—SKODA This station was built in June 2005 to replace the outdated Power Station 1. The generators are connected to the 22 kV grid through a 0.4/22 kV transformer, supplying power to 11 step-down transformers (22/0.4 kV) with capacities ranging from 100 to 250 kVA. Diesel generators at this station operate for approximately 5 h/day.
- **Diesel power station on small island**
    - 1 × 15 kVA connected directly to the 0.4 kV grid.
- Additionally, there are separate diesel power stations for specific loads, such as the Post Office, Radio Station, District Committee, Hospital, and Lighthouse Station. These stations operate independently or serve as backup power sources.
- Table 5.10 provides a summary of the power supply sources on Ly Son Island:

Table 5.10 Power supply system on Ly Son Island

Source	Location	Capacity (kVA)	Year of Operation	Notes
Ly Son diesel station	An Hai	1 × 380	1999	22 kV
		2 × 457	2001	22 kV
		2 × 250	2001	22 kV
		2 × 975	2005	22 kV
Diesel station on small island	Small island	1 × 15	2000	0.4 kV
Ly Son post office	Post office	2 × 8	1995	Backup
Radio station	Radio station	1 × 5	Before 1995	Backup
District committee	District committee	1 × 3	Before 1995	Backup
Health station	Health station	1 × 5	1996	Backup
Lighthouse station	Lighthouse station	1 × 5	1994	Independent
Radar station 505	Radar station 505	40	Before 1995	Independent
Large radar station	Large radar station	4 × 40	1998	Independent
	Total		**1256**	

(b) **Electrical grid system**
- The medium voltage (22 kV) power line system has a total length of 3,815 m, using M50 wire.
- The low voltage (0.4 kV) power line system has a total length of 9.44 km, using CV-50 wire (Tables 5.11 and 5.12).

(c) **The system of substations**

(d) **Total power consumption of the entire island**

Based on the data provided by the RISO forecast and the analysis from the IPSYS program, the day and night load forecast for the entire island has been derived (Figure 5.10).

(e) **Wind–diesel power generation system**
- As previously discussed, the wind potential at Ly Son Island has been analyzed. The consideration of integrating wind power into the energy mix is crucial in order to extend the electricity usage duration for consumers on the island (currently 5 h/day and to reduce diesel fuel costs for the diesel-powered stations.
- To integrate wind power, various scenarios need to be established and analyzed. In this context, RISO proposes four options for comparison:
  - **Option 1**: Use only the existing diesel sources.
  - **Option 2**: The existing diesel sources combined with two 350 kW WTs or three 350 kW WTs.

Table 5.11 The power line system on Ly Son Island

Lines	Conductor	Length (m)	Note
Medium voltage lines 22 kV	M50	**3,185**	
Ly Son diesel–An Hai–An Vinh		3,185	
Low voltage lines 0.4 kV		9,483	
Ly Son diesel–An Hai 5	CV-50	1,100	
An Hai 1	CV-50	720	
An Hai 4	CV-50	500	
An Vinh 5	CV-50	670	
An Vinh 2	CV-50	812	
An Vinh 6	CV-50	400	
An Vinh 1	CV-50	1,520	
An Vinh 3	CV-50	472	
An Vinh 4	CV-50	280	
Ly Son diesel–An Hai 2	CV-50	992	
An Hai 3	CV-50	972	
Small island	CV-50	1,000	

Table 5.12 The transformer substation system on Ly Son Island

Station	Capacity (kVA)	Voltage (kV)	P_{max} (kW)	Note
Ly Son step-up transformer	1,770	22/0.4	700	
Step-down transformer	1,910	22/0.4	1,244	Surcharge load 10%
An Hai 1	250	22/0.4	220	
An Hai 2	160	22/0.4	102	
An Hai 3	250	22/0.4	160	
An Hai 4	100	22/0.4	64	
An Hai 5	100	22/0.4	64	
An Vinh 1	250	22/0.4	160	
An Vinh 2	160	22/0.4	90	
An Vinh 3	160	22/0.4	102	
An Vinh 4	160	22/0.4	90	
An Vinh 5	160	22/0.4	102	
An Vinh 6	160	22/0.4	90	

- **Option 3**: Replace one 680 kW diesel generator and add three 350 kW WTs.
- **Option 4**: Replace two 680 kW diesel generators and add three 350 kW WTs.
- Based on the IPSYS simulation program, the results for the above options are shown in Table 5.13.

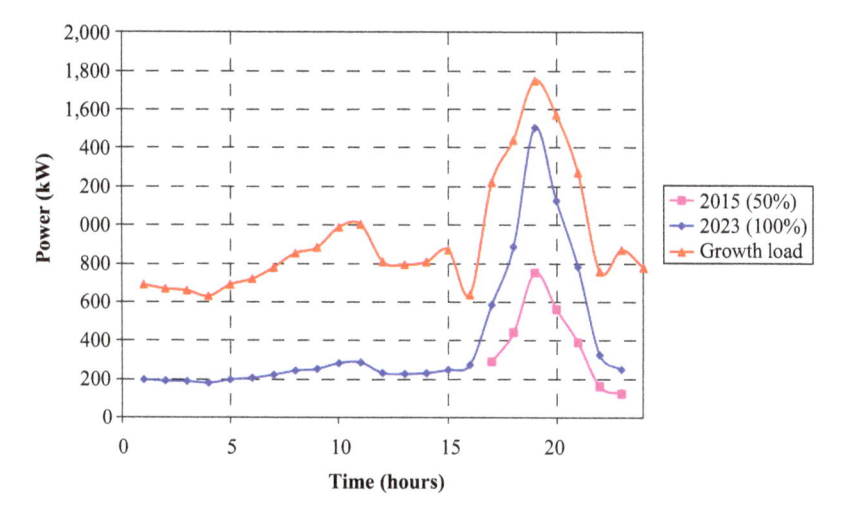

Study and application of microgrid systems 197

Figure 5.10 Day and night load curve for Ly Son Island

Table 5.13 Power supply options for Ly Son Island

Scenario	Diesels	WTGs	Fuel used	Wind energy share	Wind energy usage	Dump load share
1	2 × 680 kW Skoda 1 × 360 kW IVECO	0	100%	0%	0%	0%
2a	2 × 680 kW Skoda 1 × 360 kW IVECO	2 × 350 kW	83%	16%	99%	0.2%
2b	2 × 680 kW Skoda 1 × 360 kW IVECO	3 × 350 kW	79%	24%	84%	4%
3	1 × 680 kW Skoda 1 × 680 kW (new) 1 × 360 kW IVECO	3 × 350 kW	72%	24%	87%	3%
4	2 × 680 kW (new) 1 × 360 kW IVECO	3 × 350 kW	67%	25%	94%	1%

- In option 4, replacing two 680 kW diesel generators and adding three new 350 kW WTs will reduce the diesel consumption to 67% of the current level. Given a diesel consumption rate of 0.25 kg/kWh and an annual electricity consumption of 5 GWh for the entire island, this could save 400 tonnes of diesel per year (Figure 5.11).

5.8.1.3 Calculation of reliability indicators for option 4
(a) **Wind turbine specifications from the manufacturer**
   - Rated power: 350 kW
   - Cut-in wind speed: 4 m/s

198 *Clean energy in South-East Asia*

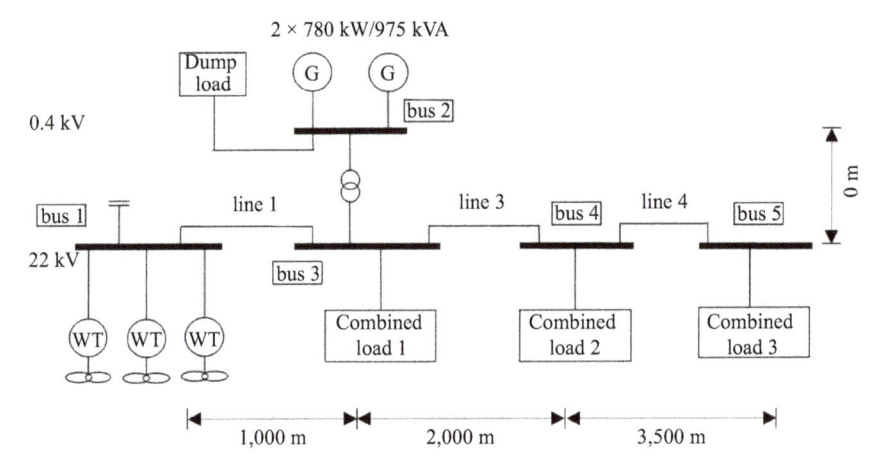

Figure 5.11 Electrical supply diagram for Ly Son according to option 4

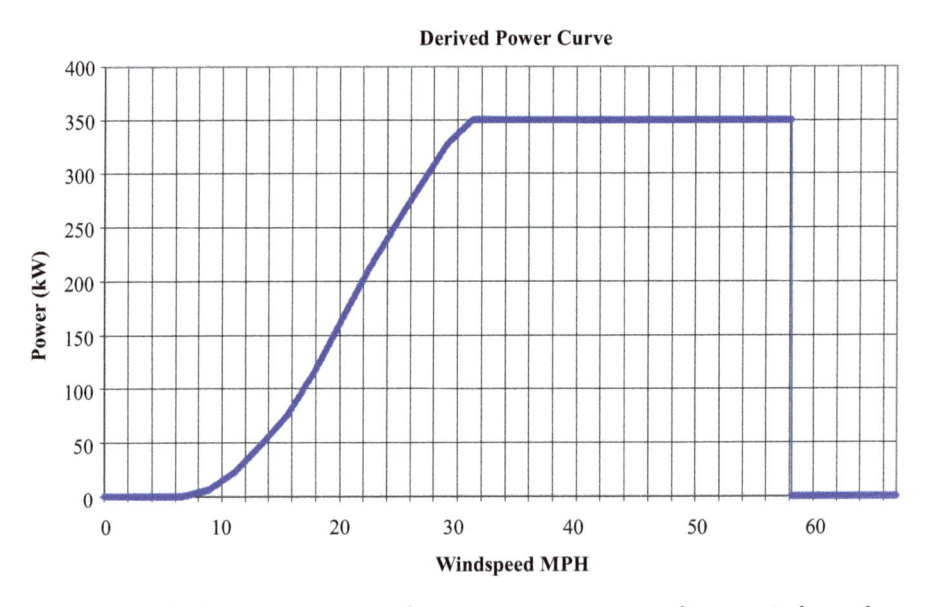

Figure 5.12 Power generation characteristic curve according to wind speed

– Rated wind speed: 14 m/s
– Cut-out wind speed: 25 m/s (Figure 5.12 and Table 5.14)

Based on statistical wind data for one year from the meteorological station on Ly Son Island, we set up the power states and the probabilities of these states for each WT and the combination of three turbines as shown in Table 5.15.

Table 5.14 Power generation according to wind speed

Wind speed (m/s)	Wind speed (mph)	Power (kW)
0.00	0.00	0.00
1.00	2.20	0.00
2.00	4.50	0.00
3.00	6.70	0.00
4.00	8.90	6.00
5.00	11.20	23.00
6.00	13.40	48.00
7.00	15.70	76.00
7.30	16.37	87.50 = 0.25Pr
8.00	17.90	115.00
9.00	20.10	162.00
9.30	20.77	175.00 (0.5Pr)
10.00	22.40	210.00
11.00	24.60	250.00
11.33	25.34	262.50 = 0.75Pr
12.00	26.80	288.00
13.00	29.10	327.00
14.00	31.30	350.00
15.00	33.60	350.00
16.00	35.80	350.00
17.00	38.00	350.00
18.00	40.30	350.00
19.00	42.50	350.00
20.00	44.70	350.00
21.00	47.00	350.00
22.00	49.20	350.00
23.00	51.40	350.00
24.00	53.70	350.00
25.00	55.90	350.00 = Pr
26.00	58.20	0.00
27.00	60.40	0.00
28.00	62.60	0.00
29.00	64.90	0.00
30.00	67.10	0.00

Table 5.15 Power generation and power generation probabilities for one wind turbine according to wind speed

No.	Wind speed range (m/s)	Power range	Probability	Converted power
1	$V < 4$ & $V > 25$	$P = 0$	33.5%	0
2	$4 \leq V < 7.3$	$0 < P < 0.25$Pr	33%	0
3	$7.3 \leq V < 9.3$	0.25Pr $\leq P < 0.5$Pr	14.5%	0.5Pr
4	$9.3 \leq V < 11.33$	0.5Pr $\leq P < 0.75$Pr	8%	0.75Pr
5	$11.33 \leq V < 25$	0.75Pr $\leq P \leq$ Pr	11%	Pr

Pr—Rated power of wind turbine; Pr = 350 kW

Combined with the failure probability of a WT, which is 0.03, we have the probability table of power output states for the combination of three WTs at different wind speeds as shown in Table 5.16.

From Table 5.16, we have the summary table of the output power of the three WT system and the probability of each state (Table 5.17).

(b) **Diesel generator and transformer parameters (source 1)**

Equivalent parameters for diesel generators and transformers as a unified source

The diesel generator and transformer components are combined into a single source element with the following reliability parameters (Table 5.18):

Table 5.16 Power output and probability of power output for three wind turbines by wind speed

No.	Wind speed range (m/s)	Combination of 3 WT	State probability	Power output state
1	$V < 4$; $V > 25$; $4 \leq V < 7.3$	–	0.665	0
2	$7.3 \leq V < 9.3$	0 Wind turbine operating	0.000003915	0 Pr
		1 Wind turbine operating	0.000379755	0.5 Pr
		2 Wind turbines operating	0.012278745	1 Pr
		3 Wind turbines operating	0.132337585	1.5 Pr
3	$9.3 \leq V < 11.33$	0 Wind turbine operating	0.00000216	0
		1 Wind turbine operating	0.00020952	0.75 Pr
		2 Wind turbines operating	0.00677448	1.5 Pr
		3 Wind turbines operating	0.07301384	2.25 Pr
4	$11.33 \leq V < 25$	0 Wind turbine operating	0.00000297	0 Pr
		1 Wind turbine operating	0.00028809	1 Pr
		2 Wind turbines operating	0.00931491	2 Pr
		3 Wind turbine operating	0.10039403	3 Pr

Table 5.17 Summary of output power and probability of output power for three wind turbines

No.	State power capacity, Pr	State power, kW	Operating probability, F_i
1	0	0	0.665009045
2	0.5	175	0.000379755
3	0.75	262.5	0.000209520
4	1	350	0.012566835
5	1.5	525	0.139112065
6	2	700	0.009314910
7	2.25	787.5	0.073013840
8	3	1,050	0.100394030

Table 5.18 Diesel generator and transformer parameters

No.	Equipment name	Average failure frequency (times/year)	Repair time (h)
1	22 kV transformer	0.02	90
2	Diesel generator	0.05	60

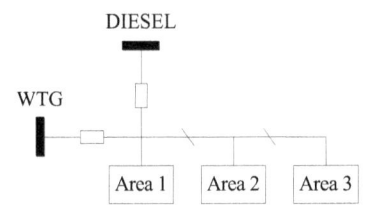

Figure 5.13 Regional division of Ly Son power supply system

- **Failure intensity of source 1:**

$$\lambda N1 = \lambda_{Transformer} + \lambda_{Gennerator} = 0.02 + 0.05 = 0.07 (\text{occurrences/year}) \tag{5.8}$$

- **Average recovery time:**

$$t_{scN1} = \frac{\lambda_{Transformer} \cdot t_{Transformer} + \lambda_{Gennerator} \cdot t_{Gennerator}}{\lambda_{NI}}$$

$$= \frac{0.02 \cdot 90 + 0.05 \cdot 60}{0.07} = 68.6 \text{ h} \tag{5.9}$$

(c) **Zonal division based on reliability indicators**

Diagram of regional division based on reliability indicators is shown in Figure 5.13.

The electrical loads on Ly Son Island can be divided into three load groups as illustrated in Table 5.19.

5.8.1.4 Calculation of the reliability indicator—load interruption time

- **Assumptions for calculation**:
  - The line protection system has a failure rate of $\lambda = 4$ (failures per 100 km per year), with a fault repair time of $r = 12$ h per repair.
  - The disconnecting switch typically has a fault operation time of $r = 2$ h per fault operation.

202   *Clean energy in South-East Asia*

Table 5.19 Electrical loads in different areas on Ly Son Island

Area	Component	P_{max} (kW)	Load profile by level			
			Level 1	Level 2	Level 3	Level 4
Area 1	An Hai 3	160	280	150	90	50
	An Hai 4	64				
	An Hai 5	64				
	Time period		0–3	3–7	7–9	9–24
Area 2	An Hai 1	220	510	270	170	80
	An Hai 2	102				
	An Vinh 5	102				
	An Vinh 6	90				
	Time period		0–4	4–8	8–10	10–24
Area 3	An Vinh 1	160	440	240	140	70
	An Vinh 2	90				
	An Vinh 3	102				
	An Vinh 4	90				
	Time period		0–3	3–8	8–11	11–24

- **Lengths of equivalent grid segments for Area 1:**
  - Segment 1: 1 km
  - Segment 2: 2 km
  - Segment 3: 3.5 km
- **Structure matrices setup**

$$D = \begin{bmatrix} 0 & 1 & 0 \\ 1 & 0 & 1 \\ 0 & 1 & 0 \end{bmatrix}; S = \begin{bmatrix} 1 & 0 & 0 \\ 1 & 1 & 0 \\ 1 & 1 & 1 \end{bmatrix}; AS = \begin{bmatrix} 0 & 1 & 1 \\ 0 & 0 & 1 \\ 0 & 0 & 0 \end{bmatrix};$$

$$B = \begin{bmatrix} 0 & 1 & 1 \\ 0 & 0 & 2 \\ 0 & 0 & 0 \end{bmatrix}; AK = \begin{bmatrix} 0 & 1 & 1 \\ 0 & 0 & 1 \\ 0 & 0 & 0 \end{bmatrix};$$

(5.10)

- **Calculation for Area 1: situations causing power outage in Area 1**
  - **Situation 1—fault occurrence in Area 1:**
    The equivalent length of Area 1 is 1 km with a failure rate of $\lambda = 4$ (failures per 100 km per year). The fault repair time is $r = 12$ h per repair.

$$T_{ND1.1} = R_{pd}(1.1)\lambda 1 + [1 - AS(1.1)]\frac{\tau_{11}}{T}\lambda_{1r1} = 4 \cdot \frac{1}{100} \cdot 12$$

$$= 0.48 \text{ h} \tag{5.11}$$

  - **Situation 2—fault occurrence in Area 2:**
    Area 2 (KV2) is not on the connection line between Area 1 and the source. Therefore, when a fault occurs, it will only cause a power outage in Area 1

during the fault isolation operation time.

$$T_{ND1.2} = R_{pd}\,(1.2)\lambda 2 + [1-AS(1.2)]\,\frac{\tau_{12}}{T}\,\lambda_{2r2} = 2 \cdot \frac{2}{100} = 0.04 \text{ h}$$

(5.12)

- **Situation 3—fault occurrence in Area 3**:
  Area 3 is not on the connection line between Area 1 and the source. Therefore, when a fault occurs, it will only cause a power outage in Area 1 during the fault isolation operation time.
  As in Situation 2, the fault isolation operation time for KV3 will be r_{bd} = **2 h per fault operation**.
  Thus, in this situation, the outage time is also the time required for the fault isolation operation, which is **2 h**.

$$T_{ND1.3} = R_{pd}\,(1.3)\lambda 3 + [1-AS(1.3)]\,\frac{\tau_{13}}{T}\,\lambda_{3r3} = 2 \cdot \frac{3.5}{100} = 0.07 \text{ h}$$

(5.13)

- **Situation 4—fault occurrence at the diesel generator—transformer set (diesel − transformer)**:

To calculate the outage time during the occurrence of a fault in the diesel generator − transformer set, we will first determine the duration of the power outage caused by this specific fault.

$$\tau_{1,N} = \sum_{i=1}^{8} \tau_{1,N,i} \cdot F_i$$

(5.14)

in which: $\tau_{1,N,I}$ is the power outage time corresponding to the ith state of the three WT system combination. F_i is the probability corresponding to the i state of the combination (Table 5.20).

Table 5.20 Power outage time in Area 1

No.	State power	Probability for the corresponding state	$\tau_{1,N,i}$	$\tau_{1,N,I} \cdot F_i$
1	0	0.665009045	24	15,960,217.08
2	175	0.000379755	3	0.001139265
3	262.5	0.000209520	3	0.00062856
4	350	0.012566835	0	0
5	525	0.139112065	0	0
6	700	0.009314910	0	0
7	787.5	0.073013840	0	0
8	1,050	0.100394030	0	0
	Total	$\tau_{1,N} = \sum_{i=1}^{8} \tau_{1,N,i} \cdot F_i$		15.961

204   Clean energy in South-East Asia

$$T_{ND1.N} = R_{pd}\,(1.N)\,\lambda\,N + [1-AS(1.N)]\,\frac{\tau_{1N}}{T}\,\lambda_{NrN}$$

$$= \frac{15{,}961{,}984.91}{24} \cdot 0.07 \cdot 68.6 = 3.1937\,\text{h} \tag{5.15}$$

The total power outage time for all of Area 1 is calculated by summing the weighted outage times for each state. The formula is:

$$T_{ND1} = T_{ND1.1} + T_{ND1.2} + T_{ND1.3} + T_{ND1.4} = 0.48 + 0.04 + 0.07 + 3.1937$$

$$= 3.7837\,(\text{h})$$

$$\tag{5.16}$$

- **Calculation for Area 2: situations causing power outage in Area 2**
  - **Situation 1—fault occurrence in Area 1**:

    The equivalent length of Area 1 is 1 km, with a failure rate of $\lambda_0 = 4$ (failures per 100 km per year). The repair time for each fault is $r = 12$ h per repair. Area 1 lies on the connection line between the wind power source combination and the diesel generator combination with Area 2. Therefore, a fault in Area 1 would have a significant impact on Area 2.

    To calculate the power outage time for Area 2 due to a fault in Area 1, we would need to account for:

    1. The failure rate (λ_0) and repair time (r) for Area 1.
    2. The system configuration and the length of Area 1 influencing Area 2.

    Since Area 1 is connected to Area 2, the power outage in Area 1 would propagate to Area 2 during the fault repair time, affecting the operation of the entire system. The time required for fault repair in Area 1 would determine the outage time for Area 2.

$$T_{ND2.1} = R_{pd}\,(2.1)\lambda 1 + [1-AS(2.1)]\,\frac{\tau_{2.1}}{T}\,\lambda_{1r1} = 4 \cdot \frac{1}{100} \cdot 12 = 0.48\,(\text{h})$$

$$\tag{5.17}$$

  - **Situation 2—fault in Area 2**

    The equivalent length of Area 2 is 2 km, with a failure rate of $\lambda_0 = 4$ (failures per 100 km per year), the repair time for each fault is $r = 12$ h per repair.

$$T_{ND2.2} = R_{pd}\,(2.2)\lambda 2 + [1-AS(2.2)]\,\frac{\tau_{2.2}}{T}\,\lambda_{2r2} = 4 \cdot \frac{2}{100} \cdot 12$$

$$= 0.96\,(\text{h})$$

$$\tag{5.18}$$

- **Situation 3—fault in Area 3**

   Area 3 is not located on the line connecting Area 2 to the power source, so when a fault occurs, it only causes a power outage in Area 2 during the fault isolation time.

$$T_{ND2.3} = R_{pd}\ (2.3)\lambda 3 + [1-AS(2.3)]\ \frac{\tau_{23}}{T}\ \lambda_{3r3} = 2 \cdot \frac{3.5}{100} = 0.07\ (\text{h})$$

(5.19)

- **Situation 4—fault in the diesel power source—transformer combination**

   First, we calculate the power outage time during the occurrence of the fault:

$$\tau_{2,N} = \sum_{i=1}^{8} \tau_{2,N,i} \cdot F_i \tag{5.20}$$

where $\tau_{2,N,i}$ is the power outage time corresponding to the ith state of the three-wind farm combination. F_i is the probability corresponding to the i state of the combination.

This power outage time corresponds to the period when the output power of the wind farm combination is less than the total power demand of both Area 1 and Area 2.

To determine the shortage of power, the total power demand of both Area 1 and Area 2 is as shown in Tables 5.21 and 5.22.

Table 5.21   Total power consumption of Area 1 and Area 2

Time (h)	0–3	3–4	4–7	7–8	8–9	9–10	10–24
Power (kW)	790	660	420	360	260	220	130

Table 5.22   Power outage time in Area 2

No.	Power demand	State probability	$\tau_{2,N,i}$	$\tau_{2,N,i} \cdot F_i$
1	0	0.665009045	24	15.96021708
2	175	0.000379755	10	0.00379755
3	262.5	0.000209520	8	0.00167616
4	350	0.012566835	8	0.10053468
5	525	0.139112065	4	0.55644826
6	700	0.009314910	3	0.02794473
7	787.5	0.073013840	0	0.00000000
8	1,050	0.100394030	0	0.00000000
Total		$\tau_{2,N} = \sum_{i=1}^{8} \tau_{2,N,i} \cdot F_i$		**16.650**

$$T_{ND2.N} = R_{pd}\,(2.N)\lambda N + [1-AS(2.N)]\,\frac{\tau_{2N}}{T}\,\lambda_{NrN}$$

$$= \frac{16{,}650{,}618.46}{24} \cdot 0.07 \cdot 68.6 = 3.3315\ (\text{h}) \tag{5.21}$$

The total power outage time for the entire Area 2 is:

$$T_{ND2} = T_{ND2.1} + T_{ND2.2} + T_{ND2.3} + T_{ND2.4} = 0.48 + 0.96 + 0.07 + 3.3315$$

$$= 4.8415\,(\text{h})$$

$$\tag{5.22}$$

- **Calculation for Area 3: fault scenarios in Area 3**
  - **Situation 1—fault in Area 1**
    The equivalent length of Area 1 is 1 km, with a failure rate of $\lambda_0=4$ (failures per 100 km per year), and the repair time for each fault is $r=12$ h per repair. Since Area 1 is located on the line connecting the wind power source combination and the diesel power source combination to Area 2, a fault in Area 1 significantly impacts).

$$T_{ND2.1} = R_{pd}\,(2.1)\lambda 1 + [1-AS(2.1)]\,\frac{\tau_{2.1}}{T}\,\lambda_{1r1} = 4\cdot\frac{1}{100}\cdot 12$$

$$= 0.48\ (\text{h})$$

$$\tag{5.23}$$

  - **Situation 2—fault in Area 2**
    The equivalent length of Area 2 is 2 km, with a failure rate of $\lambda_0=4$ (failures per 100 km per year), and the repair time for each fault is $r=12$ h per repair. Since Area 2 is located on the line connecting the wind power source combination and the diesel power source combination to Area 3, a fault in Area 2 significantly impacts Area 3.

$$T_{ND3.2} = R_{pd}\,(3.2)\lambda 2 + [1-AS(3.2)]\,\frac{\tau_{3.2}}{T}\,\lambda_{2r2} = 4\cdot\frac{2}{100}\cdot 12$$

$$= 0.96\ (\text{h})$$

$$\tag{5.24}$$

  - **Situation 3—Fault in Area 3**

$$T_{ND3.3} = R_{pd}\,(3.3)\lambda 3 + [1-AS(3.3)]\,\frac{\tau_{33}}{T}\,\lambda_{3r3} = 4\cdot\frac{3.5}{100}\cdot 12$$

$$= 1.68\ (\text{h})$$

$$\tag{5.25}$$

  - **Situation 4—Fault in the diesel power source and transformer combination**

First, we calculate the power outage time during the occurrence of the fault.

$$\tau_{3,N} = \sum_{i=1}^{8} \tau_{3,N,i} \cdot F_i \tag{5.26}$$

where $\tau_{3,N,i}$ is the power outage time corresponding to the ith state of the three wind power systems combination. F_i is the probability associated with the ith state of the combination. This outage time refers to the period when the power output from the wind power system combination is less than the total power demand of the three areas (Area 1, Area 2, and Area 3). To determine the duration of the power shortage, the total power demand for the three areas (Area 1, Area 2, and Area 3) is as shown in Tables 5.23 and 5.24.

$$T_{ND3.N} = R_{pd}(3.N)\lambda N + [1 - AS(3.N)] \frac{\tau_{3N}}{T} \lambda_{NrN}$$

$$= \frac{17{,}840{,}697.88}{24} \cdot 0.07 \cdot 68.6 = 3.569.6 \text{ (h)} \tag{5.27}$$

The total power outage time for the entire Area 3 is:

$$T_{ND3} = T_{ND3.1} + T_{ND3.2} + T_{ND3.3} + T_{ND3.N} = 0.48 + 0.96 + 1.68 + 3.5696$$

$$= 6.6896 \text{(h)}$$

(5.28)

5.8.2 Summary

For the Ly Son power grid, which is an independent power system with a wind power source running parallel to a diesel power source, the impact of the wind

Table 5.23 Total power consumption of Area 1 and Area 2

Time (h)	0–3	3–4	4–7	7–8	8–9	9–10	10–11	11–24
Power (kW)	1230	900	660	600	400	360	270	200

Table 5.24 Power outage time in Area 3

No.	State power, kW	State probability, F_i	$\tau_{2,N,i}$	$\tau_{2,N,i} \cdot F_i$
1	0	0.665009045	24	15,960,217.08
2	175	0.000379755	24	0.00911412
3	262.5	0.000209520	11	0.00230472
4	350	0.012566835	10	0.12566835
5	525	0.139112065	8	1.11289652
6	700	0.009314910	4	0.03725964
7	787.5	0.073013840	4	0.29205536
8	1,050	0.100394030	3	0.30118209
Total		$\tau_{3,N} = \sum_{i=1}^{8} \tau_{3,N,i} \cdot F_i$		17.840

source on the overall grid reliability is quite significant. Specifically, when a fault occurs in the power supply section—the diesel and transformer combination—the load areas will be powered by the wind energy source. The randomness of wind speed in the wind power system is addressed through a four-state power generation model of the wind system, each corresponding to different probabilities.

Chapter 5 provides an overview of MG systems, including their definition, objectives, key components, and the application of this technology in the development of renewable energy as well as cloud-based systems. This chapter highlights the crucial role of hybrid wind–diesel systems in ensuring continuous power supply to remote areas, particularly in Vietnam and globally. The implementation of MG technology not only optimizes the use of renewable energy but also enhances the stability of the power system and reduces dependence on non-renewable energy sources.

Both domestic and international studies are discussed to explore solutions for optimizing power supply, improving power quality, and increasing wind energy usage. A notable example is the hybrid wind–diesel system on Phu Quy Island, which was selected as the primary research subject. The control and operation solutions for this system have been thoroughly analyzed to ensure stability and optimal efficiency in providing power to residents and supporting economic activities on the island.

The chapter also introduces the power model applied on Phu Quoc Island, where projects such as the 110 kV underground cable from Ha Tien to Phu Quoc have helped ensure stable power supply, promoted socio-economic development, enhanced competitiveness, and protected the environment. Notably, Phu Quoc has implemented a simulation of the optimal EV charging scenario with two test scenarios. The first scenario simulates the charging process without using a coordination algorithm, gradually increasing the number of EVs and tracking changes in voltage, frequency, and grid power. The second scenario applies a coordination algorithm to monitor the number of EVs charging and the operation of charging stations, thus improving coordination and optimizing grid power.

Furthermore, the power system on Ly Son Island, Quang Ngai, is an independent system with a wind power source operating alongside a diesel power source. This combination has improved the reliability of the power system, particularly in case of failure in the power supply sources—the MF diesel and transformer units. In such cases, the wind power source is activated to supply electricity to load areas, ensuring the continuous operation of the system. The randomness of wind speed is addressed through a model with four power output states of the WT, each with different probabilities, ensuring stable operation under all conditions.

In conclusion, the development and application of MG systems, particularly hybrid wind–diesel systems, not only offer economic benefits but also contribute to the sustainable use of renewable energy. This chapter lays the foundation for further research aimed at improving operational performance and expanding the application of MG systems in the future. Additionally, it opens new avenues for enhancing the reliability and efficiency of independent power grids in island areas. The optimization of EV charging stations on Phu Quoc also helps integrate renewable energy with modern infrastructure, ensuring sustainable energy management for the future.

Further reading

[1] Ngo Dang Luu, Nguyen Hung, Le Anh Duc and Long D. Nguyen, (2022), "Model and Simulate Electricity Spot Prices Using the Skew-Normal Distribution", Ministry of Culture, Sports and Tourism.

[2] Ngo Dang Luu, Nguyen Hung, Le Anh Duc and Long D. Nguyen, (2022), "Lithium-Ion Battery Pack with Fault", Ministry of Culture, Sports and Tourism.

[3] Ngo Dang Luu, Nguyen Hung, Le Anh Duc and Long D. Nguyen, (2022), "Anomaly Detection in Industrial Machinery Using Three-Axis Vibration Data", Ministry of Culture, Sports and Tourism.

[4] Ngo Dang Luu, Nguyen Hung, Le Anh Duc and Long D. Nguyen, (2022), "Predict Battery State of Charge Using Machine Learning", Ministry of Culture, Sports and Tourism.

[5] Ngo Dang Luu, Nguyen Hung, Le Anh Duc and Long D. Nguyen, (2022), "MATLAB Implementation of Harmonic Analysis of Time Series (HANTS)", Ministry of Culture, Sports and Tourism.

[6] Ngo Dang Luu, Nguyen Hung, Le Anh Duc and Long D. Nguyen, (2022), "Phasor Measurement Unit (PMU)-based Fault Analysis", Ministry of Culture, Sports and Tourism.

[7] Ngo Dang Luu, Nguyen Hung, Le Anh Duc and Long D. Nguyen, (2022), "Power System State Estimation with Phasor Measurements", Ministry of Culture, Sports and Tourism.

[8] Ngo Dang Luu, Nguyen Hung, Le Anh Duc and Long D. Nguyen, (2022), "Hybrid AC/DC & DC Microgrid Test System Simulation", Ministry of Culture, Sports and Tourism.

[9] Ngo Dang Luu, Nguyen Hung, Le Anh Duc and Long D. Nguyen, (2022), "Fixed vs Tracking Solar Panel Output Comparison in Simulink", Ministry of Culture, Sports and Tourism.

[10] Ngo Dang Luu, Nguyen Hung, Le Anh Duc and Long D. Nguyen, (2022), "Load Frequency Control Model in MATLAB SIMULINK", Ministry of Culture, Sports and Tourism.

[11] Ngo Dang Luu, Nguyen Hung, Le Anh Duc and Long D. Nguyen, (2022), "Privacy Preserving Smart Metering based on Randomized Response Model", Ministry of Culture, Sports and Tourism.

[12] Ngo Dang Luu, Nguyen Hung, Le Anh Duc and Long D. Nguyen, (2023), "DC Microgrid Demo", Ministry of Culture, Sports and Tourism.

[13] Ngo Dang Luu, Nguyen Hung, Le Anh Duc and Long D. Nguyen, (2023), "Artificial Neural Network Based MPPT of 2000 W PV System", Ministry of Culture, Sports and Tourism.
[14] Vietnam's Eighth National Power Development Plan (PDP VIII)
[15] Environmental Protection Law: Law No. 72/2020/QH14, signed on 17/11/2020.
[16] Law on Water Resources: Law No. 17/2012/QH13, signed on 21/6/2012, effective from 01/1/2013.
[17] Planning Law: Law No. 21/2017/QH14, passed by the 14th National Assembly on 24/11/2017.
[18] Amendment Law No. 28/2018/QH14: Amendments and supplements to several articles of 11 laws related to planning, passed on 15/6/2018.
[19] Amendment Law No. 35/2018/QH14: Amendments and supplements to several articles of 37 laws related to planning, passed on 20/11/2018.
[20] Decree No. 29/2011/NĐ-CP: Issued by the Government on 18/2/2011, regulating strategic environmental assessment, environmental impact assessment, and environmental protection commitments.
[21] Decree No. 23/2006/NĐ-CP: Issued by the Government on 03/3/2006, on the implementation of the Law on Forest Protection and Development.
[22] Decree No. 46/2012/NĐ-CP: Issued by the Prime Minister on 22/5/2012, amending and supplementing several articles of:
[23] Decree No. 35/2003/NĐ-CP (dated 04/4/2003) detailing the implementation of several articles of the Law on Fire Prevention and Fighting.
[24] Decree No. 130/2006/NĐ-CP (dated 08/11/2006) on compulsory fire and explosion insurance policies.
[25] Decree No. 32/2006/NĐ-CP (2006): Application of the Sorensen Coefficient (S) for comparing species composition in the study area with neighboring areas.
[26] Decree No. 18/2015/NĐ-CP (Article 12) and Circular No. 27/2015/TT-BTNMT (Article 7): Provisions on consultation activities during the implementation of the Environmental and Social Impact Assessment (ESIA).
[27] Circular No. 27/2015/TT-BTNMT: Issued on 29/5/2015 by the Ministry of Natural Resources and Environment (MONRE) on Strategic Environmental Assessment, Environmental Impact Assessment, and Environmental Protection Plans (Appendix 2.3, Chapters 1, 3, 4, and 5).
[28] Circular No. 26/2011/TT-BTNMT: Issued on 18/7/2011 by MONRE, detailing certain provisions of Decree No. 29/2011/NĐ-CP on strategic environmental assessment, environmental impact assessment, and environmental protection commitments.
[29] Circular No. 12/2011/TT-BTNMT: Issued on 14/4/2011 by MONRE, on the management of hazardous waste.
[30] Directive No. 08/2006/CT-TTg: Issued by the Government on 08/3/2006, on urgent measures to prevent illegal deforestation, forest burning, and unauthorized logging.

[31] Ngo Dang Luu, Nguyen Hung, Nguyen Anh Tam, and Long D. Nguyen, (2021), "Wireless Power Transfer simulation for EV", Ministry of Culture, Sports and Tourism.

[32] Ngo Dang Luu, Nguyen Hung, and Long D. Nguyen, (2021), "Power flow calculation of the electric power system by MATLAB", Ministry of Culture, Sports and Tourism.

[33] Ngo Dang Luu, Nguyen Hung, Nguyen Anh Tam, and Long D. Nguyen, (2022), "Distance Relay Protection in AC Microgrid", Ministry of Culture, Sports and Tourism.

[34] Ngo Dang Luu, Nguyen Hung, Nguyen Anh Tam, and Long D. Nguyen, (2022), "Analysis of Solar Photovoltaic System Sunlight Shadowing", Ministry of Culture, Sports and Tourism.

[35] Ngo Dang Luu, Nguyen Hung, Le Anh Duc and Long D. Nguyen, (2022), "Estimate Frequency Response at the Command Line", Ministry of Culture, Sports and Tourism.

Chapter 6
Research and application of IoT technology in sustainable energy development

6.1 Introduction to wireless sensor systems based on remote management and their role in renewable energy development

The Internet of Things (IoT) and artificial intelligence (AI) are two core technologies driving the Fourth Industrial Revolution. IoT provides an integrated platform that connects sensor networks and communication technologies to collect data from industrial objects, thereby delivering useful information to users. However, conventional IoT systems are limited to data collection and are unable to automatically analyze or make decisions based on this data.

On the other hand, AI, particularly machine learning (ML) solutions, requires vast amounts of data for training and development. When these two technologies are combined, the AIoT (AI + IoT) system emerges, enabling IoT systems not only to collect data but also to use AI to analyze, make assessments, and optimize operational efficiency. AIoT promises to offer significant potential value in industrial, energy, and service sectors.

In the context of renewable energy development, particularly solar energy, Vietnam is witnessing strong growth. According to the Vietnam Electricity Corporation, by December 2020, the total solar power capacity reached 19,400 MWp, accounting for 25% of the total system capacity, with rooftop solar systems contributing 50% of the solar power production and 10% of the national grid capacity.

This boom presents numerous opportunities but also challenges, including:

- Energy monitoring for investors and management units.
- Evaluating the operational efficiency of the system for economic and statistical purposes.
- Detecting faults for timely maintenance and repairs.

Among these issues, detecting and addressing faults in photovoltaic (PV) panels is crucial. PV panels are the core components of solar energy systems, and any fault here reduces system efficiency, causing electricity production loss and impacting economic performance. However, traditional fault detection methods, which rely on parameters from inverters, often take between 3 to 19 days, resulting in delays in fault rectification.

A wireless sensor system integrated with AIoT capable of remote monitoring and real-time analysis would be an optimal solution. With the support of ML algorithms, this system can detect and alert faults as soon as they occur, thus reducing operational and maintenance costs while improving solar energy system efficiency. This system not only helps optimize renewable energy sources but also promotes sustainable development in the energy sector.

6.1.1 IoT applications in solar power monitoring

Currently, IoT applications in solar power monitoring are also being widely researched. In these studies, authors use Arduino microcontrollers in combination with sensors for current, voltage, light intensity, and temperature to collect operational data from PV panels. The data is then transmitted to a cloud server and the ThingSpeak application via communication modules that use wireless technologies such as WiFi or Zigbee. Researchers at the SenSIP Center at the University of Arizona have conducted a series of studies using a smart monitoring device (SMD) integrated with IoT for monitoring PV arrays and solving maintenance issues such as fault detection, classification, and customizing connections between panels.

6.1.2 Studies on monitoring and analyzing the operational status of photovoltaic panels

The analysis of PV panel operation is based on information such as current, voltage, and power output. Methods for analyzing the operational status of PV panels include: model-based diagnostic methods (MBDM), real-time diagnostic methods (RDM), output signal analysis (OSA), and machine learning-based techniques (MLT).

MBDM methods use environmental parameters like solar radiation and temperature to input into PV models built on theoretical principles. These models' outputs are compared with actual operational data to assess the status. The comparison can be made by evaluating discrepancies between voltage, current, and power parameters in theoretical models and actual values, or by setting thresholds for characteristic ratios between model results and actual values.

In contrast to MBDM, RDM analyzes parameters collected from PV panels and sets warning thresholds through pre-programmed rules and knowledge. These rules for comparison and threshold-setting may still be derived from theoretical PV model analysis or from operational characteristics. However, since RDM does not require the calculation of a parallel PV model during data collection, it is more easily applied than MBDM.

OSA methods transform PV panel parameter signals using mathematical transformations to extract signal features and assess conditions based on these features. Transformation methods can include normal distribution, quantum probability models, and Wavelet transformations. The assessment of PV panel conditions is done by comparing and setting thresholds for the transformed signal feature values with values in the steady state. OSA methods are advantageous as they do not require environmental sensors, as only the panel output needs to be analyzed. Moreover, the accuracy of OSA methods is superior to RDM methods. However, OSA methods are more computationally complex, and transformations using signal processing algorithms like Wavelet can introduce some delay.

MLT apply AI or ML models to train and learn the features of the PV panel's operation under both steady and fault conditions. The training data is obtained through real-world data or results from PV model simulations. The training data must include both environmental input parameters and PV panel output parameters under all operational conditions, not just the steady-state condition. The trained ML model can analyze and assess the PV panel's operational status in real time based on input parameters like solar radiation intensity and temperature. Studies on MLT methods will be presented in Section 6.1.3.

6.1.3 Studies on the application of MLT in monitoring and analyzing the operational status of photovoltaic panels

The general aim of MLT is to use ML and AI algorithms to learn from an available data source and then establish correlations between input parameters and desired outputs. Training data can be constructed by collecting real-world data or using simulation results. Most algorithms used in monitoring and analyzing the operational status of PV panels are supervised learning algorithms, meaning the training data must be labeled—recorded with known outcomes through expert knowledge or other analysis algorithms. Commonly applied methods include: artificial neural networks (ANN), support vector machines (SVM), decision trees (DT), and fuzzy logic.

ANN methods use neural network structures to link input parameters with output results. The relationship between inputs and outputs is expressed through functions and weights at each neuron in the network structure. During training, the weights are adjusted to make the network's output as close as possible to the labeled data in the training source. ANN structures can be used to analyze the direct relationship between panel output parameters and operational status using feedforward networks, Bayesian neural network (BNN), convolutional neural network (CNN), Euclidean neural network (ENN), probabilistic neural network (PNN) or replace the PV model with ANN to analyze the relationship between environmental parameters (such as radiation intensity and temperature) and PV output values.

SVM is an ML method that divides data based on the inner product to maximize the margin between the dataset and the dividing line. SVM machines are often used for learning and classifying the operational conditions of PV panels based on operational data or other transformed analytical results.

DT methods perform threshold analysis based on statistical results from training data. During training, threshold values are selected to divide the data into the largest possible components, and these thresholds are used to classify data after training. Like SVM, DT models typically use PV panel operational parameters as input. The accuracy of a DT model depends on the size (the number of checks) of the DT. A larger tree size increases accuracy but also affects computation time and memory usage.

Fuzzy Logic methods analyze the operational status of PV panels by comparing input values with predefined standard values based on fuzzy logic rules. For example, one study used the Takagi–Sugeno–Kahn rule for analyzing collected data against predicted results based on environmental parameters such as radiation intensity and temperature. Another study combined a NARX neural network with a Sugeno fuzzy classifier to detect faults and analyze the operational status of PV panels using environmental parameters and panel output values.

6.1.4 Research objective

This research aims to develop an IoT system for monitoring solar energy systems with the capability to analyze the operational status of PV panels.

6.1.5 Research content

Building on the previous studies, the topic proposes a model for monitoring a solar energy system using IoT and analyzing the condition of PV panels based on ML algorithms.

The IoT monitoring system consists of three components: the monitoring unit, the server, and the user application. In the monitoring unit, system parameters of the solar power system, including current intensity, AC and DC voltage at both sides of the system, and solar radiation intensity at the installation site, are collected through sensors and processed by a central microcontroller. These data are then sent to the Server via the communication function block. On the server, functions for data transmission, reception, and storage are performed, allowing the data to be stored in databases (DBs) to meet the data requirements of both the monitoring unit and the user application. In addition to data transmission, the Server also performs analysis and provides assessments of the PV panel's status based on the parameters sent by the monitoring unit, using ML models. These include an ANN model that simulates the PV panel and a comparison evaluation model that investigates two ML methods: k-means and SVM. The evaluation results are stored in the incident DB on the server. The user application includes mobile apps and desktop software to help customers monitor system parameters. Additionally, the software grants management and technical staff the ability to set up and customize system parameters, ensuring accurate monitoring and supervision. The system block diagram is depicted in Figure 6.1.

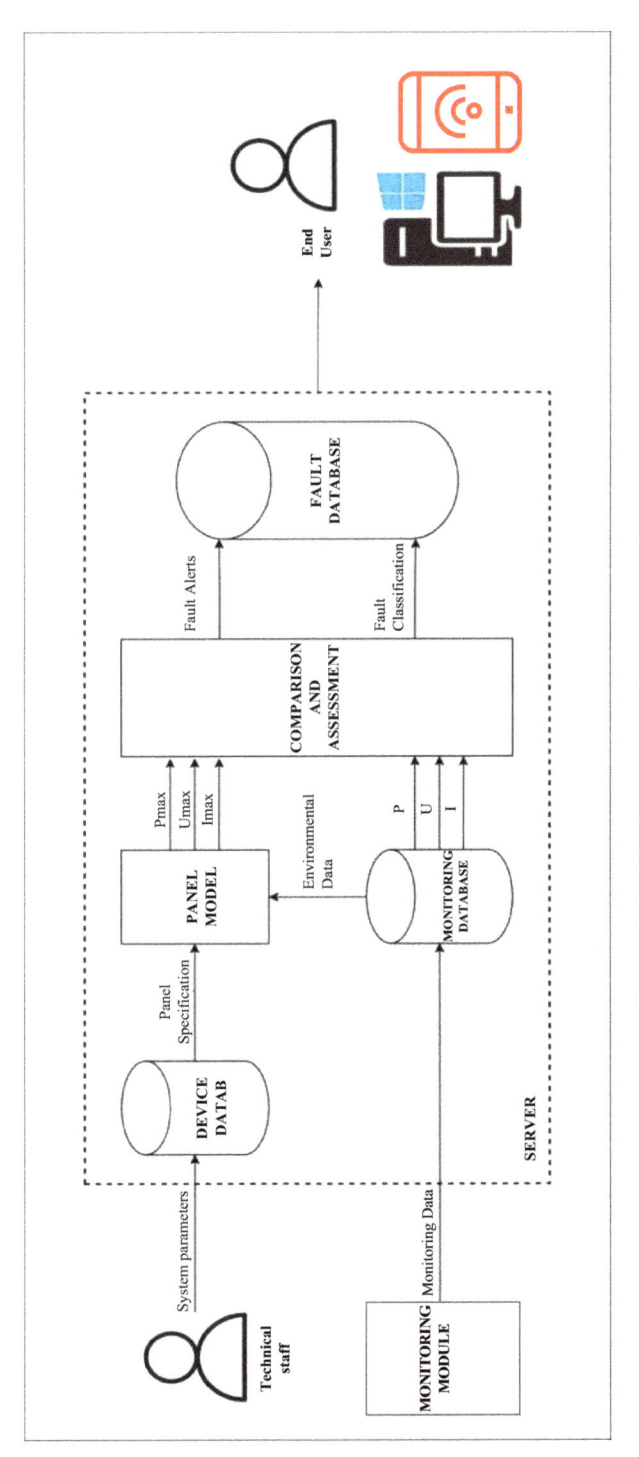

Figure 6.1 Block diagram of the proposed system

6.2 Types of sensors and communication technologies for monitoring and managing renewable energy systems

In today's era, renewable energy has become an essential part of sustainable energy solutions. Energy sources such as solar, wind, hydro, and biomass are increasingly being utilized. To ensure the efficiency and stability of the operation of these energy systems, sensors and communication technologies play a crucial role in monitoring and management.

The sensor and communication systems not only record important parameters but also allow for automation of the operation process, minimizing human intervention while enhancing the accuracy in optimizing the system. This section will focus on two key aspects: sensors and communication technologies in renewable energy systems.

6.2.1 Sensors in renewable energy

Sensors are devices responsible for collecting data from the environment and transmitting it to a processing system or control center. In renewable energy systems, sensors play a critical role in monitoring the factors that directly influence energy production. Below are some common types of sensors.

6.2.1.1 Solar energy sensors

In solar energy systems, PV panels are the main components for capturing and converting sunlight into electrical energy. Solar energy sensors measure the amount of light absorbed, the temperature of the panel, and the level of energy generated. These parameters help determine the system's operational status and detect early abnormalities such as performance degradation due to dirt, damaged panels, or connectivity issues. Additionally, temperature and humidity sensors are used to monitor weather conditions, which helps optimize the positioning and tilt angle of the panels to enhance energy absorption.

6.2.1.2 Wind sensors

In wind energy systems, sensors that measure wind speed and direction are essential for adjusting the position of wind turbines to maximize energy capture. Information from these sensors allows turbines to automatically adjust the angle of the blades, improving electricity generation efficiency and reducing the risk of damage when wind speeds are too high. Vibration and temperature sensors are also used to monitor mechanical components such as shafts and bearings, helping to detect wear or overload issues early.

6.2.1.3 Environmental sensors

In addition to specialized sensors, environmental sensors such as temperature, humidity, and air pressure sensors are integrated to monitor the operating conditions of the system. This information allows the system to self-adjust to maintain stable performance and extend the lifespan of the equipment.

6.2.2 Communication technologies in renewable energy management

Communication technologies are crucial for connecting sensors to the management and control systems. Data collected from sensors needs to be transmitted quickly and accurately to support timely decision-making. Below are common communication technologies.

6.2.2.1 WiFi and mobile networks

WiFi is a widely used wireless communication technology due to its accessibility and high data transmission speed. In renewable energy systems, WiFi often connects sensors to a central management system, especially in small-scale projects or residential areas.

Mobile networks, including 4G and 5G, are also widely applied. With the advantage of wide coverage, mobile networks work effectively in remote areas where installing wired infrastructure is challenging.

6.2.2.2 Zigbee and Bluetooth

Zigbee is a low-power wireless communication protocol that allows multiple devices to connect within the same network. It is an ideal choice for small-scale renewable energy systems that require remote monitoring and control with minimal power consumption.

Bluetooth is also used in small-scale systems, with the advantage of connecting devices over short distances while saving energy. Bluetooth is commonly used in personal monitoring devices or local systems.

6.2.2.3 IoT and specialized communication protocols

The IoT is becoming a trend in the development of smart renewable energy systems. Sensors and devices connected within an IoT system can exchange real-time data with each other and the central control system. Through IoT, renewable energy systems can be remotely managed, automatically adjust performance, and generate detailed reports on operational conditions.

In addition to popular communication technologies, there are specialized protocols such as Modbus, CAN bus, and Industrial IoT protocols. These protocols are designed to ensure high reliability and are suited for harsh industrial environments, where stability and resistance to interference in data transmission are critical.

6.2.3 Benefits of sensor and communication systems in renewable energy management

The application of sensors and communication technologies in renewable energy systems offers many practical benefits. One of the most important benefits is the ability to monitor the system in real time. Continuous data collection from sensors allows for the detection of small changes in environmental conditions and operational status, enabling timely adjustments to maintain optimal performance.

Moreover, sensor and communication systems help minimize human intervention. Instead of manually inspecting and adjusting equipment, the system can automatically manage operations through pre-programmed processes, saving time and effort while reducing errors. Finally, these systems play a vital role in extending the lifespan of renewable energy equipment. Continuous monitoring allows for the early detection of issues such as wear or damage, enabling timely maintenance before serious failures occur.

Sensors and communication technologies are key elements in the development and management of renewable energy systems. With continuous monitoring and intelligent management capabilities, they help optimize performance, minimize human intervention, and extend equipment lifespan. In the future, as IoT technology and advanced communication protocols continue to develop, renewable energy systems will become more efficient and sustainable.

6.3 Application of wireless technology and sensor systems in monitoring and predicting the performance of renewable energy sources and some practical examples related to wireless sensor systems in the renewable energy sector

PV panels are the most important devices in solar energy systems, serving as the components that convert solar radiation into electrical energy for consumption. A PV panel consists of multiple connected PV cells, with each cell containing a p–n junction as shown in Figure 6.2. When sunlight strikes the PV panel, the PV cells absorb energy from the photons in the sunlight. If the energy level exceeds the bandgap energy of the semiconductor material, it causes polarization due to the movement of electrons across the p–n junction, following the PV principle. Connecting the two ends of the p–n junction in the PV cell/panel generates an electric current that flows through the circuit.

The characteristics and operational parameters of a solar panel are typically represented through the *I–V* or *P–V* characteristic curves provided by the

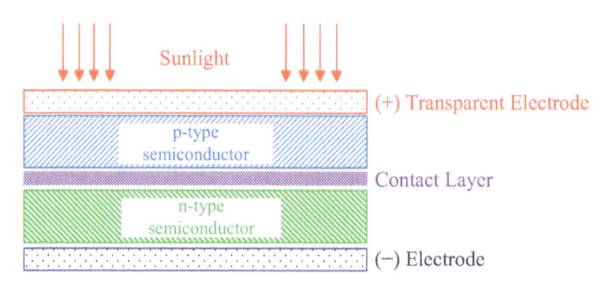

Photovoltaic Cell Structure Diagram

Figure 6.2   Structure of a photovoltaic panel

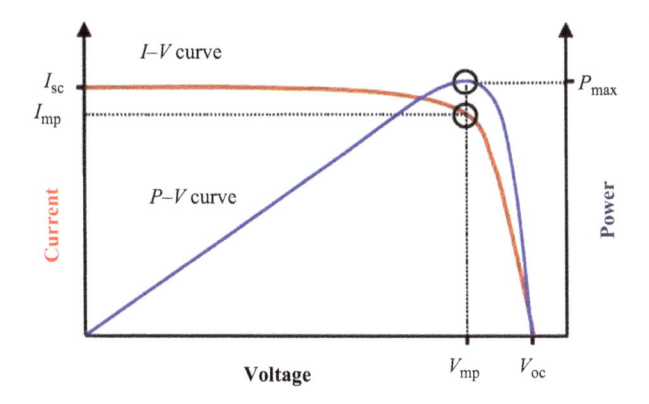

Figure 6.3 I–V (red) and P–V (blue) characteristic curves of a photovoltaic panel

manufacturer, as shown in Figure 6.3. Additionally, the manufacturer must supply specific parameters of the panel as given below.

- **Panel material**
  - **Nominal power**: The maximum power the panel can generate under standard test conditions (STC).
  - **Voltage and current values at maximum power**: The voltage and current at which the panel operates at its maximum power (nominal power) under STC conditions.
  - **Open-circuit voltage**: The voltage across the panel when it is in an open-circuit state under STC.
  - **Short-circuit current**: The current generated by the panel when it is in a short-circuit state under STC.
  - **Temperature coefficient of current**: The rate at which the short-circuit current decreases as the panel's operating temperature changes compared to the STC.

- **Standard test conditions**
  - **Radiation intensity**: 1,000 W/m^2
  - **Operating cell temperature**: 25 °C

6.3.1 Failure cases in photovoltaic panels

Like any other system, a PV system can experience unexpected failures during operation. These failures arise due to malfunctions or degradation of internal components within the system. Figure 6.4 from the article illustrates the types of failures in a PV system, categorized according to the system diagram into: failures due to the array of PV panels, failures in the DC side of the system, and failures in the AC side of the system.

222　*Clean energy in South-East Asia*

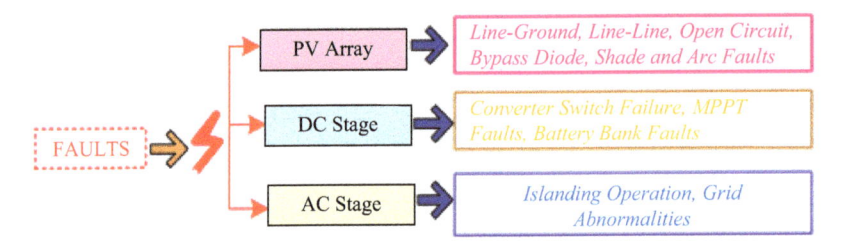

Figure 6.4　Types of faults in solar power systems

Figure 6.4 illustrates several failure cases in the PV panel array, including:

- **Line-line**: A failure due to a short circuit between two different PV panels, which may occur within the same array or across different arrays.
- **Shading**: A failure caused by uneven solar radiation intensity hitting the panels in the array due to external factors.
- **Open circuit**: A failure caused by the PV panel array being open-circuited.

Additionally, other cases such as degradation due to the aging of the panels and degradation of the wiring can be considered.

6.3.2　Internet of Things

IoT is a system platform capable of connecting devices through wireless communication technologies, which allows for the provision of data and operational information of component devices.

The structure of an IoT system can be divided into various models. A synthesis of these IoT structural models is presented in Table 6.1. However, the model of an IoT system can be simplified and generalized into three layers: the physical layer, the network layer, and the application layer. Figure 6.5 details each layer as follows:

- **Physical layer**: Includes the devices participating in the system and sensor systems that collect the necessary data.
- **Network layer**: Performs functions such as data processing, storage, and communication, including the server that contains the DB and processing algorithms.
- **Application layer**: Comprises software and applications on user devices such as smartphones, computers, laptops, or online websites that transmit information obtained from the collected data.
- **Results achieved**
- **Theoretical calculation model**

The model was calculated and tested for operational results based on the parameters of the Redsun P618-25W PV panel, with the specifications presented in Table 6.2.

Research and application of IoT technology in SE development 223

Table 6.1 Some IoT models

Model	Structure
European FP7	Tree structure: • Root: devices integrating communication technologies • Trunk: components responsible for communication and security • Leaves: potential applications derived from collected data
ITU	• Sensing layer • Access layer • Network layer • Middle ware layer • Application layer
IoT Forum	• Applications • Transportation • Processors
Qian Xiaocong, Zhang Jidong [42]	• Application layer • Transportation layer • Perception layer
Kun Han, Shurong Liu, Dacheng Zhang, and Ying Han	• Near-field communication technology • Connection management devices • High-speed wireless networks

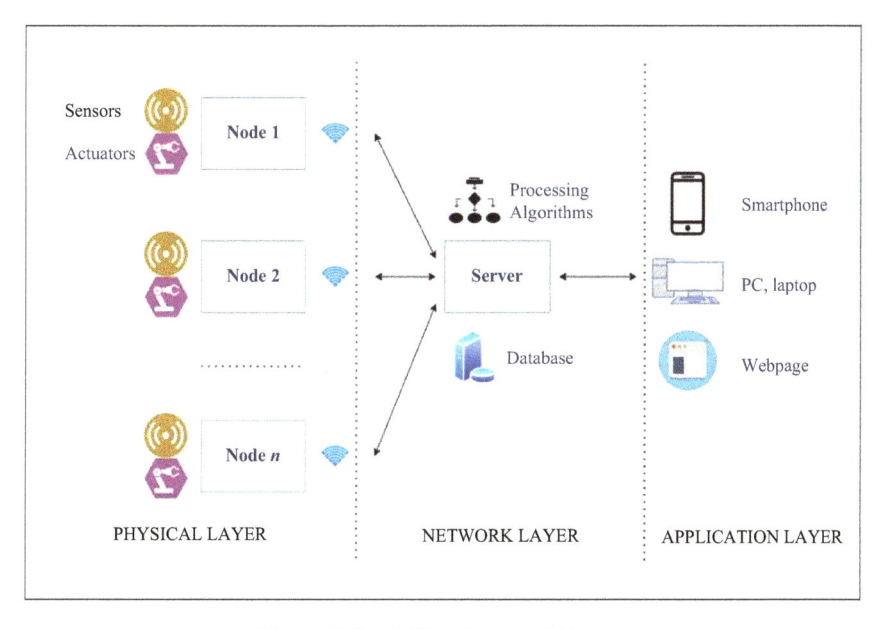

Figure 6.5 IoT system architecture

Table 6.2 Specifications of the P618-25W solar panel used for simulation

Technical specifications	Value
Maximum power (P_{max})	25 W
Type of solar cell	Poly
Open-circuit voltage	21.69 V
Short-circuit current	1.55 A
Voltage at maximum power	17.35 V
Current at maximum power	1.44 A
Temperature coefficient	−0.46%/°C

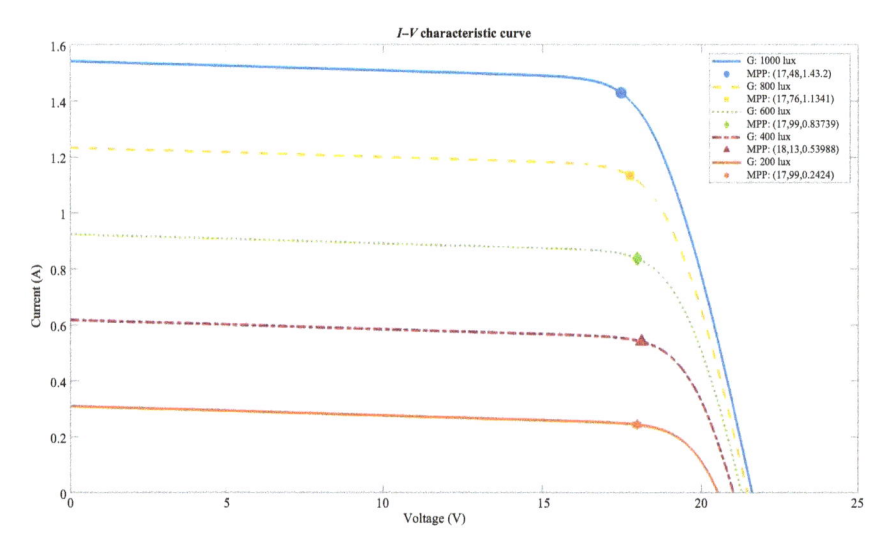

Figure 6.6 I–V characteristic curve

Based on the formulas and algorithm diagrams presented in Section 3.1, the PV panel model was programmed and tested using MATLAB® software. The results of the model include the I–V characteristic curve and the P–V characteristic curve, along with the position of the maximum power point (MPP) under radiation intensities of 200, 400, 600, 800, and 1,000 W/m², as shown in Figures 6.6 and 6.7.

This simulation model was then implemented in a server environment to calculate the position of the MPP = [V_{mp}, I_{mp}] for radiation intensity levels ranging from 0 to 1,000 W/m². The resulting MPP points are represented in the I–V space, as shown in Figure 6.6. It can be observed that the MPP points tend to move toward the point (0 V, 0 A) at low radiation intensity values before stabilizing. This occurs because, in the model, at low radiation conditions, the difference between voltage and current is very small, causing the Newton–Raphson algorithm to converge in

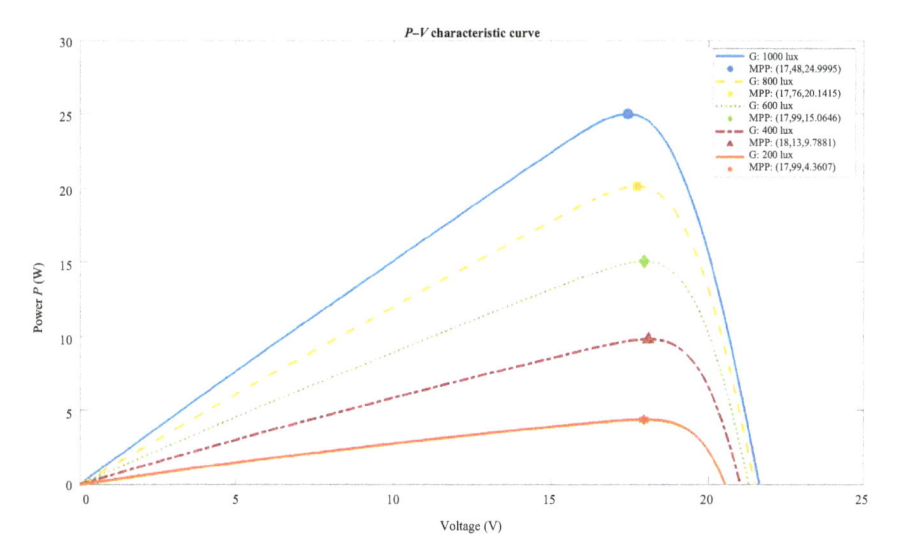

Figure 6.7 P–V characteristic curve

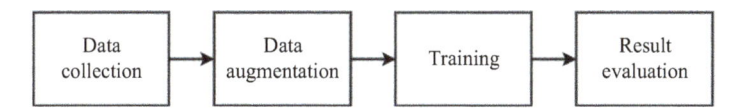

Figure 6.8 Steps in building a machine learning model

the early calculation iterations. This issue needs to be considered when incorporating the MPP points into the ML model for training.

6.3.3 Machine learning model for fault classification
6.3.3.1 Steps to build the classification model
The steps to build the ML model are presented as shown in Figure 6.8.

The data collection process was carried out through simulations using the Simulink® software of MATLAB to model fault scenarios in a solar power system with **Np** parallel-connected arrays, each consisting of **Ns** series-connected solar panels. Data augmentation was implemented by increasing the number of points for each fault scenario and adding a case to represent fault scenarios that had not yet been considered. The increase in the number of points for each fault scenario was achieved by introducing random errors into the current or voltage parameters of the initial points collected from the simulation. These added errors represent two cases: one substitutes fault points under radiation intensity conditions adjacent to the levels already considered, and the other accounts for errors that could arise during real-world data collection or from the computational results of the solar panel model when applying the fault classification model in **MBDM** or **RDM**.

Two classification models using ML algorithms, ***k*-means** and **SVM**, were developed on the Google Colab platform using the **Python** language and leveraging the **sklearn** libraries. These libraries provide tools for building models, training them, and presenting the results post-training to evaluate the classification performance of the model.

6.3.3.2 Fault simulation

Simulation method

To generate data for the surveyed faults, a PV panel system connected in a 8×2 configuration (2 rows, 8 panels in series) was simulated using MATLAB software. The type of PV panel is not critical, as after surveying the current and voltage parameters, they will be normalized according to the MPP parameters. However, for reference purposes, the specifications of the panels used in the MATLAB simulation are presented in Table 6.3.

The PV panel simulation module in the Simulink tool of MATLAB is shown in Figure 6.9. The module has two inputs for the radiation intensity and cell temperature parameters, and two outputs for the positive and negative terminals to connect with other blocks in the simulation system. To obtain the current and voltage parameters of the system, the PV panel module is connected to corresponding measurement blocks provided by Simulink. Figure 6.9 illustrates the system connection in MATLAB/Simulink, where the panel modules are connected in an 8×2 configuration, along with functional blocks that collect and output data on the system's voltage, current, and power.

The system is simulated for two fault cases: short circuit and open circuit, under the following conditions:

- **Short circuit**: Voltage is zero.
- **Open circuit**: Current is zero.
    - **Open-circuit case**
        In the open-circuit scenario of a string of panels, the system is assumed to consist of only eight series-connected panels, with the open-circuited string being removed from the system.

Table 6.3 Technical specifications of the Sunpower 245 W

Model	Sunpower SPR-E20-245
Rated power	245 W
Open-circuit voltage	48.8 V
Short-circuit current	6.43 A
Voltage at maximum power point (MPP)	40.5 V
Current at MPP	6.05 A
Current temperature coefficient	0.07 (%/°C)
Number of series-connected cells	72

Under STC conditions – 1,000 W/m^2, 25°C.

- **Other faults**
  Other faults are incorporated into the simulation by adding connections or adjusting abnormal parameters (as shown in Figure 6.10(a) and (b)).
- **Study conditions**
  The faults are investigated under various radiation intensity levels, including 200, 400, 600, 800, and 1,000 W/m^2, while also considering the variation in the number of panels affected by faults within the system.

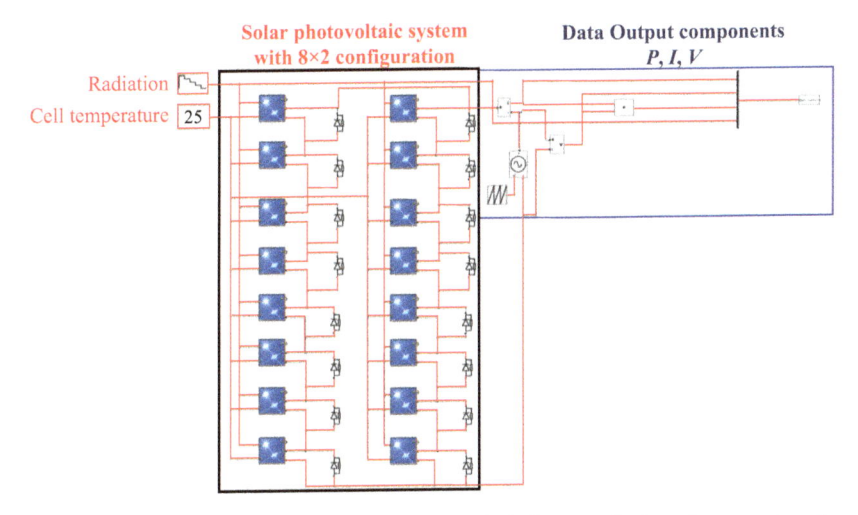

Figure 6.9 Connecting modules in Simulink for simulation data retrieval

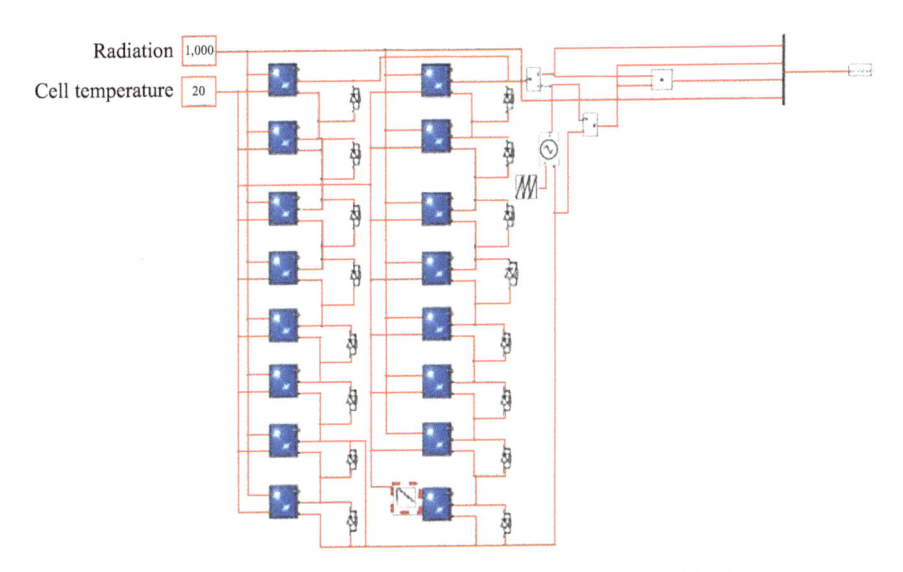

Figure 6.10 Simulation of faults in the solar power system: (a) shading and (b) string–string fault

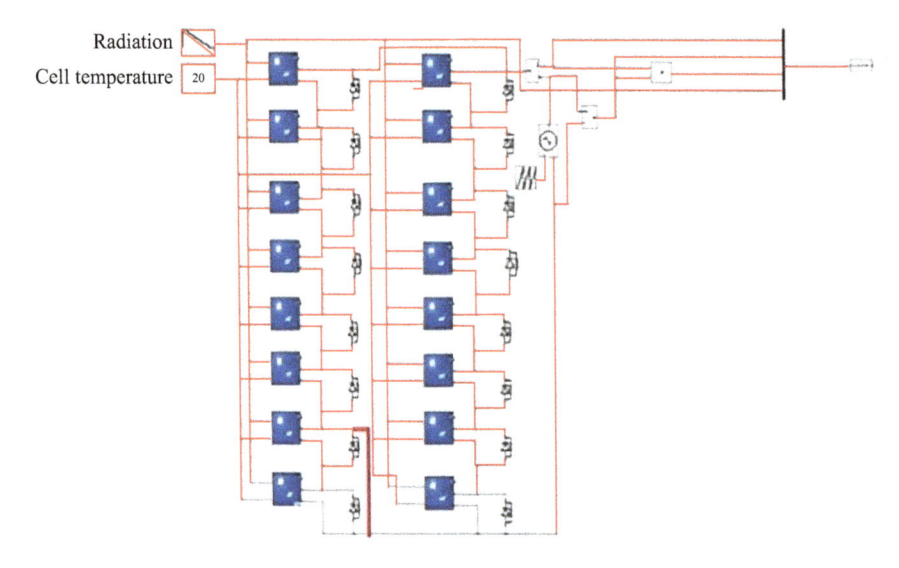

Figure 6.10 (Continued)

Simulation results

Figures 6.9 and 6.10 show of the system for two fault cases: wire-to-wire and shading, with a radiation intensity of 1,000 W/m² and the number of affected panels being 1 and 6, respectively.

- **Wire-to-wire fault case**
  - When the number of faulty panels is small ($n = 1$), the change in the *I–V* curve is negligible.
  - When the number of faulty panels increases ($n = 6$), the change becomes more pronounced, reflected in a voltage loss compared to the normal state of the system.
- **Shading fault case**
  - For a small number of affected panels, the MPP is influenced differently depending on the shading radiation intensity level:
    - When the radiation difference is small (e.g., 800 W/m² compared to 1,000 W/m²):
      - The *I–V* curve is slightly distorted.
      - The MPP decreases slightly in current.
    - When the radiation difference is large (e.g., 400 W/m² compared to 1,000 W/m²):
      - The *I–V* curve is more significantly distorted.
      - The MPP shifts toward a lower voltage.
      - When the number of shaded panels increases:

Research and application of IoT technology in SE development 229

- The MPP tends to decrease in current while maintaining the voltage level.
- The greater the difference between shading intensity and the normal radiation intensity, the more significant the current reduction.

The number of data points from the fault cases forms the dataset for evaluating the two ML algorithms, as shown in Table 6.4.

6.3.4 Machine learning model training results
6.3.4.1 *K*-means algorithm model

Figure 6.11 illustrates the classification results of the ML model using the *k*-means algorithm.

- Figure 6.11(a) shows the augmented dataset.
- Figure 6.11(b) displays the classification results after training, with corresponding fault labels as shown in Table 6.5.

Table 6.4 Composition of the training dataset

Fault type	Quantity
Shading	2,000
Open circuit in one string	500
Wire-to-wire	3,500

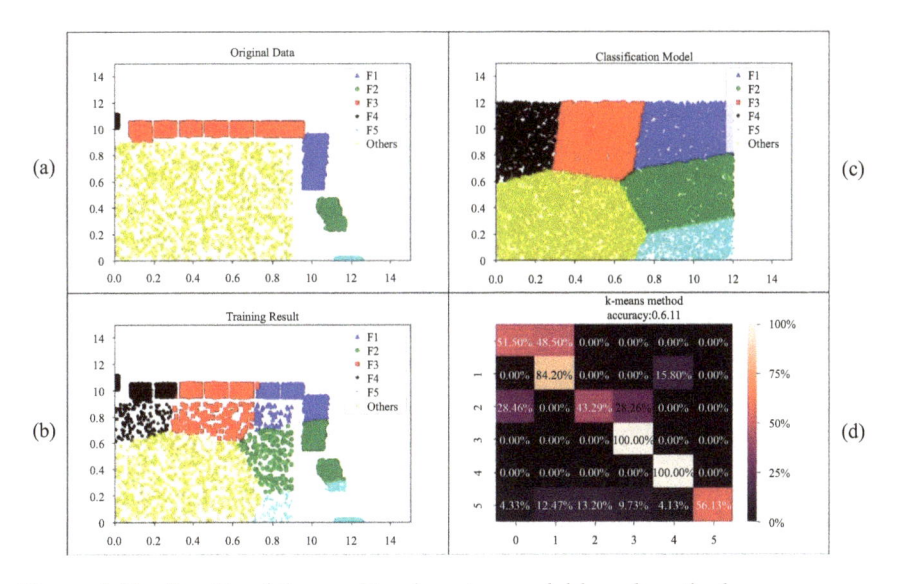

Figure 6.11 Results of the machine learning model based on the k-means algorithm. (a) Initial dataset, (b) training results, (c) model in the I–V space domain, and (d) fault classification matrix.

Table 6.5 Symbols for fault cases

Fault case	Symbol	Label
Shading	F1	0
Open circuit in one string	F2	1
Wire-to-wire	F3	2
Short circuit	F4	3
Open circuit	F5	4
Other faults	Others	5

The training results indicate that the k-means model did not perform well in classifying fault cases:

- **Overlapping classification**: Data points from different fault types overlap.
- **Misclassification**: There is significant confusion between fault types.
- Figure 6.11(c) shows the predicted results of the model in the normalized I–V parameter space.
    - The results reveal that the separating lines between fault types are far from the actual separation points, causing misclassification.
- Figure 6.11(d) presents the confusion matrix for fault classification, showing an overall accuracy of 61.1%.

1. **Highest accuracy prediction case:**
    - **Short circuit and open circuit**: Accuracy reaches 100%.
2. **Misclassification cases**:
    - **Wire-to-wire and open circuit on one string**: The model frequently confuses these two fault types.
    - **Open circuit on one string**: Although the accuracy is relatively high (84.20%), 48.50% of shading cases were misclassified as open circuit on one string.
3. **Remaining cases**: Accuracy is low, ranging from 40% to 60%.

The k-means algorithm is ineffective in comprehensively classifying fault cases. This may be due to:

- **Overlapping data distribution** in the parameter space.
- The **incompatibility** of the k-means algorithm with the complexity of the problem

6.3.4.2 SVM algorithm model

Figure 6.12 illustrates the training and classification results of the ML model using the SVM algorithm.

- Figure 6.12(a) displays the initial dataset.
- Figure 6.12(b) shows the training results of the model.

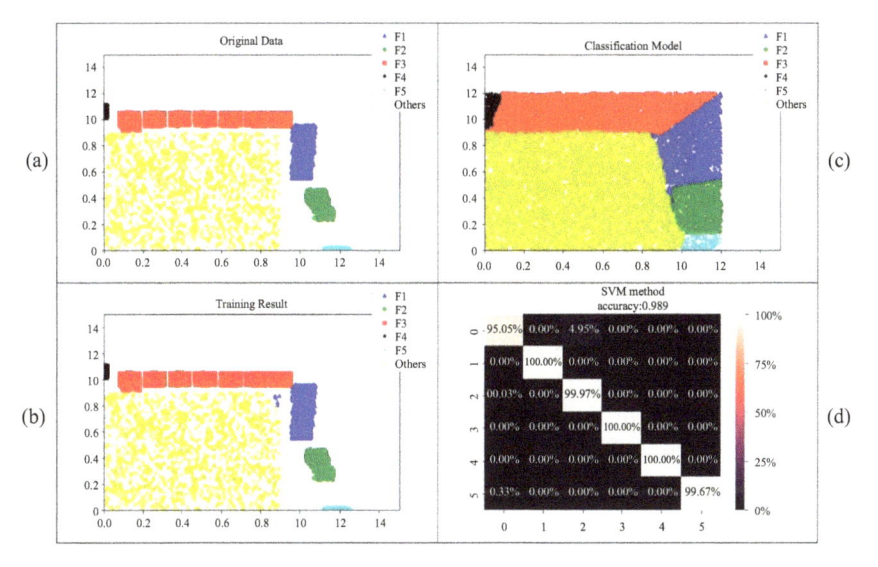

Figure 6.12 *Results of the machine learning model based on the SVM algorithm. (a) Initial dataset, (b) training results, (c) model in the I–V space domain, and (d) fault classification matrix.*

The results demonstrate that the SVM model achieves high classification performance:

- **Data fitting**: The model classifies effectively and is well aligned with the characteristics of the initial dataset.
- Figure 6.12(c) illustrates the classification results in the parameter space.
  - The separating hyperplane between the fault groups closely follows the boundary data points, creating a clearer and more accurate classification threshold compared to the *k*-means algorithm.
- Figure 6.12(d) presents the fault classification confusion matrix, showing an overall model accuracy of 98.9%.

1. **Classification performance**
   - All fault cases achieved an accuracy of over 95%.
   - **Short circuit and open circuit**: 100% accuracy.
2. **Advantages over *k*-means**
   - The separating boundary is more distinct, reducing confusion between fault groups.
   - Effectively classifies difficult cases due to the SVM's ability to optimize the separating margin in the parameter space.

The SVM algorithm outperforms *k*-means in classifying fault cases, especially when the data has overlap. With a high overall accuracy of 98.9%, it is the optimal choice for classifying fault cases in the surveyed system.

6.3.4.3 Comparison of the two models

Table 6.6 provides a detailed comparison of accuracy and F1-score for each type of fault between the two ML models based on the k-means and SVM algorithms.

Comparison results
1. **Overall accuracy**
   - **SVM**: 98.9%
   - **K-means**: 61.1% The SVM model significantly outperforms the k-means model, demonstrating better classification effectiveness.
2. **Data characteristics and classification performance**
   - The fault data in the normalized I–V parameter space tends to concentrate in distinct rectangular regions rather than being widely distributed.
   - SVM leverages this characteristic by using the optimal separating margin to create more precise decision boundaries, whereas k-means relies solely on the distance to cluster centroids, leading to confusion between fault groups.
3. **Advantages of SVM**
   - **Superior clustering capability**: By optimizing the separating margin, SVM reduces confusion and ensures accurate classification across different fault groups.
   - **Avoidance of bias**: Unlike k-means, which can be influenced by subjective clustering thresholds, SVM allows for more objective and consistent classification.

The superiority of the SVM model over k-means lies in its ability to cluster based on the maximum margin boundary, which is particularly effective with

Table 6.6 Comparison of results between two machine learning models based on k-means and SVM algorithms

		k-Means (%)	SVM (%)
	Model accuracy	61.1	98.9
Accuracy of each fault case	Shading	50	97
	Open circuit in one string	41	100
	Wire-to-wire	58	99
	Short circuit	64	100
	Open circuit	93	100
	Other faults	72	100

concentrated data types. This makes SVM a more reliable choice for fault classification, ensuring higher accuracy and reliability.

The results of the study and comparison between the two ML models, k-means clustering and SVM, show that SVM significantly outperforms k-means in identifying and classifying fault conditions in PV panels. With an overall accuracy of up to 98.9%, the SVM model demonstrates its capability to efficiently handle concentrated and complex data characteristics, ensuring higher reliability compared to the k-means model. This confirms the potential of SVM for use in diagnosing and managing solar energy systems.

6.3.5 Development of an IoT-based solar energy monitoring system

6.3.5.1 Proposed IoT monitoring system

The proposed IoT-based solar energy monitoring system is designed to effectively and conveniently monitor and manage the operational parameters of the system. The system uses electronic components to measure key parameters, including:

- Current, voltage, and power of the solar panel array (DC side)
- Current, voltage, and power of the system (AC side)

These parameters will be displayed on-site, allowing users to monitor them in real time.

In addition, the system integrates IoT technology, enabling the creation of a data processing platform that provides necessary information to users via a mobile application. The information includes the status of the solar panel array, power quality, and other relevant metrics, ensuring accuracy, convenience, and ease of use. The system is designed to help users comprehensively and effectively monitor the operational state of the solar energy system (Figure 6.13).

6.3.5.2 Design of the monitoring and data collection unit

Based on the parameters of the solar energy system under consideration, the design process involves selecting the corresponding and appropriate measurement components.

As illustrated in Figure 6.14, the monitoring unit consists of the following functional blocks:

- Microcontroller
- DC data collection block
- AC data collection block
- Data transmission block
- Display block

The DC and AC data collection blocks are responsible for collecting current, voltage parameters from both sides of the solar energy system, as well as the light radiation intensity. These parameters are gathered and processed by the

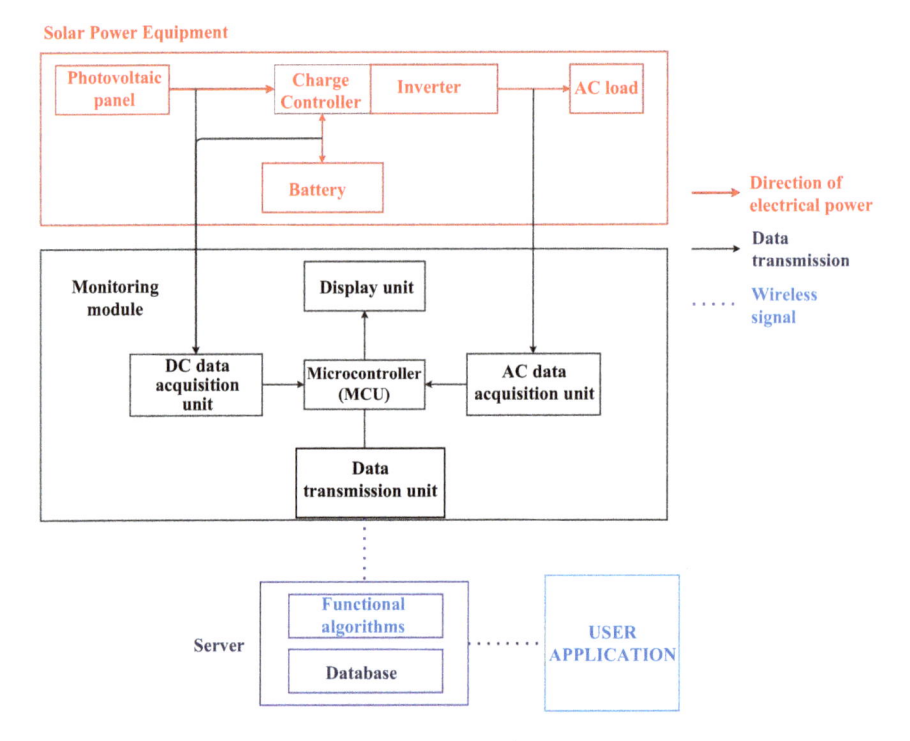

Figure 6.13   Proposed IoT system

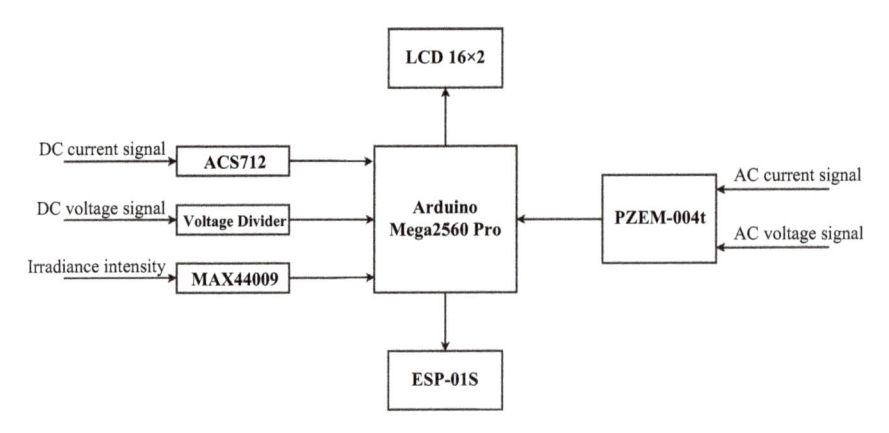

Figure 6.14   Components corresponding to functional blocks

microcontroller, which then sends the data to the server via the data transmission block. The display block is responsible for showing the real-time monitoring results at the installation location of the monitoring unit. The components corresponding to each functional block are shown in Figure 6.14.

The sensors for the DC and AC data collection blocks of the solar energy system use the ACS712-20 A current sensor, the MAX44009 light intensity sensor, and the PZEM–004T AC power measurement module.

The ACS712-20A current sensor is a Hall-effect sensor that outputs a voltage value ranging from 0 to 5 V, which is proportional to the current being measured. The current intensity (I_p) to be measured is calculated using the formula where V_{out} is the output voltage of the sensor, and the sensitivity depends on the type of sensor. The ACS712-20A has a sensitivity of 100 mV/A and a measurement range from −10 A to 10 A.

$$I_p = \frac{V_{out} - 2.5 \text{ V}}{\text{Sensitivity}} \tag{6.1}$$

The MAX44009 sensor is a light intensity sensor integrated with I2C communication protocol, operating at a voltage of 3.3–5 V_{DC}. The sensor is capable of measuring light intensity in the range of 0.045–188,000 lux.

The microcontroller block uses the Arduino Mega2560 Pro board in combination with the ESP-01S WiFi module for data transmission. The display block uses a 16×2 LCD with I2C integration.

6.3.5.3 Server

The server of the system is designed with customizable functional APIs and a DB to support the effective management and monitoring of solar energy system parameters. The DBs on the server are built using MongoDB software and include the following main components:

- **Information DB**
  - Stores instantaneous parameters sent from the monitoring unit.
  - Supports real-time display and monitoring on the user application.
- **Graph DB**
  - Stores all parameter data along with timestamps.
  - Serves the function of data storage and graph plotting for analysis on the user application.
- **User DB**
  - Stores individual user IDs and passwords.
  - Ensures security and supports the management of separate users via login functionality.
- **Device DB**
  - Stores detailed information of PV panels in the system.
- Supports the analysis of panel performance when integrating ML models.

The detailed structure of the components within the DBs on the server is presented in Table 6.7.

6.3.5.4 API functions for data communication

The IoT system uses JSON strings to communicate data between components, which include label-value pairs. In this context:

- **Label**: Defined according to the label of the parameter in the DB.
- **Value**: The data collected from the monitoring unit.

Each JSON string will also include a label code to differentiate the specific function that the server needs to perform. The functions built on the server and the communication objects are presented in Table 6.8.

Table 6.7   Structure of databases used on the server

Information database	Graph database	User database	Device database
Station name	Station name	User	Module ID
Authorization code	Module current	Password	Short-circuit current
Radiation intensity	Module voltage		Open-circuit voltage
Module current	Module power		Nominal power
Module voltage	Battery voltage		MPP voltage
Module power	Operating status		MPP current
Battery voltage	Fault type		Temperature coefficient
AC voltage	AC power		
AC current	Update hour		
AC power	Update date		
Power factor			
Module ID			
Operating status			
Fault type			

Table 6.8   List of functional APIS

No.	Function	Client
1	Update monitoring parameters	Monitoring unit
2	Create dataset (new station)	Admin
3	Register	Computer software
4	Log in	User application
5	Retrieve monitoring parameters	User application
6	Retrieve data for graph plotting	User application
7	Input photovoltaic panel information	Computer software
8	Input installation information	Computer software
9	Save information in the database	Computer software

Research and application of IoT technology in SE development 237

6.3.6 Mobile application

The mobile application is designed to display solar energy system monitoring information in an intuitive and user-friendly manner. This information is presented in both tabular form and graphs, allowing users to track and analyze data related to the operation of PV panels.

In addition to displaying information, the application also features user role-based access control to ensure that each user can only access the pages corresponding to their permissions. Specifically, information pages are divided based on specific monitoring stations, and each user is allowed to access the relevant stations.

6.3.6.1 User access rights

- **User1**: Only has access to information from Station 1.
- **User2**: Only has access to information from Station 2.
- **Admin**: Has access to all monitoring stations and can edit or manage system information.

6.3.6.2 Application interface flowchart

The interface flowchart of the application is shown in Figure 6.15. The application interface is organized into separate information pages for each monitoring station. Each page follows a common layout, but the user's access is restricted to the relevant information pages according to their role:

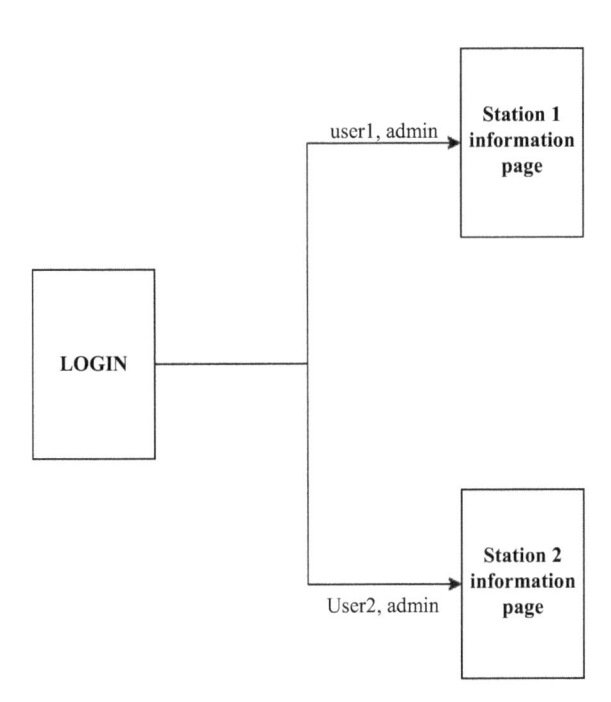

Figure 6.15 Flowchart of interface and access rights in the user application

1. **Main page**: Provides options for the user to choose "Station 1," "Station 2," or "User Management" (for Admin).
2. **Station 1 and station 2 information pages**: Displays data and parameters such as current, voltage, power of the panels and system, in both tabular and graphical formats.
3. **User management page (admin)**: Allows the Admin to add, delete, edit users, and manage access permissions.

This application simplifies the process of monitoring and managing the solar energy system while ensuring security and clear role-based access to information.

6.3.7 Mobile application and computer software interface

6.3.7.1 Mobile application: monitoring information interface

The mobile application interface is designed to provide real-time monitoring information for the solar energy system, while also displaying graphs that represent the power trends of the PV panel system over time. This enables users to gain a visual understanding of the system's performance and identify any abnormalities or potential issues during operation.

- **Real-time display**: Monitoring parameters such as current, voltage, and power are continuously updated and displayed directly on the user interface.
- **Trend graphs**: A section of the interface will display graphs (e.g., power vs. time), allowing users to easily track the changes in power production from the PV panels over a specified period. This feature not only helps identify variations by the hour, day, or week but also assists in analyzing performance trends.

The monitoring data will be presented in the form of line or bar graphs, with clearly marked time intervals, enabling users to observe fluctuations in the system over a selected timeframe.

6.3.7.2 Computer software: data management and backup functions

The computer software will support more in-depth management functions compared to the mobile application, aimed at making it easier for system administrators to monitor and manage the PV panel system. The main functions are as follows:

- **Entering PV panel information**: The software will feature a function that allows users (primarily admins) to input detailed information about the PV panel system into the system, such as:
  – Panel ID
  – Rated power
  – Parameters for voltage, current, and power
  – Information about the type of panel, operational status, and other technical specifications. This information will be stored in a DB on the server and can be edited or updated as changes occur.

- **Monitoring data backup**: The software will include a backup feature that allows data from monitoring devices to be uploaded to the server for long-term storage. This ensures that data is protected from loss and can be easily retrieved when needed. The backed-up data will include all monitored parameters such as current, voltage, power of the PV panels, and the system's operational status.

This backup functionality will support the automation of the backup process (on a periodic basis) and allow users to download or restore data as necessary.

6.3.7.3 System overview

- The mobile application will enable users to monitor and track the real-time status of the solar energy system easily, with trend graphs for power over time, providing a smooth and intuitive user experience.
- The computer software will offer more robust management features, enabling the input of PV panel information and the backup of monitoring data to protect information, facilitating long-term system maintenance and operation.

Together, these two systems will form a comprehensive monitoring solution, helping to manage and optimize the performance of the solar energy system.

6.3.7.4 Computer software

Requirements analysis
The functional requirements and design approach for the solar energy monitoring software are based on the features outlined in Table 6.9.

Login functionality
In the login dialog, the user enters their ID and corresponding password, along with the server address and port number. The application will send a login request to the server and receive a JSON string containing an "access" label and a "status" label. The "access" label indicates the login account's access rights, including admin privileges for management and user rights for monitoring individual nodes. The

Table 6.9 List of required functions in the software

Required function	Design method
Data transmission protocol between the application and server	Build a database, API, and JSON string for communication with the server.
Login, registration, and user authorization functionality	Build the interface, algorithm, API, and JSON string for communication with the server.
Display monitoring parameters	Create a monitoring interface, algorithm, and API.
Plotting function	Create a graph plotting interface, algorithm, and API.
System status check and alert function	Design check and alert functions based on available algorithms for system status.
Data storage function	Store data in .txt or Excel files.

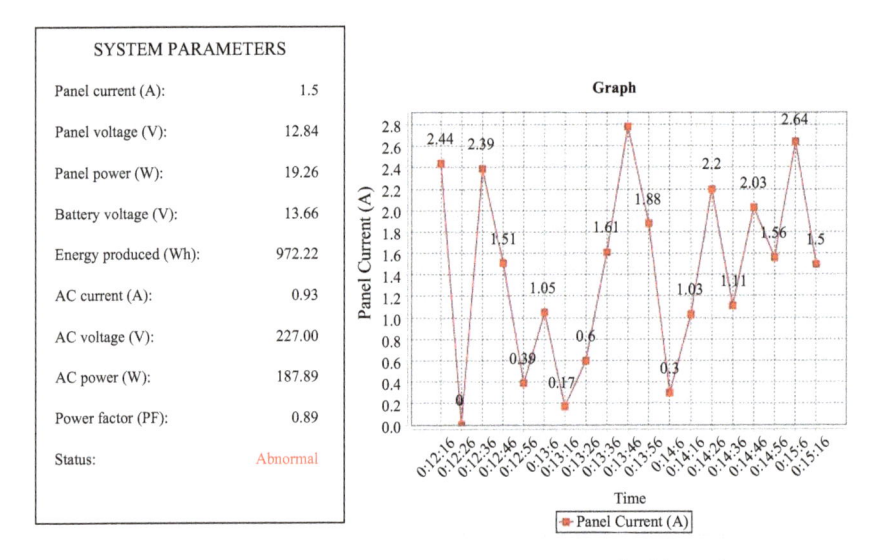

Figure 6.16   Administrator's monitoring dashboard

"status" label provides the server's response indicating whether the login process was successful or failed.

In addition to the SIGN IN button, which initiates the login request, there is a SIGN UP button used to begin the registration process and display the registration dialog box.

Monitoring function design

During the login process, the application will send a code to the server to retrieve the panel parameters, which will create two threads: a **data thread** (for retrieving and displaying parameter values) and a **chart thread** (for drawing graphs on the interface). Here, a thread refers to creating a data flow from the management software, using a socket to send requests to the server to receive information for displaying the parameters and generating the graph.

The management interface will differ depending on whether the user is accessing it with admin rights or as a regular user.

Figure 6.16 shows the management interface dashboard, which includes system parameters, graphs, installation buttons, parameter buttons, and a node list. This allows the manager to monitor user activities.

In Figure 6.17, the user interface of the solar energy monitoring application, some functions are restricted and reserved for system administrators. These functions are as follows:

1. **Install button**
   - When the "Install" button is pressed, the administrator gains access to the Installation Parameters dialog box. Here, the administrator can enter key installation parameters for the PV panels, including:

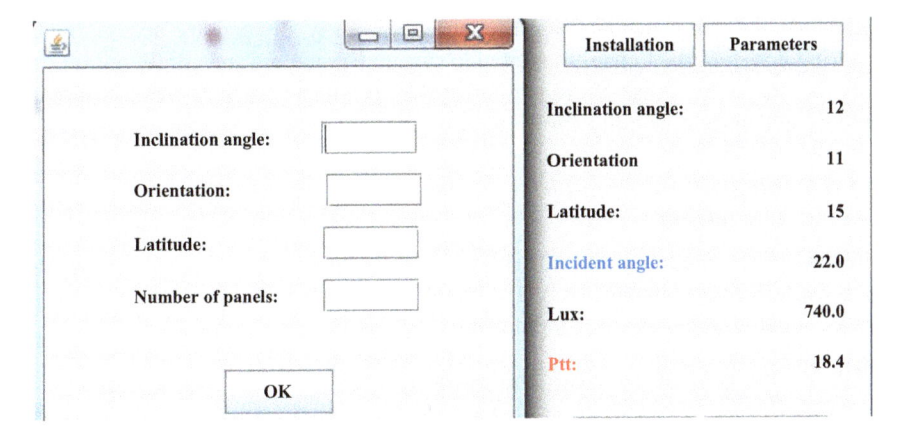

Figure 6.17   System parameters box

- **Tilt angle**: The angle at which the panels are installed relative to the ground, affecting the efficiency of solar energy collection.
- **Orientation**: The direction the panels face such as toward the South (in the Northern Hemisphere) to optimize energy collection.
- **Latitude**: The geographical latitude of the installation site, which helps calculate the precise solar light angle.
- **Number of panels**: Information on how many panels are installed in the system, which helps estimate the total power capacity the system can generate.

These parameters are crucial for optimizing solar system performance, so only the administrator has the rights to access and modify them.

2. **Change panel type button**
   - This button is located within the parameters section, allowing the user to select the type of panels used in the system. There are two options:
     - **Mono panels**: Monocrystalline PV panels, which typically offer higher efficiency.
     - **Poly panels**: Polycrystalline PV panels, which are cheaper but less efficient than mono panels.
   - The choice of panel type will impact the system's power output and efficiency, so only the administrator can change these settings to ensure system accuracy.
3. **Station selection list**
   - The station selection list is only available to the administrator. This ensures that only authorized personnel with management rights can access and modify information related to different stations within the monitoring system.

Thus, the functions on this interface are clearly divided by access rights, ensuring that important changes related to system parameters and configuration can only be made by those with administrative privileges, which helps maintain security and system stability for the solar energy monitoring system.

6.3.8 Integration of machine learning models into the IoT system

Integrating ML models into the IoT system aims to enhance the monitoring and fault classification capabilities of PV panels. The system involves two main processes added to the server for monitoring and classification purposes.

6.3.8.1 Process of creating the panel model

- This process is triggered when the user enters the panel parameters from the computer software. These parameters are then sent to the server through a predefined API.
- The panel parameters are stored in the panel parameter DB on the server. Upon receiving the information, the server will initiate a child process to simulate the panel data based on the obtained parameters.
- The panel simulation method is based on techniques presented in Section 6.3.8.1. This simulated data helps create an equivalent model of the panel, which predicts the system's performance and operational condition.
- The panel model is stored under a panel code name, and this code is updated in the DB for future reference during the next step—the operation status check.
- Once the panel model is created, the child process will stop.

6.3.8.2 Process of checking the panel's operational status

- After the panel model is created and stored, the system uses this model to check the panel's actual operational status. Operational parameters (such as current, voltage, power, and other data from the monitoring system) are sent from the monitoring units to the server.
- The server uses the created model to compare and analyze the real-time parameters with the simulation, aiming to detect faults or performance issues in the system.
- This process utilizes ML models to classify the panel's condition, helping provide timely alerts for faults or performance-impacting factors in the solar energy system.

Figure 6.18 illustrates the process of building the equivalent panel model on the server. This process allows the solar energy IoT monitoring system to perform monitoring, classification, and fault prediction functions automatically and more accurately, thanks to the analysis capabilities of ML models.

By integrating ML models into the IoT system, solar energy monitoring and maintenance become smarter, enabling users to easily track and detect issues early. This, in turn, enhances the operational efficiency of the system.

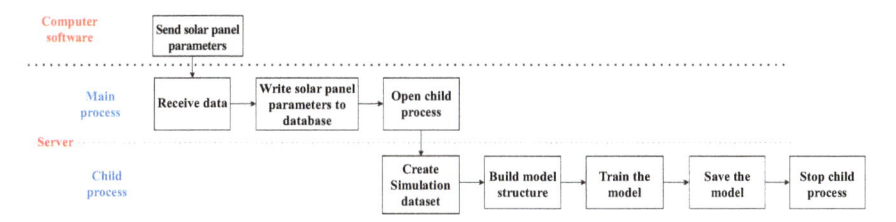

Figure 6.18 Process of building the equivalent solar panel model on the server

The IoT-based solar energy monitoring system has been fully developed with three main components: the monitoring unit, the server, and the user application, along with processes for data linking, transmission, and communication. Important operational parameters of the system, such as voltage, current, power of the panels, and environmental conditions (radiation intensity), are collected via electronic sensors and transmitted to the server.

The server acts as the central station, transmitting data between the monitoring unit and the user application, while also storing information in the necessary DBs. Additionally, the server integrates an ML model to assess and classify the operational status of the solar energy system, helping detect faults and optimize the performance of PV panels.

The user application on smartphones enables users to monitor the system's operational parameters, while the computer software not only provides monitoring functions but also supports management features such as daily data retrieval, modification of panel parameters used in the system, and other features to assist in the operation and maintenance process.

In conclusion, the developed IoT-based solar energy monitoring system tightly connects hardware, software, and data components, offering an effective solution for monitoring and managing the operation of the solar energy system.

6.3.9 *Experimental results*

6.3.9.1 Fault diagnosis process

After integrating ML models into the solar energy monitoring system, the DB is updated with information about the system's operational status. This enhances the system's ability to detect and classify faults during the monitoring of parameters such as current, voltage, and power.

Figure 6.19 illustrates the monitoring results of these parameters from the DB on December 28, 2021, from 3:00 PM to 12:00 AM the following day. From this chart, it is clearly visible that from 3:00 PM to around 5:00 PM, the system experienced a fault, as the actual voltage was lower than the calculated voltage. Additionally, the fault classification results fluctuated between two levels: 3 and 6, corresponding to a "wire-to-wire short circuit" and "other faults."

However, due to the system configuration, which includes two parallel panels, the "wire-to-wire" fault classification was inaccurate. This indicates that the system

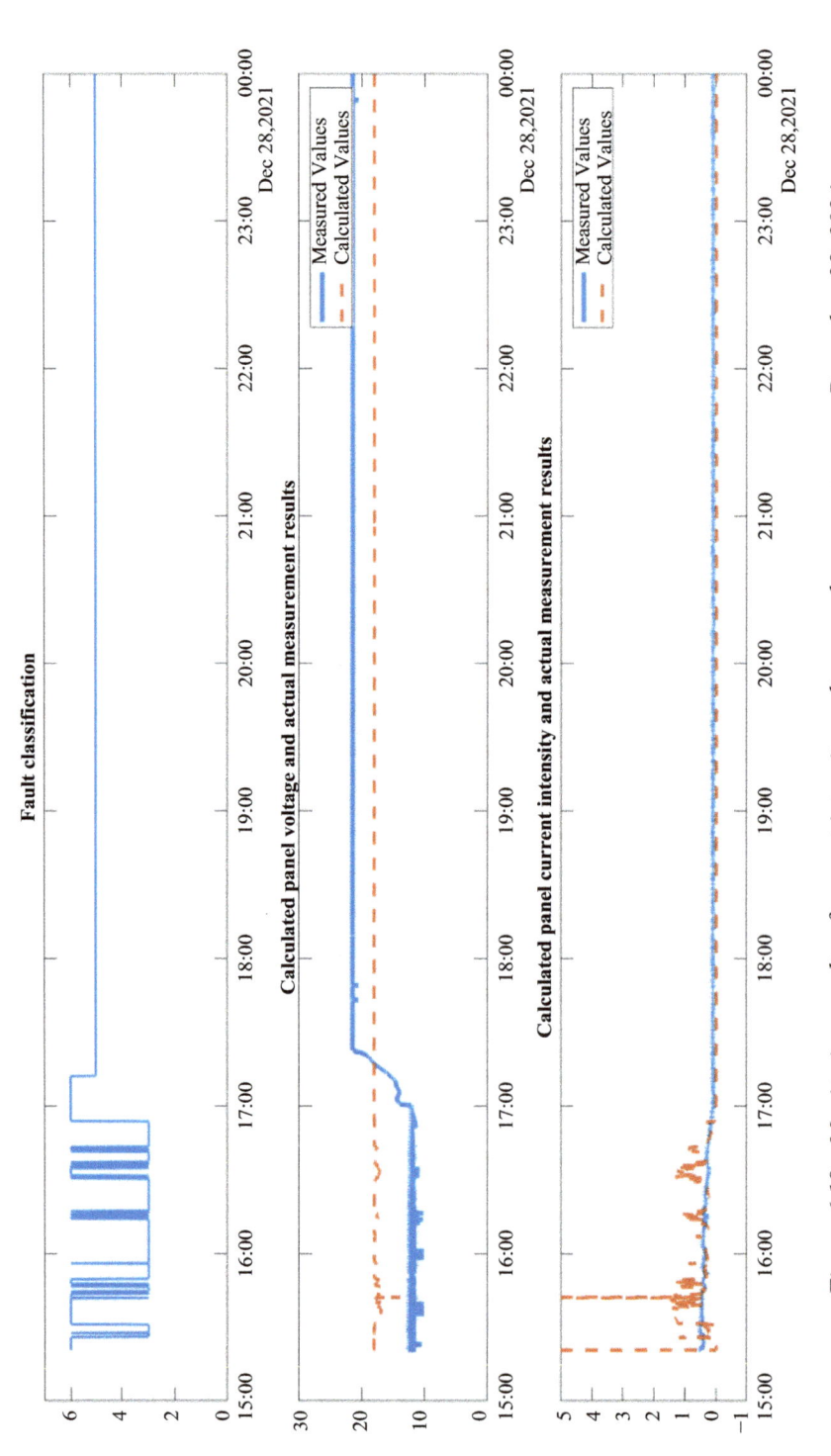

Figure 6.19 Monitoring results of current intensity, voltage, and system status on December 28, 2024

is more likely experiencing issues related to power degradation or the resistance of the output connection wires. These types of faults were not part of the research scope of the current study.

From 5:00 PM onward, the system's status was classified at level 5, corresponding to an "open-circuit" condition. This result was accurate because it reflects the absence of solar radiation during that time, leading to no power generation.

The fault diagnosis process of the solar system demonstrated the accuracy of fault classification using the ML model, while also highlighting potential issues that require further monitoring and analysis in future studies. This approach enables the system to automatically and accurately detect operational problems, helping to improve the overall performance and maintenance of the solar energy system.

6.3.9.2 Fault classification results

The fault classification results can be further examined through Figure 6.20, where they are compared with the normalized current and voltage values calculated at the MPP. It is observed that in certain sections, the fault classification identifies a "wire-to-wire short circuit" when the ratio $I/I_{mpp} >= 1$ and $V/V_{mpp} < 1$. The error in these cases may arise due to the calculated MPP current values at low irradiance levels. During the training process, the low irradiance regions were "relaxed" and adjusted to match high irradiance points. This is an issue that needs to be addressed to improve the system's accuracy.

Moreover, since the calculated current values turn out to be negative and very small, along with the fact that the open-circuit current is not 0 A, the normalized current also becomes negative, and this value is amplified beyond the range used by the fault classification model presented in the error reference section. Even though the classification still correctly identifies an "open-circuit" condition, this issue must be addressed to enhance the reliability of the system in future applications.

6.3.9.3 Key improvements

- **Irradiance-based calibration**: The model's sensitivity to low irradiance levels should be improved, ensuring that low irradiance points do not disproportionately affect the MPP calculations.
- **Handling negative current values**: Measures should be taken to prevent negative values from distorting the normalization process and affecting the classification accuracy.

These adjustments will help enhance the system's ability to classify faults correctly, especially in varied environmental conditions.

The IoT system was tested for operation from December 8–12, 2020, with the solar power system installed at a building in Ho Chi Minh City. The data collected by the monitoring device was transmitted to a server and stored in a pre-established DB. After integrating ML models and adding the capability to assess and classify the operational status of the system, the model was tested again on December 28, 2021.

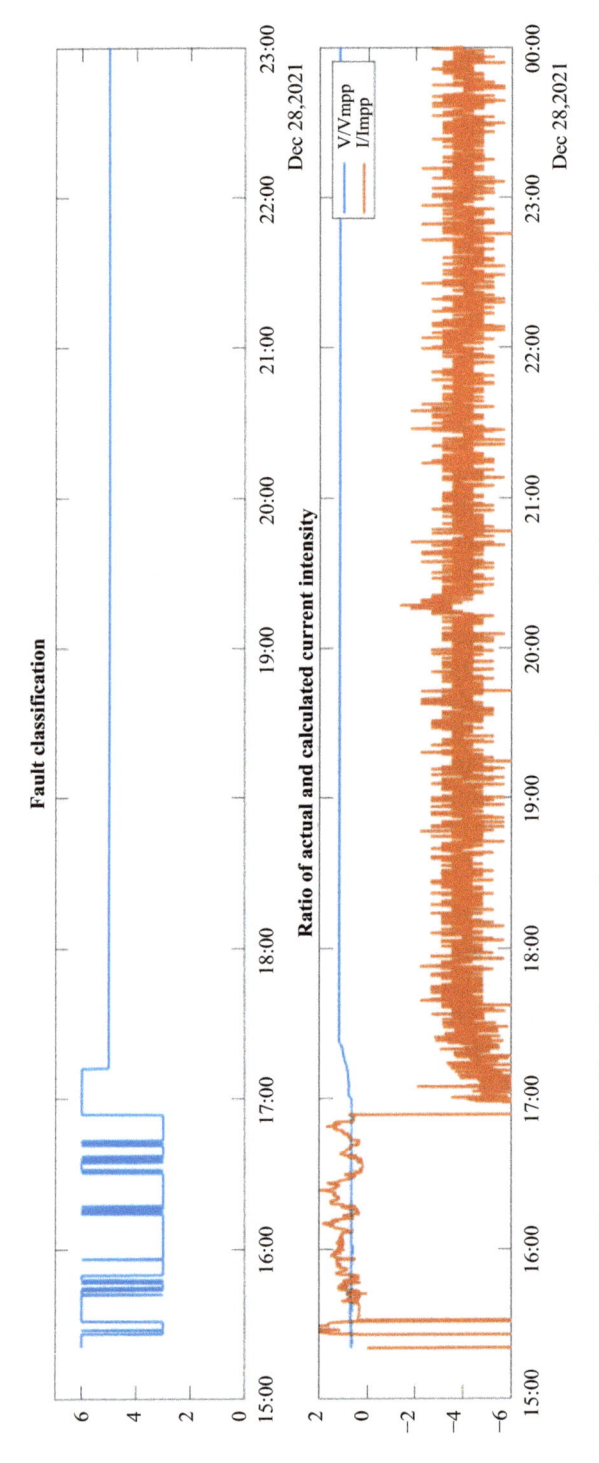

Figure 6.20 Classification results corresponding to normalized current intensity and voltage

6.3.10 Achievements

The IoT system for monitoring solar power parameters has been successfully developed and operates effectively with the function of alerting on the operational status of the PV panels. The system employs ML algorithms to simulate the panels and classify panel faults. The main components of the system include: a monitoring device for collecting solar power parameters, a server for data storage and communication between the monitoring device and the mobile application. This system allows users to remotely monitor solar power parameters in real time via a mobile app. Additionally, a DB has been set up to store the operational parameters of the solar power system on the server. From this DB, combined with simulated solar radiation calculations and panel models, the system can compute the power output at any given time, providing a basis for evaluating the system's performance and developing AI algorithms for monitoring and managing solar power systems.

The system has potential applications for enhancing the management capabilities of residential solar power systems in Vietnam, especially in supporting maintenance through the evaluation of panel performance. When combined with geographical data on radiation intensity provided by weather data centers, the system can assist in evaluating performance during the installation and operation of solar power systems.

6.3.11 Challenges and limitations

One of the system's major limitations is that during the testing phase, it was only deployed under local conditions. Therefore, the system's stability and load-bearing capacity have not been fully tested.

In practice, monitoring the parameters of the PV panels is carried out by inverters provided by the manufacturers. Furthermore, as the system's capacity increases, the voltage and current requirements for the sensors also rise, leading to increased costs for the monitoring system, which may make the system economically inefficient. As such, there needs to be a balance between the need for independent data and the economic viability of the system's implementation.

Additionally, the research is limited to common PV panel faults, with many other faults yet to be investigated. The ability to identify fault locations is also limited. To complete the analysis and fully determine the operational status of the PV panels, further research and the integration of other methods are needed.

6.3.12 Recommendations and development directions

To address the above challenges and limitations, several development and follow-up research directions are proposed, including:

- **Integration with existing solar power systems**: To avoid redundant functionality, it is essential to use market-available devices such as data collection nodes. However, this requires agreements between parties and the synchronization of data transmission processes between devices when connected to the server.

- **Combining with infrared fault detection methods for PV panels**: Infrared imaging techniques can identify fault locations within PV panel arrays. Integrating these methods with the system will enhance and complete the analysis of PV panel performance.
- **Research and application of new technologies**: The fields of AI, ML, and related industries are rapidly evolving. Large platforms such as AWS and Google have developed tools that assist in integrating and deploying ML models into computing systems. Applying these new technologies can improve the system's effectiveness and expand its monitoring and management capabilities.

Further reading

[1] Vietnam's Eighth National Power Development Plan (PDP VIII).
[2] Environmental Protection Law: Law No. 72/2020/QH14, signed on 17/11/2020.
[3] Law on Water Resources: Law No. 17/2012/QH13, signed on 21/6/2012, effective from 01/1/2013.
[4] Planning Law: Law No. 21/2017/QH14, passed by the 14th National Assembly on 24/11/2017.
[5] Amendment Law No. 28/2018/QH14: Amendments and supplements to several articles of 11 laws related to planning, passed on 15/6/2018.
[6] Amendment Law No. 35/2018/QH14: Amendments and supplements to several articles of 37 laws related to planning, passed on 20/11/2018.
[7] Decree No. 29/2011/NĐ-CP: Issued by the Government on 18/2/2011, regulating strategic environmental assessment, environmental impact assessment, and environmental protection commitments.
[8] Decree No. 23/2006/NĐ-CP: Issued by the Government on 03/3/2006, on the implementation of the Law on Forest Protection and Development.
[9] Decree No. 46/2012/NĐ-CP: Issued by the Prime Minister on 22/5/2012, amending and supplementing several articles of:
[10] Decree No. 35/2003/NĐ-CP (dated 04/4/2003) detailing the implementation of several articles of the Law on Fire Prevention and Fighting.
[11] Decree No. 130/2006/NĐ-CP (dated 08/11/2006) on compulsory fire and explosion insurance policies.
[12] Decree No. 32/2006/NĐ-CP (2006): Application of the Sorensen Coefficient (S) for comparing species composition in the study area with neighboring areas.
[13] Decree No. 18/2015/NĐ-CP (Article 12) and Circular No. 27/2015/TT-BTNMT (Article 7): Provisions on consultation activities during the implementation of the Environmental and Social Impact Assessment (ESIA).
[14] Circular No. 27/2015/TT-BTNMT: Issued on 29/5/2015 by the Ministry of Natural Resources and Environment (MONRE) on Strategic Environmental

Assessment, Environmental Impact Assessment, and Environmental Protection Plans (Appendix 2.3, Chapters 1, 3, 4, and 5).
[15] Circular No. 26/2011/TT-BTNMT: Issued on 18/7/2011 by MONRE, detailing certain provisions of Decree No. 29/2011/NĐ-CP on strategic environmental assessment, environmental impact assessment, and environmental protection commitments.
[16] Circular No. 12/2011/TT-BTNMT: Issued on 14/4/2011 by MONRE, on the management of hazardous waste.
[17] Directive No. 08/2006/CT-TTg: Issued by the Government on 08/3/2006, on urgent measures to prevent illegal deforestation, forest burning, and unauthorized logging.
[18] Ngo Dang Luu, Nguyen Hung, Nguyen Anh Tam, and Long D. Nguyen, (2021), "Wireless Power Transfer simulation for EV", Ministry of Culture, Sports and Tourism.
[19] Ngo Dang Luu, Nguyen Hung, and Long D. Nguyen, (2021), "Power flow calculation of the electric power system by MATLAB", Ministry of Culture, Sports and Tourism.
[20] Ngo Dang Luu, Nguyen Hung, Nguyen Anh Tam, and Long D. Nguyen, (2022), "Distance Relay Protection in AC Microgrid", Ministry of Culture, Sports and Tourism.
[21] Ngo Dang Luu, Nguyen Hung, Nguyen Anh Tam, and Long D. Nguyen, (2022), "Analysis of Solar Photovoltaic System Sunlight Shadowing", Ministry of Culture, Sports and Tourism.
[22] Ngo Dang Luu, Nguyen Hung, Le Anh Duc and Long D. Nguyen, (2022), "Estimate Frequency Response at the Command Line", Ministry of Culture, Sports and Tourism.

Chapter 7
Impacts of policies and mechanisms on renewable energy development

7.1 Overview of Vietnam's power system: structure and challenges in ensuring adequate, stable, and safe electricity supply

7.1.1 Structure of Vietnam's power system

Vietnam's power system has been developed to ensure energy security and meet the increasing electricity demands of the economy and society. The system encompasses a diverse range of energy sources, including:

- Renewable energy: Wind power (onshore and offshore), solar power, biomass, and hydropower.
- Fossil energy: Coal-fired power plants, gas-fired power plants.
- Emerging energy sources: Hydrogen, green ammonia, and energy storage technologies like pumped-storage hydropower and battery storage.

Power generation development is strategically distributed across regions to optimize local energy potential, reduce transmission losses, and ensure stable electricity supply.

7.1.2 Solar power

Vietnam is focusing on the following strategies to enhance its power system:

- Diversifying power sources: Enhancing energy autonomy and reducing reliance on imported fuels.
- Prioritizing renewable energy: Accelerating the development of wind, solar, and biomass power to increase the share of clean energy in total capacity.
- Developing new energy and storage solutions: Promoting hydrogen production, green ammonia, and deploying energy storage technologies as costs become competitive.
- Energy transition: Gradually reducing the share of coal-fired power, replacing it with biomass and LNG, and aiming to eliminate coal use by 2050.
- Infrastructure investment: Developing synchronized infrastructure, including grid systems and LNG import facilities, to optimize capacity allocation and system operation.

7.1.3 Challenges in ensuring electricity supply

Despite significant progress in power generation development, Vietnam faces several challenges:

- Rapidly increasing electricity demand: The fast-growing economy requires ever-larger electricity capacity.
- Infrastructure limitations: The existing grid system needs upgrades to accommodate and transmit renewable energy sources.
- Dependence on imported fuels: Reliance on imported fuels like LNG puts pressure on costs and energy security.
- Regional imbalances: Uneven distribution of power sources across regions, particularly in the North, risks electricity shortages during peak periods.
- Environmental impacts: Power development, especially hydropower and thermal power, must consider forest conservation, water resources, and emissions reduction.

7.1.4 Strategic solutions

To overcome these challenges, Vietnam needs to:

- Increase investment in renewable energy and energy storage technologies.
- Promote rooftop solar power and self-consumption models.
- Develop new power sources for both domestic consumption and export.
- Upgrade and expand the grid system.
- Accelerate the transition from fossil fuels to clean and sustainable energy sources.

Vietnam is undergoing a critical energy transition, with substantial potential for renewable energy and supportive development policies to ensure stable and sustainable electricity supply in the future.

7.2 Vietnam's regulations on grid connection for wind and solar power projects: capacity, voltage, and frequency requirements

7.2.1 Regulations on connecting wind and solar power projects to the grid

7.2.1.1 Wind power

Under the 2004 Electricity Law, approved by the National Assembly, power plants are allowed to connect to the national grid. Chapter VI outlines the rights and responsibilities of electricity companies and consumers. Article 39 states that power plant operators are entitled to connect their plants to the national grid, provided they meet technical and regulatory conditions for grid connection.

Circular No. 32/2010/TT-BCT, issued by the Ministry of Industry and Trade on July 30, 2010, details conditions and procedures for grid connection to the distribution system. Chapter 5, Section 5.1 defines the grid connection point, and

Section 5.2 specifies connection conditions. Decision No. 18/2008/QD-BCT outlines grid connection requirements for small renewable energy power plants.

Accordingly, the connection point is the location where the seller's electrical system connects to the buyer's distribution system, agreed upon by both parties and aligned with the approved power development plan.

For cases where grid connection to a 110 kV voltage level or unsuitable connection to the approved power development plan is proposed, the investor must submit a proposal to the provincial People's Committee for review by the Electricity Regulatory Authority of Vietnam (ERAV) and approval from the Ministry of Industry and Trade.

The seller is responsible for investing in, operating, and maintaining the grid and transformer system from the power plant to the buyer's grid connection point in compliance with standards and legal regulations.

Grid connection procedure
To connect to the grid, an agreement must be reached with EVN (the grid operator and electricity buyer). Investors must submit:
A grid connection application dossier.
A schematic diagram of key electrical equipment beyond the connection point.
Technical documentation on the equipment to be connected, project completion schedule, and economic-technical data of the connection project.

EVN will review and approve the connection plan within 30 days upon receiving valid documents. If approved, an agreement is formalized within 10 working days following EVN's leadership approval.

Draft regulations on wind power projects
Draft regulations provide favorable terms for wind power investors regarding grid connection. EVN is responsible for investing in the transmission line from the wind power plant's substation to the national grid. Additionally, EVN is required to prioritize mobilization and purchase of all electricity generated by grid-connected wind power projects.

7.2.1.2 Solar power
Vietnam has not yet issued detailed guidelines on technical requirements for grid connection of solar power plants. Current criteria are based on:
Circular No. 39/2015/TT-BCT on Distribution Power Systems (issued November 18, 2015).
Circular No. 25/2016/TT-BCT on Transmission Power Systems (issued November 30, 2016).

7.2.1.3 Technical requirements for medium-voltage solar power plants
Reactive power and voltage adjustment capability
- Adjust reactive power continuously within a power factor range of 0.95 (leading/lagging) at the connection point when generating at least 20% of rated active power, with grid voltage within the normal operational range.

254 *Clean energy in South-East Asia*

- Reduce reactive power generation or absorption proportionally for output below 20% of rated active power.
- Maintain voltage adjustment accuracy within ±0.5% of the nominal voltage, completing the adjustment within 2 min.

Voltage range compliance

Solar power plants must operate within specified voltage ranges at the connection point, maintaining grid stability during transient conditions.

Further sections can be added to complete specific voltage and frequency-related requirements.

$$\text{Meanwhile}: T_{\min} = 4 \times U - 0,6 \tag{7.1}$$

$T_{\min}$ (s): Minimum time duration for maintaining power generation.
U (pu): Actual voltage at the point of connection, expressed in per-unit (pu).

Voltage requirements for solar power plants

- For voltages ranging from 0.9 pu to below 1.1 pu, the solar power plant must continuously maintain power generation.
- For voltages ranging from 1.1 pu to below 1.15 pu, the solar power plant must sustain power generation for a duration of 3 s.
- For voltages ranging from 1.15 pu to below 1.2 pu, the solar power plant must sustain power generation for a duration of 0.5 s.

Harmonic distortion and voltage imbalance

The solar power plant must ensure that the negative sequence component of the phase voltage at the point of connection does not exceed 1% of the nominal voltage.

The solar power plant must be capable of withstanding a negative sequence component of the phase voltage at the point of connection up to 3% of the nominal voltage for 110 kV voltage levels or up to 5% of the nominal voltage for voltage levels below 110 kV.

Voltage flicker requirements

Under normal operating conditions, the voltage flicker level at all points of connection must not exceed the specified limits, as shown in Table 7.1.

Meanwhile

- Short-term voltage flicker level (P_{st}): The measured value over a 10-min interval using standard measuring equipment as specified in IEC 868. P_{st} 95%

Table 7.1 *Voltage flicker levels for distribution grids*

Voltage level	Permissible flicker levels
110 kV	$P_{st95\%} = 0.80$
	$P_{lt95\%} = 0.60$
Medium and low voltage	$P_{st95\%} = 1.00$
	$P_{lt95\%} = 0.80$

Impacts of policies and mechanisms on renewable energy development 255

represents the threshold value of P_{st} such that, during 95% of the measurement period (at least one week) and at 95% of the measurement locations, the P_{st} does not exceed this value.
- Long-term voltage flicker level (P_{lt}): Calculated from 12 consecutive P_{st} measurements (over a 2-h interval) using the following formula:

$$P_{lt} = \sqrt{\frac{1}{2} * \sum_{j=1}^{12} P_{stj}^2} \qquad (7.2)$$

P_{lt} *95% definition*
P_{lt} 95% is the threshold value of P_{lt} such that, over 95% of the measurement time (at least one week) and 95% of the measurement locations, P_{lt} does not exceed this value.

Requirements for short-circuit current and fault clearing time
The maximum allowable short-circuit current and fault clearing time are specified in Table 7.2.

7.2.1.4 Technical requirements for solar power plants connected to the transmission grid

Requirements for reactive power and voltage regulation capability

- Solar power plants must have the capability to regulate reactive power and voltage as follows:
  – If the active power output of the plant is greater than or equal to 20% of the rated active power and the voltage is within the normal operating range, the plant must be able to continuously adjust reactive power within a power factor range of 0.95 (corresponding to reactive power generation) to 0.95 (corresponding to reactive power absorption) at the point of connection for the rated power output.
  – If the active power output of the plant is less than 20% of the rated power, the plant may reduce its ability to absorb or generate reactive power in accordance with the characteristics of the generating units.
  – If the voltage at the point of connection is within ±10% of the nominal voltage, the plant must be capable of adjusting the voltage at the point of connection with a deviation not exceeding ±0.5% of the nominal voltage

Table 7.2 *Maximum permissible short-circuit current and fault clearing time*

Voltage level	Maximum short-circuit current	Fault clearing time	Equipment withstand duration
22–35 kV	25	500	1
110 kV	31.5	150	1

(relative to the set voltage value) across the entire allowable operating range of the generating units, completing the adjustment within 2 min.
- If the voltage at the point of connection fluctuates beyond ±10% of the nominal voltage, the plant must be capable of adjusting reactive power at a minimum rate of 2% of the rated reactive power for every 1% voltage variation at the point of connection.

- Voltage range and operational time requirements:
  - The solar power plant, at all times while connected to the grid, must be capable of maintaining power generation corresponding to the voltage range at the point of connection for the following durations:
  - For voltage below 0.3 pu, the minimum duration is 0.15 s.
  - For voltage between 0.3 pu and below 0.9 pu, the minimum duration is determined by the following formula:

$$\text{Meanwhile}: T_{min} = 4 \times U - 0,6 \tag{7.3}$$

T_{min} (s): Minimum duration for maintaining power generation.
U (pu): Actual voltage at the point of connection, expressed in per-unit (pu).

- Voltage range and operational time requirements:
  - For voltages from 0.9 pu to below 1.1 pu, the solar power plant must maintain continuous power generation.
  - For voltages from 1.1 pu to below 1.15 pu, the solar power plant must sustain power generation for 3 s.
  - For voltages from 1.15 pu to below 1.2 pu, the solar power plant must sustain power generation for 0.5 s.

Requirements for negative sequence components

The solar power plant must ensure that the negative sequence component of the phase voltage at the point of connection does not exceed 1% of the nominal voltage.

The solar power plant must be capable of withstanding a negative sequence component of the phase voltage at the point of connection up to 3% of the nominal voltage for voltage levels of 220 kV and above.

Voltage flicker requirements

The voltage flicker caused by the solar power plant at the point of connection must not exceed the specified values as shown in Table 7.3.

Table 7.3 Voltage flicker levels for transmission grids

Voltage level	$P_{lt95\%}$	$P_{st95\%}$
220 kV, 500 kV	0.6	0.8

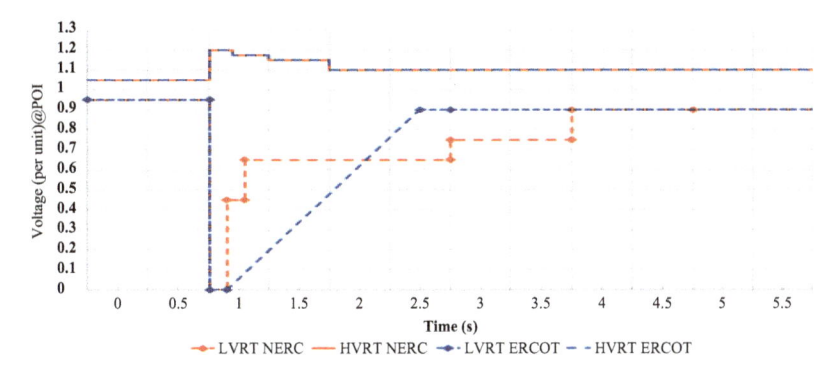

Impacts of policies and mechanisms on renewable energy development 257

Figure 7.1 Graph representing voltage according to the VRT standards of ERCOT and NERC

P_{lt}95%: The threshold value of P_{lt} such that, during 95% of the measurement period (at least one week) and at 95% of the measurement locations, P_{lt} does not exceed this value.

P_{st}95%: The threshold value of P_{st} such that, during 95% of the measurement period (at least one week) and at 95% of the measurement locations, P_{st} does not exceed this value.

Introduction to Voltage Ride-Through (VRT) capability of inverters/turbines
As defined by NERC/ERCOT (North American Electric Reliability Corporation/ Electric Reliability Council of Texas), VRT capability refers to the ability of inverters or turbines to remain connected to the grid, continue supplying active and reactive power, and meet grid requirements while the voltage at the point of connection fluctuates according to specified characteristic curves.

According to the Vietnamese definition, VRT is the capability of wind and solar power plants to maintain power generation at all times while connected to the grid, corresponding to the voltage range at the point of connection, for the duration specified in Figure 7.1.

Voltage ride-through capability is divided into two smaller capabilities: low-voltage ride-through (LVRT) and high-voltage ride-through (HVRT). Detailed information on these VRT capabilities will be presented in the following sections.

7.2.1.5 Analysis of HVRT and LVRT by ERCOT and NERC

ERCOT and NERC define the normal operating voltage range of the system at the point of connection as 0.95–1.05 pu, as shown in Figure 7.2. The system is in HVRT mode if the voltage at the point of connection is no lower than 1.1 pu. Similarly, the system is in LVRT mode if the voltage at the point of connection does not exceed 0.9 pu.

258 *Clean energy in South-East Asia*

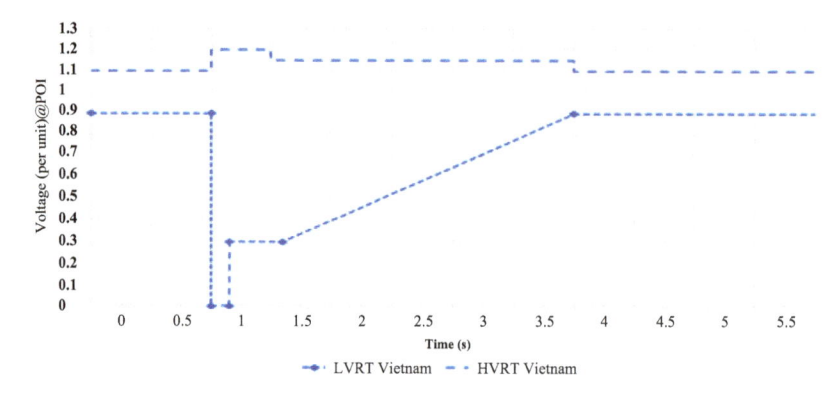

Figure 7.2 Graph representing voltage according to the VRT standards of Vietnam

When the voltage at the point of connection exceeds the normal operating voltage, i.e., 1.05 pu as shown in Figure 7.2, the inverter and turbine must detect the voltage change and comply with the following HVRT requirements:

- **Reactive power consumption during high voltage transients**: The inverter and turbine must consume reactive power to maintain the voltage at the point of connection at 1.1 pu. In this case, the inverter and turbine typically consume maximum reactive power. Note that the reactive power before the HVRT mode should be greater than the reactive power in the HVRT mode, i.e., $Q1 > Q2$.
- **Maintain power consumption according to design power**: The inverter and turbine must not enter a momentary cessation mode (no immediate cessation of power generation).

When the voltage at the point of connection is lower than the normal operating voltage, i.e., 0.95 pu, the inverter and turbine must detect the voltage change and comply with the following LVRT requirements:

- **Reactive power generation during low voltage transients**: The inverter and turbine must generate reactive power to maintain the voltage at the point of connection at 0.9 pu. In this case, the inverter and turbine typically need to generate maximum reactive power. Note that the reactive power before entering LVRT mode must be smaller than the reactive power in LVRT mode, i.e., $Q1 < Q2$.
- Power consumption must not decrease by more than 10% of the design power.
- No immediate cessation of power generation: The inverter and turbine must avoid entering a momentary cessation mode (no immediate cessation of power generation).

7.2.1.6 Analysis of LVRT and HVRT in Vietnam

According to Article 42 of Circular No. 25/2016/TT-BCT and Circular No. 30/2019/TT-BCT issued by the Ministry of Industry and Trade, wind and solar

Impacts of policies and mechanisms on renewable energy development 259

power plants, at all times when connected to the grid, must have the capability to maintain operation and generation corresponding to the voltage range at the point of connection for the required duration, as shown in Figure 3.2.

For voltages below 0.3 pu, the minimum duration of operation is 0.15 s.

For voltages from 0.3 pu to below 0.9 pu, the minimum duration is calculated using the following formula:

$$T_{min} = 4 \times U - 0.6 \tag{7.4}$$

where T_{min} (s) represents the minimum duration for maintaining power generation, and U (pu) is the actual voltage at the connection point, measured in per-unit (pu) values.

- Voltage from 0.9 pu to below 1.1 pu: Wind and solar power plants must maintain continuous operation and generation.
- Voltage from 1.1 pu to below 1.15 pu: Wind and solar power plants must maintain operation and generation for at least 3 s.
- Voltage from 1.15 pu to below 1.2 pu: Wind and solar power plants must maintain operation and generation for at least 0.5 s.

Thus, the current HVRT and LVRT requirements in Vietnam focus on the grid connection time of inverters and turbines, without specific requirements for active power and reactive power. As we know, HVRT and LVRT requirements are intended to provide grid support from wind and solar power plants during high and low voltage events. Therefore, HVRT and LVRT requirements must also have specific reactive power provisions for inverters and turbines, similar to the standards of ERCOT and NERC.

Additionally, the working voltage range during HVRT and LVRT in Vietnam is currently defined as 0.9–1.1 pu. This is inappropriate because inverters and turbines cannot distinguish between normal operating conditions and VRT modes. Moreover, the normal operating voltage range defined in Circular No. 30/2019/TT-BCT differs from the range specified in the National Standard TCVN 12230:2018. In this standard, the normal operating voltage is 0.95–1.05 pu.

7.2.1.7 Requirements for solar power systems connected to low-voltage distribution grids

- **Connection power**
    - The total installed capacity of the solar power system connected to the low-voltage side of the substation must not exceed the installed capacity of the substation.
    - Solar power systems with an installed capacity of 20 kWp or less can be connected to a single-phase or three-phase grid as agreed with the distribution company and the electricity retail company.
    - Solar power systems with an installed capacity of 20 kWp or more must be connected to a three-phase grid.

- At all times when connected to the grid, solar power systems must have the ability to maintain power generation for a minimum duration corresponding to the operating frequency ranges as specified in Table 7.4.

When the system frequency exceeds 50.5 Hz, solar power systems with a capacity of 20 kWp or more must reduce the active power output according to the following formula:

$$\Delta P = 20 \times Pm \times \frac{50.5 - f_m}{50} \quad (7.5)$$

where

ΔP: The reduction in active power output (MW);
P_m: The active power output corresponding to the time before the power reduction is implemented (MW);
f_n: The system frequency before the power reduction is implemented (Hz).

- The solar power system must have the ability to continuously operate and generate power within the voltage ranges at the connection point as specified in Table 7.5.
- The solar power system connected to the low-voltage grid must not inject reactive power into the grid and must operate in reactive power consumption mode with a power factor (cosφ) greater than 0.98.
- The solar power system must not cause direct current (DC) injection into the distribution grid that exceeds 0.5% of the rated current at the connection point.
- The solar power system connected to the low-voltage grid must comply with the voltage, phase balance, harmonic distortion, voltage flicker, and grounding requirements specified in Articles 5, 6, 7, 8, and 10 of this Circular.

Table 7.4 Minimum operating time for power generation corresponding to system frequency ranges

System frequency range	Minimum operating time
48–49 Hz	30 min
49–51 Hz	Continuous operation
51–51.5 Hz	30 min

Table 7.5 Minimum operating time for power generation corresponding to voltage ranges at the point of connection

Voltage at point of connection	Minimum operating time
Less than 50% of nominal voltage	Not required
50–80% of nominal voltage	2 s
85–110% of nominal voltage	Continuous operation
110–120% of nominal voltage	2 s
Greater than 120% of nominal voltage	Not required

Impacts of policies and mechanisms on renewable energy development 261

- The solar power system must be equipped with protection devices to ensure the following requirements:
  - Automatically disconnect from the distribution grid when an internal fault occurs within the solar power system;
  - Automatically disconnect when a loss of power from the distribution grid occurs and not feed power into the grid when the distribution grid is down;
  - Not automatically reconnect to the grid unless the following conditions are met:
- The grid frequency must be maintained within the range of 48–51 Hz for at least 60 s;
- The voltage of all phases at the connection point must be maintained within 85–110% of the rated voltage for at least 60 s.
  - For a solar power system connected to a three-phase low-voltage grid, the customer requesting connection must agree on and align the protection system requirements with the distribution power company, which must include at least the protections specified in Points a, b, and c of this Clause, overvoltage protection, undervoltage protection, and frequency protection.

7.2.2 Simulation of the impact of renewable energy systems on the operating parameters of the neighboring grid at the connection point

7.2.2.1 Introduction to PSS/ADEPT software

PSS/ADEPT (power system simulator/advanced distribution engineering productivity tool) is a utility software for simulating electrical systems and analyzing power grids with the following functions:

1. Load flow analysis: Distribution of power within the network.
2. Fault analysis: Short-circuit calculation at one or multiple points (fault, fault all analysis).
3. Motor starting analysis: Analyzing the startup of motors.
4. CAPO optimization: Optimization for the installation of fixed and adjustable compensation capacitors.
5. Harmonic analysis: Analyzing harmonic distortions within the system.
6. Protective coordination: Ensuring the coordination of protection devices within the grid.
7. TOPO analysis: Optimal opening point analysis.
8. Grid reliability analysis distribution reliability analysis [DRA]: Analyzing the reliability of the power grid.

In the current electrical grid calculation and analysis software, Shaw Power Technologies, Inc.'s PSS/ADEPT software is widely used (Figure 7.3). Each version, depending on the user's requirements, is sold with a hardware key for use on a single machine or networked machines. The version for single machines, along with the hardware key, can only run on one computer.

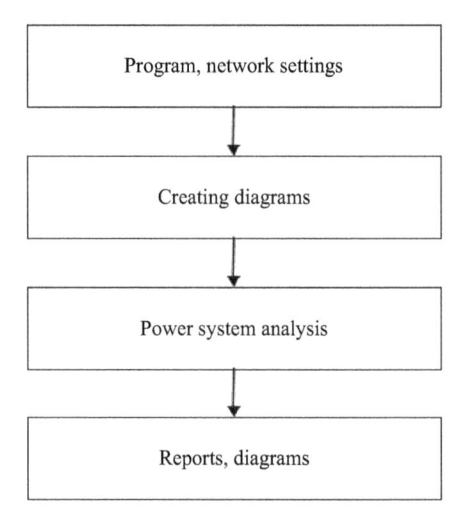

```
┌─────────────────────────┐
│ Program, network settings│
└───────────┬─────────────┘
            ▼
┌─────────────────────────┐
│    Creating diagrams    │
└───────────┬─────────────┘
            ▼
┌─────────────────────────┐
│  Power system analysis  │
└───────────┬─────────────┘
            ▼
┌─────────────────────────┐
│    Reports, diagrams    │
└─────────────────────────┘
```

Figure 7.3 Implementation diagram of PSS/ADEPT

PSS/ADEPT software (Shaw Power Technologies, Inc.) is a powerful tool for electrical grid analysis and calculation, applicable to both high-voltage and low-voltage power grids. It can handle an unlimited number of nodes, making it highly suitable for widespread use in power companies.

PSS/ADEPT is developed for engineers and technical staff in the electricity industry. It is used as a tool for designing and analyzing distribution grids. PSS/ADEPT also allows users to design, edit, and analyze grid diagrams and models in a highly visual manner using a graphical interface with an unlimited number of nodes.

7.2.2.2 Simulating the distribution grid on PSS/ADEPT software

Setting parameters for the grid network: Before creating diagrams, conducting analysis, or calculating a project, the parameters must be set. PSS/ADEPT allows users to independently configure parameters based on user profiles. The Construction Dictionary (PTI.CON) library in PSS/ADEPT is an ASCII-formatted file that provides data for the system, including impedance, wire specifications, and transformer parameters.

Users can also create customized library files for wires, transformers, etc., suited for the electricity grid of Vietnam. These files, with a specific extension, can be drafted using any word processing application such as Word, Notepad, and WordPad.

7.2.2.3 Simulating a real project

According to the project plan of the investor, the Dam An Khe Solar Power Plant officially connected to the national grid in 2018 with a generating capacity of

8.5 MW, and its capacity was increased to 43 MW in 2019. Therefore, the evaluation of the electrical losses in the grid, considering the participation of the Dam An Khe Solar Power Plant, is calculated using the electrical source and load forecast data for the area in 2018 and 2019.

We will use **PSS/ADEPT 5.0** software to simulate the impact of the Dam An Khe Solar Power Plant on the electricity grid of Duc Pho District and the 110 kV grid in Quang Ngai Province.

Phase 1: Evaluating the impact of the Dam An Khe Solar Power Plant with a generating capacity of 8.5 MW on the distribution grid of Duc Pho District. Aside from the Dam An Khe Solar Power Plant, there are no other power sources in the Duc Pho District. The area is supplied by the 110/22 kV Duc Pho substation with a transformer capacity of 25 MVA.

Phase 2: Evaluating the impact of the Dam An Khe Solar Power Plant with a generating capacity of 43 MW on the 110 kV grid of Quang Ngai Province.

7.2.2.4 Status of the Quang Ngai Province electrical grid

500 kV Grid: Quang Ngai Province is primarily powered by the national grid through the 500 kV Doc Soi substation located in the province. The 500 kV Da Nang–Doc Soi–Pleiku transmission line, with a length of 298 km and ACSR-4 × 330 conductors, supplies power to the 500 kV Doc Soi 500/220 kV—450 MVA substation, with a 90.3 km section running through Quang Ngai Province.

220 kV Grid: Currently, Quang Ngai Province has three 220 kV substations: the Doc Soi 220/110 kV substation (63+125 MVA), the Dung Quat 220/110 kV substation (125 MVA), and the Quang Ngai 220/110 kV substation (1×125 MVA). These provide the main electricity supply to the 110 kV substations in the area, and also connect to the 110 kV substations in Tam Ky and Ky Ha in Quang Nam Province, and the Hoai Nhon substation in Binh Dinh Province.

As for the 220 kV transmission lines, the province has the following:

- The two-circuit 220 kV Doc Soi–Son Ha line (operating at 110 kV) with a length of 47 km, receiving 110 kV power from the Dakdrinh Hydropower Plant, connecting to the 110 kV busbar at the Doc Soi substation.
- The double-circuit 220 kV line connecting Doc Soi–Tam Ky, with the Tam Ky 220 kV substation linking to the Song Tranh Hydropower Plant and connecting to the 500 kV Da Nang substation.

110 kV Grid: Quang Ngai Province receives electricity from the Central Region's grid through nine 110 kV substations with a total capacity of 345 MVA, including substations such as:

- Doc Soi 500 kV substation (450 MVA)
- Dakdrinh Hydropower (2×62.5 MW)
- Dung Quat Oil Refinery power plant (108 MW, often generating 50–60 MW)
- Quang Ngai City's diesel power plants (2×2,100 kW)
- Other power sources such as Dakdrinh Hydropower (125 MW), Ca Du Hydropower (2.6 MW), Song Rieng Hydropower (2.9 MW), Ha Nang Hydropower (11 MW), and Nuoc Trong Hydropower (16.5 MW).

7.2.2.5 Status of the Duc Pho District electrical grid

Currently, Duc Pho District is mainly supplied by the 110/22 kV Duc Pho substation and the 110/22 kV Mo Duc substation. The 22 kV feeder lines currently meet the load demand of the district and surrounding areas. By 2025, the 110/22 kV Pho Minh substation will require investment and improvements, including the construction of new transmission lines as follows:

- Upgrading the 35 kV Mo Duc–Ba To line shared with the newly built 22 kV line (feeder 471 Mo Duc), supplying power to the Pho Phong Industrial Zone, with a conductor size of AC150 and a length of 7 km, connected to feeder 478 Mo Duc.
- Upgrading the conductor size from 3M-50 to 3ACKP120 over 15 km from post 01/TRC to 136/XT 471/110 kV Duc Pho, supplying the areas of Pho Khanh, Pho Cuong, and TT Duc Pho, connected with feeders 477 Duc Pho and 476, 478 Mo Duc.
- Upgrading the conductor size from 3M-35 to 3ACKP-150 over 3.15 km from Pho Thuan to the Pho Chau 5 substation on feeder 473/110 kV Duc Pho, supplying the areas of Pho Thanh, Pho Chau, connected with feeders 472 and 475 at the 110 kV Duc Pho substation.
- New 22 kV feeder from the 110 kV Duc Pho substation, with a conductor size of 3AC150, 3 km in length, connected to the grid in Binh Dinh Province and the feeder 472 Duc Pho.
- New 22 kV feeders from the Pho Minh 110/22 kV substation, connected to feeders 471 and 477 at the existing Duc Pho substation, to reduce the supply radius for these feeders.
- By 2025, additional branch improvements will be necessary in Duc Pho District to ensure a safe and reliable power supply for the communes.

Duc Pho receives electricity from the 110/22 kV substations in the area, with feeders such as 471, 476, 477, 478 110/35/22 kV Mo Duc substation (Table 7.6).

7.2.2.6 Simulation of the power grid in the area before and after the Dam An Khe Solar Power Plant

The Dam An Khe Solar Power Plant project is planned to be installed in Phu Khanh commune, Duc Pho District, Quang Ngai Province. The plant will be developed in two phases:

Table 7.6 Specifications of main distribution lines in Duc Pho

No.	Transmission line	Cable type	Supply radius (km)	P_{max} (kW)	$\Delta U\%$ (%)
I	**Duc Pho (25 MVA)**				
1	Line 471	ACX-120	7.50	3,500	1.12
2	Line 473	ACX-120, AC-150	13.10	4,600	2.13
3	Line 475	AC-120	5.20	4,500	1.04
4	Line 477	AC-95	7.50	3,900	1.38

Phase 1: The Dam An Khe Solar Power Plant will install 10 MWp, corresponding to a grid power output of approximately 8.5 MW, and will be operational by the end of 2018. During Phase 1, a 22 kV transmission line approximately 0.2 km long will be constructed to connect the solar power plant to the 22/0.4 kV substation at Sa Huynh Heritage Site. The transmission line will operate at a voltage of 22 kV during this phase.

Phase 2: The Dam An Khe Solar Power Plant will install an additional 40 MWp, increasing the total capacity to 50 MWp, with a grid power output of approximately 42 MW, operational by May 2019. After the installation of Phase 2, the entire power output from the plant will be transferred to a 110 kV voltage level for connection to the power grid. Consequently, the connection method for the Dam An Khe Solar Power Plant in Phase 2 will be considered for integration into the existing 110 kV TBA at Duc Pho.

Based on the projected power output of the Dam An Khe Solar Power Plant, the project aims to calculate and evaluate the impact of the plant on the local power system in two phases:

Phase 1: Evaluate the impact of the Dam An Khe Solar Power Plant with a grid power output of approximately 8.5 MW on the distribution grid in the Duc Pho District area. Due to the large distribution grid volume, the study focuses on evaluating the plant's impact on two medium-voltage lines (473 and 475) fed from the 110/22 kV substation in Duc Pho.

The system node is chosen at the 110 kV substation in Duc Pho, with circuit breakers used to connect/disconnect the Dam An Khe Solar Power Plant from the grid. The PSS/ADEPT 5.0 software is then run to simulate the following grid configurations:

- Grid simulation without the Dam An Khe Solar Power Plant under maximum load conditions.
- Grid simulation with the Dam An Khe Solar Power Plant under maximum load conditions.

Phase 2: Evaluate the impact of the Dam An Khe Solar Power Plant with a grid power output of approximately 43 MW on the 110 kV power grid in Quang Ngai Province. The system node is chosen at the 220 kV Doc Soi substation, with circuit breakers used to connect/disconnect the Dam An Khe Solar Power Plant from the grid. The **PSS/E 30.2** software is then run to simulate the following grid configurations:

- Grid simulation without the Dam An Khe Solar Power Plant under maximum load conditions in both the rainy and dry seasons.
- Grid simulation with the Dam An Khe Solar Power Plant under maximum load conditions in both the rainy and dry seasons.

7.2.2.7 Results of the power grid simulation in the project area
Phase 1: Evaluation of the plant's impact on the distribution grid in the project area

The results of the grid simulation in Phase 1 will provide insight into the changes in key parameters such as voltage levels, power flow, and the impact of the added solar power generation on the local distribution grid. These results will help assess the ability of the grid to handle the additional load and power flow under maximum load conditions with and without the solar power plant connected (Figure 7.4).

Based on the results from the simulation using the PSS/E software, it was observed that the load-carrying capacity of the lines when the Dam An Khe Solar Power Plant is connected to the distribution grid at the 22 kV busbar connection point of the 22/0.4 kV substation at Sa Huynh Heritage Site indicates that the current load on several nearby medium-voltage lines, such as line Pho Thanh 9-Pho Thanh 3, line Pho Thanh 3-TDC Đong Muoi, line Pho Thanh 16-Pho Thanh 7, and line Pho Thanh 7-DT Sa Huynh, increases but not significantly. Therefore, it does not affect the load-carrying capacity of the lines (all lines are carrying load below 60%), ensuring that the local grid continues to operate normally.

The addition of another power source to the Duc Pho distribution grid has caused some of the main trunk lines to reduce their load-carrying current such as DZ TC22 Duc Pho-CVD. An Khe, line CVD.An Khe-Pho Thanh 9, line Pho Thanh 17-XD Hang 473, and line XD Hang 473-TC22 Đuc Pho.

The detailed comparison of the load-carrying capacity of some of these lines is shown in Table 7.7.

7.2.2.8 Assessment of the impact of the solar power plant on voltage losses

From the results of the simulation assessing the impact on voltage quality at the nodes of the grid before and after the connection of the Dam An Khe Solar Power Plant to the local grid with maximum load conditions, it was observed that after the connection, the addition of a new power supply to the distribution grid reduces the voltage losses at all nodes of the Duc Pho District distribution grid. The results comparing the voltage difference at specific nodes are as shown in Table 7.8.

The findings indicate significant voltage fluctuations in current levels before and after the commissioning of the power plant (Figure 7.5).

7.2.2.9 Simulation of reactive power compensation to reduce losses for the Go Cong town 22 kV distribution network, Tien Giang Province

This section presents the optimal reactive power compensation solution for the 22 kV distribution line 471 in Go Cong Town, Tien Giang Province, with the aim of reducing energy losses in the grid using the PSS/ADEPT software with the CAPO (capacitor placement optimization) module. CAPO assists in identifying the optimal capacitor placement locations to minimize power losses on the electrical grid. The compensation capacitors can be either fixed or dynamic capacitors.

7.2.2.10 Current situation
The 471GC line originates from the 110 kV/22 kV substation—MBA T1-40MVA, with a length of 7.11 km.

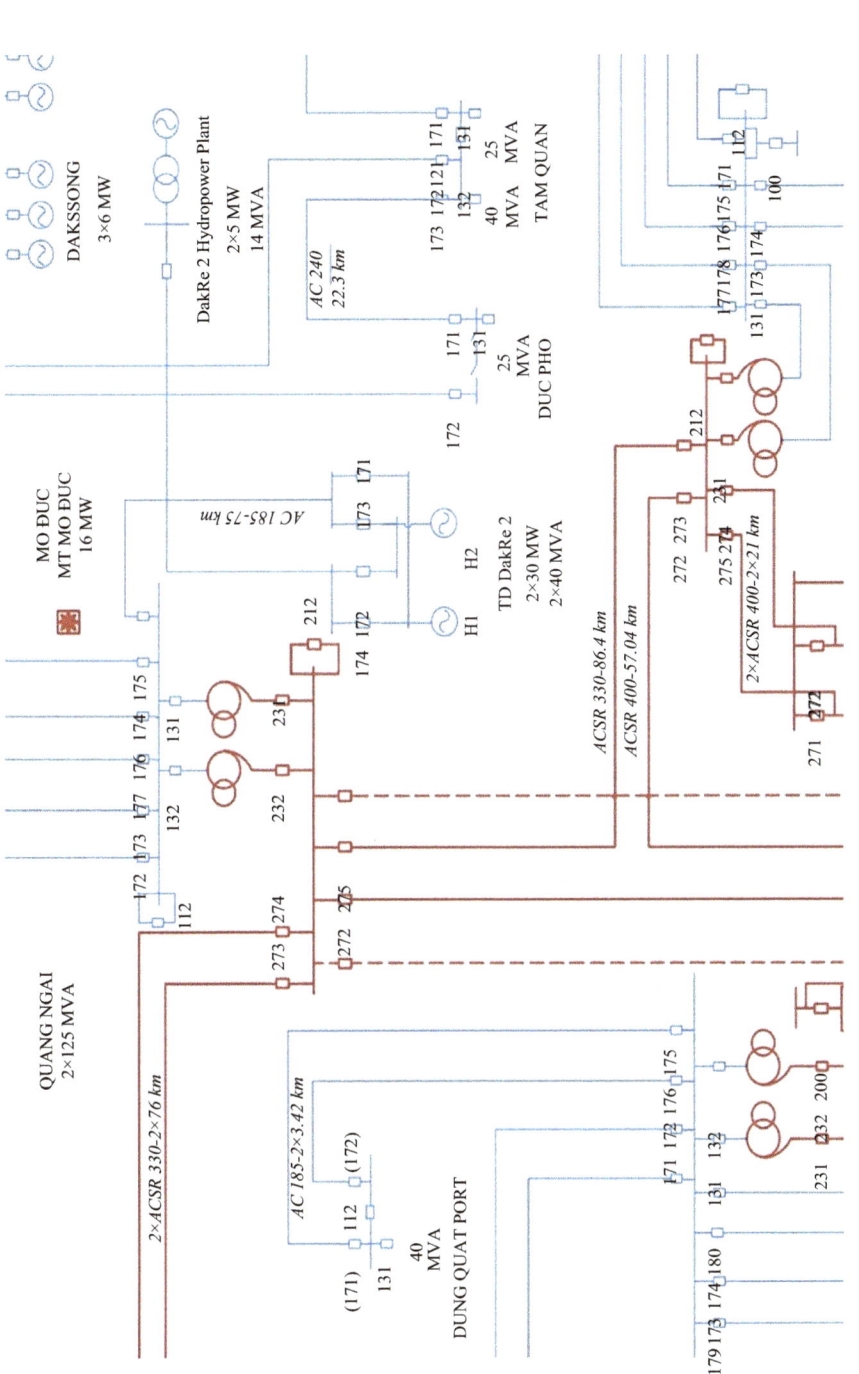

Figure 7.4 110 kV power grid diagram of Quang Ngai Province

Table 7.7 Comparison of load-carrying capacity of some medium-voltage lines with and without power plants

From point	To point	I_d (A) without PV	I_d (A) with PV	Load difference (A)
TC22_DucPho	CVD.AnKhe	102.28	93.52	−8.76
CVD.AnKhe	PhoThanh9	92.49	83.73	−8.75
PhoThanh9	NODE5	9.33	18.35	9.02
PhoThanh3	TDCDongMuoi	12.55	21.55	9.00
TDCDongMuoi	NODE3	14.49	23.47	8.98
NODE3	NODE1	19.25	5.20	−14.05
NODE1	PhoThanh13	6.26	17.23	10.96
PhoThanh16	PhoThanh7	4.32	19.14	14.82
PhoThanh7	DT_KCSaHuynh	1.07	22.33	21.26
NODE3	PhoThanh1	33.71	19.33	−14.38
PhoThanh1	PhoThanh17	36.96	22.58	−14.38
PhoThanh17	XDHung473	38.90	24.52	−14.38
XDHung473	TC22_DucPho	39.97	25.59	−14.38

Table 7.8 Comparison of voltage deviations at selected nodes in the distribution network

From point	To point	I_d (A) without PV	I_d (A) with PV	Load difference (A)
TC22_DucPho	CVD.AnKhe	102.28	93.52	−8.76
CVD.AnKhe	PhoThanh9	92.49	83.73	−8.75
PhoThanh9	NODE5	9.33	18.35	9.02
PhoThanh3	TDCDongMuoi	12.55	21.55	9.00
TDCDongMuoi	NODE3	14.49	23.47	8.98
NODE3	NODE1	19.25	5.20	−14.05
NODE1	PhoThanh13	6.26	17.23	10.96
PhoThanh16	PhoThanh7	4.32	19.14	14.82
PhoThanh7	DT_KCSaHuynh	1.07	22.33	21.26
NODE3	PhoThanh1	33.71	19.33	−14.38
PhoThanh1	PhoThanh17	36.96	22.58	−14.38
PhoThanh17	XDHung473	38.90	24.52	−14.38
XDHung473	TC22_DucPho	39.97	25.59	−14.38

- The total capacity on the line is:
  - Active power P (kW): 4,580 kW
  - Reactive power Q (kVAr): 1,430 kVAr
- Losses on the line are:
  - Active power P (kW): 110.698 kW
  - Reactive power Q (kVAr): 217.903 kVAr

Impacts of policies and mechanisms on renewable energy development 269

Comparison of Voltage Loss

[Chart: Node Voltage (V) Without Solar Power Plant vs Node Voltage (V) With Solar Power Plant across nodes PhoThanh3, TDCDongMuoi, NODE3, NODE1, PhoThanh13, PhoThanh2, PhoThanh14, PhoThanh16, PhoThanh7, DT_KCSaHuynh, PhoThanh1, PhoThanh17]

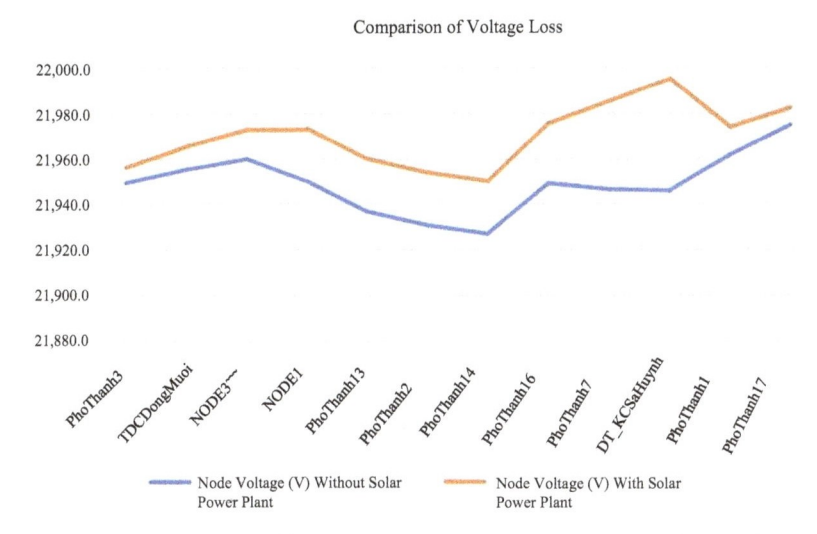

Figure 7.5 Comparison of load-carrying capacity of several medium-voltage lines with and without a solar power plant

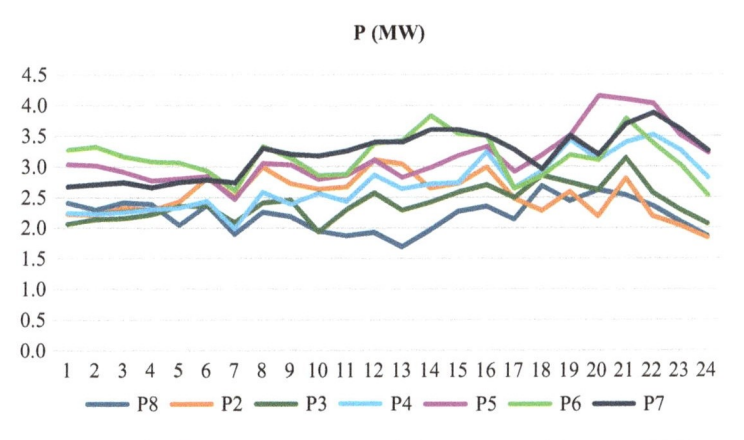

P (MW)

[Chart showing P8, P2, P3, P4, P5, P6, P7 across hours 1–24]

Figure 7.6 Power consumption (P) diagram of feeder 471GC

- Current grid compensation:
  – Total compensation capacity: 300 (kVAr)
  – Compensation location: Only at IT6 position at node GC55, with a fixed capacitor of 300 (kVAr).
- According to the power consumption chart (*P*) of the line as shown in Figure 7.6, the power consumption from Monday to Saturday is unstable and does not follow a consistent pattern.
- According to the reactive power chart (*Q*) of the line as shown in Figure 7.7, the reactive power over the same period across the days of the week is

270 *Clean energy in South-East Asia*

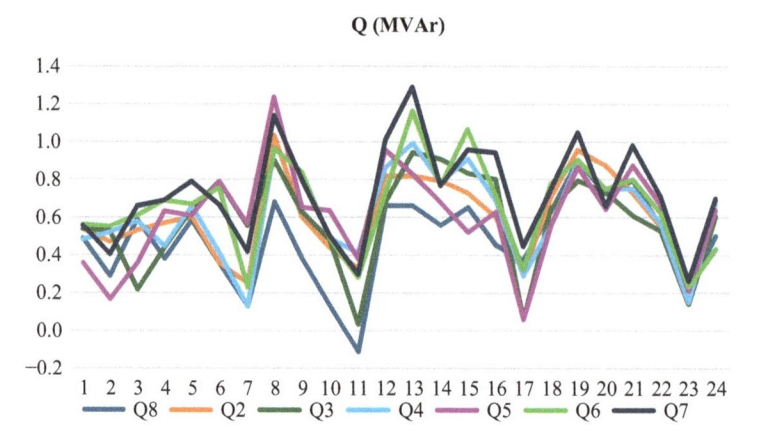

Figure 7.7 Reactive power (Q) diagram of feeder 471GC

Table 7.9 Initial power losses of feeder 471GC

Feeder name	P (kW)	Q (kVAr)	Active power loss (kW)	Reactive power loss (kVAr)
471GC	4,580	1,430	110.697	217.903

unstable and does not sufficiently ensure reactive power compensation on the grid, with some instances of overcompensation.
– The initial power loss of the 471GC line is shown in Table 7.9.
– Power loss results: From the power loss results, it is evident that the 471GC line experiences significant active and reactive power losses. Therefore, reactive power compensation for the 471GC line is necessary.

7.2.2.11 Calculation of optimal reactive power compensation for the 22 kV distribution line 471GC using the CAPO compensation model

- Selection of capacitor bank size for low-voltage transformer stations:
  – Based on the statistics of the distribution grid on the 471GC line, there are many single-phase and three-phase transformer stations with a rated capacity of $S\_nom < 75$ kVA. These mainly serve residential loads that have a high power factor ($\cos\varphi$). Therefore, if reactive power compensation is implemented for single-phase and three-phase transformer stations with $S\_nom < 75$ kVA, the compensating capacity would be low, but the number of capacitor units to be installed would be large, leading to high installation costs. Hence, reactive power compensation at

Impacts of policies and mechanisms on renewable energy development 271

- transformer stations with S_nom < 75 kVA would not be economically effective.
- It is therefore proposed to only perform reactive power compensation at three-phase transformer stations with S_nom > 75 kVA.
• Selection of capacitor bank size for low voltage:
  - For low-voltage compensation, the fixed capacitor bank size module is chosen as 30 kVAr, and the adjustable capacitor bank module is chosen as 15 kVAr.
• Selection of capacitor bank size for medium voltage:
  - Currently, the minimum capacity for a medium-voltage capacitor bank unit is 300 kVAr. Therefore, the medium-voltage capacitor bank size module is chosen as 300 kVAr.
• Economic compensation calculation approaches:
  - When calculating economic compensation options, it is assumed that the existing capacitors on the grid are retained and additional compensation is added.
• Selection of optimal compensation parameters for medium and low voltage:

The following parameters are assumed:

  - Voltage limits: high = 1.1 pu and low = 0.9 pu
  - Power factor, $\cos\varphi = 0.95$

• Compensation methods:
  - Optimal fixed compensation calculations for the medium voltage side are performed using the CAPO program for the 471GC line.

The position and capacity of the compensation capacitors after running the CAPO calculation are detailed in Table 7.10.

Calculation of losses for the 471GC feeder after implementing fixed medium voltage reactive power compensation. The total compensation capacity and the reduction in losses compared to before compensation for the feeder are shown in Table 7.11.

Perform overvoltage checks in the minimum load mode after implementing fixed medium voltage reactive power compensation. The results show that the voltage at all nodes is within the permissible limits. Therefore, the capacity of the fixed medium voltage compensating capacitors is acceptable.

Table 7.10 Location and capacity of capacitors after running CAPO

No.	Compensation location	Fixed (kVAr)	Dynamic (kVAr capacity)
1	GC43	300	0
2	GC192	300	0
Total		**600**	

Calculation of optimal dynamic reactive power compensation on the medium voltage side using the CAPO program for the 471GC feeder

The locations and capacities of the compensating capacitors after running the CAPO algorithm are shown in Table 7.12.

Calculate the losses for the 471GC feeder after implementing dynamic medium voltage reactive power compensation. The total compensation capacity and losses are reduced compared to the pre-compensation state, as shown in Table 7.13.

Table 7.11 Total compensation capacity and loss reduction after implementing fixed medium-voltage compensation on 471GC line

Line name	Medium-voltage fixed compensation capacity (kVAr)	Loss before compensation		Loss after medium-voltage fixed compensation		Loss reduction after medium-voltage fixed compensation	
	Q_b (kVAr)	ΔP (kW)	ΔQ (kVAr)	ΔP (kW)	ΔQ (kVAr)	ΔP (kW)	ΔQ (kVAr)
471	600	110.697	217.903	109.113	212.776	1.584	5.127

Table 7.12 The locations and capacities of the compensating capacitors after running the CAPO algorithm

No.	Compensation location	Fixed (kVAr capacity)	Dynamic (kVAr capacity)
1	GC43		300
2	GC192		300
Total			600

Table 7.13 Total compensation capacity and loss reduction after implementing dynamic medium-voltage compensation for feeder 471GC

Feeder name	Dynamic medium-voltage compensation capacity (kVAr)	Losses before compensation		Losses after medium-voltage fixed compensation		Loss reduction after dynamic medium-voltage compensation	
	Q_b (kVAr)	ΔP (kW)	ΔQ (kVAr)	ΔP (kW)	ΔQ (kVAr)	ΔP (kW)	ΔQ (kVAr)
471	600	110.697	217.903	107.876	208.817	2.821	9.086

Perform over-voltage testing under minimum load conditions after implementing dynamic medium voltage reactive power compensation. The test results show that the voltages at the nodes are within the permissible limits. Therefore, the capacity of the dynamic medium voltage capacitors is acceptable.

Calculate the optimal fixed and dynamic reactive power compensation for the 471GC feeder using the CAPO program

The location and capacity of the capacitors after running the CAPO calculation are shown in Table 7.14.

Calculate the losses for the 471GC feeder after implementing both fixed and dynamic medium voltage reactive power compensation. The total compensation capacity and losses have decreased compared to before the compensation, as shown in Table 7.15.

Perform overvoltage checks under minimum load conditions after implementing both fixed and dynamic medium voltage reactive power compensation. The test results show that the voltage at all nodes is within the permissible limits. Therefore, the capacities of the fixed and dynamic medium voltage compensating capacitors are acceptable.

Calculate the optimal fixed reactive power compensation on the low-voltage side using the CAPO program for the 471GC feeder

The location and capacity of the compensating capacitors after running the CAPO calculation are shown in Table 7.16.

Table 7.14 Location and compensation capacity after running CAPO

No.	Compensation location		Fixed compensation capacity (kVAr)	Dynamic compensation capacity (kVAr)
	Fixed	Dynamic		
1	GC43		300	0
2		GC192	0	300
Total			600	

Table 7.15 Total compensation capacity and loss reduction after implementing fixed and dynamic medium voltage compensation of the 471GC feeder

Feedline name	Medium voltage fixed and dynamic compensation capacity (Q_b)	Loss before compensation (ΔP, ΔQ)		Loss after compensation (ΔP, ΔQ)		Loss reduction after compensation (ΔP, ΔQ)	
	Q_b (kVAr)	ΔP(kW)	ΔQ(kVAr)	ΔP(kW)	ΔQ(kVAr)	ΔP(kW)	ΔQ(kVAr)
471	600	110.697	217.903	107.876	208.817	2.821	9.086

Table 7.16 Location and capacitor size after running CAPO

Feedline	Number of units	Fixed compensation capacity	Dynamic compensation capacity
471GC	16	480	0
Total		**480**	

Table 7.17 Total compensation capacity and loss reduction after implementing fixed low voltage compensation for feedline 471GC

Feedline	Fixed low voltage compensation capacity Q_b (kVAr)	Loss before compensation ΔP(kW)	Loss before compensation ΔQ(kVAr)	Loss before compensation ΔP(kW)	Loss before compensation ΔQ(kVAr)	Loss reduction after compensation ΔP(kW)	Loss reduction after compensation ΔQ(kVAr)
471	480	110.697	217.903	104.679	197.369	6.018	20.534

Table 7.18 Locations and capacitor compensation capacities after running CAPO

Feedline	Number of units	Fixed compensation capacity	Dynamic compensation capacity
471GC	46		690
Total			**690**

Calculate the losses for the 471GC feeder after implementing fixed low-voltage reactive power compensation. The total compensation capacity and losses have decreased compared to before compensation, as shown in Table 7.17.

Perform over-voltage testing under minimum load conditions after implementing fixed low-voltage compensation. The test results show that the voltage at the nodes is within the permissible limits. Therefore, the capacity of the fixed low-voltage capacitors is acceptable.

Calculate the optimal dynamic compensation for the low-voltage side using the CAPO program for the 471GC feeder

The position and capacity of the capacitors after running the CAPO model are shown in Table 7.18.

Calculate the losses for the 471GC feeder after implementing dynamic low-voltage compensation. The total compensation capacity and losses have decreased compared to the pre-compensation state, as shown in Table 7.19.

Perform overvoltage check under minimum load condition after reactive power compensation at low voltage. The results of the check show that the voltage at all nodes is within the permissible limits. Therefore, the capacity of the dynamic reactive power compensators at low voltage is acceptable.

Table 7.19 Total compensation capacity and loss reduction after implementing dynamic low-voltage compensation for feedline 471GC

Feedline	Dynamic low-voltage compensation capacity	Losses before compensation		Losses after dynamic low-voltage compensation		Loss reduction after dynamic low-voltage compensation	
	Q_b (kVAr)	ΔP(kW)	ΔQ(kVAr)	ΔP(kW)	ΔQ(kVAr)	ΔP(kW)	ΔQ(kVAr)
471	600	110.697	217.903	103.040	191.886	7.657	26.017

Table 7.20 Locations and capacitor compensation capacities

Feedline	Number of units		Fixed compensation	Dynamic compensation
	Fixed	Dynamic		
471GC	16	14	480	210
	Total			610

Table 7.21 Total compensation capacity and loss reduction after fixed and dynamic low-voltage compensation of feedline 471GC

Feedline	Total low-voltage compensation capacity	Losses before compensation		Losses after fixed and dynamic low-voltage compensation		Loss reduction after fixed and dynamic low-voltage compensation	
	Q_b (kVAr)	ΔP(kW)	ΔQ(kVAr)	ΔP(kW)	ΔQ(kVAr)	ΔP(kW)	ΔQ(kVAr)
471	1135	110.697	217.903	103.040	191.886	7.657	26.017

Optimal fixed and dynamic reactive power compensation calculation at low voltage using the CAPO program for feeder 471GC

The location and capacity of the compensating capacitors after running the CAPO program are shown in Table 7.20.

Calculate the losses for feeder 471GC after implementing both fixed and dynamic reactive power compensation at low voltage. The total compensating capacity and the losses have decreased compared to the pre-compensation state of the feeder, as shown in Table 7.21.

Perform overvoltage checks under minimum load conditions after fixed and dynamic reactive power compensation at low voltage.

The test results show that the voltage at the nodes is within the permissible limits. Therefore, the capacity of the fixed and dynamic reactive power compensation at low voltage is acceptable.

- Comparison of optimal compensation calculation methods
  - Based on the results of the loss reduction from the compensation methods, it is observed that all compensation methods lead to a reduction in losses compared to the pre-compensation state. Additionally, the power factor and voltage at the nodes are within the permissible limits. Therefore, all compensation methods meet the technical criteria, and the option that provides the highest profit should be considered the most reasonable solution.
  - Each calculation method provides the following results: total fixed and adjustable compensation capacity, active power loss, reactive power loss across the entire feeder, and the amount of power loss reduction compared to the initial losses.
  - From these, the total current value of the operating and installation costs for the compensation capacitors is calculated:

$$C = Q_b^{cd}\left(q_o^{cd} + N_e \cdot C_{bt}^{cd}\right) + Q_b^{dc}\left(q_o^{dc} + N_e \cdot C_{bt}^{dc}\right) \tag{7.6}$$

where

Q_b^{cd}, Q_b^{dc}: Fixed and adjustable compensation capacity (kVAr);
q_o^{cd}, q_o^{dc}: Investment cost for fixed and adjustable compensation capacitors (VND/kVAr);
C_{bt}^{cd}, C_{bt}^{dc}: Maintenance cost per year for fixed and adjustable compensation capacitors (VND/year.kVAr);

The time conversion to the current time is:

$$N_e = \sum_{n=1}^{N} \left[\frac{1+i}{1+r}\right]^n \tag{7.7}$$

where
$N = 8$: Calculation period (years);
$R = 0.12$: Discount rate for the calculation;
$i = 0.05$: Inflation rate.

$$N_e = \sum_{n=1}^{N}\left[\frac{1+i}{1+r}\right]^n = \sum_{n=1}^{N=8}\left[\frac{1+0,05}{1+0,12}\right]^n = 6{,}049208$$

The present value of the benefits from installing the capacitor banks is determined as follows:

$$B = (\Delta P \times gp + \Delta Q \times gq) \times N_e \times T \tag{7.8}$$

where
$\Delta P = \Delta P$ (before compensation) − ΔP (after compensation): The reduction in active power loss compared to before compensation (kW);
$\Delta Q = \Delta Q$ (before compensation) − ΔQ (after compensation): The reduction in reactive power loss compared to before compensation (kVAr);
$gp = 1.358$: The cost of active power consumption (VND/kWh);
$gq = 7.938$: The cost of reactive power consumption (VND/kVArh);
$T = 8{,}760$: The working hours of the capacitor banks (hours/year);

Impacts of policies and mechanisms on renewable energy development 277

- Present value of profit:

$$NPV = B - C \tag{7.9}$$

- Substitute the values into the expression above to determine values for *B*, *C*, and *NPV* (net present value) of the optimal compensation options for line 471GC, as shown in Table 7.22.
- Comparison of profit value, *NPV* = *B* − *C* for the compensation options of the 471GC line, it is observed that the fixed low-voltage compensation option has the highest *NPV*. Therefore, it is recommended to select the fixed low-voltage compensation combined with adjustment for the 471GC distribution grid.

7.2.2.12 Simulation of the Tuy Phong Wind Power Plant project

Simulate the connection of the Tuy Phong Wind Power Plant (30 MW)—Binh Thuan with the local 110 kV grid using PSS/E software—Delimit the influence area of the Tuy Phong Wind Power Plant. PSS/E software is used to simulate the operating conditions studied in this project, initially to delimit the impact range of the wind power plant on the local electrical grid (Table 7.23).

The Tuy Phong Wind Power Plant, with a capacity of 30 MW, based on simulations and operational surveys over the past years, only affects a few neighboring nodes. The simulation results of the wind power plant's impact on the voltage at the local grid nodes near the connection point for some characteristic modes are presented in Figure 7.8.

7.2.3 PSCAD software

PSCAD is developed by scientists from MANITOBA (Manitoba HVDC Research Centre Inc., Canada) as part of an effort to create a unified, modern, object-oriented, and non-human-centric simulation and modeling language in the field of Electrical Engineering. As an object-oriented language, PSCAD defines each subsystem, such as synchronous and asynchronous generators, capacitors, inductors, overhead lines, underground cables, and other components like circuit breakers, and regulators, as a class. Behavior can be described using equations (algebraic, differential, etc.) or algorithms. PSCAD supports event-based behavior descriptions, which makes it easy to model discrete-event systems and hybrid systems. A class can inherit from another class, thereby inheriting all the characteristics of that class, including behavior.

PSCAD can simulate the operation dynamics of both real-time systems and allows users to directly intervene in control processes to predict all possible scenarios that the system may encounter. The simulation results of the model are very intuitive and easy to understand. Moreover, since it does not require defining input-output relationships, this model can be used for various purposes, whether it is calculating the power dynamics of a generator unit or modeling the

Table 7.22 Comparison of optimal compensation methods for feedline 471GC

No.	Compensation methods	Compensation capacity (kVAr) Fixed LV	Adjustable LV	Fixed MV	Adjustable MV	Power loss ΔP (kW)	ΔQ (kVAr)	Cost (C) (VND)	Benefit (B) (VND)	Net present value ($NPV = B - C$) (VND)
1	Initial power loss					110.697	217.903			
2	Fixed MV compensation			600	0	109.113	212.776	164,515,277.9	3,813,681,210.32	3,649,165,932.4
3	Adjustable MV compensation			0	600	107.876	208.817	402,337,055.2	0	−402,337,055.23
4	Combined fixed and adjustable MV compensation			300	300	107.876	208.817	164,515,277.6	3,813,681,210.32	3,649,165,932.4
5	Fixed LV compensation	480	0			104.679	197.369	154,324,802.6	9,401,942,686.95	9,247,617,884.37
6	Adjustable LV compensation	0	690			103.040	191.886	305,867,440.1	0	−305,867,440.12
7	Combined fixed and adjustable LV compensation	480	210			103.040	191.886	247,957,692.4	9,401,942,686.95	9,153,984,994.54

Table 7.23 Summary of calculation results for characteristic operating modes

Characteristic mode	Power flow (MVA)/power loss (MW) on transmission line					Node voltage (kW)				
	Electric power system (HTĐ-Luong Son)	Electric power system (HTĐ-Phan Ri)	Luong Son–Phan Ri	Phan Ri–Tuy Phong	Tuy Phong–Vinh Hao	Vinh Hao–Ninh Phuoc	Luong Son	Phan Ri	Tuy Phong	Vinh Hao	Ninh Phuoc
1	$\frac{8.1-j0.2}{\approx 0}$	$\frac{7.2-j0.2}{\approx 0}$	$\frac{7.1+j0}{\approx 0}$	$\frac{-11.9-j0.6}{0.1}$	$\frac{18.1+j1.8}{\approx 0}$	$\frac{16-j1.9}{0.2}$	109.5	109.4	109.8	109.4	108.3
2	$\frac{21.2+j1.5}{0.2}$	$\frac{24.8+j2}{0.4}$	$\frac{14.6+j1.3}{\approx 0}$	$\frac{18.2+j1.5}{0.1}$	$\frac{18.1+2.0}{\approx 0}$	$\frac{16.0+j2.0}{0.2}$	108.7	108.0	107.5	107.1	106.0
3	$\frac{24.5+j3.3}{0.3}$	$\frac{28.9+j4.4}{0.5}$	$\frac{17.5+j2.9}{0.1}$	$\frac{23.4+j3.9}{0.1}$	$\frac{28.1+4.3}{0.1}$	$\frac{17.3+j3}{0.2}$	108.3	107.4	106.6	106.0	104.7
4	$\frac{26.7+j3.3}{0.4}$	$\frac{31.8+j4.3}{0.5}$	$\frac{19.6+j2.7}{0.1}$	$\frac{28.3+j3.8}{0.2}$	$\frac{28.1+j3.9}{0.1}$	$\frac{17.3+j2.6}{0.2}$	108.2	107.2	106.4	105.7	104.5

Notes:

1. Mode when the wind power plant generates maximum power into the grid via the communication line (at 17:00).
2. Mode when the wind power is not operational (at 17:00).
3. Mode when the load receives maximum power from the power system with the wind power plant operational (at 21:00).
4. Mode when the load receives maximum power when the wind power plant is not operational (at 21:00).

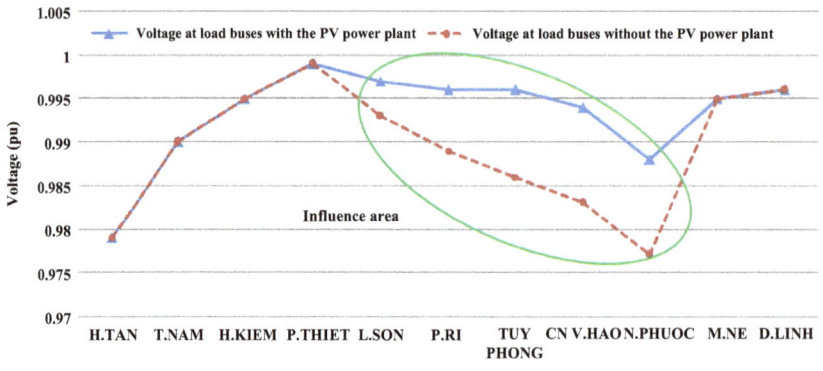

Figure 7.8 Simulation of the impact of the PV power plant on the bus voltages in the grid area

transient processes of rectifier-inverter currents. PSCAD software tools provide a powerful and convenient computing environment for applications in the technical field.

7.2.3.1 Overview and modeling of power loss problem using PSCAD program

Electric power generated by photovoltaic panels is first converted by an inverter and then transmitted through a system. The components of this system have resistance and reactance, which will cause active power loss due to resistance and reactive power loss due to reactance. Below is a detailed schematic diagram describing the rooftop solar power system connected to the grid for a typical project in Long An Province (Figure 7.9).

The main power losses in the grid-connected rooftop solar power system include: losses in the inverter, losses in the cables (CXV 0.6/1 kV 3 × 50 + 1 × 35 mm^2, CXV 0.6/1 kV 3 × 240 mm^2, CXV 0.6/1 kV 3 × 300 + 1 × 300 mm^2), and power losses in the transformer.

7.2.3.2 Modeling power losses on the PSCAD software

The detailed circuit diagram of the grid-connected rooftop solar power system for a project is redesigned and simulated on the PSCAD software.

Power loss results for the project considering grid integration with a weak grid: The design and simulation of the system's circuit diagram include: nine inverters (inverters), with the AC output connected to the busbar of the low-voltage distribution panel via CXV 0.6/1 kV 3 × 50 + 1 × 35 mm^2 cables. The cables connecting from the main ACB of the low-voltage distribution panel to the ACB of the substation use CXV 0.6/1 kV 3 × 240 mm^2 and CXV 0.6/1 kV 3 × 300 + 1 × 300 mm^2 cables. Additionally, the cable section from the main ACB of the low-voltage distribution panel to the ACB of the substation is also connected in parallel with a 600 kVAr low-voltage compensation station. All of these low-voltage circuits are

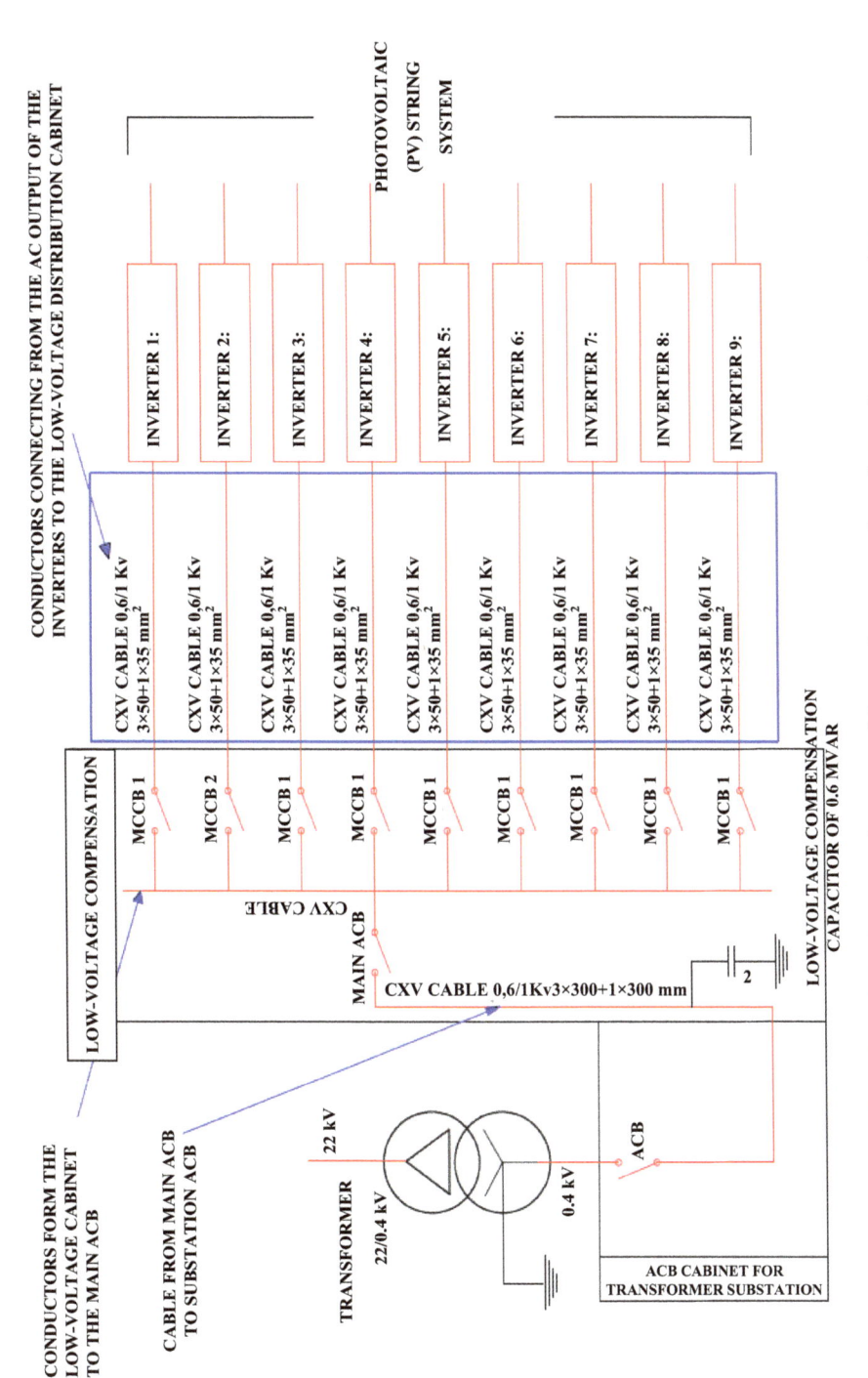

Figure 7.9 Detailed schematic diagram of the grid-connected rooftop solar power system

282　Clean energy in South-East Asia

stepped up through a 22/0.4 kV transformer with a capacity of 1,250 kVA to increase the voltage to 22 kV and integrate with the distribution grid of the electricity sector (at this point, the grid's power source is considered to be less than 24 MW—weak grid). The simulation results for each operational mode of the inverter are as follows:

- When the inverter operates in the weak grid mode at a temperature of 50 °C, the calculated power loss results on the PSCAD software are shown in Table 7.24.

When surveyed under this operational mode, the simulation results from PSCAD calculate a power loss of 132.7 kW (corresponding to a system efficiency of 85.26% for the grid-connected rooftop solar power system when integrated into a weak grid at a temperature of 50°C). The largest power loss occurs in the inverter, with a power dissipation of 107.73 kW. The greatest power loss in the cables is observed in the CXV 0.6/1 kV 3 × 50 + 1 × 35 mm^2 cable at inverter 6, with a power dissipation of 3.13 kW (Table 7.25).

7.2.3.3 Comments
Through the analysis of power losses under different operational modes of the project, it is observed that:

- The grid-connected rooftop solar power system operates most efficiently under weak grid conditions when the inverter temperature is 30°C (at this point, the system loss calculated using PSCAD is 30.3 kW, corresponding to an efficiency of 96.63%).
- When the system operates under strong grid conditions and the inverter temperature reaches 60°C, the power loss is at its highest (the system loss calculated using PSCAD in this case is 298.9 kW, corresponding to an efficiency of 66.79%).
- There is a difference in power losses between weak and strong grids, with the loss in the weak grid being lower than in the strong grid (however, the difference is not significant). This indicates that the system performs better under weak grid conditions.
- As the temperature increases, the power loss in the inverter becomes larger. A comparison of power losses in the inverter and cables also shows a significant difference as the temperature rises (Figure 7.10).

7.2.3.4 Evaluation of the impact of the Fujiwara Solar Power Plant on the power grid of Binh Dinh Province
Current status of power sources in Binh Dinh Province
Currently, Binh Dinh Province has six hydroelectric power plants (HEPPs) that are concentrated in the two districts of Tay Son and Vinh Thanh.

Vinh Son Hydroelectric Power Plant with a capacity of (2×33) MW, which was commissioned at the end of 1994 and early 1995. It feeds into the 110 kV grid system through the 110 kV Vinh Son–Hoai Nhon–Tam Quan–Duc Pho and Vinh Son–Vinh Son 5–Tra Xom–Don Pho transmission lines.

Table 7.24 Project losses in a weak grid at 50 °C

INVT	P1 (kW)	Q1 (kVAR)	P2 (kW)	Q2 (kVAR)	P3 (kW)	Q3 (kVAR)	P4 (kW)	Q4 (kVAR)	P5 (kW)	Q5 (kVAR)	P6 (kW)	Q6 (kVAR)	ΔP21 (kW)	ΔQ2 (kVAR)	ΔP43 (kW)	ΔQ43 (kVAR)	ΔP54 (kW)	ΔQ54 (kVAR)	ΔP65 (kW)	ΔQ65 (kVAR)
1	88.03	0.00	87.14	0.00									0.89	0.00						
2	88.03	0.00	86.74	0.00									1.29	0.00						
3	88.03	0.00	86.36	0.00									1.67	0.00						
4	88.03	0.00	85.08	0.00									2.95	0.00						
5	88.03	0.00	85.47	0.00	774.50	0.00	774.20	0.00	773.70	682.20	767.30	641.40	2.56	0.00	0.30	0.00	0.50	−497.20	640	40.80
6	88.03	0.00	84.90	0.00									3.13	0.00						
7	88.03	0.00	85.31	0.00									2.72	0.00						
8	88.03	0.00	86.15	0.00									1.88	0.00						
9	88.03	0.00	86.56	0.00									1.47	0.00						

Note: Bold values indicate significant results or key data points relevant to the analysis.

Table 7.25 Losses results in the project

Weak grid	Item	30°C	40°C	50°C	60°C
Weak grid	Total loss (kW)	30.30	107.10	132.70	296.10
	Efficiency (%)	96.63	88.10	85.26	67.10
Strong grid	Total loss (kW)	35.40	111.40	136.90	298.90
	Efficiency (%)	96.07	87.62	84.79	66.79

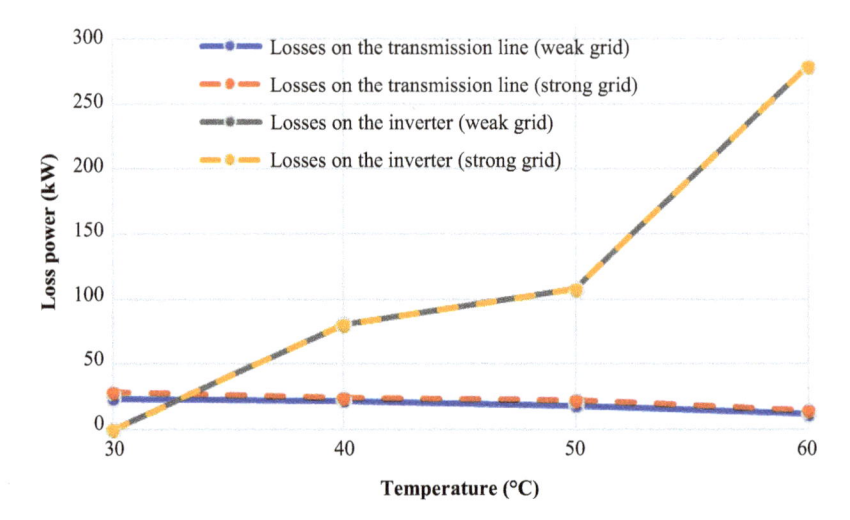

Figure 7.10 Comparison of loss power on the transmission line and inverter as the temperature increases.

Vinh Son 5 Hydroelectric Power Plant with a capacity of (2×28) MW, which was commissioned in 2014. It connects to the 110 kV grid system via the 110 kV Vinh Son 5–Phu My, Vinh Son 5–Tra Xom–Don Pho, and Vinh Son 5–Vinh Son transmission lines.

Tra Xom Hydroelectric Power Plant with a capacity of (2 × 20) MW, which began operation in March 2013. It connects to the 110 kV grid system through the 110 kV Tra Xom–Vinh Son 5 and Tra Xom–Don Pho–Nhon Tan–Quy Nhon transmission lines.

In addition, there are three small HEPPs feeding into the medium-voltage grid, including:

Dinh Binh Hydroelectric Power Plant (3 × 9.9 MW) in Vinh Thanh District, which feeds into the medium-voltage grid after the 110 kV Don Pho substation.

Tien Thuan Hydroelectric Power Plant (2 × 10 MW) and Van Phong Hydroelectric Power Plant (2 × 6 MW) in Tay Son District.

Currently, Binh Dinh Province has six medium and small HEPPs with a total capacity of 289.7 MW, meeting the load demand of the province and the regional grid during the rainy season (Table 7.26).

Table 7.26 List of power plants in Binh Dinh Province

No.	Power plant name	Location	Number of units	Installed capacity (MW)	Generation voltage (kV)	Grid connection voltage (kV)
1	Vinh Son Hydropower Plant	Vinh Thanh	2	66	13.8	110
2	Hydropower Plant Vinh Son 5	Vinh Thanh	2	28	11	110
3	Hydropower Plant Tra Xom	Vinh Thanh	2	20	6.3	110
4	Hydropower Plant Dinh Binh	Vinh Thanh	3	9.9	6.3	35.22
5	Hydropower Plant Tien Thuan	Tay Son	2	10	6.3	22
6	Hydropower Plant Van Phong	Tay Son	2	6	6.3	22

Table 7.27 Transformer parameters and operating conditions of 220 kV substations

No.	Substation name	Transformer	Rated capacity (MVA)	Rated capacity (kV)	Maximum power (MW)	Load (%)
1	Quy Nhon	AT1	125	230 /121/ 23	55	46.3%
		AT2	250	230 /121/ 23	114	48.0%

The province is served by the 220 kV Quy Nhon substation, which has a capacity of (125–250) MVA, supplying power to 110 kV substations through various 110 kV transmission lines. The 220 kV Quy Nhon substation ensures the power supply for the entire province of Binh Dinh.

7.2.3.5 Current status of the transmission network in Binh Dinh Province

The transmission network in Binh Dinh Province operates at 220 kV and 110 kV voltage levels. The provincial grid currently imports electricity from the 500 kV Pleiku substation and the An Khe Hydroelectric Power Plant to the 220 kV Quy Nhon substation and various 110 kV substations in the region.

The 220 kV Quy Nhon substation, located in the Phu Tai Industrial Zone in Quy Nhon City, has a capacity of (125–250) MVA and supplies electricity to 110 kV substations in Binh Dinh Province. The substation operates under a moderate load (P_{max} = 169 MW) after the replacement of transformer AT2, upgrading its capacity from 125 to 250 MVA (Table 7.27).

Evaluation of the ability to maintain power generation at Fujiwara Solar Power Plant—Binh Dinh during grid faults

One of the requirements for the solar power plant to connect to the 110 kV grid is its ability to continue generating electricity in the event of a short circuit on the grid, with a fault clearing time of less than 150 ms. This means that if the fault is cleared within 150 ms, the solar power plant should be able to continue operating and assist the grid in returning to a stable state.

To assess the ability of the Fujiwara Solar Power Plant in Binh Dinh to maintain power generation during short circuits on the 110 kV grid, we examine different types of faults, including a single-phase-to-ground fault and a two-phase fault (Figure 7.11). The analysis is conducted at three locations:

- At the 110 kV busbar of the plant (point of connection with the grid).
- At the busbar of Phuoc Son substation.
- At the busbar of Quy Nhon 2 substation (located further from the plant).

A single-phase short circuit occurred at the 110 kV busbar of the plant (Phase A) while operating at a power output of 42 MW. At the moment of the short circuit, the fault current in the affected phase increased significantly compared to the other two phases. The voltage on the faulted phase drops to zero, while the voltage on the other two phases decreases. When the line voltage drops below 0.85 pu, the LVRT protection system will activate and control the injection of reactive power into the grid to assist in stabilizing and raising the grid voltage during the short circuit event. At the point of 150 ms, when the fault is cleared and the grid voltage is restored, the plant continues to generate power, returning to stable operation, ensuring voltage and power output meet the requirements as per Circular 39/2015/TT-BCT of Vietnam.

7.2.4 Simulation of voltage drop and short circuit at the rooftop solar power plant in Long An Province

7.2.4.1 Selection of DC wiring

Each solar panel has two output terminals (+ and −), with each wire having a length of about 1.2 m. According to the solar panel manufacturer's catalog, the wiring used to connect the panels together has a cross-sectional area of 4 mm^2. At both ends of the wire, there are QC 4.10-35 connectors that allow for easy manual connection to another panel. However, specialized tools are required to disconnect them.

In the proposed wiring diagram, 14–16 panels are connected in series to form a string. For the project, the wiring used to connect the strings to the inverter is H1Z2Z2-K-4-1.5kV DC cable (4 mm^2). The following is the DC wiring schematic for the project (Figure 7.12).

7.2.4.2 Voltage drop check for DC cables

The design of the DC system and the cable parameters, as shown in the wiring diagram used in the project, were simulated to calculate the voltage drop using ETAP software. The results of this simulation are shown in Table 7.28.

Figure 7.11 Single-phase short circuit

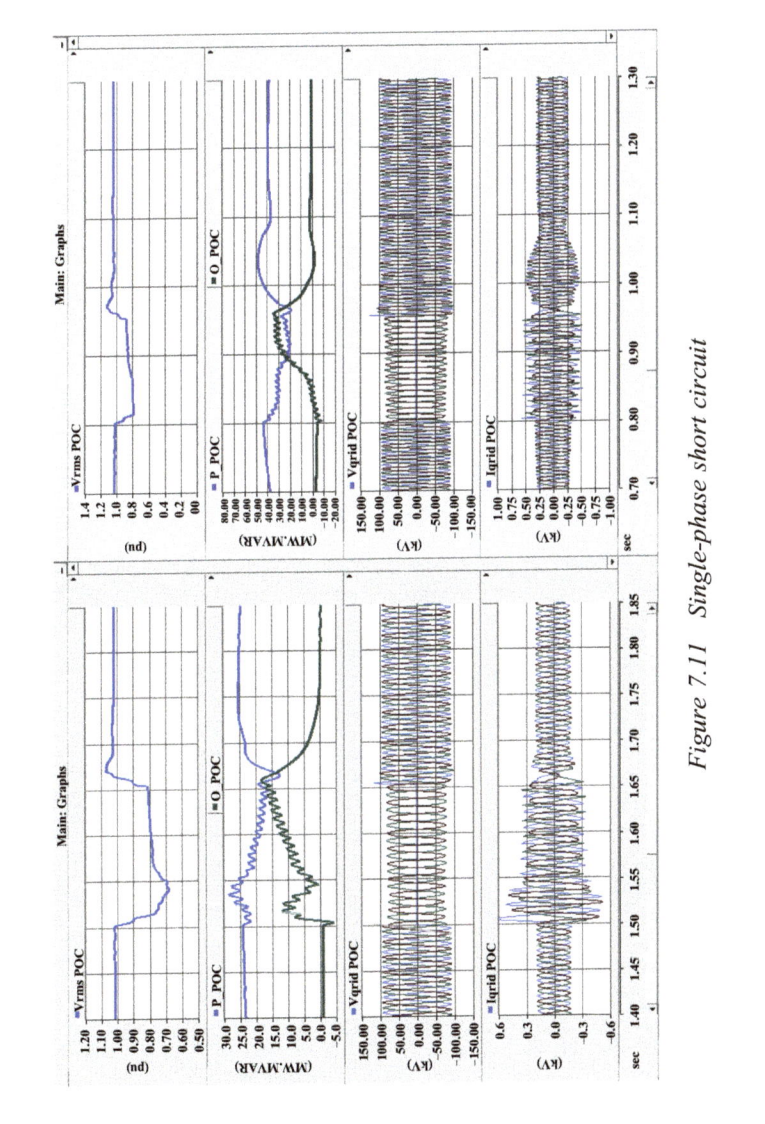

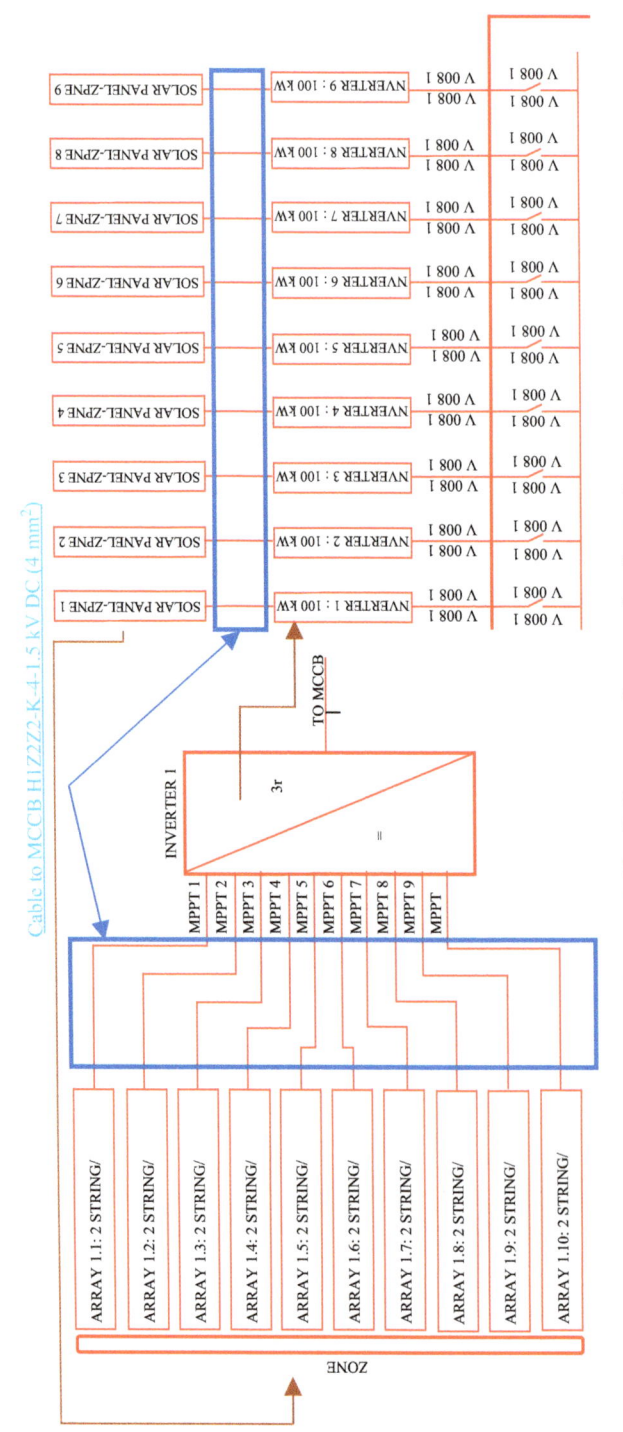

Figure 7.12 DC wiring diagram for the project

Table 7.28   Calculation results of DC conductor voltage drop for the project

Workshop	Maximum wire length (m)	Resistance (Ω/km)	Resistance (Ω)	Current (A)	Voltage drop (V)	String voltage (V)	Voltage drop (%)
1	46		0.21022		1.18163	660.8	0.1788
2	46	4.57	0.21022	8.64	1.18163	660.8	0.1788
3	56		0.25592		2.21115	660.8	0.3346

7.2.4.3   Evaluation of the selection of DC cables

According to the TCVN 9207:2012 standard, the allowable voltage drop on low-voltage transmission lines should be less than 5%. From the calculation results for the voltage drop on the DC cables used in the project, the longest cable length from the string of photovoltaic panels to the inverter is 56 m, and the calculated voltage drop is 0.33%, which is the highest voltage drop observed. This value is much lower than the allowable limit of 5% stipulated by the standard.

Thus, the use of the H1Z2Z2-K-4-1.5 kV DC cables (4 mm^2) from the strings to the inverter will ensure safe operation and durability under the environmental conditions of the project site throughout the lifetime of the solar power plant.

7.3   Vietnamese electricity market: structure of the electricity market and electricity transmission pricing model

7.3.1   Problem statement

The development of the electricity market in Vietnam began early, with the establishment of the Electricity Market Board under EVN in 2004. With the assistance of TransGrid (Australia), in 2005–2006, EVN developed and presented the Internal Electricity Market Regulations to the Ministry of Industry (now the Ministry of Industry and Trade), which became the official basis for operation starting in January 2007.

The significant benefit of electricity market reforms is that they simultaneously achieve both goals: bringing electricity prices closer to long-term marginal costs and creating competitive pressures to minimize costs across all stages of the electricity industry. Previously, vertically integrated regulatory structures, no matter how well-designed, could only achieve one of these goals. Competition can increase labor productivity in the electricity industry by up to 60% and reduce generation costs by up to 40%. This explains why electricity market reform is an inevitable trend in the electricity sectors of many countries worldwide. Even nations that initially faced setbacks in their reforms do not return to the previous vertically integrated model.

In line with global trends, Vietnam's electricity sector reform is strongly reflected in the electricity law of 2004, which outlines the development process for the electricity market in Vietnam in three phases:

- Phase 1: Vietnam's competitive generation market (VCGM) (2005–2014)
- Phase 2: Vietnam's wholesale electricity Market (VWEM) (2015–2022)
- Phase 3: Vietnam's retail electricity market (VREM) (after 2022)

In the electricity market, prices are a critical factor influencing the behaviors and strategies of participants. Buyers always desire lower electricity prices, while sellers want higher prices for greater profits. Therefore, for the market to benefit society and operate efficiently, participants must develop appropriate pricing strategies.

A key component of the electricity price is the transmission price, which is considered a common cost for all electricity market participants. Since transmission is a monopoly, the state must manage it to ensure fair and reasonable transmission pricing that balances the needs of all market participants.

The main question now is how to calculate transmission prices that are appropriate for Vietnam at the present and in the future. This is an ongoing topic of concern for the electricity industry and researchers in Vietnam.

7.3.2 Vietnam's wholesale electricity market

7.3.2.1 Objectives of the VWEM

The objectives of the VWEM are to:

- Ensure a stable and adequate electricity supply at reasonable prices without causing significant disruptions to electricity production and business activities.
- Ensure sustainable development of the electricity sector.
- Attract investment from all economic sectors, both domestic and international, into the electricity sector, while reducing the state's direct investment in electricity.
- Improve competitiveness and ensure fairness, equity, and transparency in electricity transactions and operations.

7.3.2.2 Principles for building VWEM

The principles for building the VWEM include:

- Simplicity, feasibility, and appropriateness: The system must be simple and feasible, avoiding the creation of complex mechanisms that hinder implementation. It must also consider the unique conditions of Vietnam's electricity industry.
- Building on the advantages and addressing the challenges: The market must build upon the strengths and address the challenges of the previous competitive generation market in Vietnam.
- Minimizing system-wide electricity costs: One of the key principles is to minimize the overall system costs, optimizing scheduling and dispatch while taking system security constraints into account. This also includes the effective

mobilization of ancillary services like frequency control and voltage regulation.
- Effective pricing mechanisms: The pricing mechanism must reflect the true cost of electricity purchase at each location and during each trading period, sending the correct price signals.
- Encouraging effective investment: A critical goal of the Vietnamese electricity market is to attract investments in new power generation sources.
- Transparency and efficiency in operations: The VWEM aims for effective and transparent operation of the electricity system and market.
- Enhancing competition: The market structure must support multiple buyers and sellers of electricity, as well as service providers.
- Proper risk allocation: One principle of the competitive wholesale electricity market is to ensure that risks are appropriately allocated among participants.
- Maximizing market participation: One limitation of the previous competitive generation market was that about 41% of the system's installed capacity did not participate in the market. This resulted in market prices not reflecting the true marginal cost of electricity and limited transparency. All power plants with an installed capacity above 30 MW, including strategic multi-purpose hydropower plants and BOT plants, should participate in the market.

The VWEM's aim is to foster a competitive, transparent, and efficient electricity market in Vietnam that encourages investment and meets the country's growing energy needs while minimizing costs for consumers.

7.3.2.3 VWEM structure

The detailed design of the VWEM has been approved under Decision No. 8266/QD-BCT dated August 10, 2015, by the Minister of Industry and Trade (Figure 7.13). According to the decision, the market is named as follows:

- In Vietnamese: Thị trường bán buôn iện cạnh tranh Việt Nam
- In English: Vietnam's wholesale electricity market
- Abbreviation: VWEM

7.3.2.4 Participants in the VWEM

In terms of structure, participants in the VWEM are classified into three main groups (Figure 7.14):

- Electricity sellers
- Electricity buyers
- Service providers

7.3.2.5 Day-ahead market

a) Fully centralized market: National system and market operator (NSMO) operates the VWEM under a fully centralized market mechanism. All electricity generated or received from the transmission grid must be traded through the day-ahead market.

292   *Clean energy in South-East Asia*

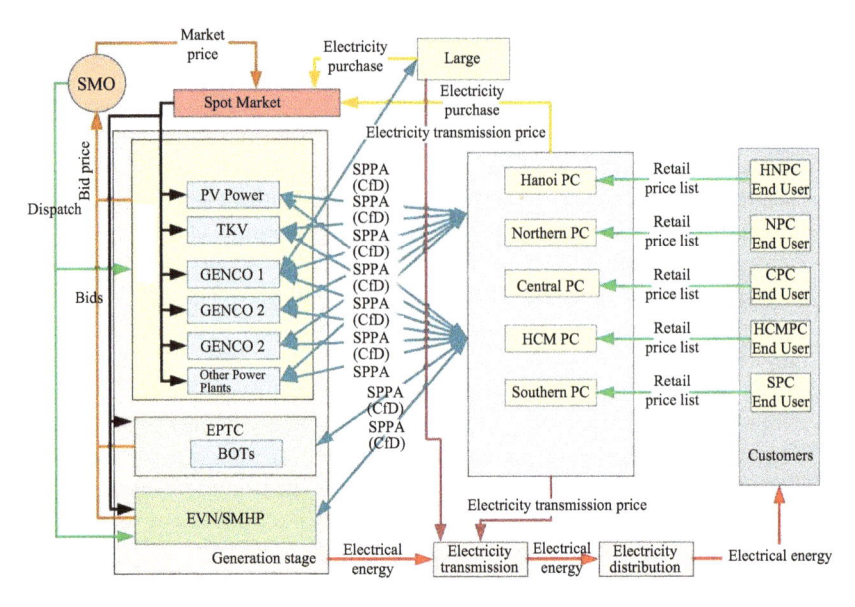

Figure 7.13   Structure of Vietnam's wholesale electricity market (VWEM)

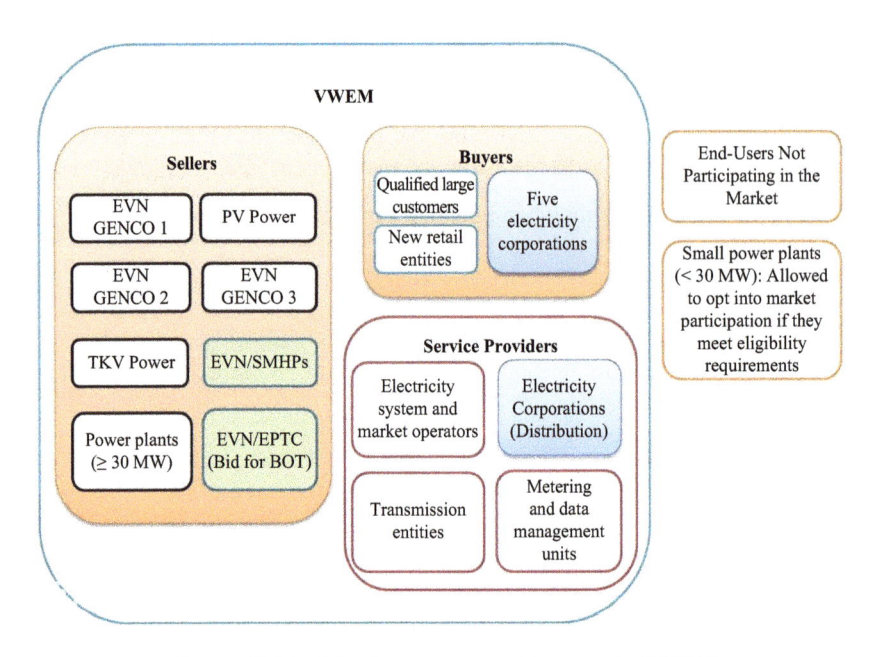

Figure 7.14   Member units participating in VWEM

b) Cost-based bidding market: NSMO operates the VWEM under a cost-based bidding market mechanism at the start of VWEM, with plans to transition to a freely priced energy market.
c) Trading nodes: NSMO must allocate each generator unit, ancillary service provider, and load (wholesale purchase point) connected to the transmission grid to a trading node. For each trading node, NSMO must allocate one or more meters to measure the electricity generated or received during each trading cycle.
d) Trading day: The trading day begins at 12:00 AM (midnight) and ends at 12:00 AM the following day.
e) Trading cycle: A trading cycle is the time period for payment calculations, which is 30 min.
f) Dispatch cycle: The dispatch cycle is the time interval between two optimal dispatch steps. NSMO must operate the VWEM with N dispatch cycles in each trading cycle, where $N \geq 1$ and is an integer. The dispatch cycle is $30/N$ min. At the start of full VWEM implementation, the dispatch cycle is 30 min ($N = 1$). When the new MMS system operates, the dispatch cycle is 5 min for long-term VWEM ($N = 1$).

7.3.2.6 Legal framework analysis in electricity market development

The documents analyzed focus on factors related to electricity transmission pricing methods.

a) **Electricity law**

Several aspects related to electricity transmission pricing methods can be highlighted:
- Article 8 requires the development of local and national electricity planning, approved by the competent state authority, and compliance by organizations and individuals with the approved electricity development plans.
- Article 9 stipulates that the Ministry of Industry (now the Ministry of Industry and Trade) is responsible for submitting the national electricity development plan, mostly covering transmission activities, to the prime minister for approval, publication, and oversight.
- Article 15 specifies that transmission lines and substations must meet technical-economic standards and operate optimally to ensure stable, safe, and continuous electricity supply while minimizing energy losses.
- Articles 26 and 27 address the quality of electricity supply, interruptions, and reductions in supply, specifying penalties or compensation for service failures or interruptions.

Regarding pricing policy, Articles 29–31 establish principles for state management of electricity transmission pricing, specifically:
- Investments in the electricity sector should be profitable under efficient environmental and economic conditions, ensuring the rights and legitimate interests of electricity companies and customers.

- A rational price structure should gradually eliminate cross-subsidies between customer groups.
- The price framework for electricity transmission, system operation, market transactions, and ancillary service prices must be developed by electricity companies, with ERAV assessing and presenting it for Ministry of Industry and Trade approval.
- Article 39 outlines the rights and obligations of electricity generation units, including the right to connect to the national grid and meet technical conditions.
- Article 40 specifies the rights and obligations of the transmission company:
  - The right to establish transmission tariffs for approval.
  - The obligation to ensure the safe, stable, and continuous operation of the transmission network and equipment.
  - The obligation to invest in the development of the transmission system to meet the transmission needs as per the national grid plan.

b) **Transmission license**

Aspects related to electricity transmission pricing methods include:
Article 27 stipulates:
- Transmission companies are not allowed to apply electricity tariffs or prices without ERAV's approval.
- Transmission prices are only related to costs outlined in the overall grid plan and financial activities approved by ERAV.
- The transmission license proposes pricing that aligns with the principles and methods of electricity transmission pricing and matching connection contracts.
- Transmission system operators (TNO) must submit required documents for ERAV's review and approval of transmission pricing, along with an explanation of proposed prices consistent with legal requirements. Regarding revenue risks, the license includes compensation for transmission customers in the event of faults or failure to meet voltage, frequency, output, or other standards, causing harm to customers, and other conditions per the connection contract.

For transmission network expansion and upgrades, Section 7.3.2.6 establishes the obligation to expand and upgrade the transmission grid according to the grid plan and the annual list of activities proposed by ERAV.

The license also anticipates TNO purchasing previously owned equipment as part of the electricity generation license when connecting power plants to the transmission system. In these cases, Section 7.3.2.6 requires TNO to pay reasonable compensation to the asset owners, including previous and current electricity costs, construction, installation, and maintenance costs for transmission assets, plus a reasonable return on investment.

c) **Transmission grid regulations for competitive market**

The key relationship between transmission grid regulations and electricity transmission pricing is identifying financial penalties for TNO if they fail to meet established standards. In various transmission pricing methods, failure to meet standards is viewed as providing lower-than-standard service quality as defined by grid regulations, and customers must be compensated for receiving substandard service.

Transmission grid regulations set objectives for TNO, including:

- Providing transmission asset services to the NSMO.
- Maintaining transmission assets to ensure reliable, high-quality electricity transmission.
- Scheduling repairs to minimize impacts on system operation.

These goals must ensure that TNO meets the standards in grid regulations while aligning with the transmission pricing methods, ensuring clarity without unnecessary complexity.

In cases where NSMO needs specific objectives, these include:

- Maintaining real-time electricity supply security.
- Operating the real-time system to meet demand.
- Minimizing electricity costs for end customers.
- Ensuring electricity supply standards and goals.

d) **General comments on the institutional framework for electricity market development**

From the above analysis, the key conclusions are:

- The Vietnamese electricity sector clearly exhibits centralized planning characteristics, with electricity entities responsible for investing in compliance with the approved national grid plan.
- Transmission customers do not have the freedom to decide where to build power plants or distribution companies, as locations are specified in the national grid plan.
- Transmission pricing mechanisms are not decided by transmission customers but are set during the electricity system planning process.
- System expansion is part of the national grid plan, with decisions about what to expand and how it is implemented falling outside TNO's discretion.
- Transmission pricing methods must ensure the legitimate rights of both the company and customers under efficient conditions.
- Transmission pricing must be determined using the standard method approved by ERAV.
- Transmission pricing must be based on proposals from TNO to ERAV, with proposals adhering to ERAV's procedures; ERAV must assess and present the pricing for Ministry of Industry and Trade approval.
- TNO must compensate customers when incidents cause customer harm.

7.3.2.7 Suitability of the transmission pricing method for the electricity market in Vietnam

a) **Electricity law**

The new transmission pricing method focuses on regulatory aspects related to the development of a competitive electricity market based on a centralized planning mechanism. This method aligns with the goal of ensuring operational efficiency for transmission companies. Therefore, transmission companies are responsible for investing in and constructing the lines to the meter for the electricity purchaser. The transmission pricing method primarily focuses on determining the transmission price to recover all electricity costs and allowed profits. This pricing method encourages the economical and efficient use of electricity, boosts production, and enhances the competitiveness of companies through economic mechanisms while also aligning with international best practices. Finally, the National Power Development Plan (NPDP) defines the economic efficiency of expanding the transmission grid. As such, the economic efficiencies reflected in the NPDP foster effective capital recovery mechanisms, and customers are free to choose the location of their units. Consequently, the transmission price will provide economic signals to these customers.

b) **Electricity operation license**

The electricity operation license in the field of electricity transmission requires TNO to invest in expanding and upgrading the transmission network to ensure the system can receive power from power plants, transmit electricity to other transmission systems, and distribute electricity to the distribution systems connected to the transmission grid in line with the NPDP.

Regarding tariffs, for general transmission electricity costs and related services, the transmission company proposes a price to recover these electricity costs and submits it to the regulatory authority for approval, in accordance with the approved transmission pricing method.

c) **Grid regulations**

VCGM has yet to finalize terms regarding the ownership of the TNO and the NSMO. For VCGM, the separation of the transmission and generation stages, as well as the separation of investment responsibilities and the establishment of distinct pricing mechanisms for each of the transmission and distribution stages, is crucial. In the competitive wholesale electricity market, distribution and retail activities are separated from the transmission process. This aligns with the transmission pricing method, where the price is directly related to transmission operations. The transmission pricing method will establish distinct prices for the different responsibilities of the TNO and the NSMO. This ensures that other activities are not redundantly factored into the transmission pricing calculation. The responsibilities of the NSMO include:

- Maintaining the security of the electricity supply system in real-time.
- Operating the electrical system to meet electricity demands.
- Minimizing electricity costs for the supply to end customers.
- Meeting electricity supply quality objectives.

Impacts of policies and mechanisms on renewable energy development 297

The responsibilities of the TNO include:

- Providing transmission network services through the NSMO.
- Maintaining the transmission network to ensure reliability, transfer capacity, and availability.
- Planning transmission network shutdowns to minimize electricity operation costs.

7.3.2.8 Suitability of the transmission pricing method for the electricity market in Vietnam

a) Electricity law

The new transmission pricing method focuses on regulatory aspects related to the development of a competitive electricity market based on a centralized planning mechanism. This method aligns with the goal of ensuring operational efficiency for transmission companies. Therefore, transmission companies are responsible for investing in and constructing the lines to the meter for the electricity purchaser. The transmission pricing method primarily focuses on determining the transmission price to recover all electricity costs and allowed profits. This pricing method encourages the economical and efficient use of electricity, boosts production, and enhances the competitiveness of companies through economic mechanisms while also aligning with international best practices. Finally, the NPDP defines the economic efficiency of expanding the transmission grid. As such, the economic efficiencies reflected in the NPDP foster effective capital recovery mechanisms, and customers are free to choose the location of their units. Consequently, the transmission price will provide economic signals to these customers.

b) Electricity operation license

The electricity operation license in the field of electricity transmission requires TNO to invest in expanding and upgrading the transmission network to ensure the system can receive power from power plants, transmit electricity to other transmission systems, and distribute electricity to the distribution systems connected to the transmission grid in line with the NPDP.

Regarding tariffs, for general transmission electricity costs and related services, the transmission company proposes a price to recover these electricity costs and submits it to the regulatory authority for approval, in accordance with the approved transmission pricing method.

c) Grid regulations

VCGM has yet to finalize terms regarding the ownership of the TNO and the NSMO. For VCGM, the separation of the transmission and generation stages, as well as the separation of investment responsibilities and the establishment of distinct pricing mechanisms for each of the transmission and distribution stages, is crucial. In the competitive wholesale electricity market, distribution and retail activities are separated from the transmission process. This aligns with the transmission pricing method, where the price is directly related to transmission operations. The transmission pricing method will establish distinct prices for the different responsibilities of the TNO and the NSMO. This ensures that other activities are not redundantly factored into the transmission pricing calculation. The responsibilities of the SMO include:

- Maintaining the security of the electricity supply system in real time.
- Operating the electrical system to meet electricity demands.
- Minimizing electricity costs for the supply to end customers.
- Meeting electricity supply quality objectives.

The responsibilities of the TNO include:

- Providing transmission network services through the NSMO.
- Maintaining the transmission network to ensure reliability, transfer capacity, and availability.
- Planning transmission network shutdowns to minimize electricity operation costs.

7.3.3 Current transmission pricing method in Vietnam
7.3.3.1 Existing network

The average transmission price for year N is determined based on the total allowed transmission revenue for that year and the entities required to pay transmission costs.

a) Transmission price

The average transmission price for year N is calculated using the following formula:

$$g_{TT_N} = \frac{G_{TT_N}}{\sum_{i=1}^{n} A_{GN_{i,N}}^{DB}} \tag{7.10}$$

in which
G_{TT_N} is the total allowed transmission revenue for year N (VND)
$A_{GN_{i,N}}^{DB}$ is the total forecasted electricity exchanged by unit i at all delivery points in year N (kWh)

b) Total transmission revenue

The total allowed transmission revenue for year N (G_{TT_N}) includes the following components:
The allowed capital electricity cost ($C_{CAP_{TT_N}}$),
The allowed operation and maintenance (O&M) electricity cost ($C_{OM_{TT_N}}$),
The revenue adjustment component for transmission electricity for year $N-2$ (CL_{N-2}),
which is determined by the following formula:

$$G_{TT_N} = C_{CAP_{TT_N}} + C_{OM_{TT_N}} + CL_{N-2} \tag{7.11}$$

in which
$C_{CAP_{TT_N}}$ is the total allowed capital electricity cost for transmission in year N (VND)

Impacts of policies and mechanisms on renewable energy development 299

$C_{OM_{TT_N}}$ is the total allowed O&M electricity cost for transmission in year N (VND)
CL_{N-2} is the difference in electricity costs and transmission revenue for year $N-2$ adjusted into the total allowed transmission revenue for year N (VND)

c) Total electricity delivered and received

The total forecasted electricity delivered and received by unit i at all delivery points in year N (kWh) for units that are required to pay the transmission electricity costs.

d) Total capital electricity costs

The total capital electricity cost for transmission in year N ($C_{CAP_{TT_N}}$) is determined according to the following formula:

$$C_{CAP_{TT_N}} = C_{KH_N} + C_{LVDH_N} + LN_N \quad (7.12)$$

in which

C_{KH_N} is the total depreciation costs of fixed assets in year N ( )
C_{LVDH_N} is the total interest costs on long-term loans and other costs for borrowing capital, payable in year N for transmission assets ( )
LN_N is the allowed transmission profit in year N ( )

e) Total electricity O&M costs

The total electricity O&M costs for transmission in year N ($C_{OM_{TT_N}}$) are determined according to the following formula:

$$C_{OM_{TT_N}} = C_{VL_N} + C_{TL_N} + C_{SCL_N} + C_{MN_N} + C_{K_N} \quad (7.13)$$

in which:

C_{VL_N} is the total material electricity costs in year N (VND)
C_{TL_N} is the total electricity salary costs in year N (VND)
C_{SCL_N} is the total major repair electricity costs in year N (VND)
C_{MN_N} is the total external service electricity costs in year N (VND)
C_{K_N} is the total other electricity costs in year N (VND)

f) Electricity cost and transmission revenue differential

The difference between electricity costs and transmission revenue in year $N-2$ is adjusted into the total transmission revenue for year N, as determined by the following formula:

$$CL_{TT(N-2)} = \left[G^D_{TT(N-2)} - G^{TH}_{TT(N-2)} + \Delta C_{N-2} + SV_{N-2}\right] * (1 + I_{N-1}) \quad (7.14)$$

where:

$G^D_{TT(N-2)}$ is the total approved transmission revenue for year $N-2$
$G^{TH}_{TT(N-2)}$ is the total actual transmission revenue for year $N-2$ (according to financial reports)
ΔC_{N-2} is the total difference between actual and approved electricity costs for year $N-2$
SV_{N-2} is the total electricity costs incurred for disaster recovery and handling force majeure incidents; total depreciation costs, interest costs, and profit from equity investments for reasonable investments incurred in year $N-2$

I_{N-1} is the average interest rate for deposits in Vietnamese Dong

g) Group of electricity costs to determine total capital electricity costs

Total depreciation costs for fixed assets in year N (C_{KH_N})
Total long-term loan interest and financing fees to be paid in year N (C_{KH_N})
Allowed profit for year N, which is determined by the following formula:

$$LN_N = V_{CSH,N} \times ROE_N \tag{7.15}$$

where:
$V_{CSH,N}$ is the average of estimated equity as of December 31st of year $N-1$ and estimated equity as of December 31st of year N (VND)
ROE_N is the return on equity ratio

h) Group of electricity costs to determine total O&M electricity costs

Total material costs for electricity in year N (C_{VL_N}) are determined by the following formula:

$$C_{VL_N} = C_{VL_N}^{DD} + C_{VL_N}^{TBA} + C_{VL_N}^{MBA} \tag{7.16}$$

where:
$C_{VL_N}^{D}$ is the total material costs for transmission line electricity in year N (VND)
$C_{VL_N}^{TBA}$ is the total material costs for transformer station electricity in year N (VND)
$C_{VL_N}^{MBA}$ is the total material costs for transformer electricity in year N (VND)

- Total labor costs for electricity in year N include the total labor costs and any other wage-related electricity costs.
- Total repair costs for electricity in year N are determined by the following formula:

$$C_{SCL_N} = C_{SCL_N}^{D} + C_{SCL_N}^{TBA} + C_{SCL_N}^{MBA} + C_{SCL_N}^{PT} \tag{7.17}$$

where:
$C_{SCL_N}^{DD}$ is the total major repair costs for transmission line electricity in year N (VND)
$C_{SCL_N}^{TBA}$ is the total major repair costs for transformer station electricity in year N (VND)
$C_{SCL_N}^{MBA}$ is the total major repair costs for transformer electricity in year N (VND)
$C_{SCL_N}^{PT}$ is the total major repair costs for auxiliary and service facilities in year N (VND)

- Total external service costs for electricity in year N (VND)
- Total other electricity costs in year N (VND)

Comments

- The transmission electricity price is low and applies a unified price every year.
- It does not encourage investors to invest capital into the transmission grid projects.

- The transmission and distribution sectors still have monopolies, maintaining control over the electricity industry.
- Distribution companies are not allowed to choose their electricity suppliers to reduce costs.

7.3.3.2 New investment grid

a) Calculation of the electricity sales price for capital recovery

The electricity sales price for capital recovery is calculated to ensure the repayment of all loans (within the credit term) and the full recovery of the project's capital within the project's lifespan or the BOT or BLT term. Therefore, the electricity sales price is calculated based on the total profit over the entire project's operational period, discounted to a present value of zero ($NPV = 0$).

b) Formula for calculating electricity sales price for capital recovery and transmission electricity price

To ensure the feasibility of the investment project comprehensively, the high-voltage grid structure for transmitting electricity from power plants to consumption locations must also be considered. The issue is: what is the minimum electricity sales price at the consumption point or the annual transmission grid rental fee that will ensure the project's feasibility, ensure full repayment of all loans (within the credit term), and the recovery of the full capital within the project's lifespan or the BOT or BLT term (build-operate-transfer)? Therefore, the analysis of the electricity sales price for capital recovery at the consumption point or transmission electricity price needs to be calculated.

In other words, the calculation of the electricity sales price for capital recovery and the transmission electricity price is applied based on financial indicators. This explains why new investment projects adopt this calculation method.

$$NPV = (-C_0 + V) + \sum_{t=1}^{n} P_B(A_t - \Delta A)(1+i)^{-t}$$

$$- \sum_{t=1}^{n}(P_M A_t + C_{VH} + G_t + L_t + T_t)(1+i)^{-t} \qquad (7.18)$$

$$NPV = 0 \Rightarrow P_B = \frac{(C_0 - V) + \sum_{t=1}^{n}(P_M A_t + C_{VH} + G_t + L_t + T_t)(1+i)^{-t}}{\sum_{t=1}^{n}(A_t - \Delta A)(1+i)^{-t}}$$

$$(7.19)$$

$$\Rightarrow \Delta P = P_B - P_M \qquad (7.20)$$

where

C_0 is the Investment capital of the project (USD)
V is the loan capital (USD)
P_B is the electricity selling price (US cents/kWh)
P_M is the electricity purchase price (US cents/kWh)

ΔP is the transmission electricity price (US cents/kWh)
A_t is the purchased electricity (kWh)
ΔA is the loss of electricity (kWh)
C_{VH} is the annual electricity management and operation cost [the annual O&M of the transmission network is usually set at 2.5% of the total investment capital (avh = 0.025)] (USD)
G_t is the loan principal repayment (USD)
Lt is the loan interest repayment (USD)
T is the income tax (USD)
i is the discount rate (%)

c) **Remarks**
- The transmission electricity price is high and applies a fixed rate for a certain amount of capacity.
- This stimulates investment in the transmission network infrastructure.
- New investment projects are still dependent on the operation of the national electricity transmission system.
- It creates pressure for all power plants under the state-owned electricity company to improve operational efficiency and reduce costs in order to supply electricity to the grid.

7.3.3.3 Comparison and analysis of the two methods

Both methods for calculating the transmission electricity price are highly feasible and easy to apply to the current conditions in Vietnam, according to the roadmap towards an electricity market. However, each method has its own advantages and disadvantages, which are analyzed as follows:

a) **Existing network**

Advantages

- Requires minimal changes to the current industry structure.
- Easy to implement and has a high success rate.
- In the short term, there is little impact on distribution and retail companies, allowing these companies more time to improve their financial and management capabilities in preparation for future competition.
- The transmission electricity price is low.

Disadvantages

- Initial competition is limited.
- Distribution companies cannot select electricity suppliers to reduce electricity costs.
- There is still the perception that the electricity purchasing process is not entirely transparent.
- Ignores transmission-related constraints.

Impacts of policies and mechanisms on renewable energy development 303

b) **New investment network**
Advantages
- Simple and easy to implement based on financial indicators.
- Attracts capital investment.
- The transmission electricity price is aligned with the needs of investors.

Disadvantages
- Difficult to implement, suitable only for a single new investment project.
- Raises the electricity price.
- Ignores transmission-related constraints.

7.3.4 Transmission electricity price model in Vietnam's competitive wholesale electricity market

This section presents the results of a study on the transmission electricity price model based on the optimal node electricity price for the 500 kV power grid market of the Power Transmission Company 4, Vietnam. Based on this, the feasibility of applying this model to the competitive wholesale electricity market in Vietnam in the coming years is proposed.

The simulation diagram for the electricity market of the 500 kV power grid, consisting of 17 nodes of Power Transmission Company 4, a subsidiary of the National Power Transmission Corporation, Vietnam, is modeled using the Power World Simulator V19 software, as shown in Figure 7.15.

When operating according to the original construction plan, the 17-bus power system remains stable. The voltage at all buses stays within permissible limits. Additionally, the nodal prices at the buses under normal loading conditions are illustrated in Figure 7.18.

Simulation results for the 17-node power market under normal load conditions

When the load operates according to the initial construction plan, the 17-node power system remains stable (Figure 7.16). The voltage at all nodes is maintained within the allowable limits (Tables 7.29–7.32).

The nodal price at the buses under normal load conditions can be represented as shown in Figure 7.17.

The simulation results for the electricity market considering transmission limits and losses show that

When the load at node 7 (Phu Lam node) is increased from 190 MW to 370 MW, the 17-node electricity market undergoes changes. The simulation results, as shown in Figure 7.18, reveal the following:

Transmission limit violations: The transmission line between node 5 (Nha Be) and node 7 (Phu Lam) reaches 88% of its transmission limit, and the line between node 6 (Song May) and node 9 (Tan Dinh) reaches 83% of its capacity.

Price fluctuations: The increase in load causes fluctuations in the nodal prices at various points in the system, as the increased demand leads to transmission congestion and higher costs in certain regions.

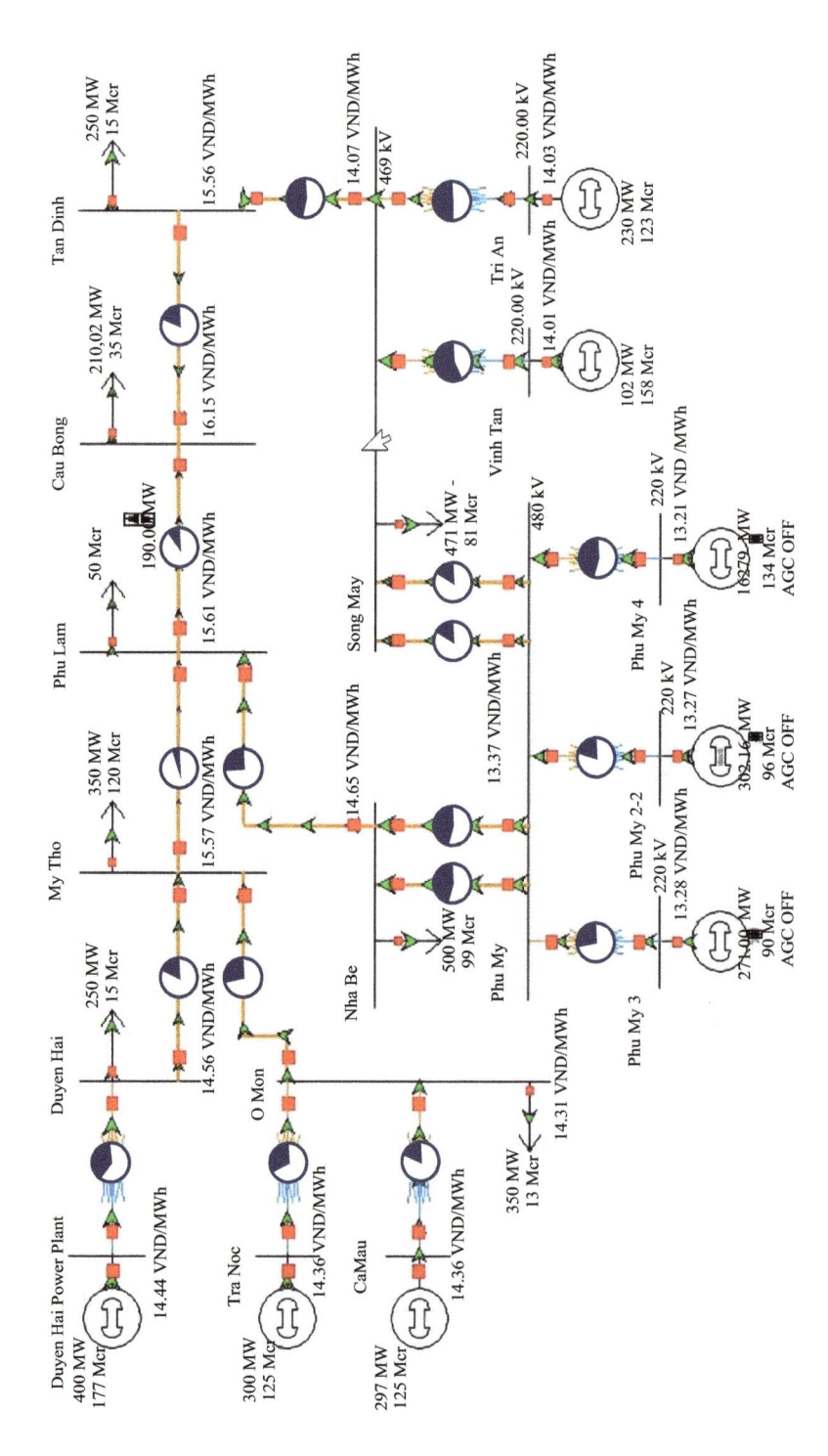

Figure 7.15 The 17-node electricity market of power transmission company no. 4 under the national power transmission corporation, Vietnam

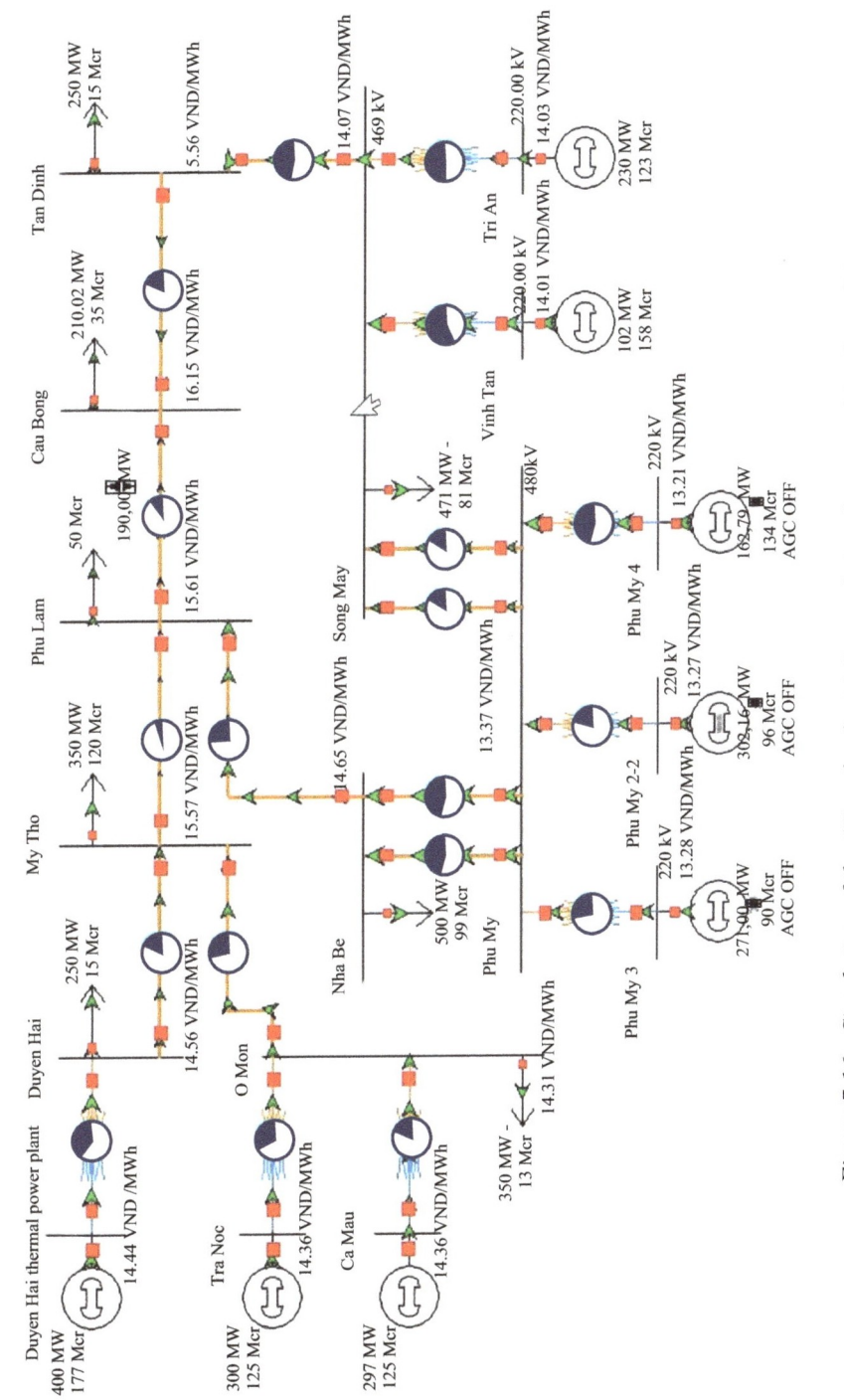

Figure 7.16 Simulation of the 17-node electricity market under normal load conditions

Table 7.29   System node parameters for the 17-node electrical system

Node	Name	Voltage (kV)	Voltage (PU)	Actual voltage (kV)	P (MW)	Q (Mvar)
1	Phu My 2-2	220	1	220	–	–
2	Phu My 3	220	1	220	–	–
3	Phu My 4	220	1	220	–	–
4	Phu My	500	0.96034	480.168	–	–
5	Nha Be	500	0.91971	459.857	500	199
6	Song May	500	0.9376	468.799	471	−81
7	Phu Lam	500	0.8934	446.701	190	50
8	Cau Bong	500	0.87737	438.684	210.02	35
9	Tan Dinh	500	0.89341	446.704	250	45
10	Tri An	220	1	220	–	–
11	My Tho	500	0.8986	449.299	350	120
12	Duyen Hai	500	0.92618	463.09	250	45
13	O Mon	500	0.94315	471.575	350	13
14	Tra Noc	220	1	220.001	–	–
15	Ca Mau	220	1	220.001	–	–
16	Duyen Hai Thermal Power Plant	220	1	220	–	–
17	Vinh Tan	220	1	220	–	–

Table 7.30   Line parameters for the 17-node electrical system

From node	To node	R	X	B	Length (km)	Limit (MVA)
Phu My 2-2	Phu My	0	0.0586	0	1.1	1.100
Phu My 3	Phu My	0	0.0586	0	2.45	1.000
Phu My 4	Phu My	0	0.0586	0	1.58	1.100
Nha Be	Phu My	0.01	0.00023	0.00023	42.628	1.000
Nha Be	Phu My	0.01	0.00023	0.00023	42.628	1.000
Song May	Phu My	0.017	0.00025	0.00021	65.947	1.000
Song May	Phu My	0.017	0.00025	0.00021	65.947	1.000
Nha Be	Phu Lam	0.01	0.00023	0.00023	16.7	390
Tan Dinh	Song May	0.01	0.00022	0.00023	41.198	500
Song May	Tri An	0	0.0586	0	11.62	500
Song May	Vinh Tan	0	0.0586	0	145.59	750
Phu Lam	Cau Bong	0.02	0.00023	0.00023	30.127	1,000
My Tho	Phu Lam	0.017	0.00025	0.00021	77	1,000
Cau Bong	Tan Dinh	0.01	0.00023	0.00023	21.015	1,000
My Tho	Duyen Hai	0.017	0.00025	0.00021	113.318	1,000
My Tho	O Mon	0.017	0.00025	0.00021	79.75	300
Duyen Hai thermal power plant	Duyen Hai	0	0.0586	0	2.1	300
Tra Noc	O Mon	0	0.0586	0	7.09	300
Ca Mau	O Mon	0	0.0586	0	80.135	2,000

Impacts of policies and mechanisms on renewable energy development 307

Table 7.31 Generator parameters of the 17-bus electrical system

Node	Name	P_{Gen} (MW)	Q_{Gen} (Mvar)	P_{min} (MW)	P_{max} (MW)
1	Phu My 2-2	302.46	95.84	100	1.050
2	Phu My 3	271	90.25	100	900
3	Phu My 4	462.79	134.39	100	645
10	Tri An	230	123.1	50	500
14	Tra Noc	300	125.22	50	200
15	Ca Mau	297	124.66	100	1,800
16	Duyen Hai	400	177.43	50	3,000
17	Vinh Tan	402	157.81	50	750

Table 7.32 Results of the 17-bus power system under normal load

Node	Name	Voltage (PU)	Node price (VND/MWh)
1	Phu My 2-2	1	13.27
2	Phu My 3	1	13.28
3	Phu My 4	1	13.21
4	Phu My	1.00345	13.37
5	Nha Be	0.97431	14.65
6	Song May	0.98538	14.07
7	Phu Lam	0.95198	15.61
8	Cau Bong	0.93484	16.15
9	Tan Dinh	0.94771	15.56
10	Tri An	1	14.03
11	My Tho	0.95424	15.57
12	Duyen Hai	0.94625	14.56
13	O Mon	0.99143	14.31
14	Tra Nóc	1	14,36
15	Ca Mau	1	14,36
16	Duyen Hai	1	14.44
17	Vinh Tan	1	14.01

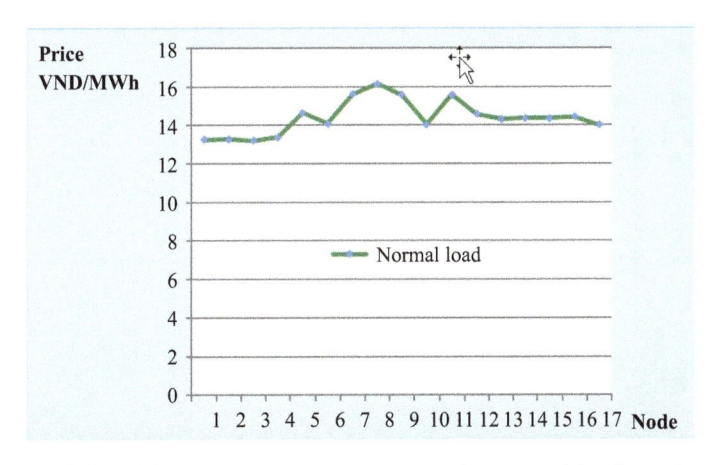

Figure 7.17 Nodal price representation under normal load conditions

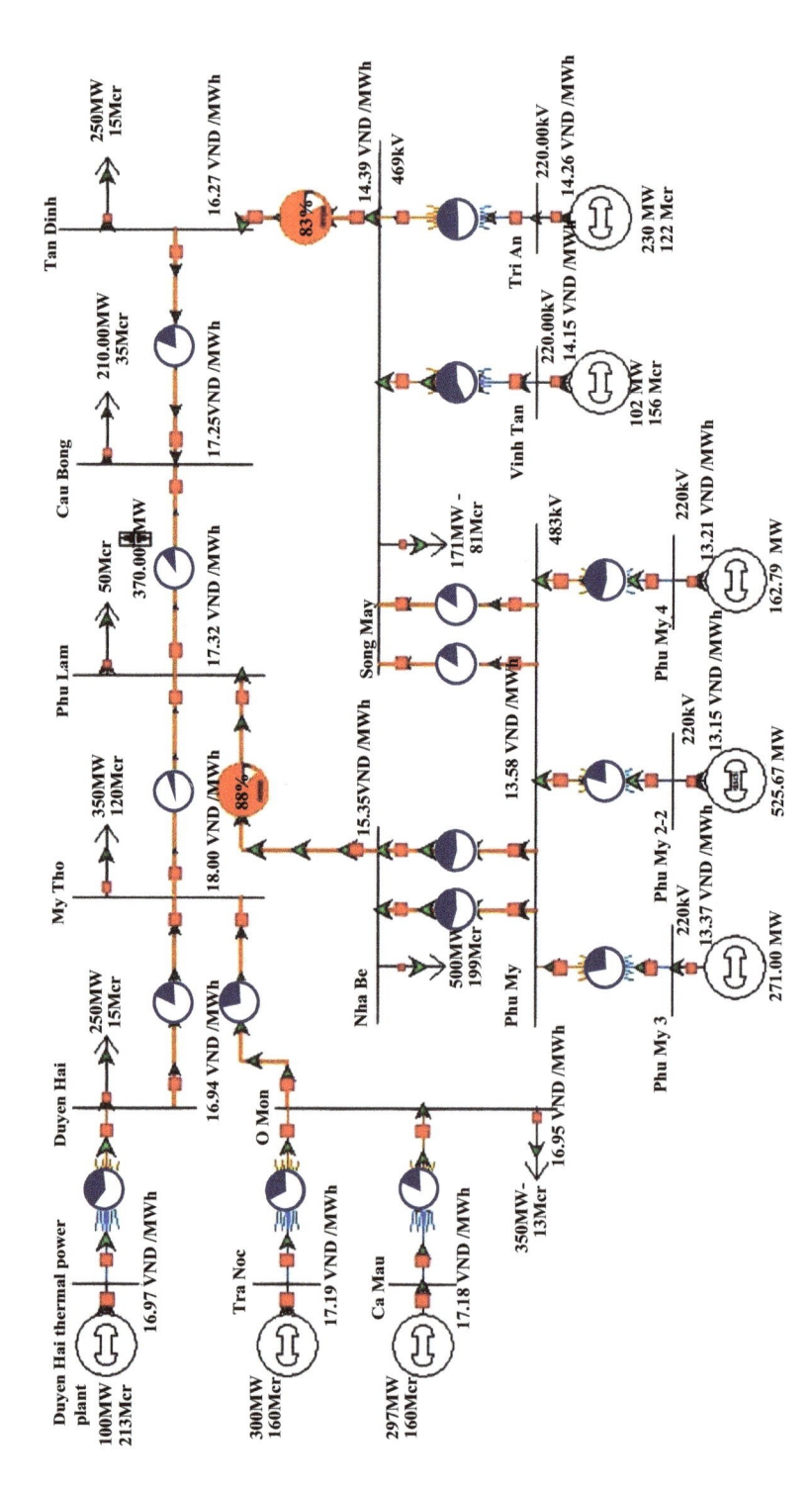

Figure 7.18 Simulation of the 17-node electricity market with transmission constraints and losses

Impacts of policies and mechanisms on renewable energy development 309

Table 7.33   Results of the 17-bus power system under transmission limitations and losses

No.	Name	Voltage (PU)	Node price (when transmission limited and losses) (VND/MWh)
1	Phu My 2-2	1	13.27
2	Phu My 3	1	13.28
3	Phu My 4	1	13.21
4	Phu My	1.00345	13.37
5	Nha Be	0.97431	14.65
6	Sông May	0.98538	14.07
7	Phú Lam	0.95198	15.61
8	Cau Bong	0.93484	16.15
9	Tan Dinh	0.94771	15.56
10	Tri An	1	14.03
11	My Tho	0.95424	15.57
12	Duyen Hai	0.94625	14.56
13	O Mon	0.99143	14.31
14	Tra Noc	1	14.36
15	Ca Mau	1	14.36
16	Duyen Hai thermal power plant	1	14.44
17	Vinh Tan	1	14.01

Voltage stability: Despite these changes, the voltage at all nodes remains within the acceptable stability limits, ensuring that the system remains operational and does not experience voltage instability.

These findings demonstrate the impact of increased load on the transmission system and the need for careful management of power flow to avoid congestion while maintaining system stability (Table 7.33).

The nodal prices at the buses, considering the transmission limits and losses, can be represented as shown in Figure 7.19.

Simulation results considering a generator fault at the Tri An node
When the generator at the Tri An node experiences a fault and stops generating power, the electricity market for the 17-node system is disrupted. The simulation results are shown in Figure 7.20: In this case, the likelihood of violating the transmission limits increases on the line from node 5 (Nha Be node) to node 7 (Phu Lam node) by 87%, and on the line from node 6 (Song May node) to node 9 (Tan Dinh node) by 85%. Consequently, the node prices at various locations fluctuate and rise at several nodes, as shown in Table 7.34.

The node prices at the busbars when the Tri An generator experiences a fault can be represented as shown in Figure 7.21.

Simulation results considering transmission line congestion
When transmission line congestion is considered, the power flow and market dynamics are significantly impacted. The simulation results reflect the changes in

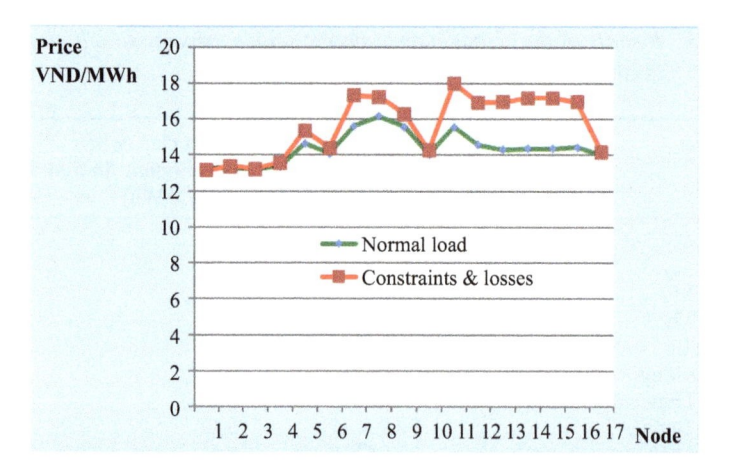

Figure 7.19 Node, price representation under normal load and during constraints and losses (price in VND/MWh)

voltage stability, node prices, and the overall transmission network performance under congestion conditions. As congestion occurs, power is constrained in certain areas, causing an increase in the price at the affected nodes.

These effects can be illustrated by the adjusted voltage levels and node prices across the system, showing how congestion affects both supply and demand in the electricity market. The results highlight potential limitations in the grid's ability to handle high loads, necessitating optimization of the transmission system to prevent price spikes and ensure stability.

The detailed simulation results will be shown in the corresponding figures and tables, which provide a visual representation of these changes under different congestion scenarios.

When a transmission line fault occurs from node 13 (O Mon node) to node 11 (My Tho node), the 17-node electricity market is disrupted, and the simulation results are shown in Figure 7.22. As a result, there is a transmission limit violation on the line from node 5 (Nha Be node) to node 7 (Phu Lam node) at 101%, and the line from node 6 (Song May node) to node 9 (Tan Dinh node) at 95%. Consequently, the node prices at the affected nodes also fluctuate, as shown in Table 7.35.

The node prices at the affected nodes due to transmission line congestion can be represented as shown in Figure 7.23.

Transmission pricing in the simulation of the 17-node power system
From the results obtained above, the transmission price can be calculated by the price differential between the nodes as follows:

- In the case where transmission limits and losses are considered, the transmission price is given as shown in Table 7.36.

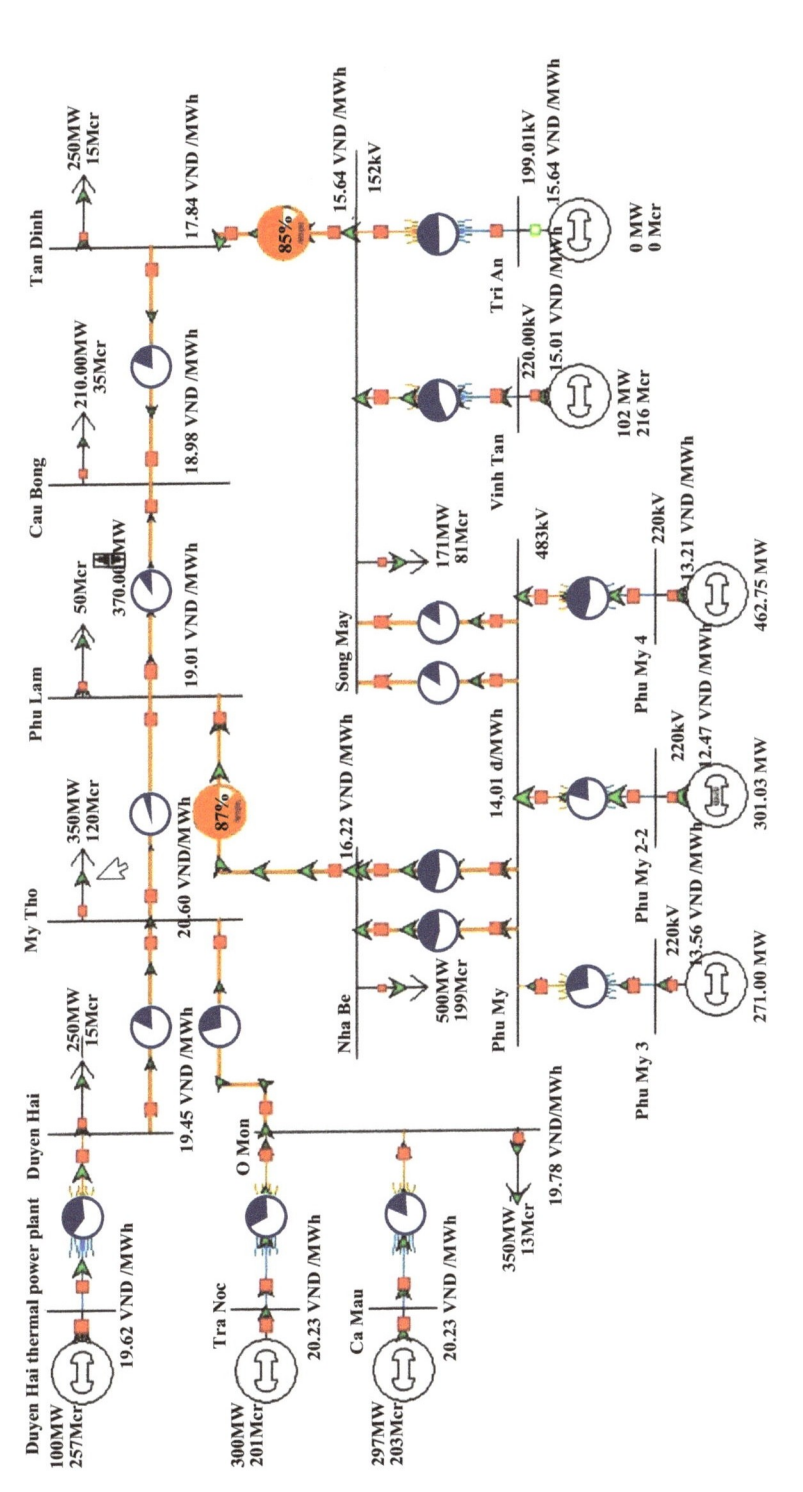

Figure 7.20 Simulation of a 17-node power market with generator failure at Tri An node

312 *Clean energy in South-East Asia*

Table 7.34 Results of the 17-bus power system under Trị An power plant generator failure

Node	Name	Voltage (PU volt)	Node price (when generator failure) (VND/MWh)
1	Phu My 2-2	1	12.47
2	Phu My 3	1	13.56
3	Phu My 4	1	13.21
4	Phu My	0.95208	14.01
5	Nha Be	0.89926	16.22
6	Song May	0.90473	15.64
7	Phu Lam	0.84988	19.01
8	Cau Bong	0.83752	18.98
9	Tan Dinh	0.85648	17.84
10	Tri An	0.90473	15.64
11	My Tho	0.85188	20.6
12	Duyen Hai	0.88092	19.45
13	O Mon	0.89791	19.78
14	Tra Noc	1.00001	20.23
15	Ca Mau	1.00001	20.23
16	Duyen Hai thermal power plant	1.00001	19.62
17	Vinh Tan	1	15.01

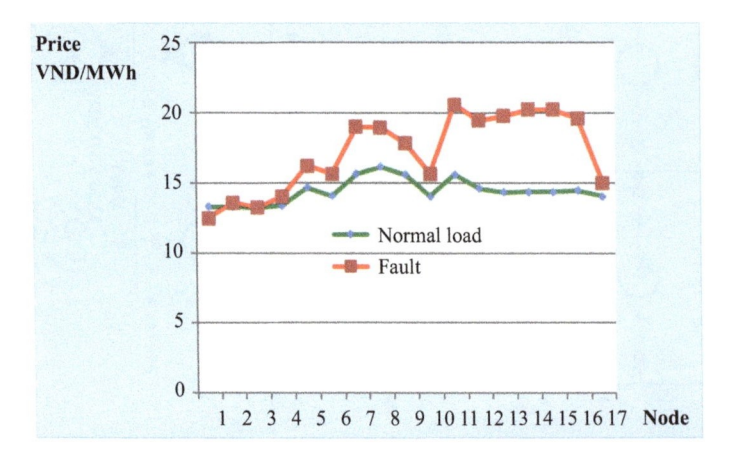

Figure 7.21 Node, price representation under normal load and during a fault

The transmission price, considering transmission limits and losses, can be represented as shown in Figure 7.24.

In the case of considering the Trị An generator failure, the transmission price is as shown in Table 7.37.

The transmission price considering the Tri An generator failure can be represented as shown in Figure 7.25.

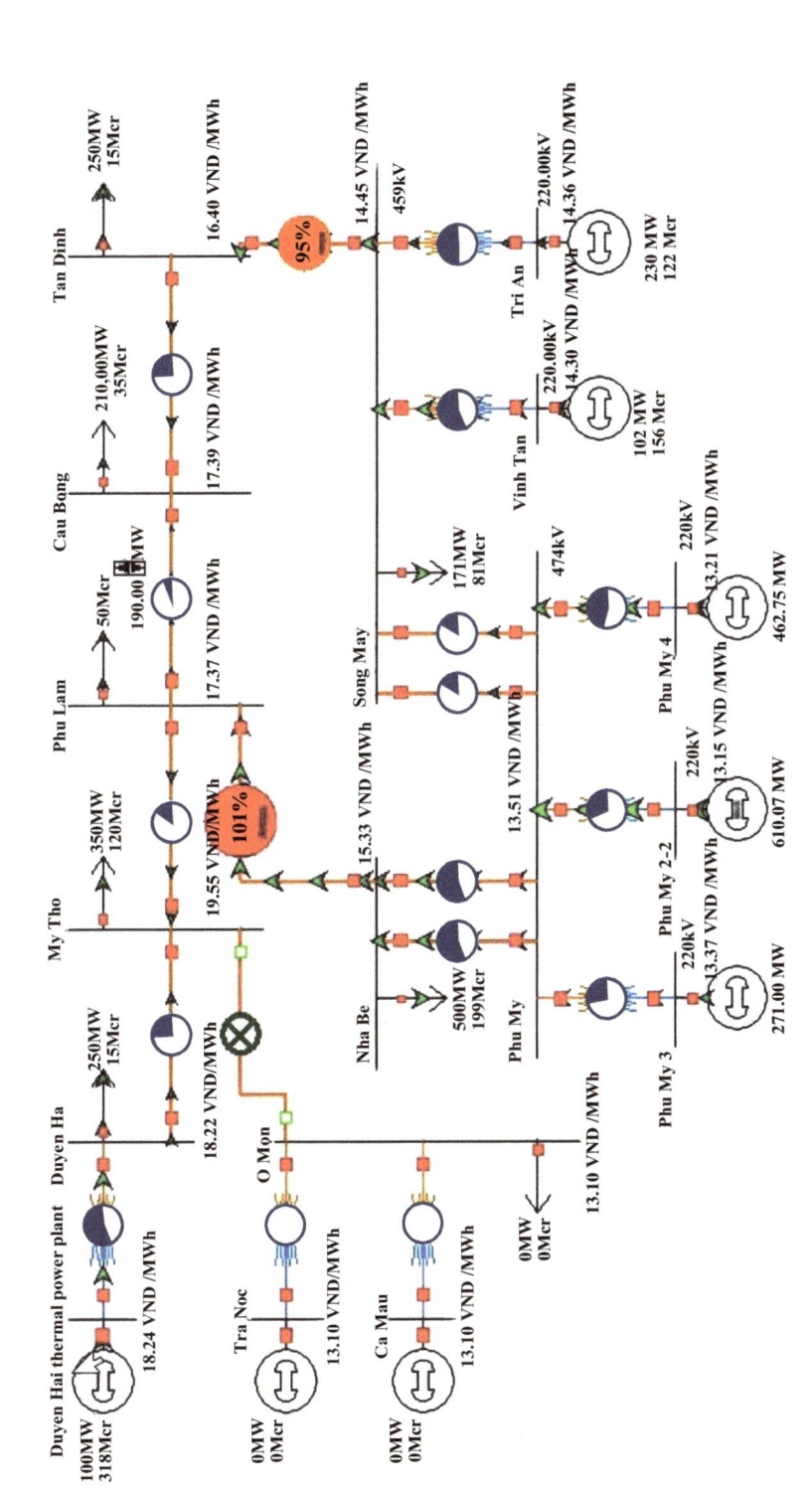

Figure 7.22 Simulation of the 17-node electricity market during a circuit congestion

Table 7.35 Results of the 17-bus power market under transmission line congestion

Node	Name	Voltage (PU)	Node price (when transmission congestion) (VND/MWh)
1	Phu My 2-2	1	13.10
2	Phu My 3	1	13.34
3	Phu My 4	1	13.21
4	Phu My	0.94752	13.51
5	Nha Be	0.89414	15.33
6	Song May	0.91775	14.45
7	Phu Lam	0.8439	17.37
8	Cau Bong	0.84115	17.39
9	Tan Dinh	0.8649	16.40
10	Tri An	1	14.36
11	My Tho	0.79922	19.55
12	Duyen Hai	0.82987	18.22
13	O Mon	–	13.10
14	Tra Noc	–	13.10
15	Ca Mau	–	13.10
16	Duyen Hai thermal power plant	1	18.24
17	Vinh Tan	1	14.30

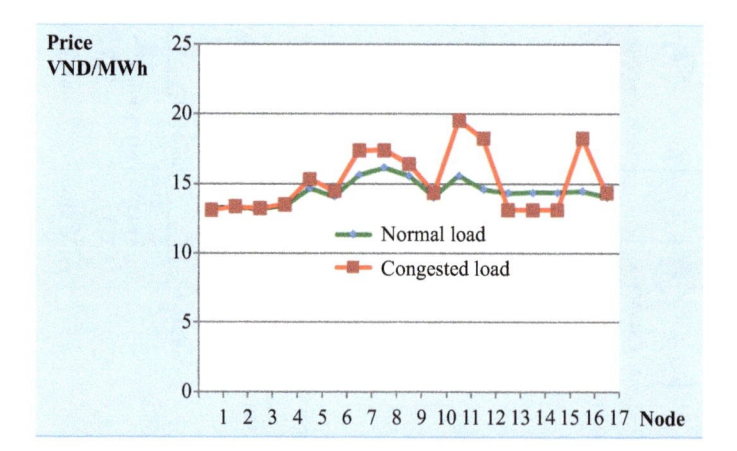

Figure 7.23 Representation of node prices under normal load and circuit congestion conditions

The transmission price considering congestion can be represented as shown in Table 7.38.

The transmission price considering congestion can be represented as shown in Figure 7.26.

Impacts of policies and mechanisms on renewable energy development 315

Table 7.36 Transmission electricity prices considering transmission limitations and losses

No.	From node	To node	Transmission electricity price (VND/MWh)
1	Phu My 2-2	Phu My	0.11
2	Phu My 3	Phu My	0.13
3	Phu My 4	Phu My	0.22
4	Nha Be	Phu My	1.18
5	Nha Be	Phu My	1.18
6	Song May	Phu My	0.57
7	Song May	Phu My	0.57
8	Nha Be	Phu Lam	1.42
9	Tan Dinh	Song May	1.43
10	Song May	Tri An	0.09
11	Song May	Vinh Tan	0.16
12	Phu Lam	Cau Bong	0.09
13	My Tho	Phu Lam	0.64
14	Cau Bong	Tan Dinh	0.69
15	My Tho	Duyen Hai	0.61
16	My Tho	O Mon	0.91
17	Duyen Hai thermal power plant	Duyen Hai	0.07
18	Tra Noc	O Mon	0.07
19	Ca Mau	O Mon	0.11

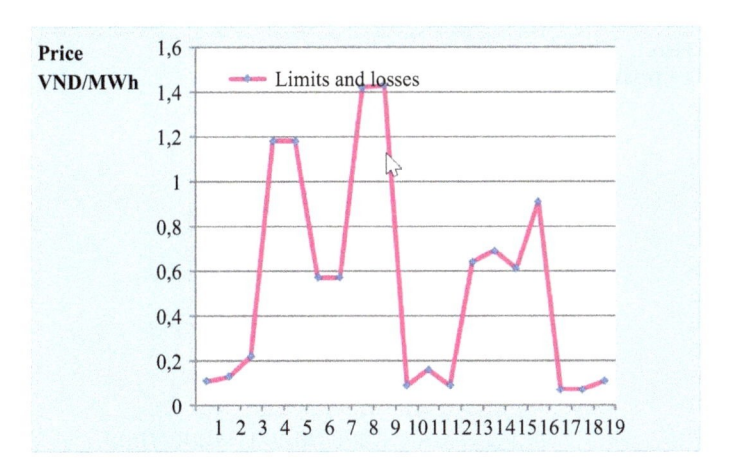

Figure 7.24 Transmission price representation considering transmission limits and losses

Table 7.37 Transmission electricity prices considering the Tri An generator fault

No.	From node	To node	Transmission electricity price (VND/MWh)
1	Phu My 2-2	Phu My	0.07
2	Phu My 3	Phu My	0.10
3	Phu My 4	Phu My	0.18
4	Nha Be	Phu My	1.09
5	Nha Be	Phu My	1.09
6	Song May	Phu My	0.56
7	Song May	Phu My	0.56
8	Nha Be	Phu Lam	1.22
9	Tan Dinh	Song May	1.36
10	Song May	Tri An	0.07
11	Song May	Vinh Tan	0.12
12	Phu Lam	Cau Bong	0.23
13	My Tho	Phu Lam	1.11
14	Cau Bong	Tan Dinh	0.62
15	My Tho	Duyen Hai	1.07
16	My Tho	O Mon	0.90
17	Duyen Hai thermal power plant	Duyen Hai	–
18	Tra Noc	O Mon	0.09
19	Ca Mau	O Mon	0.14

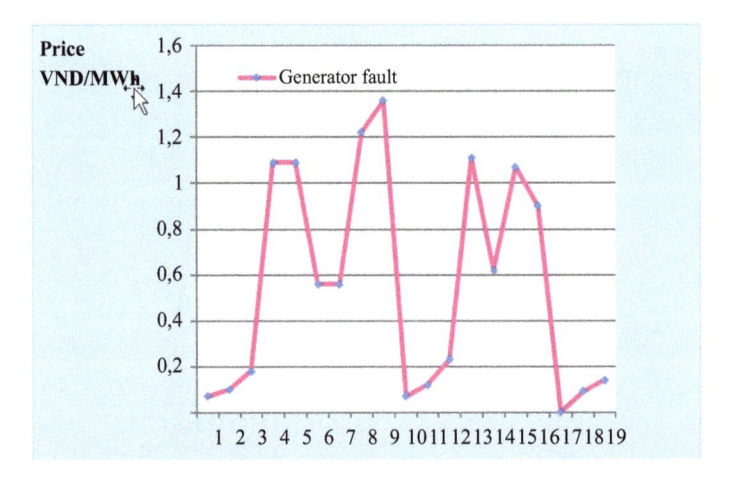

Figure 7.25 Transmission price representation considering the Tri An generator fault

Impacts of policies and mechanisms on renewable energy development 317

Table 7.38 *Transmission electricity prices considering transmission line congestion*

No.	From node	To node	Transmission electricity price (VND/MWh)
1	Phu My 2-2	Phu My	0.16
2	Phu My 3	Phu My	0.15
3	Phu My 4	Phu My	0.26
4	Nha Be	Phu My	1.29
5	Nha Be	Phu My	1.29
6	Song May	Phu My	0.60
7	Song May	Phu My	0.60
8	Nha Be	Phu Lam	1.64
9	Tan Dinh	Song May	1.50
10	Song May	Tri An	0.11
11	Song May	Vinh Tan	0.20
12	Phu Lam	Cau Bong	0.06
13	My Tho	Phu Lam	0.80
14	Cau Bong	Tan Dinh	0.77
15	My Tho	Duyen Hai	0.66
16	My Tho	O Mon	0.86
17	Duyen Hai thermal power plant	Duyen Hai	0.08
18	Tra Noc	O Mon	0.09
19	Ca Mau	O Mon	0.14

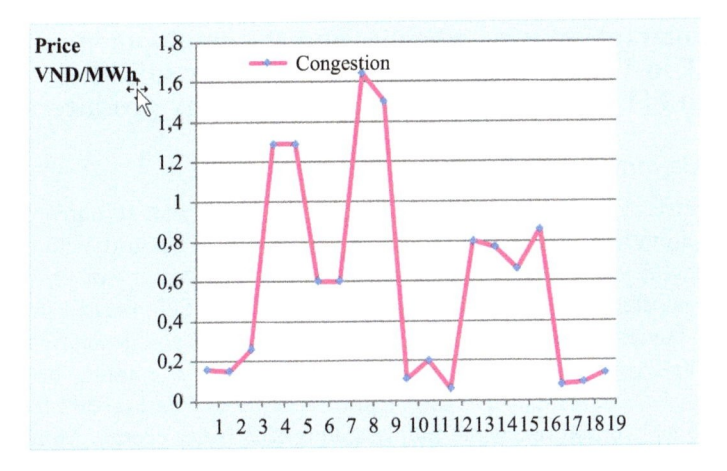

Figure 7.26 *Transmission price representation considering grid congestion*

7.3.5 Analysis of simulation results

- With the simulation results presented in Tables 7.21–7.23, the node prices can be either higher or lower than the initial price. This provides economic signals for investors participating in the electricity market.
- Node prices give market participants information about location-based pricing with transmission constraints. According to the node pricing method, each key transmission node (generator, interface, transmission line, or substation) will have a different electricity price.
- When considering transmission line constraints, optimizing power flow to the nodes becomes necessary. This optimization will change the price between nodes, and the price differences will be incorporated into the transmission pricing.

Based on the results achieved, accurately determining the transmission price will enhance transparency, with optimal node prices attracting various investments both domestically and internationally. This ensures continuous investment in the development of power generation and the grid in a competitive electricity market, which is highly practical. The study has focused on solving the following key issues:

- The transmission price results align with the developing and improving conditions of Vietnam's electricity market.
- The transmission price, as the price difference between two different nodes, affects all parties involved in the electricity market. This provides economic signals for investors in the electricity market.

7.4 Renewable energy subsidies and the development of FiT in Vietnam: calculation methods for subsidies and FiT mechanism for renewable energy producers

7.4.1 Definition of FiT

Feed-in tariffs (FiT) are electricity prices set by governments to allow electricity producers from renewable energy sources to sell power to the grid, with the aim of encouraging development and enhancing the competitiveness of these energy sources compared to traditional ones. Additionally, the FiT mechanism requires electricity buyers to purchase power at the fixed price for any generation capacity over a long-term period (usually 15–25 years) through a contract between the electricity producer and buyer. The FiT price is adjusted based on different energy types such as hydropower, wind, biomass, biogas, or solar energy. Moreover, the FiT price changes every 4–5 years. In the initial years, the FiT price is higher than the following years to incentivize renewable energy producers to develop technologies, reduce costs, and bring renewable energy prices closer to the market prices of fossil fuels.

7.4.2 FiT pricing method

7.4.2.1 Existing methods and their advantages and disadvantages

One basic method for calculating FiT is based on the lifetime cost of the project and the electricity generation output. The total lifetime cost includes the initial investment, O&M costs, fuel costs for operation, inflation factors, and an additional portion of profit for the investor. This method is simple and aligned with investment costs for each technology; however, it does not account for changes due to the specific characteristics of each energy type.

The "avoided cost" method, which is the cost of producing 1 kWh from the most expensive generation unit in the national grid, can be calculated separately for the dry and wet seasons, peak hours, average hours, and off-peak hours.

The fixed price method usually considers the market electricity price, with policymakers setting a price higher than the market price to ensure profit for investors. This method is suitable for countries with competitive markets and transparent electricity prices.

The bidding-based price determination method allows electricity project developers to submit wide-ranging bids, selecting the lowest price that meets technical requirements.

7.4.2.2 Objective function for FiT price calculation

The formula for calculating FiT is as follows:

$$\text{FiT} = \text{LCOE} + K_{\text{ln}} \tag{7.21}$$

where

$$\text{LCOE} = \frac{\sum_{t=1}^{n} \left(\frac{I_t + M_t + F_t}{(1+i)^t}\right)}{\sum_{t=1}^{n} \left(\frac{E_t}{(1+i)^t}\right)} \tag{7.22}$$

FiT price calculation method:
where

- *LCOE* (levelized cost of energy) represents the average cost of electricity generation, calculated as the ratio of total initial investment cost, reinvestment cost, O&M costs, and fuel costs over the total electricity generation output.
- K_{ln} is the profit margin for the investor.
- I_t represents the total initial and reinvestment costs in year *tt* (USD).
- M_t is the total O&M costs in year *tt* (USD).
- E_t is the electricity production in year *tt* (MWh).
- F_t is the total fuel cost in year tt (USD).
- *i* is the financial discount rate (%).
- *n* is the economic lifetime of the solar power plant (usually 25 years).
- *t* represents the year in the plant's economic life (from year 1 to year *nn*).

LCOE calculation

LCOE is the cost per unit of electricity, representing the lowest price at which electricity must be sold for the total revenue to just cover the total costs without providing any profit to the investor. The formula for calculating LCOE is defined in (7.22).

where the terms represent the costs and generation over the economic life of the plant (as detailed above).

Factors influencing FiT pricing

Several key factors influence the FiT price:

- Solar irradiance: The solar radiation levels in different areas where the solar plant is located will impact the output of the plant. Higher solar irradiance leads to higher energy generation, and lower irradiance leads to lower output.
- Grid connection conditions: The distance from the plant to the grid connection point also affects the total investment costs. If the connection is far, the costs increase.
- Investment and output variables: Both the total investment amount and the output are variables of the objective function, thus influencing the FiT price.

7.4.2.3 Simulation of FiT calculation process considering specific changing factors

Based on the practical implementation of solar energy projects in Vietnam, the paper proposes a method for calculating the FiT price using a block diagram as shown in Figure 7.27. The calculation steps involve combining the PVSys V6.62 software with an Excel program developed by the authors. The specific steps are outlined as follows:

1. Input data: Collect data on solar irradiance, grid connection conditions, and initial cost projections.

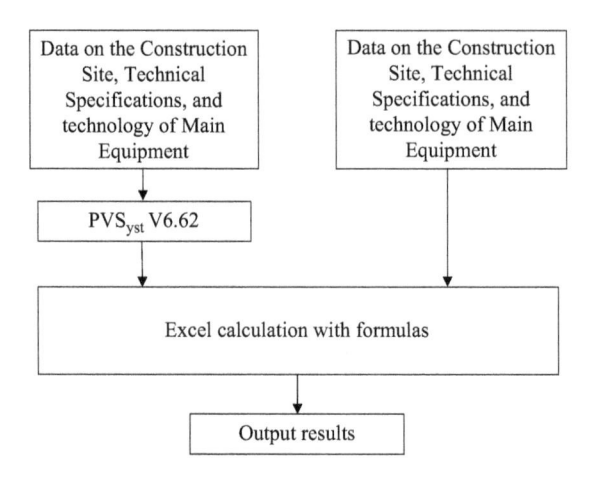

Figure 7.27 Block diagram of the simulation process

Impacts of policies and mechanisms on renewable energy development 321

2. Estimate output: Use the PVSys V6.62 software to simulate solar energy output based on irradiance data and design specifications.
3. Cost calculation: Calculate the total initial costs, annual operational and maintenance costs, and fuel costs over the plant's lifetime.
4. Apply discount rate: Discount future costs to their present value using the financial discount rate.
5. LCOE calculation: Compute the LCOE using the above formula to determine the minimum cost per unit of electricity.
6. Determine FiT: Finally, calculate the FiT price by considering the investor's required profit margin (K_{ln}) and the LCOE.

This method integrates both technical software simulations and financial calculations to determine an optimal FiT price, ensuring that the price accurately reflects the costs and profits for solar energy producers.

7.4.2.4 FiT price calculation process
Input data

The baseline scenario is based on data from a 50 MW solar power plant in Khanh Hoa Province. The following details provide context for this project:

- Total construction area: 70 hectares
- Installed capacity: 61.1 MWp
- Grid connection capacity: 50 MWac
- Technology used: Poly-Si photovoltaic panels (188,000 panels with a rated capacity of 325 Wp)
- Inverters: 50 units, each with a capacity of 1,000 kWac
- Transformers: 2 units of 1,250 kVA (0.4/22 kV), 24 units of 2,500 kVA (0.4/22 kV)
- Grid connection distance: 10 km of 110 kV lines and 22/110 kV 63 MVA substation
- Financing: 20% equity, 80% commercial loans at an annual interest rate of 8.6%
- Total investment: Approximately 1.261 trillion VND (around 51 million USD)
- Land compensation cost: 35 billion VND (around 1.4 million USD), averaging 0.5 billion VND per hectare
- Lifespan of equipment: 20–25 years, with inverters having a lifespan of around 10 years
- Reinvestment costs: Reinvestment for inverters in the 11th year, totaling 111 billion VND (around 4.4 million USD)
- Annual O&M costs: 2% of the total initial investment
- Fuel costs: Zero for solar energy
- Profit margin (K_{ln}): Assumed to be 0.3 cents (USD), based on a profit ratio after tax of approximately 3% of total revenue.

The PVSyst V6.62 software is used to simulate and calculate the output of the solar power plant (Tables 7.39 and 7.40). The software uses solar radiation data

Table 7.39 Data of representative areas and coordinates

Representative area name	Representing province	Latitude (°)	Longitude (°)
Areas 1: Northwest	Lang Son	21°51' B	106°45' D
Areas 2: Northeast	Son La	21°19' B	103°54' D
Areas 3: North Central	Quang Tri	16°44' B	107°11' D
Areas 4: Central Highlands and South Central	Khanh Hoa	12°01' B	109°09' D
Areas 5: Southern	Bac Lieu	09°15' B	105°45' D

Table 7.40 Different calculation options

No.	Option	Power plant capacity and connection details	Total investment
1	Base PA	50 MWAC capacity, 12 km of 110 kV transmission line, 1 × 22/110 kV 63 MVA substation	1.261 billion VND
2	PA1	50 MWAC capacity, 20 km of 110 kV transmission line, 1 × 22/110 kV 63 MVA substation	1.289 billion VND
3	PA2	50 MWAC capacity, 1 × 22/110 kV 63 MVA substation	1.232 billion VND

from the MeteoNorm 7.1 source, along with additional location-specific data such as:

Coordinates (latitude and longitude)

- Elevation above sea level
- Total installed capacity of the solar panels
- Technical specifications of the main equipment

Process for FiT calculation

- Step 1: Enter input data into PVSyst V6.62
  Enter geographical coordinates (longitude, latitude), elevation, and technical data of the primary equipment into the software.
- Step 2: Run PVSyst V6.62 Software
  The software calculates the electricity output for the entire year based on the input data.
- Step 3: Input investment and O&M costs into Excel
  Input all relevant cost data, including investment costs, O&M costs, and any other associated costs into an Excel spreadsheet.
- Step 4: Import electricity output data
  Enter the electricity output data from PVSyst V6.62 into Excel, along with the costs from Step 3. Then apply Formula (1) (the LCOE formula) to calculate the FiT price.
- Step 5: Continue data entry for different scenarios
  For each variable scenario (e.g., location, grid connection conditions), input the data accordingly and run the calculations to see the FiT results for each situation.

Impacts of policies and mechanisms on renewable energy development 323

Key considerations
Solar radiation data: Solar irradiance varies by location, so different areas will produce different levels of energy. Areas with higher solar radiation yield more electricity, affecting the FiT price.

Grid connection conditions: The distance between the plant and the grid connection point impacts investment costs. Longer distances result in higher costs.

Reinvestment in inverters: The inverter replacement is an essential factor in the 11th year. The reinvestment cost should be accounted for in the calculations.

Financing model: With 80% commercial loan financing, the interest rate impacts the total investment cost. The equity portion contributes to the project's capital cost, influencing the profitability.

O&M costs: These costs are estimated as a percentage of the initial investment and are crucial for determining the long-term operational viability of the plant.

Example data for base scenario

- Installed capacity: 61.1 MWp
- Annual electricity production: 85,000 MWh (as calculated by PVSyst)
- Investment costs: 1.261 trillion VND
- Annual O&M costs: 2% of the total investment
- Reinvestment in inverters (11th year): 111 billion VND
- Fuel costs: 0

Output results
Once the calculations are completed, the FiT price can be determined based on these inputs, adjusting for any changes in the solar irradiance, grid connection, and investment conditions.

The result will provide a FiT that ensures the project covers its costs and generates a reasonable return on investment for the investor, given the specific conditions of the solar plant location and technology used.

7.4.3 Results

The calculated FiT prices for different scenarios and regions are presented in Table 7.41 and visually compared through Figure 7.28. These results show the variation in FiT prices depending on several factors.

Comment 1: According to the results presented in Table 7.41 and Figure 7.29, the FiT price for large-scale solar power plants varies significantly depending on

Table 7.41 FiT price calculation results for different scenarios

No.	Representative areas	Base scenario price for each area	Option 1 price	Option 2 price
1	Areas 1: Northwest	11.23	11.44	11.01
2	Areas 2: Northeast	9.93	10.11	9.74
3	Areas 3: North Central	9.33	9.51	9.15
4	Areas 4: Central Highlands and South Central	7.83	7.97	7.68
5	Areas 5: Southern	9.19	9.36	9.02

324 *Clean energy in South-East Asia*

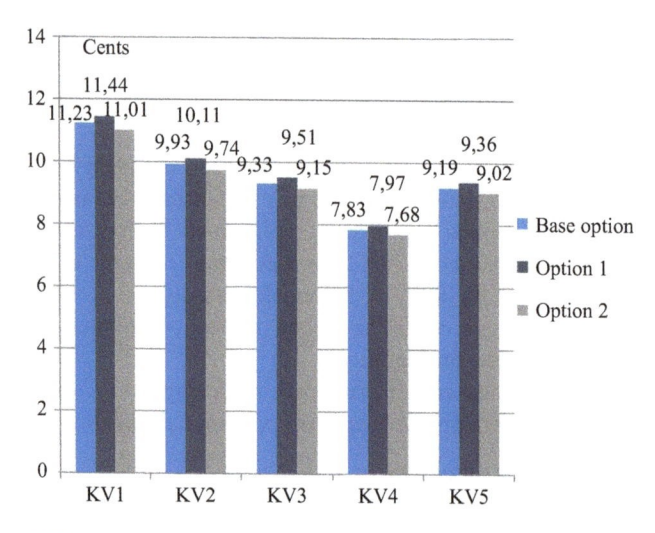

Figure 7.28 FiT price calculation table according to different options

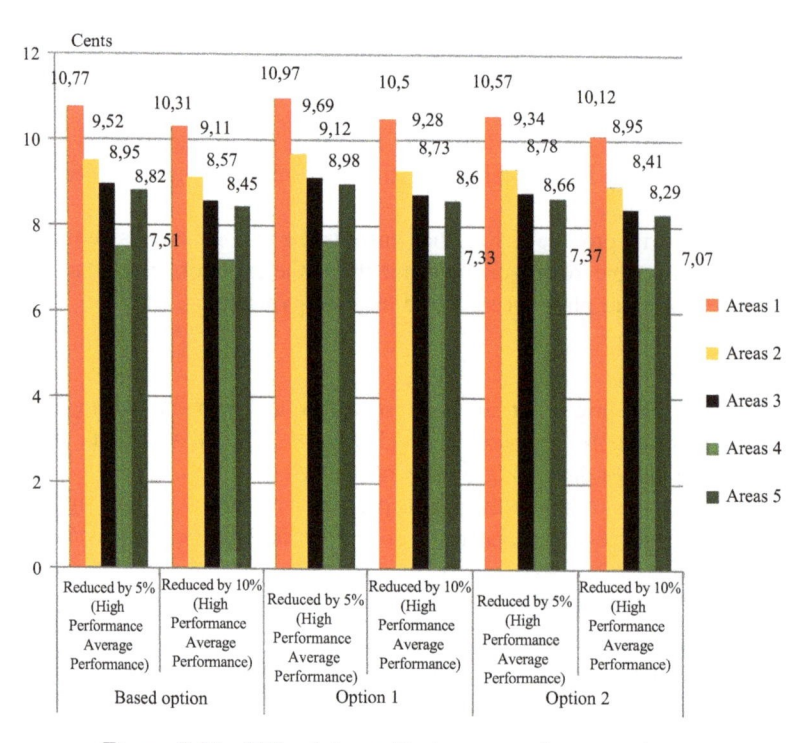

Figure 7.29 FiT pricing table in cases of price reduction

the geographic location in each region. With the current government-set FiT price of 9.35 cents/kWh, only solar power plant projects located in regions from the central area to the south are financially feasible for construction. Among these, the Central South and Central Highlands regions are the most favorable. Projects in the Northern region are not financially viable. Additionally, the FiT price is influenced by the conditions for connecting to the existing grid. If the grid connection length exceeds 20 km, only plants in the Central South and Central Highlands regions remain financially feasible. In cases where a price reduction is anticipated after bidding, the calculated FiT prices for various options are presented and compared in the graphs in Figure 5.

Comment 2: In cases where the EPC contract price (design, construction, and installation contract) is reduced by 5–10% from the estimated price, most plants in the regions would become financially viable, except for those in Region 1 (Northwest) as shown in Figure 7.29. However, this reduction introduces potential risks related to the output quality of the plant's electricity.

The FiT pricing model mentioned in the article above does not consider system constraints such as transmission line congestion or the shortage of reserve generation capacity, which could affect the stable operation of the power system. This is particularly important because solar power plants inherently have variable and unstable generation capacity depending on weather conditions. Such issues could impact the security of power system operations, potentially requiring the temporary shutdown of solar power plants.

7.4.4 FiT pricing policy

Vietnam's electricity sector is undergoing a transformation, with significant changes since 2017. These changes have driven exceptional growth through the development of renewable energy sources, such as wind and solar power, from 2018 to 2021.

According to data from Vietnam electricity (EVN), by the end of 2020, the total installed solar capacity reached 19.4 GWp, of which 9.3 GWp came from over 100,000 rooftop solar systems, with the remainder from solar power plants. By the end of October 2021, wind power capacity was expected to increase by approximately 5.7 GW. Thus, the total installed capacity of variable renewable energy sources was about 28%, excluding hydropower.

However, Vietnam's electricity system is still primarily dependent on fossil fuel-based power (64% of the total, with coal-fired thermal power contributing 123 billion kWh (~50%) and gas-fired thermal power contributing 34.7 billion kWh (~14%)). The favorable results mentioned above were driven by the impact of FiT policies for renewable energy during this period. Specifically:

7.4.4.1 FiT pricing policy for wind power

The first FiT policy for wind power was issued in 2011 with a price of 78 USD/MWh. However, this price was considered commercially unfeasible, resulting in an installed capacity of only 135 MW by 2017. Notably, all three projects implemented during this period applied special financial mechanisms rather than relying entirely on the FiT price.

To promote wind power development in Vietnam, the prime minister issued FiT2 in November 2018, with a price of 85 USD/MWh for onshore wind projects and 98 USD/MWh for offshore wind projects (Decision No. 39/2018/QD-TTg). This decision quickly had a positive impact on the market, with an estimated 5,886 MW of wind power expected to be operational by the end of 2021, 43 times the total installed capacity in 2017.

7.4.4.2 FiT pricing policy for solar power

The first FiT policy for solar power was issued in 2017, setting a price of 9.35 US cents/kWh, with a deadline for grid connection by June 30, 2019. This policy triggered the market, leading to approximately 4.5 GW of solar power being connected during this period.

The second FiT policy set prices of 7.09, 7.69, and 8.38 US cents/kWh, respectively, for ground-mounted, rooftop, and floating solar power plants, with a grid connection deadline of December 31, 2020.

7.4.5 Challenges and solutions

The above policies, which apply to electricity purchases from renewable energy sources for 20 years, have had a positive impact on investors, helping the market mature quickly. However, this has raised concerns that purchasing renewable energy at attractive FiT prices could lead to an increase in retail electricity prices and a supply-demand imbalance.

According to experts, Vietnam should learn from other countries to promote the sustainable development of the domestic renewable energy market. In terms of mechanisms, there needs to be a gradual shift from the FiT policy to competitive bidding and renewable energy pricing.

7.4.6 Benefits of competitive bidding

Competitive bidding mechanisms are being implemented in over 60 countries due to the following benefits:

1. Rational development planning: Effectively integrates renewable energy into the grid.
2. Cost savings: The price of purchasing renewable energy approaches market prices.
3. Ensures timely project completion: Supports a fair energy transition process.
4. Increases transparency: Clear power purchase agreements (PPAs) reduce financial and legal risks.

Additionally, competitive bidding helps achieve other development goals, such as creating jobs, promoting technology, and contributing to national energy security.

7.4.7 Policy recommendations

It is necessary to set a cap on prices and pre-establish installed capacities, while also implementing clear commitment mechanisms in PPAs, specifying the responsibilities of the parties, ensuring project progress, and striving for a green economy and carbon neutrality by 2050.

7.5 Subsidy mechanism: limitations, analysis of subsidy mechanisms, and recommendations to promote renewable energy development in Vietnam

7.5.1 Introduction

Renewable energy plays a crucial role in Vietnam's sustainable economic development strategy. The government has implemented several subsidy mechanisms to promote renewable energy projects, aiming to mitigate the impact of climate change, reduce carbon emissions, and enhance national energy security. However, current mechanisms still have many shortcomings, which reduce their effectiveness. This article will analyze the existing subsidy mechanisms, highlight their limitations, and propose solutions to improve their effectiveness.

7.5.2 Existing subsidy mechanisms and their importance

- **Direct subsidies**
  - Financial support to offset investment costs for renewable energy projects.
  - Provides incentives for investors to participate in the sector despite high initial costs.

- **Feed-in tariff**
  - Fixed price for purchasing electricity from renewable energy projects for 20 years.
  - Ensures stable income for producers.
  - For example, in 2017, the FiT for solar power was 9.35 US cents/kWh, which helped trigger 4.5 GW of capacity by mid-2019.

- **Tax incentives**
  - Tax exemptions or reductions for renewable energy projects.
  - Reduces operational costs and increases investment attractiveness.

7.5.2.1 Limitations of current subsidy mechanisms

- **Management challenges**
  - Monitoring and allocating subsidies is complex due to the large number of projects.
  - Risks of fraud or inefficient use of resources.

- **Lack of transparency**
  - Subsidy processes are unclear, leading to a lack of trust from investors.
  - There is insufficient publicly available information on subsidy policies.
- **Limited appeal**
  - Current subsidy levels are not attractive enough compared to profits from fossil fuels.
  - Not sufficient to maintain competition and strong growth in the renewable energy market.
- **Economic balance**
  - Excessive subsidies for renewable energy may strain the national electricity budget.
  - There is a risk of increased retail electricity prices for consumers.

7.5.2.2 Relevant legal documents

- **General policy**
  - Decision No. 2068/QD-TTg (2015): Renewable energy development strategy through 2030.
  - Resolution No. 55-NQ/TW (2020): National energy development strategy.
- **Wind energy**
  - Decision No. 39/2018/QD-TTg: FiT for onshore wind at 8.5 US cents/kWh, offshore at 9.8 US cents/kWh.
- **Solar energy**
  - Decision No. 11/2017/QD-TTg: Initial FiT for grid-connected solar power at 9.35 US cents/kWh.
  - Decision No. 13/2020/QD-TTg: Revised FiT applicable until December 31, 2020, at 8.38 US cents/kWh.

7.5.2.3 Impact analysis

- **Renewable energy development**
  - Significant growth: By 2021, 28% of installed capacity came from solar and wind energy.
  - The FiT mechanism played a critical role, especially in regions with high renewable potential.
- **Market challenges**
  - Over-reliance on FiT can lead to supply-demand imbalances.
  - Some renewable energy projects have delayed grid connection due to rapid development.
- **Economic and environmental benefits**
  - Reduced emissions and supported the goal of net-zero emissions by 2050.
  - Created jobs and promoted technological advancements in the renewable energy sector.

7.5.2.4 Proposed solutions

- **Increase transparency and management**
  - Build a public database to monitor subsidies.
  - Implement independent audits to ensure transparency.

- **Transition to competitive bidding**
  - Replace FiT with competitive bidding mechanisms.
  - Price renewable energy based on market fluctuations, as implemented in Chile and Denmark.

- **Public–private partnerships (PPP)**
  - Encourage PPP models to share risks and responsibilities in investment.

- **Enhance the capacity of regulatory bodies**
  - Train regulatory authorities to improve oversight and enforcement of subsidy policies.
  - Utilize modern tools in subsidy management.

- **Learn from international experiences**
  - Study and adopt competitive bidding models, such as those used for offshore wind projects in Denmark.
  - Implement CO_2 quotas and green certificates to encourage sustainable energy use.

The current renewable energy subsidy mechanisms in Vietnam have contributed to strong development but need reform to ensure long-term sustainability. Shifting to competitive bidding, improving transparency, and encouraging PPPs will enhance subsidy effectiveness. By addressing these challenges, Vietnam can achieve its goal of sustainable energy development while contributing to environmental protection and economic growth in the future.

7.6 Impact of renewable energy development on environmental, economic, and social policies in Vietnam

7.6.1 Introduction

RE has become an essential part of the sustainable development strategy for many countries, especially Vietnam. Faced with increasing environmental issues such as air pollution, climate change, and the depletion of natural resources, developing RE is not only an option but also an urgent requirement. This study explores the impacts of RE on the environment, economy, and society in Vietnam, emphasizing its crucial role in sustainable development.

7.6.2 Environmental impact

7.6.2.1 Reducing pollution

The development of RE is an effective solution to reduce carbon emissions and pollutants compared to fossil fuel usage. Studies show that the renewable energy sector can significantly reduce air pollution, improving the quality of life for communities. The use of solar, wind, and hydropower energy not only limits greenhouse gas emissions but also reduces negative health effects such as respiratory and cardiovascular diseases.

A clear example of this positive impact is the rapid development of wind and solar power plants in Vietnam in recent years. These projects not only generate clean energy but also bring environmental benefits to the surrounding areas.

7.6.2.2 Protecting natural resources

RE helps reduce dependence on non-renewable resources such as oil and coal. Energy sources like wind, solar, and hydropower ensure a sustainable energy supply while preserving natural resources for future generations.

The development of RE projects encourages individuals and businesses to participate in environmental protection activities, thereby raising awareness about resource conservation. Additionally, RE helps alleviate pressure on natural ecosystems, contributing to the protection of biodiversity and ecological balance amidst current climate change.

7.6.3 Economic impact

7.6.3.1 Job creation

The renewable energy sector has the potential to create numerous job opportunities in fields such as production, installation, and maintenance of energy systems. Reports show that RE development not only generates millions of new jobs but also stimulates economic growth in both rural and urban areas.

Industries related to RE, such as solar and wind equipment manufacturing and energy storage technology, are also growing rapidly, opening up opportunities for investors and businesses.

7.6.3.2 Attracting investment and development

Both the government and businesses can attract investments in RE projects, contributing to economic growth and infrastructure development. Investing in RE not only ensures a sustainable energy supply but also creates opportunities to improve transportation, electricity transmission, and other utilities. This will stimulate economic activities in neighboring areas and create favorable conditions for business development.

Increasing investment in RE also enhances Vietnam's competitiveness in the global market, particularly as the demand for clean energy grows.

7.6.4 Social impact

7.6.4.1 Raising public awareness

The development of RE can enhance public awareness of environmental protection and encourage more sustainable lifestyles. The use of renewable energy brings not only economic benefits but also educates the community on the importance of environmental conservation. Through educational programs and community activities, people will gain a better understanding of the impacts of fossil fuels and the value of RE.

This shift in awareness may lead to positive changes in consumer behavior, encouraging people to use energy more efficiently and support local RE projects.

7.6.4.2 Improving quality of life

The use of RE not only brings environmental benefits but also ensures a stable and sustainable electricity supply for communities. Clean energy from RE reduces electricity shortages and powers economic, educational, and daily activities. This contributes to improving the quality of life for people, especially in rural and remote areas.

Moreover, with the development of RE, people can access new and advanced technologies, creating more opportunities for individual and community development.

Developing renewable energy can bring many benefits to the environment, economy, and society in Vietnam, laying the foundation for sustainable development in the future. These positive impacts help address environmental issues, stimulate economic growth, and improve people's quality of life.

In the current context, accelerating the development of RE is necessary for Vietnam to meet its energy needs and build a green, sustainable, and environmentally friendly economy. The government, businesses, and communities need to work closely together to achieve these goals, thus creating a bright future for the country.

7.7 Solutions for renewable energy development in Vietnam: supportive policies, market mechanisms, and technological solutions

Supportive policies are key to creating a favorable environment for RE development. To ensure that RE can be widely and effectively implemented, the government needs to build a robust and cohesive legal framework. The enactment of clear laws and regulations to encourage the use of RE is essential. This may include setting specific targets for RE production and establishing approval processes for projects. A strong legal framework ensures transparency for investors and minimizes risks during the investment process. For example, setting clear technical and safety standards for RE projects can help avoid issues during implementation.

To promote RE development, financial incentives such as tax exemptions, subsidies, or capital support for both producers and consumers are necessary. These programs not only reduce initial installation costs for RE systems but also increase access to this energy source for the public. Additionally, training programs and technical support should be implemented to help the public understand the benefits of RE and how to use it effectively.

One of the most effective ways to encourage RE investment is to apply an FiT system. This policy ensures that RE producers receive a fixed price for energy over a certain period, providing financial stability to invest in RE infrastructure. In addition to reducing risks for investors, the fixed price system makes RE more competitive compared to traditional energy sources. Furthermore, developing comprehensive national energy production strategies is crucial, especially in the context of dwindling fossil fuel resources. These plans should prioritize renewable energy sources and outline steps to transition from fossil fuels, enhancing energy security and promoting sustainable development.

Market mechanisms play a significant role in creating a competitive and efficient RE market. The implementation of carbon taxes or emissions trading systems is an effective way to encourage reductions in greenhouse gas emissions. This policy makes RE more attractive than fossil energy. By taxing high-emission sources, the government can encourage businesses and consumers to switch to cleaner energy alternatives. PPAs are important tools to ensure a stable market for RE. Establishing contracts between energy producers and buyers (such as electricity companies) will ensure financial stability for RE projects, reduce risks for investors, and create a more competitive environment in the energy sector. Competitive bidding for new RE projects is also an effective method to encourage efficiency and cost savings during project development. Contractors can compete to provide the best solution at the lowest cost, benefiting both investors and end consumers.

Additionally, the renewable energy certificate (REC) system allows companies to meet emission requirements and enhance their sustainable image. Building a trading system for certificates representing the environmental benefits of RE production will encourage more businesses to join the RE market.

Technological solutions are key to enhancing the efficiency and feasibility of RE systems. Investment in research and development to improve existing technologies is necessary. Developing new technologies such as solar panels and wind turbines will help increase energy efficiency and reduce production costs. Technological innovation not only helps lower prices but also improves the performance of RE systems. Deploying smart grid technology will help better integrate RE sources into the energy system. Smart grids can monitor and adjust energy consumption in real-time, enhancing reliability and efficiency in energy distribution, thereby reducing power outages and better meeting energy demands. Additionally, the development and application of advanced energy storage technologies are essential to ensure stable energy supply. Solutions such as lithium-ion batteries and pumped hydro storage will help store excess energy from renewable sources, providing power during low production periods and ensuring energy security for the entire system. Finally, using software and energy optimization systems is crucial. Energy management systems will help households, businesses, and industries use RE more efficiently by providing real-time data on energy consumption, allowing users to adjust their behavior accordingly.

Impacts of policies and mechanisms on renewable energy development 333

By focusing on three key areas—supportive policies, market mechanisms, and technological solutions—Vietnam can effectively develop its RE sector. This approach not only helps meet the growing energy demand but also reduces greenhouse gas emissions and promotes sustainable economic growth. Implementing these solutions will enhance energy security and position Vietnam as a leader in RE in the region. Furthermore, RE development will contribute to environmental protection, natural resource conservation, and drive the green transition in a global context.

7.8 Summary and future research directions

Vietnam is at a critical juncture for renewable energy development to meet increasing energy demands, reduce dependence on fossil fuels, and achieve sustainable development goals. However, the current centralized power system, with its asynchronous expansion, faces significant challenges in ensuring stable, secure, and sustainable electricity supply. Initial results from supportive policies, such as FiT mechanisms, renewable energy subsidies, and interconnection regulations, have been promising. Yet, issues with transparency, efficiency, and adaptability to market changes remain.

To accelerate renewable energy development, a comprehensive approach is needed. Policies must be strengthened with clear legal frameworks, flexible financial incentives, and effective market mechanisms, such as competitive bidding, PPAs, and RECs. In parallel, technological solutions like smart grids, advanced energy storage, and energy optimization are essential to improve system efficiency and stability.

Open research directions

1. **Integrating renewable energy into smart grids**: Research and application of technologies such as IoT and AI to optimize grid operation with high shares of renewable energy.
2. **Economic models for renewable energy**: Analyze the effectiveness of flexible FiT mechanisms, competitive bidding, and carbon taxes in the context of Vietnam's economy.
3. **Social and environmental studies**: Assess the impact of renewable energy projects on local communities, including job creation, environmental improvements, and quality of life enhancements.
4. **Energy storage optimization**: Develop energy storage technologies suitable for Vietnam, such as lithium-ion batteries, thermal energy storage systems, and pumped hydroelectric storage.
5. **Demand-side management**: Encourage research on demand response programs to reduce system load during peak hours.

Comprehensive solutions: The combination of these research efforts will enable Vietnam not only to achieve its sustainable energy goals but also to position itself as a regional and global leader in renewable energy.

Further reading

[1] Vietnam's Eighth National Power Development Plan (PDP VIII).
[2] Environmental Protection Law: Law No. 72/2020/QH14, signed on 17/11/2020.
[3] Law on Water Resources: Law No. 17/2012/QH13, signed on 21/6/2012, effective from 01/1/2013.
[4] Planning Law: Law No. 21/2017/QH14, passed by the 14th National Assembly on 24/11/2017.
[5] Amendment Law No. 28/2018/QH14: Amendments and supplements to several articles of 11 laws related to planning, passed on 15/6/2018.
[6] Amendment Law No. 35/2018/QH14: Amendments and supplements to several articles of 37 laws related to planning, passed on 20/11/2018.
[7] Decree No. 29/2011/NĐ-CP: Issued by the Government on 18/2/2011, regulating strategic environmental assessment, environmental impact assessment, and environmental protection commitments.
[8] Decree No. 23/2006/NĐ-CP: Issued by the Government on 03/3/2006, on the implementation of the Law on Forest Protection and Development.
[9] Decree No. 46/2012/NĐ-CP: Issued by the Prime Minister on 22/5/2012, amending and supplementing several articles of:
[10] Decree No. 35/2003/NĐ-CP (dated 04/4/2003) detailing the implementation of several articles of the Law on Fire Prevention and Fighting.
[11] Decree No. 130/2006/NĐ-CP (dated 08/11/2006) on compulsory fire and explosion insurance policies.
[12] Decree No. 32/2006/NĐ-CP (2006): Application of the Sorensen Coefficient (S) for comparing species composition in the study area with neighboring areas.
[13] Decree No. 18/2015/NĐ-CP (Article 12) and Circular No. 27/2015/TT-BTNMT (Article 7): Provisions on consultation activities during the implementation of the Environmental and Social Impact Assessment (ESIA).
[14] Circular No. 27/2015/TT-BTNMT: Issued on 29/5/2015 by the Ministry of Natural Resources and Environment (MONRE) on Strategic Environmental Assessment, Environmental Impact Assessment, and Environmental Protection Plans (Appendix 2.3, Chapters 1, 3, 4, and 5).
[15] Circular No. 26/2011/TT-BTNMT: Issued on 18/7/2011 by MONRE, detailing certain provisions of Decree No. 29/2011/NĐ-CP on strategic environmental assessment, environmental impact assessment, and environmental protection commitments.
[16] Circular No. 12/2011/TT-BTNMT: Issued on 14/4/2011 by MONRE, on the management of hazardous waste.
[17] Directive No. 08/2006/CT-TTg: Issued by the Government on 08/3/2006, on urgent measures to prevent illegal deforestation, forest burning, and unauthorized logging.

[18] Ngo Dang Luu, Nguyen Hung, Nguyen Anh Tam, and Long D. Nguyen, (2021), "Wireless Power Transfer simulation for EV", Ministry of Culture, Sports and Tourism.
[19] Ngo Dang Luu, Nguyen Hung, and Long D. Nguyen, (2021), "Power flow calculation of the electric power system by MATLAB", Ministry of Culture, Sports and Tourism.
[20] Ngo Dang Luu, Nguyen Hung, Nguyen Anh Tam, and Long D. Nguyen, (2022), "Distance Relay Protection in AC Microgrid", Ministry of Culture, Sports and Tourism.
[21] Ngo Dang Luu, Nguyen Hung, Nguyen Anh Tam, and Long D. Nguyen, (2022), "Analysis of Solar Photovoltaic System Sunlight Shadowing", Ministry of Culture, Sports and Tourism.
[22] Ngo Dang Luu, Nguyen Hung, Le Anh Duc and Long D. Nguyen, (2022), "Estimate Frequency Response at the Command Line", Ministry of Culture, Sports and Tourism.

Chapter 8
Investment procedures for renewable energy projects: a case study in Vietnam

8.1 Investment policies, mechanisms, and competition among power producers in Vietnam

The expansion of renewable energy (RE) sources in Vietnam has driven private investment through the independent power producer (IPP) model, increasing private ownership of installed capacity from 20% in 2020 to 30% in 2018. As the RE market grows, Vietnam must support fair competition between private developers and state-owned enterprises (SOEs), particularly with the dominant Electricity of Vietnam (EVN). Measures include refining competitive bidding frameworks, officially operating a competitive wholesale electricity market, privatizing EVN's generation corporations, and enhancing the independence of the National Load Dispatch Center (A0). These steps aim to level the playing field between EVN and IPPs.

In the past decade, Vietnam has successfully attracted foreign direct investment (FDI). Transitioning to clean energy will further support its goal of becoming a leading regional manufacturing hub. However, building an effective competitive electricity market requires overcoming several challenges, including:

- Upgrading transmission systems for RE integration
  - **Context**: By 2045, variable RE could account for over 44% of installed capacity, necessitating significant investment (~$85 billion from 2021 to 2045) in transmission infrastructure.
  - **Recommendations**
    o Allow private sector investment in transmission grids.
    o Develop clear legal frameworks to ensure feasibility and cost-effectiveness.

- Limitations in public–private partnerships (PPP) for clean energy
  - **Context**: Despite the PPP law allowing energy project investments, its application is limited, especially for small-scale RE projects.
  - **Recommendations**
    o Enhance the PPP legal framework to increase flexibility and attract investments in large-scale projects.
    o Address risks related to guarantees and foreign currency conversion.

- **Dispute resolution and land access**
  - **Context**: Transparent and independent regulation is critical for resolving disputes and ensuring fair outcomes.
  - **Recommendations**
    - Strengthen the Electricity Regulatory Authority of Vietnam (ERAV) for fair dispute resolution.
    - Simplify land access and clearance processes, empowering provinces to facilitate project implementation.
- **Clean energy commitments and enhanced FDI competitiveness**
  - **Context**: Vietnam can capitalize on the global trend of carbon emission reduction to enhance its competitive edge.
  - **Recommendations**
    - Provide clean, stable, and cost-effective electricity to support Vietnam's industrial ambitions, targeting 30% of GDP from manufacturing by 2030.

8.1.1 Key policy recommendations

8.1.1.1 Ensuring accurate price signals in the wholesale electricity market

- **Objectives**
  - Encourage grid-friendly operations.
  - Improve sector efficiency and capital allocation.
- **Solutions**
  - Improve pricing mechanisms in the wholesale electricity market to deliver clear and timely signals.
  - Use pricing tools to encourage RE development and usage.

8.1.2 Strengthening the independence of ERAV

- **Objectives**
  - Ensure ERAV decisions are fair, transparent, and free from political interference.
- **Solutions**
  - Enhance ERAV's authority and establish arbitration mechanisms for power purchase agreements (PPAs).
  - Adopt international best practices in dispute resolution to build investor confidence.

8.1.3 Enhancing national load dispatch center (A0) independence

- **Objectives**
  - Ensure transparency and fairness in system management and coordination.

- **Solutions**
  - Plan national load dispatch center (A0)'s separation from EVN.
  - Unbundle production, distribution, and retail operations within EVN to reduce conflicts of interest.

8.1.4 Promoting private investment in transmission infrastructure

- **Objectives**
  - Mobilize non-government capital for transmission grid expansion and upgrades.
  - Align infrastructure upgrades with RE development plans.
- **Solutions**
  - Amend legal frameworks to facilitate private investment in transmission networks.
  - Implement the new PPP law, develop concession agreements, and promote competitive bidding.

8.1.5 Establishing a transparent competitive bidding framework

- **Objectives**
  - Ensure a level playing field for private and SOEs.
- **Solutions**
  - Develop detailed guidelines for competitive bidding in RE projects.
  - Incorporate risks related to land-use certification into bidding frameworks.
  - Ensure transparency throughout all bidding phases to attract investors.

Vietnam's efforts to develop RE not only foster industry growth but also solidify its position as an attractive global investment destination.

8.1.6 Solutions for creating a level playing field in the clean energy infrastructure sector in Vietnam

8.1.6.1 Current situation

- **Market structure**: Vietnam's energy market is partially competitive, with dominance by EVN and SOEs such as PetroVietnam and Vinacomin.
- **Ownership proportion**: As of 2018, EVN and its subsidiaries controlled 58% of electricity output, while private investors (IPPs) accounted for only 31% of total installed capacity by 2020, primarily through RE projects (wind and solar).
- **SOE advantages**: SOEs benefit from strong ties with the Ministry of Industry and Trade, easier access to land, preferential financing from state-owned banks, and streamlined project approval processes.
- **Regulatory framework**: IPPs rely on non-negotiable PPAs with EVN, and there is no established framework for direct transactions between IPPs and industrial or commercial users.

8.1.6.2 Barriers to competition
- **SOE priority**: Current policies tend to favor SOEs, limiting the effective participation of private sector players.
- **Lack of transparency in approvals**: RE project approval procedures vary across provinces, leading to inconsistencies and opportunities for project speculation.
- **Dependence on EVN**: IPPs are entirely dependent on EVN for project development and operation, reducing autonomy and competitive dynamics.

8.1.6.3 Proposed solutions
- **Legal and regulatory reforms**
    - Develop and implement a transparent and well-defined competitive bidding mechanism, with clear processes for selecting RE project investors based on financial and technical capabilities.
    - Establish a legal framework for direct transactions between IPPs and industrial or commercial consumers.
    - Introduce uniform regulations on project approval authority across government levels.

- **Support for IPPs**
    - Encourage deeper IPP participation in the wholesale electricity market with pricing reflecting real-world conditions.
    - Provide financial support through tax incentives or access to preferential financing sources.

- **Privatization and improved governance of SOEs**
    - Accelerate the privatization of EVN's generation companies, ensuring a gradual reduction in EVN's involvement in non-essential segments.
    - Enhance corporate governance to mitigate risks of market power abuse or anti-competitive domestic pricing practices.

- **Learning from international experience**
    - Integrate RE projects into the market with pricing designs that adjust according to market signals, ensuring long-term benefits for investors.
    - Leverage technical support from international organizations such as the World Bank in developing bidding strategies and fostering the development of a competitive electricity market.

8.1.6.4 Expected benefits
- **Create a fair and transparent investment environment**: Attract additional resources from the private sector.
- **Ensure stability in renewable energy supply**: Meet the growing sustainable energy needs of the economy.
- **Reduce public financial burden:** Promote the development of a competitive electricity market.

8.2 Aspects of implementing wind and solar energy projects in Vietnam: capital mobilization policies and electricity pricing

8.2.1 Promoting the development of independent renewable energy systems

For independent RE systems (off-grid), the government needs to support advanced models for widespread deployment, particularly in remote areas. This requires creating a comprehensive support mechanism to ensure the system's sustainable operation, from design, construction, and management to local workforce training.

Prime Minister Decision No. 2068/QD-TTg dated November 25, 2015, approving the "Renewable Energy Development Strategy to 2030, Vision to 2050," outlines the following:

- **By 2030**
  - Develop independent RE sources to electrify rural areas, reduce poverty, and promote socio-economic development. The goal is to have most rural households electrified by 2020, and by 2030, the majority will use clean and hygienic energy.
  - Invest in the development of grid-connected RE plants that are economically viable and promote long-term technology.

8.2.2 Characteristics and policy requirements for renewable energy development

RE is highly dependent on natural conditions (water, sunshine, wind, geography), technology, and costs. Vietnam needs to:

- Apply support mechanisms such as quotas, fixed pricing, bidding, and certification.
- Introduce policies to limit the use of fossil fuels.

8.2.3 Mechanisms and policies to support investment

- **Quota mechanism**: Set a ratio for RE production/consumption to encourage competition among technologies and reduce costs.
- **Flexible electricity pricing**: Determine preferential pricing for each type of RE technology, ensuring investors earn reasonable profits while minimizing risks.
- **Competitive bidding**: Select RE projects with the lowest prices, ensuring minimal price subsidies and long-term stability.
- **Issuance of certificates**: Exempt taxes or provide deductions for investors in RE, effectively combining with other mechanisms.
- **Offset payments**: Apply to customers who generate and consume their own electricity from renewable sources, reducing the burden on the national grid.

8.2.4 Preferential policies

- **Taxation**: Exempt import taxes on equipment and production materials; reduce corporate income tax.
- **Land use**: Exempt or reduce land usage fees for RE projects.
- **Research and technology**: Provide funding from research funds and support technological improvements.

8.2.5 Developing human resources

- Strengthen the management capacity for the development of RE at all levels.
- Support universities and training institutions in developing curricula and teaching materials for RE.

8.2.6 Environmental protection policies

- Apply environmental fees on fossil fuels, allocating a portion to support RE development through the Sustainable Energy Development Fund.

To develop RE sustainably, Vietnam needs to flexibly combine various support mechanisms, encourage investment, enhance capacity, and implement comprehensive policies aligned with global trends.

8.2.7 Factors affecting household investment intentions in the rooftop solar energy sector

Identifying factors influencing household investment intentions in rooftop solar energy: There are many research models on investment in the rooftop solar energy sector. By synthesizing the previous studies, the author selects various factors and groups them based on similar characteristics (even if they are labeled differently) to construct a theoretical framework regarding the factors influencing investment intentions in the rooftop solar energy sector. Table 8.1 summarizes the factors influencing investment intentions in the rooftop solar energy sector as presented by researchers and demonstrated through practical examples.

After synthesizing, grouping, and adjusting the factors, the key determinants of investment intentions in the rooftop solar energy sector have been narrowed down to ten main factors. These factors have been mentioned by at least two studies.

Using the expert method (which includes scientists and managers in the field of electricity generation), the author has selected and formed groups of variables influencing households' investment intentions in rooftop solar energy systems. The factors agreed upon by 8 out of ten experts will be retained, while those that do not meet this threshold will be discarded. In cases where experts remove or add new factors, the author will conduct interviews to understand the reasons for such adjustments during the survey process. It is worth noting that a household's investment intention in rooftop solar energy systems is not only influenced by

Table 8.1 Summary table adjusting factors affecting investment intentions in rooftop solar power

No.	Factor	Theoretical basis	Adjustment name
1	Risk acceptance	Stefan Poier (2021) Stevens and associates (2018) Pham Hong Manh, Duong Van Son (2020) Dinh Thi Trang and associates (2021)	Riskiness
2	Willingness to protect the environment	Stefan Poier (2021) S.A. Malik, A.R. Ayop (2020) Braito et al. (2017) Anke Jackson et al. (2019) Schwartz (1977) Le Tran Thanh Liem, Pham Ngoc Nhan (2020) Dinh Thi Trang and associates (2021)	Environmental protection perspective
3	Social impact	Ramos, A. Gago, X. Labandeira, P. Linares, 2015 Rai và Robinson (2015) Vasseur và Kemp (2015) Schwartz (1977) Ingo Kastner, Paul C. Stern (2015) Dinh Thi Trang and associates (2021)	Social consumption trends
4	Family members' opinions	Stefan Poier (2021) Wasi and Carson (2013) Ondraczek (2013) Pham Hong Manh, Duong Van Son (2020)	Family members
5	Government policies	Christine and Chernyakhovskiy (2014) Hansen and associates (2015) International Energy Agency (2014) Ingo Kastner, Paul C. Stern (2015) Pham Hong Manh, Duong Van Son (2020)	Government policies

(Continues)

Table 8.1 (Continued)

No.	Factor	Theoretical basis	Adjustment name
6	Demand and motivation: personal preferences	Ondraczek (2013) Pham Hong Manh, Duong Van Son (2020) A. Ramos, A. Gago, X. Labandeira, P. Linares, (2015)	Demand and motivation: personal preferences
7	Installation requirements: availability of goods, new technology, warranty, benefits, ease of use	Anke Jackson et al. (2019) S.A. Malik, A.R. Ayop (2020) Schelly (2014) Ondraczek (2013) Vasseur and Kemp (2015) Le Tran Thanh Liem, Pham Ngoc Nhan (2020) Dinh Thi Trang and associates (2021) Pham Hong Manh, Duong Van Son (2020)	Product characteristics
8	Product characteristics	S.A. Malik, A.R. Ayop (2020) Rai and Robinson (2015) Braito et al. (2017) Wasi and Carson (2013) Anke Jackson et al. (2019) Schelly (2014) Vasseur and Kemp, 2015 Schwartz (1977) Black et al. (1985) Ingo Kastner, Paul C. Stern (2015) Christine and Chernyakhovskiy (2014) Ondraczek, 2013 Hansen and associates (2015) Schwartz (1977) Dinh Thi Trang and associates (2021) Black et al. (1985) Le Tran Thanh Liem, Pham Ngoc Nhan (2020)	Economic feasibility
9	Household income	Dinh Thi Trang and associates (2021) Ondraczek (2013)	Household income
10	Demographics	Rai and Robinson (2015) Schelly (2014)	Demographics

family members but also by those who have had direct experience, information channels, and acquaintances.

8.3 Procedure for building and developing renewable energy projects in Vietnam

Implementing a RE project involves several critical stages. Below are the basic steps:

- **Site Identification and feasibility study**
  - Identify a suitable location for project implementation.
  - Conduct a feasibility study to assess business potential, profitability, and related factors such as market conditions, energy production costs, and legal regulations.

- **Project design**
  - Based on the feasibility study results, proceed with detailed project design.
  - Define the technology to be used, develop an implementation plan, analyze costs, assess risks, and create a project management plan.

- **Registration and permitting**
  - Complete the legal procedures by registering and applying for necessary permits from the relevant authorities.
  - Permits include: registration certificate, construction permit, operating license, and other related procedures.

- **Financial mobilization**
  - Ensure funding for project implementation from sources such as bank loans, investment capital, international funding, or joint venture partnerships.

- **Construction and operation**
  - Once legal procedures are completed and sufficient funding is raised, begin the construction phase of the project.
  - After the project is completed, move to the operational and maintenance phase to ensure optimal performance.

- **Monitoring and performance evaluation**
  - Regularly monitor and evaluate the operational performance of the project.
  - Timely adjustments should be made as necessary to optimize performance and ensure sustainability.

This procedure ensures that the project is implemented effectively, meets legal requirements, and achieves both economic and environmental goals.

8.4 Technical criteria in the construction and development of renewable energy projects

The technical criteria in the construction and development of RE projects include the following:

- **Energy efficiency**: This criterion evaluates how effectively energy from renewable sources such as wind, solar, hydro, etc., is harnessed to generate electricity. The higher the energy efficiency, the more effective the project is, providing greater environmental benefits.
- **Reliability**: This criterion assesses the reliability of the electrical system, including its ability to minimize faults and the occurrence of failures. RE projects with high reliability help ensure a stable power supply, meeting users' demands.
- **Lifespan and maintenance**: This criterion evaluates the lifespan of the system and its periodic maintenance requirements to ensure stable and efficient operation. The lifespan and maintenance needs of RE projects vary depending on the type of technology used.
- **Energy storage and management**: This criterion evaluates the ability of RE systems to store and manage energy, including the capability to store energy and redistribute it when needed.
- **Economic efficiency**: This criterion evaluates the economic effectiveness of the RE project, including investment costs, operation and maintenance costs, income from electricity sales, and other economic benefits.
- **Safety and environmental protection**: This criterion assesses safety factors and environmental protection during the construction, operation, and maintenance of RE projects.

8.4.1 Site selection, survey, land clearance, solar radiation calculation, and climate conditions

To implement a solar power project, the following key steps for site selection, survey, land clearance, solar radiation calculation, and climate condition assessment are as follows:

- **Site selection**: A suitable location for installing the solar power system should be selected based on factors such as available land area, topography, proximity to the substation, altitude, and the angle of inclination of the roof or surface where the system will be installed.
- **Survey**: After identifying the location, a survey is conducted to evaluate technical and environmental factors at the installation site, including soil properties, terrain, environment, access to the substation, weather conditions, etc.
- **Land clearance**: For solar power projects installed on land, land clearance is carried out to determine the scope of land usage and reclaim land needed for the project.

- **Solar radiation calculation**: Solar radiation is the amount of solar energy received at the installation site. Calculating solar radiation helps determine the amount of electricity the solar power system can produce at the installation location.
- **Climate conditions**: The climate conditions of the installation site also affect the system's efficiency in generating electricity. Climate conditions need to be assessed to determine how suitable the location is for system installation and to calculate the amount of solar power that can be generated.

These steps need to be performed carefully and thoroughly to ensure a successful installation.

8.4.2 Application of fuzzy logic algorithm for supplier selection of solar equipment

Fuzzy logic is an important tool for addressing problems related to ambiguity and uncertainty in decision-making. When applied to the selection of solar equipment, fuzzy logic can help determine the suitability of the equipment according to the user's needs, optimizing the efficiency and cost of the solar system.

The steps to apply fuzzy logic for selecting solar equipment are as follows:

1. **Identify evaluation parameters**: To apply fuzzy logic, the critical parameters need to be identified. These may include conversion efficiency, durability, cost-effectiveness, etc.
2. **Define fuzzy functions**: Fuzzy functions represent the ambiguity of a parameter. These functions are defined using evaluation algorithms to determine fuzzy values corresponding to various parts of the evaluation parameters.
3. **Define rules**: Rules are used to combine the input values and fuzzy functions to determine the output values. These rules can be based on expert knowledge or input data.
4. **Apply fuzzy logic algorithm**: Once the fuzzy functions and rules are defined, the fuzzy logic algorithm can be applied to decide which solar equipment best fits the usage requirements. The result will be fuzzy, allowing the decision to choose the most appropriate values that meet the necessary requirements.
5. **Test and evaluate**: after applying the fuzzy logic algorithm, testing and evaluation are necessary to ensure the accuracy and effectiveness of the algorithm. If the result does not meet expectations, adjustments to the parameters and rules are required to optimize the output.

8.4.3 Power calculation

To calculate the solar power output, the following parameters need to be determined:

1. **Surface area for installing solar panels**: This area influences the number of solar panels that can be installed.

2. **Efficiency of solar panels**: This is the ratio between the electrical energy the solar panel can produce and the solar energy received. This efficiency is typically determined based on the technical specifications of the solar panel.
3. **Solar radiation**: This is the amount of solar energy received per unit area. Solar radiation depends on the location of the solar panels, the time of day, and the weather conditions.

The formula for calculating solar power is:
Solar power = surface area × solar radiation × panel
Efficiency of solar power = surface area × solar radiation × panel efficiency

For example, if ten solar panels, each with 20% efficiency, are installed on a surface area of 20 m^2, and the solar radiation is 1,000 W/m^2, the solar power output would be:

Solar Power = 20 m^2 × 1,000 W/m^2 × 0.2 × 10 = 4,000 W

Therefore, the solar power output in this case would be 4,000 W. However, it is important to note that this value represents the maximum power under optimal conditions, and actual performance will depend on various factors such as weather conditions, installation location, tilt angle of the solar panels, and how solar radiation is measured.

8.4.4 Calculating the connection to the electrical system

To calculate the connection of the solar power system to the electrical grid, the following parameters need to be determined:

- **Solar power system capacity**: This is the maximum power that the solar power system can supply.
- **Voltage of the solar power system**: The voltage must be compatible with the electrical grid or system to which it is being connected.
- **Connection type**: Depending on the type of electrical system, the connection can either be parallel or series.
- **Controller specifications**: If a controller is used, its technical specifications need to be determined to ensure the efficient operation of the system.

The steps to calculate the connection of the solar power system to the electrical grid are as follows:

- **Step 1**: Determine the power capacity of the solar power system and the output voltage of the system.
- **Step 2**: Identify the voltage and current requirements of the electrical system that needs to be connected.
- **Step 3**: Select the appropriate type of connection (parallel or series) and calculate the output voltage and current of the solar system to match the electrical system's requirements.
- **Step 4**: Determine the size of the controller (if used) to ensure the effective operation of the system.

Investment procedures for renewable energy projects 349

- **Step 5**: Choose the appropriate connection devices (inverters, transformers, electrical panels, etc.) and install all necessary safety devices.
- **Step 6**: Perform the connection and check the operation of the system.

These steps help in efficiently calculating the connection of the solar power system to the electrical grid while ensuring safety. However, to ensure accurate results, it is important to refer to technical regulations and standards for electrical system connections.

8.4.5 System testing and project handover

Running system tests and handing over the solar power project can be done following these steps:

- **Final checks and maintenance**: Perform final checks and maintenance on the solar power system to ensure that it operates correctly and meets all technical requirements.
- **Connection verification**: Verify that the solar power system is correctly connected to the building's electrical grid, ensuring that all connecting and protection devices are properly installed.
- **Safety inspection**: Conduct a safety inspection to ensure that the system does not pose any risks to the users or the building. This includes checking wiring, grounding, and protection systems.
- **System handover**: Hand over the system to the owner or building manager, providing instructions on how to operate and maintain the system, as well as offering warranty and technical support information.
- **System monitoring (initial period)**: Monitor the system during the initial period to ensure that it operates correctly and that no issues arise. If problems occur, take immediate corrective actions.
- **Performance reporting**: Provide periodic reports on the performance of the solar power system to the owner or building manager to ensure that the system is working efficiently and without any problems.

This process ensures that the solar power system is handed over and operates correctly, meeting the energy needs of the home or building.

8.5 Estimating the economic and financial indicators of the project and identifying funding sources for solar and wind energy systems

To calculate the economic and financial indicators of the project and identify potential funding sources for the project, follow these steps:

Step 1: Identify economic and financial indicators

In this step, you need to define the basic economic and financial indicators for the project, including:

- **Initial investment costs**: This includes construction costs, equipment purchases, business registration costs, consulting fees, etc.
- **Operating costs**: These include wages, operational costs, maintenance costs, advertising costs, and other ongoing expenditures.
- **Revenue**: The amount of money earned from the project's business activities.
- **Profit**: The amount of money remaining after subtracting operating costs and taxes.

Step 2: Calculate the economic and financial indicators of the project

Once you have identified the economic and financial indicators, calculate these values using financial analysis methods such as:

- **Net present value (NPV)**: The value of the project's cash flows over time, discounted by a specific interest rate.
- **Internal rate of return**: The rate at which the present value of future cash flows equals the initial investment.
- **Payback period**: The time it takes to recover the initial investment through the project's net cash inflows.
- **Profitability index (PI)**: The ratio of the present value of future cash flows to the initial investment, showing the project's relative profitability.

These indicators help evaluate the feasibility of the project and its potential return on investment (ROI).

Step 3: Identify funding sources for the project

After calculating the economic and financial indicators, you need to find suitable funding sources for the project. Possible funding options include:

- **Own capital**: Funds provided by the project owner to finance the project. This is usually used when the owner has sufficient financial resources.
- **Bank loans**: Loans from financial institutions, such as banks, to fund the project with the agreement to repay the loan with interest over time.
- **Investment from investors**: You can raise funds by offering shares, bonds, or other financial instruments to investors in exchange for capital.
- **Government grants or incentives**: Some governments provide financial assistance, tax incentives, or subsidies for RE projects like solar and wind energy systems.

Choosing the appropriate funding source depends on various factors, including the project's size, market conditions, social and political factors, and the legal and regulatory framework in place.

By estimating the economic and financial indicators and finding the appropriate funding sources, you can ensure the success and sustainability of solar and wind energy projects.

8.5.1 Calculating the cost and benefit components of rooftop grid-connected solar systems

Cost components

- **Initial investment costs**
  - The initial investment involves purchasing and installing solar system components, such as solar panels, inverters, wiring, and mounting systems. This is typically the largest expense at the outset.

- **Operating and maintenance costs**
  - This includes costs associated with system operation, routine maintenance, repairs, and replacement of faulty components over time.

- **Electricity bills**
  - Prior to having a solar system, the household needs to pay monthly electricity bills to the utility company. After installation, the solar system can reduce the electricity consumption from the grid, leading to lower monthly electricity costs.

- **Energy storage costs**
  - In cases where the solar system does not generate enough electricity (e.g., at night or during cloudy days), a battery storage system may be needed to store excess energy for later use. The cost of purchasing and installing these storage systems should be considered.

Benefit components

- **Cost savings**
  - Households can save on their monthly electricity bills by using solar-generated electricity. In some cases, surplus electricity can be sold back to the grid, generating additional income.

- **Environmental protection**
  - Solar systems reduce dependence on fossil fuels, which lowers greenhouse gas emissions and contributes to the reduction of the greenhouse effect.

- **Energy independence**
  - Solar systems offer energy independence, reducing reliance on utility companies and insulating the household from fluctuating electricity prices.

- **Increased property value**
  - Homes with rooftop solar systems are often seen as more modern and environmentally friendly, which can increase their market value and appeal to potential buyers.

8.5.2 Calculating the economic effectiveness of a rooftop grid-connected solar system

To assess the economic effectiveness of the solar system, several key indicators need to be calculated:

Key indicators

- **Initial investment costs**
  - The upfront investment for installing the rooftop solar system, including purchasing solar panels, inverters, and other necessary equipment.
- **Electricity production**
  - The amount of solar energy (in kWh) produced annually by the system, which depends on factors like system size, location, and weather conditions.
- **Electricity price**
  - The price at which electricity is sold back to the grid, as determined by regulatory authorities, which can vary over time.
- **Operating and maintenance costs**
  - The ongoing costs to operate and maintain the solar system, including repairs, replacements, and general upkeep.
- **Grid connection costs**
  - Costs related to connecting the solar system to the grid, including any fees for integration and system configuration.

Economic indicators

- **Payback period**
  - The time it takes to recover the initial investment from the savings generated by the solar system. The shorter the payback period, the more attractive the investment.
- **Return on investment**
  - The ratio of profit to the initial investment, showing the efficiency of the investment. This is calculated by dividing the net profit (after taxes) by the initial investment cost.
- **Net present value**
  - The present value of future cash inflows and outflows from the project, discounted to account for the time value of money. A positive NPV indicates a profitable project.
- **Profitability index**
  - The ratio of the present value of future cash flows to the initial investment cost. A PI greater than 1 indicates that the project is profitable.

Other factors

- **Electricity savings**
  - The amount of money saved on electricity bills due to the use of solar energy.
- **Environmental benefits**
  - The reduction in carbon emissions and the household's contribution to sustainable energy generation.

- **Grid energy compensation**
  - The amount of electricity that can be fed into the grid and compensated by the utility, depending on local regulations and policies.

8.5.3 Calculating the financial and economic effectiveness of rooftop grid-connected solar systems

To calculate the financial and economic effectiveness of rooftop solar systems, gather the following data:

- **Installation feasibility**
  - Determine the roof area and the potential for solar panel installation, considering the roof's orientation, angle, and shading.

- **Initial investment costs**
  - Include the costs of purchasing solar panels, inverters, and all necessary installation components.

- **Operating and maintenance costs**
  - Account for ongoing maintenance, repairs, and any associated operating costs.

- **Retail electricity prices**
  - The price per kWh charged for electricity consumption by the utility.

- **Electricity production**
  - Estimate the total electricity production based on the area of the solar system, roof orientation, tilt angle, and local sunlight availability.

Key financial indicators

- **Total investment costs**
  - The total cost for purchasing and installing the solar system, including any additional components like storage systems or inverters.

- **Annual income**
  - The total revenue generated by selling excess electricity to the grid.

- **Annual electricity costs**
  - The yearly cost for electricity consumption before and after the solar system is installed.

- **Annual electricity savings**
  - The amount of money saved by using solar power instead of purchasing electricity from the grid.

- **Payback period**
  - The time required to recover the initial investment through savings on electricity bills.

- **Return on investment**
  - The ROI, calculated as the ratio of annual electricity savings to the total initial investment.

- If the ROI and payback period meet expected benchmarks, the solar investment is considered economically viable and financially effective.

8.5.4 Calculation of costs and benefits of grid-connected rooftop solar power systems

The cost components of grid-connected rooftop solar power systems include:

- **Initial investment cost**: This is the cost of purchasing and installing the components of the solar power system. Typically, the initial investment is relatively high.
- **Operation and maintenance costs**: This includes the costs for operation, maintenance, repairs, and replacement of any faulty parts.
- **Electricity costs**: Prior to the installation of the solar power system, the household needs to pay monthly electricity bills to the utility company. With a solar power system, the monthly electricity cost can be partially reduced.
- **Energy storage costs**: In cases where the solar power system does not provide enough energy, the household may need to purchase energy storage systems to ensure a continuous power supply during periods with insufficient sunlight or low solar energy production.

The benefits of grid-connected rooftop solar power systems include:

- **Cost savings**: Households can save on monthly electricity costs by using electricity generated from the solar power system. In some cases, the system may even generate income by selling surplus electricity back to the utility company.
- **Environmental protection**: The use of solar power systems helps reduce the consumption of energy from fossil fuel sources, decreasing greenhouse gas emissions and mitigating the effects of climate change.
- **Energy independence**: Solar power systems help households become independent from utility companies and shield them from fluctuations in electricity prices.
- **Increased property value**: Grid-connected rooftop solar power systems can increase the value of a property and attract modern, environmentally-conscious buyers.

8.5.5 Calculation of the financial economic efficiency of wind power systems

To calculate the financial economic indicators of a project and determine the sources of funding for the project, you can follow these steps:

Step 1: Determine the financial economic indicators of the project: In this step, you need to identify the basic financial economic indicators of the project, including:

- **Initial investment cost**: This includes the costs of construction, equipment procurement, business registration, consulting fees, etc.

- **Operating costs**: This includes salaries, operational expenses, maintenance costs, advertising costs, etc.
- **Revenue**: This is the income generated from the project's business activities.
- **Profit**: This is the remaining amount after deducting operating costs and taxes.

Step 2: Calculate the financial economic indicators of the project: Once the financial economic indicators have been identified, you need to calculate these indicators using statistical methods, financial analysis, and other relevant techniques.

Step 3: Find funding sources for the project: After calculating the financial economic indicators of the project, you need to find suitable sources of funding. Possible funding sources include:

- **Own capital**: This is the capital provided by the project owner for investment in the project.
- **Bank loans**: This is a form of borrowing from financial institutions such as banks to invest in the project and repay the loan with interest as per the agreement.
- **Fundraising from investors**: You can raise capital from investors by issuing shares, bonds, or other financial instruments.

Depending on the economic situation, market conditions, social factors, and legal policies, you can choose the most appropriate funding source for the project.

8.5.6 Calculation of the costs and benefits of grid-connected wind power systems

The cost components of grid-connected wind power systems include:

- **Initial investment cost**: The initial investment cost of a wind power system includes expenses for design, procurement, and installation of equipment, as well as construction-related costs. The initial investment cost depends on factors such as the size of the project, location, and other variables.
- **Operating and maintenance costs**: These costs involve the operation, repair, and maintenance of wind power equipment, as well as expenses related to managing the system. While typically smaller than the initial investment cost, these costs still need to be considered as part of the overall expenditure.
- **Grid operation costs**: Grid-connected wind power systems need to be integrated with the power grid, so the costs of grid operation must also be considered. These costs are generally calculated based on electricity rates and the frequency of grid usage.

Benefits: The benefits of grid-connected wind power systems include:

- Improved energy supply capabilities.
- Significant reduction in carbon emissions.
- Creation of jobs and economic growth in areas where wind power projects are developed.
- Contribution to innovation and enhancement of industrial capacity.

However, to comprehensively evaluate the costs and benefits of grid-connected wind power systems, a detailed study must be conducted for each specific case. Factors such as initial investment costs, operating and maintenance costs, grid operation costs, and economic and environmental benefits need to be calculated to inform investment decisions.

8.5.7 Calculating the effectiveness of the project

To evaluate the effectiveness of a grid-connected wind power project, it is essential to identify key economic indicators as follows:

- **Initial investment cost**: This refers to the cost of building and installing the grid-connected wind power system, including expenses for equipment procurement, design, and infrastructure construction.
- **Operating and maintenance costs**: These costs encompass the expenses related to operating the wind power system, such as electricity costs, repair, and maintenance.
- **Grid operation costs**: These include the costs associated with connecting the wind power system to the grid and other expenses incurred during the operation of the system.
- **Revenue**: Revenue is calculated based on the electricity selling price and the volume of electricity generated.
- **Environmental benefits**: Environmental benefits are calculated based on the system's ability to reduce carbon emissions and other environmental impacts.

After identifying the above indicators, the following financial performance metrics can be calculated:

1. **Payback period**: The time required to recover the initial investment cost from the profits generated by the project.
2. **Return on investment**: The ratio of net profit after taxes to the initial investment cost.
3. **Net present value**: The present value of all future cash flows from the project, discounted to their present value.
4. **Profitability index**: The ratio of the present value of all future cash inflows (after discounting) to the initial investment cost.

However, to accurately assess the economic viability of a grid-connected wind power project, factors such as operating and maintenance costs, grid operation costs, and other related expenditures must also be considered.

8.5.8 A combined approach between grid-connected solar energy and independent investment optimization: a case study in Vietnam

- In Vietnam, the demand for energy has significantly increased over the past decade due to the country's rapid economic development. To avoid greenhouse gas emissions and meet the growing energy needs, much of the future energy

supply will need to come from renewable sources, requiring a substantial shift away from the use of fossil fuels as the primary energy source.
- In the Vietnam Power Development Plan VIII (PDP VIII) with a vision toward 2050, the country commits to developing new energy sources such as solar and wind energy to build and upgrade the national power grid system.
- This study presents a technical and financial model for establishing a grid-connected solar power plant with an integrated battery energy storage system (ESS), designed based on real-world optimization problems. The proposed business model links analysis with hourly radiation intensity and considers the hourly energy demand chart for planned operations.
- The proposed hybrid system is evaluated and optimized regarding investment costs, necessary capacity, and storage capacity across different scenarios. The research results demonstrate the feasibility of the system both economically and technically under the proposed methods.
- To ensure national energy security, the PDP VIII aims to provide stable electricity at a reasonable cost, contributing to rapid and sustainable socio-economic growth. The EVN, a state-owned integrated enterprise, plays a monopolistic role in the transmission, distribution, and operation of the power system, and is the main electricity producer in the country. The Vietnamese electricity sector has undergone a significant reform process from the 1990s to 2020. Total electricity consumption has continuously increased to meet economic development requirements. During the 2030 and 2050 periods, developing an appropriate electricity planning model and simulating the national energy sources is crucial for expanding energy and reducing CO_2 emissions.
- From a business perspective, investors often rely on NPV to assess the feasibility of investing in rooftop solar energy storage systems. Vietnam is considered to have significant potential in producing sustainable energy, with clean energy playing an important role in economic development, social progress, and energy security. Advances in clean energy utilization have helped save resources, reduce environmental impacts, and enhance national energy security.
- RE is becoming more competitive than traditional energy sources due to rapidly decreasing costs, especially for wind and solar energy. Factors such as increasing capacity, significant investment, and technological advancements have accelerated the global transition from fossil fuels to RE.
- The solar power system must fully meet the energy consumption needs of enterprises. When investing in additional capacity, energy from other sources such as hydropower and thermal power may need to be supplemented, as these sources are becoming exhausted. The hybrid renewable energy system, integrating RE sources in optimal ratios, can help address these limitations.
- BESS also need thorough research to ensure solar power generation meets the energy demands of enterprises both day and night, regardless of radiation conditions. This system has many applications, ranging from delaying infrastructure upgrades, providing continuous power, improving reliability, to long-term energy management. Battery technology, due to its non-self-discharge

feature, holds great potential for long-term energy storage. However, high initial investment costs and operational expenses related to equipment such as pumps and control systems are significant challenges.
- Further research into grid-connected energy systems with integrated storage is necessary, including energy management mechanisms, performance evaluation, distribution scheduling optimization, and cost–benefit analysis.

8.5.8.1 System model

The focus is on investing in a solar energy system to assess and optimize the feasibility of the project according to each country's power development plan (e.g., Vietnam's Power Development Plan VIII—PDP VIII). The primary goal is to move toward self-production and self-consumption of electricity in industrial and residential sectors, thereby reducing dependence on the grid and decreasing CO_2 emissions.

Sustainable development trends of renewable energy and future solar energy development strategy

- Currently, integrating RE sources into national energy systems faces two major challenges: connecting the grid transportation sector with future development initiatives and addressing the issue of instability in energy supply from RE systems. To develop sustainable energy policies, it is not enough to focus solely on cost reduction, performance improvement, and the development of RE sources. Instead, there is a need to develop integrated energy system solutions and apply flexible energy technologies such as BESS.
- Solar energy technology policies, investments, and support from government agencies and international organizations have created a comprehensive legal framework for the use of this RE source. However, the reduction of solar subsidies in some countries may slow down the industry's growth. The current policy is shifting toward encouraging large-scale solar energy system installations, aiming to reverse this decline, with Vietnam being a prime example. The hybrid model of grid-connected and energy storage systems (on-grid/off-grid) provides an effective platform to meet the needs of both consumers and electricity suppliers, as illustrated in Figure 8.1. The key issue is how to optimize the investment value in solar energy systems.
- One of the critical factors in building solar energy systems is ensuring the provision of adequate capacity. To achieve this, our system model develops optimization problems that combine solar energy with grid connectivity and energy storage, aiming to maximize energy efficiency and economic effectiveness. This helps optimize the scale and operation of solar energy systems according to business models.
- In this section, we present a system model that includes solar energy components, the electrical grid, and an energy storage system to illustrate the aforementioned perspectives. We then analyze investment strategy models. The general parameters of the system are described in detail in Table 8.1.
- Based on the energy generated by the solar energy system and the energy consumption of the business, the discrepancy between the energy produced

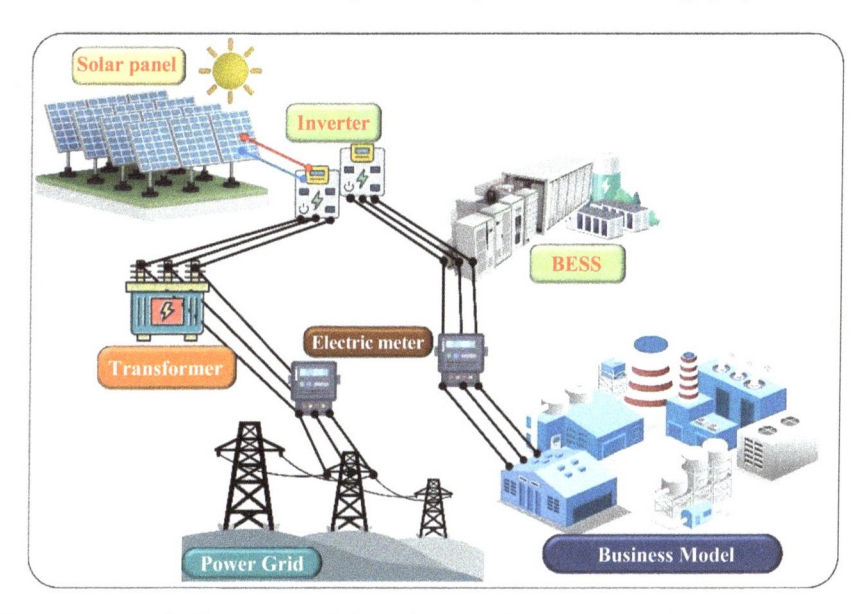

Figure 8.1 Hybrid system model combining solar energy and battery energy storage system (BESS)

and the daily energy consumption will be determined as follows:

$$F_{obj}(P_{sys}, E_{bess}) = \left| E_{generate}(P_{sys}, E_{bess}) - E_{comsume} \right| \tag{8.1}$$

where:

$$E_{generate}(P_{sys}, E_{bess}) = E_{solar}(P_{sys}) + E_{bess} \tag{8.2}$$

with:

$$E_{solar}(P_{sys}) = (t_{sun} + (12 - t_{sun})\alpha_{solar}) \times P_{sys} \tag{8.3}$$

Note:

- $E_{generate}$ is the energy produced
- $E_{consume}$ is the energy consumed
- E_{solar} is the energy from solar power
- P_{sys} is the system capacity
- α_{solar} is the solar energy coefficient
- t_{sun} is the sunlight duration

This is based on the daily operating model of solar energy to generate electricity, as shown in Figure 8.2.

The solar energy system collects energy throughout the entire daylight period (12 h from 6 AM). The period of full solar radiation (100%) is from 11 AM to 2 PM

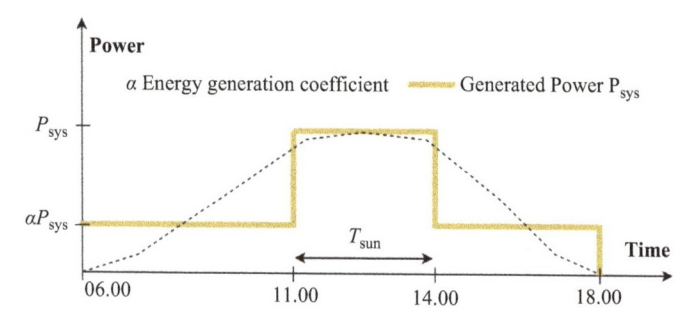

Figure 8.2 Solar energy system model operating over a single day

(3 h); during the remaining time, the solar system continues to generate energy, but at a lower radiation level ($\alpha \leq 100\%$).

$$E_{bess} = t_{charge/discharge} \times P_{bess} \tag{8.4}$$

8.5.8.2 Problem structure

The main objective of the problem is to minimize the discrepancy between the energy produced by the Hybrid Solar-BESS system and the energy required in the business model. This problem can be formulated as a minimization optimization problem as follows:

$$\text{minimize } F_{obj}(P_{sys}, E_{bess}) \tag{8.5a}$$

$$\text{subject to } 0 \leq P_{sys} \leq P_{sys}^{max} \tag{8.5b}$$

$$TC \leq TC^{max} \tag{8.5c}$$

$$E_{bess} \leq E_{consume}, E_{solar} \leq E_{consume} \tag{8.5d}$$

where:

- Equation (8.5b) represents the power limits (e.g., policy planning, land usage area), (8.5c) denotes the financial constraints for investment, and (8.5d) specifies the energy from solar or BESS, which must be less than or equal to the energy consumption.
- The optimization problem above indicates that this is a linear optimization problem with multiple variables and constraints, which can be easily solved using well-known optimization tools such as CVX (for MATLAB®) or CVXPY (for Python). Our main objective is to focus on multiple scenarios with different business models to develop a smart investment plan for the Hybrid Solar-BESS system.
- The economic value of the Hybrid Solar-BESS system is demonstrated through two cases. The first benefit arises from the solar energy produced by the solar system, which replaces the need to purchase electricity from the grid. The second benefit comes from the BESS system (in off-grid mode), which is used when there is no solar radiation (during the night).

Investment procedures for renewable energy projects 361

Table 8.2 Notations and descriptions of solar power system parameters

Notations	Describes
P_{sys} (kWp)	Power of the solar system (system capacity)
P_{model} (kW)	Power of the load (business model)
E_{solar} (kWh)	Generated energy from solar system per day
$E_{consume}$ (kWh)	Consumed energy by business model per day
E_{bess} (kWh)	Capacity of BESS system
C_{solar} ($/kWp)	Cost for building 1 unit of solar power (1 power unit = 1 kW)
C_{bess} ($/kWh)	Cost for building 1 unit of battery energy storage (1 energy unit = 1 KWh)
$Price_{solar}$ ($/kWh)	Selling price of 1 unit of solar energy generated
$Price_{grid}$ ($/kWh)	Buying price of 1 unit of energy on grid
$t_h \in \{h = 1, \ldots, 24\}$	The hth hour in a day
$t_{sun}, (t_{sun} \leq 24\ h)$	The number of hours of sunshine with full radiation each day (calculated as the mean)
α_{solar} (%)	Energy generation coefficient of solar power in the hours without full radiation, $(24 - t_{sun})$ h
$t_{operate}$	The duration of operational hours required for energy consumption per day in a business model

$$\text{NPV} = \text{Price}_{grid} \times (E^*_{generate}) = \text{Price}_{grid} \times (E^*_{solar} + E^*_{bess}) \qquad (8.6)$$

where:

- E_{solar} and E_{bess} are the total green energy produced by the solar power system and BESS, respectively, dedicated to business purposes (useful energy). This amount of energy must be sufficient to operate the business, as the selling price ($Price_{solar}$) of energy from the Hybrid Solar-BESS system is much lower than the purchasing price ($Price_{grid}$) of energy from the grid, even though the selling price of RE can reach 0 in some countries (Table 8.2).

P_{sys} (kWp): The capacity of the solar power system (system capacity).
P_{model} (kW): The power demand of the load (business model).
E_{solar} (kWh): The energy produced by the solar power system per day.
$E_{comsume}$ (kWh): The energy consumed per day according to the business model.
E_{bess} (kWh): The capacity of the BESS.
C_{solar} ($/kWp): The cost of installing one unit of solar power (1 unit of capacity = 1 kW).
C_{bess} ($/kWh): The cost of installing one unit of energy storage (1 unit of energy = 1 kWh).
$Price_{solar}$ ($/kWh): The selling price of one unit of energy produced by the solar power system.

Price$_{grid}$ ($/kWh): The purchase price of one unit of energy from the grid.
$t_h \in \{h = 1, \ldots, 24\}$: The hour of the day.
t_{sun}, ($t_{sun} \leq 24$ h): The number of hours with full solar radiation each day (calculated as the average value).
a_{solar} (%): The energy generation coefficient for the solar power system during non-full radiation hours, over the 24 h.
$t_{operate}$: The operational time required to consume energy each day in the business model.

8.5.9 Hybrid Solar-BESS system based on business models

This section proposes three common business models to leverage the potential of the proposed solar-BESS system. Each business operation requires a flexible approach to system installation.

8.5.9.1 Model 1: 24/7 service

We consider business models that operate continuously around the clock, such as industrial farming facilities, data centers, or supermarkets with cold storage. These models demand significant energy to sustain their production activities, as illustrated in Figure 8.3.

The energy usage model will be evenly distributed throughout the 24 h of each day ($t\_operate^1 = 24$ h). Therefore, the solar-BESS hybrid system will be designed to provide continuous energy throughout the company's operational hours, based on two approaches.

First, the capacity P_{sys} of the solar power system under full radiation (t_{sun}) must be greater than the capacity P_{model}. The surplus energy from the solar power system during this period will be stored in the BESS to supply energy to the business during the periods of low solar radiation ($24 - t_{sun}$).

8.5.9.2 Model 2: 8 working hours per day

Businesses typically have established operational models, which include the use of standard equipment and a workforce operating for a set number of hours. There is a

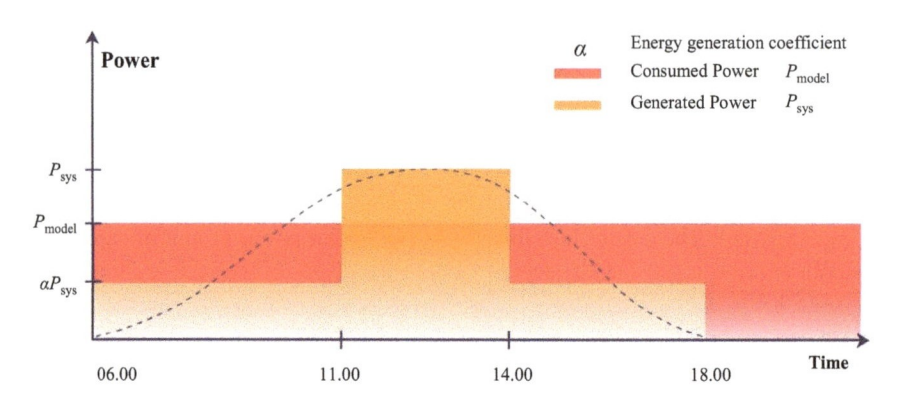

Figure 8.3 Case study of the first business model

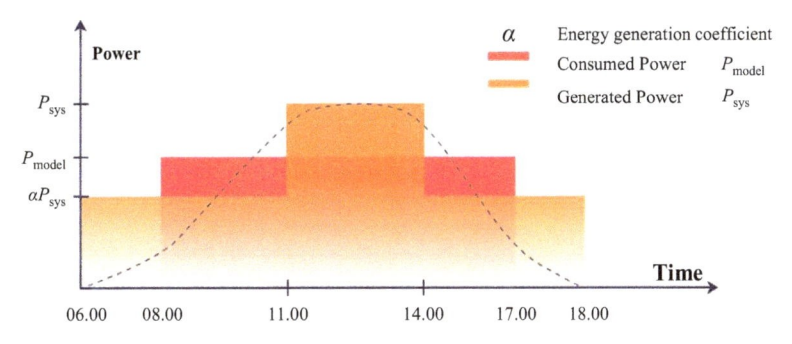

Figure 8.4 Case study of the second business model

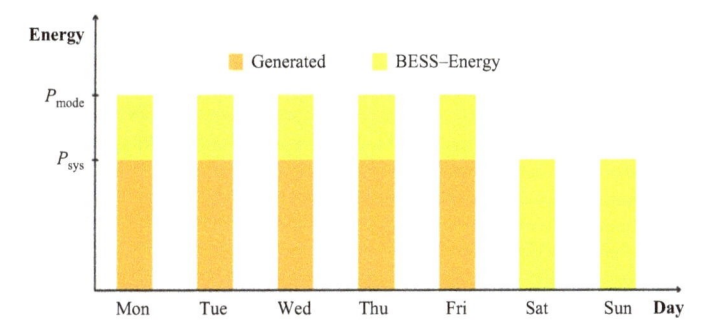

Figure 8.5 A case study of the third business model

certain energy consumption threshold that the business needs to meet within a specific period each day, for example, $t\_operate^2 = 8$ h, as illustrated in Figure 8.4.

8.5.9.3 Model 3: weekly model

Businesses often have established operational models, which include standard equipment and a workforce working five days a week, from Monday to Friday. The company operates for 8 h each working day. During the weekend, on Saturday and Sunday, the Hybrid Solar-BESS model continues to operate, storing energy in the BESS system to supply energy for the following week, as shown in Figure 8.5.

8.5.10 Some experiments: case studies in Vietnam

In this section, we present numerical results with actual values for the system parameters described in Table 8.1 for three business models. To illustrate the benefits of the proposed method, we consider four configurations related to solar and BESS systems based on on-grid and off-grid models, including: on-grid without solar and BESS, on-grid with solar but no BESS, off-grid with solar but no BESS, and off-grid with both solar and BESS. The CVXPY tool is commonly used in our simulations to solve the optimization problem and validate the results achieved through the deployment of the Hybrid Solar-BESS model.

Table 8.3 Values of general system parameters

Parameters	Values
C_{solar}	$480/kWp
C_{bess}	$240/kW
$Price_{solar}$	$0.08/kWh
$Price_{grid}$	$0.1/kWh
$t_h \in \{h = 1, \ldots, 24\}$	The hth hour in a day
$t_{sun}, (t_{sun} \leq 24\ h)$	3 h
α_{solar}	20%

Table 8.4 Values of system parameters in Model 1

Parameters	Values
$t^1_{operation}$	24 h/day
P^1_{model}	0.166 MW
P^{max}_{sys}	1 MWp

Table 8.3 provides the reference values and actual values for the common system parameters.

8.5.10.1 Experiment model 1

To illustrate the production activities in the first business model, we will conduct an experiment with an industrial farming model, specifically a pig farm in Dong Nai. This farm has a capacity of 10,000 pigs and an energy consumption of approximately 4 MWh per day. The other system parameters in this scenario are presented in Table 8.4.

Figure 8.6 and Table 8.5 provide the net values of the proposed schemes for Model 1 after subtracting the investment costs for the standalone solar power system or the Hybrid Solar-BESS system. As expected, the Hybrid Solar-BESS system achieves the best net value among the three schemes. After 10 years, the investment in the Hybrid Solar-BESS system will break even and begin generating profits. An important factor to consider is the depreciation of technology and system deployment, which will improve the payback period and the benefits associated with solar energy systems and hybrid systems.

8.5.10.2 Experiment model 2

In the second business model, we conduct an experiment with a factory that has an operational capacity of 2 MW and a working time of 8 h/day.

Table 8.6 highlights the advantages of solar energy systems and the Hybrid Solar-BESS systems. Both on-grid and off-grid platforms show that the net values

Net value of proposed methods in Model 1

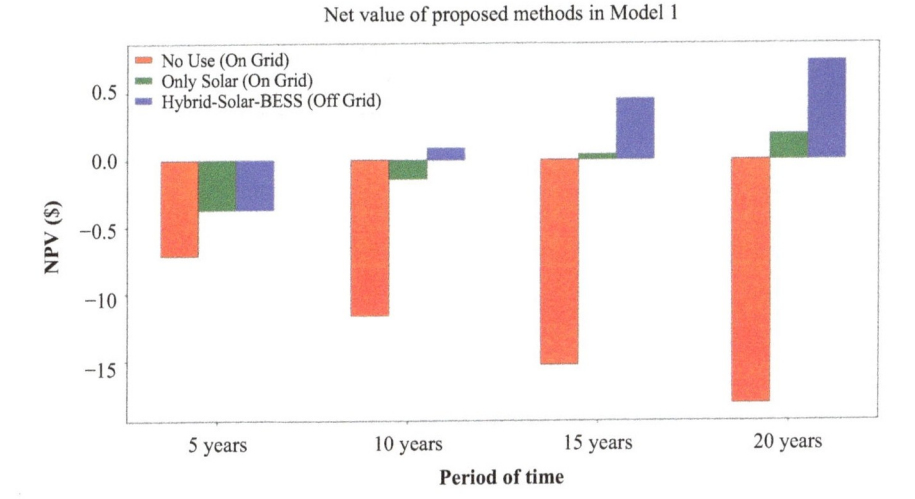

Figure 8.6 Net value of model 1 under different schemes (unit: $)

Table 8.5 Net value of Model 1 under different scenarios (unit: $)

Period	On-grid without solar (1)	On-grid without solar (2)	Off-grid (hybrid system)
5 years	$ −708, 324	$ −370, 202	$ −370, 350
10 years	$ −1, 164, 888	$ −141, 919	$ 86, 213
15 years	$ −1, 526, 702	$ 38, 987	$ 448, 026
20 years	$ −1, 813, 429	$ 182, 351	$ 734, 753

Notes: The two scenarios (1) and (2) differ in certain assumptions such as electricity pricing schemes, system configurations, or energy consumption profiles.

Table 8.6 Values of system parameters in Model 2

Parameters	Values
$t(2)$ operation	8 h/day
$P(2)$ model	2 MW
$P_{max\ sys}$	10 MWp

of these two systems outperform the other alternatives. It is important to note that the off-grid solar power system requires an initial investment and takes a long time to pay back. However, this method still provides more benefits compared to directly purchasing electricity from the grid. Overall, the use of RE systems proves to be more efficient than providing electricity from traditional fossil fuel sources.

Table 8.7 Net value of Model 2 under different schemes (unit: $)

Period	On-grid without solar	On-grid without solar	Off-grid without BESS	Off-grid without BESS
5 years	$ −128,010	$ −22,824	$ −221,080	$ −45,222
10 years	$ −210,521	$ 18,293	$ −138,568	$ 37,288
15 years	$ −275,909	$ 50,878	$ −73,180	$ 102,676
20 years	$ −327,728	$ 76,701	$ −21,362	$ 154,494

Table 8.8 Net value of Model 3 under different schemes (unit: $)

Period	On-grid without solar	On-grid without solar	Off-grid without BESS	Off-grid without BESS
5 years	$ −81,053	$ −25,536	$ −268,037	$ −193,369
10 years	$ −133,298	$ −6,548	$ −215,792	$ −141,124
15 years	$ −174,700	$ 8,498	$ −174,389	$ −99,722
20 years	$ −207,511	$ 20,422	$ −141,579	$ −66,911

8.5.10.3 Experiment model 3

In this model, we also conduct an experiment with a factory that has an operational capacity of 2 MW, operating 5 days/week and 8 h/day. Other parameters remain the same as in Model 2 (Table 8.7).

The net value of the proposed systems is presented in Table 8.8. For Model 3 (weekly model), the on-grid solar power system achieves the highest profit. In the business models outlined, particularly in this third model, the benefits of solar energy systems and BESS increase as industrial operations become more continuous and energy consumption rises. The on-grid and off-grid solar systems, along with Hybrid Solar-BESS systems, show higher efficiency compared to the traditional methods. This aligns with the current trend of energy use in Vietnam and globally.

With the growing energy demand and the depletion of fossil fuel resources, clean energy is emerging as a crucial factor in achieving sustainable development in the future. By utilizing a hybrid system combining grid-connected solar energy and off-grid solar energy, we can maximize the ROI in RE in Vietnam. This study focuses on using optimization techniques to enhance the investment efficiency in solar energy systems. It is based on the recognition of the importance of optimization models in various sectors of society. The optimization strategy described in this chapter is key to harnessing the full potential of solar energy, boosting economic profits, and protecting the environment. The optimization of investments for three different business models has demonstrated the flexibility and practicality of implementing the proposed method for various types of businesses.

8.6 Practical example for some solar and wind power projects in Vietnam

8.6.1 Overview of the project characteristics

The Solar Park 01 solar power project is invested by Hoan Cau Long An Co., Ltd, located in the Solar Park Long An Energy Zone in Binh Hoa Nam Commune, Duc Hue District, Long An Province. The solar power plant is constructed with a capacity of 50 MW and is connected to the national grid at a voltage level of 110 kV.

8.6.2 Legal basis for the estimate

8.6.2.1 Current legal documents

- Construction Law No. 50/2014/QH13 of 2014;
- Decree No. 46/2015/ND-CP dated May 12, 2015, of the Government on the management of quality and maintenance of construction works;
- Decree No. 59/2015/ND-CP dated June 18, 2015, of the Government on project investment management; Decree No. 42/2017/ND-CP dated April 5, 2017, amending and supplementing some articles of Decree No. 59/2015/ND-CP on project investment management; Decree No. 100/2018/ND-CP dated July 16, 2018, amending and supplementing, and abolishing some regulations on investment business conditions under the management of the Ministry of Construction;
- Decree No. 32/2015/ND-CP dated March 25, 2015, of the Government on the management of construction investment costs;
- Decree No. 119/2015/ND-CP dated November 13, 2015, of the Government on mandatory insurance in construction investment activities;
- Decree No. 209/2013/ND-CP dated December 18, 2013, of the Government detailing and guiding the implementation of some articles of the Value-Added Tax Law, amended and supplemented by current decrees and circulars;
- Circular No. 05/TT-BXD dated March 10, 2016, of the Ministry of Construction on guiding the determination of labor unit prices in construction investment cost management;
- Circular No. 06/2016/TT-BXD dated March 10, 2016, of the Ministry of Construction on guiding the determination and management of construction investment costs;
- Circular No. 210/2016/TT-BTC dated November 10, 2016, regulating the collection, payment, management, and use of project investment examination fees, and technical design-expenditure examination fees;
- Circular No. 329/2016/TT-BTC dated December 26, 2016, of the Ministry of Finance on the implementation of some provisions of Decree No. 119/2015/ND-CP regulating mandatory insurance in construction investment activities.

8.6.2.2 Applicable standards in the estimate

- The Construction Work Cost Estimate Standards—Part Construction published with Document No. 1776/BXD-VP on August 16, 2007, adjusted and

supplemented by Decision No. 1091/QD-BXD on December 26, 2011, Decision No. 1172/QD-BXD on December 26, 2012, Decision No. 588/QD-BXD on May 29, 2014, Decision No. 235/QD-BXD on April 4, 2017, and Decision No. 1264/QD-BXD on December 18, 2017, concerning the use of non-fired construction materials by the Ministry of Construction;
- The Construction Work Cost Estimate Standards—Part Installation published with Document No. 1777/BXD-VP on August 16, 2007, adjusted and supplemented by Decision No. 1173/QD-BXD on December 26, 2012, Decision No. 587/QD-BXD on May 29, 2014, and Decision No. 236/QD-BXD on April 4, 2017, by the Ministry of Construction;
- Decision No. 1134/QD-BXD dated October 8, 2015, on publishing the standards for machine and equipment consumption to determine the cost price for construction work;
- Decision No. 79/QD-BXD by the Ministry of Construction dated February 15, 2017, publishing the cost standards for project management and investment consulting costs in construction works;
- Decision No. 4970/QD-BCT dated December 21, 2016, by the Ministry of Industry and Trade on the publishing of specialized cost estimate standards for the installation of power lines and substations.

8.6.3 Estimate content

8.6.3.1 Estimate structure

The estimate structure for the Solar Park 01 project follows Decree No. 32/ND-CP of the Government and Circular No. 06/2016/TT-BXD, including the following items: construction costs; equipment costs; project management costs; investment consulting costs; other costs; and contingency costs.

8.6.3.2 Construction costs

The construction costs in the estimate include the costs for site leveling, the construction of main works, and auxiliary works required for the construction process. The calculation of construction costs for various work items is based on the following:

The volume according to the Technical Design documents provided by the Electricity Construction Consulting Joint Stock Company 4, March 2019 (Edition 01);
- Unit prices calculated according to the contract prices provided for the project.

8.6.3.3 Equipment costs

- Equipment costs include the costs for purchasing construction and technology equipment; costs for technology transfer training (if any); costs for installation, testing, and calibration; transportation costs, insurance; taxes, fees, and other related costs;
- The quantity for purchasing and the installation methods are based on the technical design documents provided by the Electricity Construction Consulting Joint Stock Company 4, March 2019.
- The installation, testing, and calibration costs are calculated based on the following:

- The quantities are based on the Technical Design Documentation provided by the Electric Construction Consulting Joint Stock Company 4, March 2019 (Edition 01).
- The unit prices for materials, labor, and construction machinery are calculated according to the project contract.

8.6.3.4 Unit price for procurement of equipment and materials

- The price of solar panels and inverters is based on the EPC contract price of the project, with the unit price including transportation, insurance, and other related fees.
- The prices of other equipment are based on the EPC contract price of the project, with the unit price including transportation, insurance, and other related fees.
- The exchange rate is taken from the Foreign Trade Bank's published rate on March 10, 2019: 1 USD = 23,250 VND.

8.6.3.5 Project management costs

- Project management costs are calculated based on Decision No. 79/QĐ-BXD issued by the Ministry of Construction on February 15, 2017.

8.6.3.6 Construction investment consulting costs

- Construction investment consulting costs for the project are calculated based on Decision No. 79/QĐ-BXD issued by the Ministry of Construction on February 15, 2017.

8.6.3.7 Other costs

- General item costs: Based on Circular No. 06/2016/TT-BXD.
- Design and cost estimation approval fees: According to the guidelines in Circular No. 210/2016/TT-BTC dated November 10, 2016, issued by the Ministry of Finance.
- Construction insurance costs during the construction period: As per Circular No. 329/2016/TT-BTC dated December 26, 2016, issued by the Ministry of Finance.
- Other costs: As per current regulations.

8.6.3.8 Contingency costs

- Contingency for unforeseen quantities: 5% of the total construction costs, equipment costs, project management costs, construction investment consulting costs, and other costs.
- Contingency for inflation: Calculated based on the approved total investment value.

8.6.4 Estimated cost

- The estimated cost excluding VAT is **909,811,025,000 VND**.
- Construction estimate—factory section (Tables 8.9–8.12)

8.6.4.1 Summary of estimate

Table 8.9 Summary of costs

No.	Cost Item	Unit	Formula explanation	Value before tax ($)	VAT ($)	Value after tax ($)
1	Construction cost	G_{XD}	Section 2.3	303,343,911,777	30,334,391,178	333,678,302,954
2	Equipment cost	G_{TB}	Section 2.4	387,598,390,070	38,759,839,007	426,358,229,077
3	Project management cost	G_{QLDA}	1.36% × Gxdtbpd × 0.8 × 1.35 × 2	24,357,019,495		24,357,019,495
3.1	Project management, technical consulting			16,321,500,000		16,321,500,000
3.2	Other project management costs			8,035,519,495		8,035,519,495
4	Investment consulting costs	G_{TV}		14,524,653,001	1,452,465,300	15,977,118,302
4.1	Main plant design costs		Approval for feasibility study	6,338,024,359	633,802,436	6,971,826,795
4.2	Design verification costs	GTV.3				
	• Building design verification cost		0.092% × Gxdpd	245,112,065	24,511,206	269,623,271
	• Construction estimate verification cost		0.088% × Gxdpd × 1,2	280,731,420	28,073,142	308,804,561
4.3	Supervision costs	GTV.4				
	• Construction supervision		1.522% × Gxdpd × 1,2	4,866,955,593	486,695,559	5,353,651,152
	• Equipment installation supervision		0.414% × Gtbpd × 1,2	2,793,829,565	279,382,957	3,073,212,522
5	Other costs	G_K		24,265,273,669	2,389,804,373	26,655,078,042
5.1	Construction insurance during construction	GK2	0.3% × Gxdtb	2,072,826,906	207,282,691	2,280,109,596
5.2	Technical design verification fee	GK2	0.07% × Gxdpd	187,610,928		187,610,928
5.3	Construction estimate verification fee	GK3	0.067% × Gxdpd	179,619,011		179,619,011
5.4	General cost item	GK4	Section 2.4	11,709,284,824	1,170,928,482	12,880,213,306
	Various fees					
5.5	Contractor bonding fee			550,000,000	55,000,000	605,000,000
5.6	Commitment fee			850,000,000	85,000,000	935,000,000
5.7	2-year warranty commitment fee			600,000,000	60,000,000	660,000,000
5.8	Application fee			8,115,932,000	811,593,200	8,927,525,200
6	Contingency costs	G_{DP}		75,484,333,726	7,300,943,636	82,785,277,362
6.1	Contingency for unforeseen work volume	G_{DP1}	5% × (GXD+GTB+GQLDA+GTV+GK)	37,704,462,401	3,646,824,993	41,351,287,394
6.2	Contingency for inflation	G_{DP2}	Contingency ratio according to the feasibility report	37,779,871,325	3,654,118,643	41,433,989,968
	Total			829,573,582,000	80,237,443,000	909,811,025,000

8.6.4.2 Construction cost summary

Table 8.10 Construction costs

No.	Cost category	Estimated value (excluding VAT)	VAT	Estimated value (including VAT)
I	**Main construction costs**	**263,377,918,777**	**26,337,791,878**	**289,715,710,654**
1	Construction costs	165,960,182,001	16,596,018,200	182,556,200,201
2	Procurement and Installation of Materials	97,417,736,776	9,741,773,678	107,159,510,453
II	**Auxiliary construction costs**	**39,965,993,000**	**3,996,599,300**	**43,962,592,300**
1	Electricity and water during construction phase	6,465,993,000	646,599,300	7,112,592,300
2	Temporary facilities (canteen, toilets, etc.)	6,500,000,000	650,000,000	7,150,000,000
3	Temporary access roads	7,000,000,000	700,000,000	7,700,000,000
4	Material storage warehouse (3,000 m^2)	4,500,000,000	450,000,000	4,950,000,000
5	Equipment storage yard (20,000 m^2)	6,500,000,000	650,000,000	7,150,000,000
6	Equipment staging area (containers, forklifts, trucks, loading/unloading equipment, 5,000 m^2)	6,500,000,000	650,000,000	7,150,000,000
7	Miscellaneous equipment for construction support	2,500,000,000	250,000,000	2,750,000,000
Total		**303,343,911,777**	**30,334,391,178**	**333,678,302,954**

8.6.4.3 Equipment cost summary

Table 8.11 Equipment costs

No.	Cost category	Value (excluding VAT)	VAT	Value (including VAT)
1	Equipment procurement costs	358,860,340,070	35,886,034,007	394,746,374,077
2	Equipment installation costs	26,296,800,000	2,629,680,000	28,926,480,000
3	Testing and calibration costs (included)	–	–	–
4	Transportation of imported materials	2,441,250,000	244,125,000	2,685,375,000
Total		**387,598,390,070**	**38,759,839,007**	**426,358,229,077**

8.6.4.4 General category cost summary

Table 8.12 General item costs

No.	Cost item description	Value before tax	VAT	Value after tax
1	Cost of temporary site facilities for accommodation and construction management	3,296,407,118	329,640,712	3,626,047,830
1.1	Remaining structures	3,296,407,118	329,640,712	3,626,047,830
	Remaining construction			
	Installation and testing			
2	Costs for undetermined quantities in the design phase	6,592,814,236	659,281,424	7,252,095,659
3	Remaining general item costs	1,820,063,471	182,006,347	2,002,069,818
3.1	Costs for mobilizing and demobilizing construction equipment and labor from the site	1,516,719,559	151,671,956	1,668,391,515
3.2	Costs for restoring technical infrastructure affected by construction activities	303,343,912	30,334,391	333,678,303
	Total	11,709,284,824	1,170,928,482	12,880,213,306

8.6.5 Project information summary

- The Dam Nai Wind Energy Project in Ninh Thuan Province. This document includes an analysis of wind data and an assessment of wind energy based on data collected over 24 months at the 100 m wind measurement mast in Dam Nai from May 23, 2015, to May 23, 2017. Wind speed measurements were taken at heights of 100, 80, and 50 m, while wind direction was measured at heights of 97, 77, and 47 m. The estimated energy based on the Phase 1 layout includes 3 Gamesa G114 2.0 MW turbines, totaling 6 MW, and Phase 2 consists of 12 G114 2.5 MW turbines, with a total capacity of 30 MW. Thus, the combined total capacity of both phases is 36 MW.
- Data from over 24 months of measurement shows that the average wind speed at heights of 100, 80, and 50 m is 6.53, 6.37, and 5.97 m/s, respectively. The wind rose chart aligns with the predicted wind pattern for the area, although the Northeast wind component is slightly stronger during the winter months, likely due to the surrounding terrain.
- The measured shear force is moderate to low, with a shear coefficient of 0.129 based on the power law. The air density at the project site is approximately 1.15 kg/m^3, which is within the expected range.

Investment procedures for renewable energy projects 373

- Initial analysis of turbulence levels and extreme winds indicates relatively low conditions for the project area (categorized as Level III C), although the turbulence and extreme wind analyzes are based solely on 24 months of measurement data, which may lead to an underestimation in the classification of the area.
- Based on the 24-month data, the long-term wind speed has slightly decreased, with expectations of 7.00 m/s at a height of 100 m above ground and 6.83 m/s at 80 m above ground at the wind measurement mast. These values are similar to the estimates from the previous wind data report (with a deviation of less than 1%). Further data collection at the wind measurement mast is required to confirm these projections.
- The Gamesa G114 (2.0 MW and 2.5 MW—Class II turbines, highly suitable for this area) was selected as the most cost-effective and readily available option on the market. The estimated energy based on the layout of 15 turbines within the project boundary, with a turbine spacing of approximately 7.5D along the wind direction and 2D across the wind direction to optimize energy production while minimizing land occupation (including a layout for three G114 turbines for Phase 1—limited to 6 MW—to be constructed from mid-to-late 2017, and 12 G114 turbines for Phase 2—limited to 30 MW—to be constructed from mid-2018). The estimated annual net energy production (AEP) of the 15 Gamesa G114 turbines at 80 m height (AEP) is 116.4 GWh/year (with a net capacity factor of 36.9%).
- Only a few houses/farming structures need to be relocated, though a final survey will be conducted in coordination with local authorities to confirm the classification of houses and buildings and to carry out appropriate compensation/mitigation processes.

8.6.5.1 Wind data, wind speed, and wind direction assessment

The 24-month wind data was recorded by a 100 m wind measurement mast at Dam Nai from May 23, 2015 to May 23, 2017, with a total data availability of 96.9%. Due to the rich data source and minimal elevation differences between the anemometers, with winds primarily perpendicular to the wind mast direction, the average values of both anemometers at each level, excluding periods when the anemometers were obstructed, are summarized to present an average wind speed table.

The summary of average wind speeds is presented in Tables 8.13 and 8.14.

8.6.5.2 Climate data analysis

The climate data at the wind measurement mast location indicates an average temperature of 27.5°C at both the 6 and 100 m sensor devices (with the device closer to the ground exhibiting higher variations but having the same average). The average relative humidity is 70.4% at 100 m and 70.6% at 6 m, and the average air pressure is 100.8 kPa at 6 m above ground level. All of these values allow for the calculation of the average air density at approximately 1.15 kg/m^3 throughout the measurement period at an altitude of around 15 m above sea level (ranging from 1.13 kg/m^3 in the summer months to 1.16 kg/m^3 during the winter monsoon, due to lower temperatures and strong winds).

374   *Clean energy in South-East Asia*

Table 8.13   Summary of average wind speeds

Sea	100 mA speed	100 mB speed	80 mA speed	80 mB speed	50 mA speed	50 mB speed
Measurement height	100	100	80	80	50	50
Average wind speed (m/s)	6.5260	6.5318	6.3655	6.3697	5.9705	5.9776
Monthly average wind speed	6.4740	6.4798	6.3153	6.3192	5.9243	5.9315
Median value (m/s)	5.9400	5.9540	5.8190	5.8460	5.4930	5.5110
Lowest wind speed (m/s)	0.2150	0.2060	0.2080	0.2210	0.1980	0.2120
Highest wind speed (m/s)	25.2360	25.9860	25.5760	25.5350	24.1420	23.9290
RCMC wind speed (m/s)	9.1135	9.1277	8.8881	8.8855	8.3318	8.3214
Weibull k	1.369	1.365	1.372	1.369	1.377	1.386
Weibull c (m/s)	7.1240	7.1256	6.9531	6.9513	6.5247	6.5404

Table 8.14   Summary of average wind speeds at different heights over a 24-month measurement period, including tower shading

Height	Average wind speed	Availability
100 m	6.53 m/s	96.9%
80 m	6.37 m/s	96.9%
50 m	5.97 m/s	96.9%

- **Wind speed adjustment process for long-term measurements**
  – In general, at least one full year of data from the wind measurement mast is necessary for long-term wind analysis/adjustment to ensure that seasonal patterns (i.e., summer monsoon and winter monsoon winds) are fully captured and the shear force is well understood.
  – Since 24 months of data have already been collected at the Dam Nai wind measurement mast, the wind speed has been adjusted according to long-term trends in this report.

- **Long-term data set: vortex time series (TS)**
  – Due to the lack of reliable long-term reference wind data in Vietnam (such as from automatic weather stations), the simulated wind data based on the CFSR time series analysis (virtual wind measurement mast data) obtained from Vortex is used as the reference data. This dataset spans approximately 22 years from January 1, 1995 to May 23, 2017.
  – Since the Gamesa turbines are at a hub height of 80 m, the long-term focus is primarily on the 80 m data set. Thus, the Vortex simulated data has been recalculated using the 24 months of measurement data at the 80 m mast (with

aggregated data from 80 and 77 m) from May 23, 2015 to May 23, 2017. The data spans more than 22 years from January 1, 1995 to May 23, 2017.
- Both datasets were collected over several annual time periods from May 23, 1995 to May 23, 2017, to reduce any seasonal bias. A similar process was carried out at a height of 100 m to gather relevant data.

8.6.5.3 Analysis of the overlap time between wind mast measurements and vortex TS at 80 m

At the time of writing this report, there is a 24-month overlapping period between the wind mast measurements and the Vortex CFSR TS data. During this period, a −50-min time offset has been adjusted to ensure optimal alignment. The re-simulated Vortex TS data was generated using the 24 months of measured data and includes the time adjustment. All wind direction and wind speed data from the wind mast, Vortex TS, and Vortex re-simulation data are consistent in terms of trends.

8.6.5.4 Long-term adjustment of measured wind speeds

Based on the relationship between these datasets, the long-term adjustment used the least squares linear (LLS) algorithm and matrix time series to compile the long-term dataset from May 23, 1995 to May 24, 2017 (22 years). The results are presented in Table 8.15.

Table 8.15 shows a summary of long-term wind estimates at the Dam Nai wind mast with hub height at 80 m.

The re-simulated results are of greater interest, as they have the highest correlation coefficient, and the matrix process tends to better represent wind distribution, which is crucial for energy forecasting. The long-term correlation with mesoscale time series results shows a difference of about 2–3%, but these were not considered due to shorter overlap periods. A similar process with the 100 m mast measurement dataset and the 100 m Vortex TS re-simulation gives a long-term wind speed forecast of 7.00 m/s at a height of 100 m at the wind mast location based on the matrix method.

8.6.6 Preliminary energy estimates

8.6.6.1 Wind resource grid (WRG)

Based on the 24-month measurement data from the wind mast, the Vortex re-simulated wind resource grid was created over the wide area of Dam Nai. This simulated grid is based on long-term data at a height of 80 m (combined), and the wind resource grid can be extracted at different hub heights of the turbine to generate energy estimates. Due to the relatively low shear forces, and the turbine being considered at a hub height of 80 m, all energy estimates are based on this hub height. However, the shear forces assumed by Vortex for creating the wind resource grid were not always well recorded, so several small adjustments were necessary.

8.6.6.2 Turbine selection for evaluation

- After preliminary discussions with several turbine manufacturers willing to extend their presence in Vietnam for further negotiations on pricing and

Table 8.15 Comparison of wind speed data according to the long-term method and forecast

Reference area	Long-term method	Reference period (years)	Average on-site wind speed (m/s) at 80 m	Overlap period (months)	Average overlapping wind speed (m/s) at 80 m for wind measurement Mast	Average overlapping wind speed (m/s) at 80 m by vortex TS	Forecasted Long-term wind speed (m/s) at 80 m at wind measurement mast	Correlation coefficient (R^2)
Vortex TS CSFR	LLS	22	7.53	24	6.37	6.68	6.97	0.747
	Matrix	22	7.53	24	6.37	6.68	6.96	0.747
Vortex remodeled TS (CSFR)	LLS	22	6.86	24	6.37	6.28	6.86	0.883
	Matrix	22	6.86	24	6.37	6.28	6.83	0.883

delivery times, the Gamesa G114 2.0 MW (for Phase 1) and G114 2.5 MW (for Phase 2) turbines with a rotor diameter of 114 m were selected to operate at a hub height of 80 m, arranged in a layout of 15 turbines. These turbines, classified as Type II, were deemed suitable for the project area.
- Due to the Investment Certificate allowing for a two-phase implementation, the locations of three turbines in Phase 1 (6 MW) were chosen to access the best wind resource near the AH1 highway, minimizing the need for road construction for Phase 1.
- The 12 wind turbines for Phase 2 were optimized in the southern part of the project boundary to minimize encroachment on farmland and the impact on farmhouses within the project area, while maximizing energy output. Final adjustments were made to the positions to reduce impacts on rice fields and make use of existing roads.

8.6.6.3 Project area boundaries and turbine layout diagram

There are several constraints in the development area. The following inclusions and exclusions were considered to define the "boundary range":

- **Exclusions for project development**
    - **Exclusion of 400 m minimum from houses or residential areas**: A minimum exclusion zone of 400 m from the north (due to prevailing wind direction from the north) and 500 m from the north of the area, extending up to 750 m from the village located south of the proposed turbine position. This exclusion does not consider the scattered houses/farmhouses in the rice field/agricultural area.
    - **Exclusion of 20 m from rivers/ditches**: A 20-m exclusion zone around small rivers or ditches has been considered.
    - **Exclusion of 20 m around existing roads/trails**: A 20-m exclusion zone has been established around existing roads and trails crossing the project area.
    - **Buffer zone of 180 m from existing 100 kV power lines**: A 180-m buffer zone has been considered for existing high-voltage power lines going north of the project area, as well as for the power lines passing through the project area of the potential nuclear power plant.
    - **Buffer zone of 70 m from the nuclear power plant road**: A 70-m buffer zone from the nuclear power plant's road has been considered.

- **Turbine layout**
    - A layout of **three turbines (Phase 1)** and **12 turbines (Phase 2)** has been proposed for the project area. The minimum distance between turbines is set at **7D × 2D** ($D = 114$ m), aligned both along and across the prevailing wind direction (20°).
    - The turbine positions were selected near the existing wind mast and are optimized to capture the best wind resource, maximizing energy production while minimizing uncertainty.

- The layout was slightly adjusted to minimize land encroachment by utilizing existing roads and optimizing the design of new access roads.

- **Energy estimation**
  The software **Openwind V01.04.00.1097 64 bit** was used to estimate the energy for the wind turbines:
  - **Turbine layout**: 3 RevK wind turbines (2 MW each, totaling 6 MW for Phase 1) and 12 RevS wind turbines (2.5 MW each, totaling 30 MW for Phase 2).
  - **Wind resource grid**: The wind resource grid with 100 m resolution at 80 m height above ground from Vortex (re-simulated and adjusted) (Table 8.16).

- **Turbine specifications**
  - The turbines used are detailed in the following section, with a power curve at **1.15 kg/m³**.
  - Air density is calculated as 1.15 kg/m³ (based on measured climatic data) at approximately 10 m above ground level.
  - The average turbine bottom height is 6 m above ground level.

Estimated energy: The energy estimation for the proposed **36 MW layout** (6 MW + 30 MW) is detailed below.

The wind turbine hub height is 80 m, with an air density of 1.15 kg/m³, and a loss factor of 10.16% (due to wind turbine generator, balance of plant (BoP), grid availability, power curve, etc., as detailed below) for the net capacity factor/annual net energy based on Vortex simulation with a 100 m resolution and adjusted wind resource grid.

- **Assumptions for the loss factor applied to the calculation of annual net electricity production**
  - **Turbine availability**: 97.0%
  - **BoP availability**: 99.8%
  - **Grid availability**: 98.0%
  - **Scheduled maintenance**: 99.3%
  - **Electrical efficiency**: 97.0%
  - **Power curve adjustment**: 99.4%
  - **Wind delay during high winds**: 99.9%
  - **Environmental conditions**: 99.5%
  - **Turbine shutdown**: 100.0%
  - **Blade wear & other factors**: 99.5%

Total energy capacity of Gamesa (with 10.16% loss): 89.84%
No transmission losses are considered between the field and the substation due to the short distance between them. However, the assumed losses may need to be revisited depending on the information from EVN and the connection selection, although no major changes are expected.

8.6.6.4 Uncertainty

Several sources of uncertainty may impact the electricity estimate, which could stem from measurement discrepancies and long-term forecasts (due to anemometry, long

Table 8.16 Layout and performance of wind turbines at Dam Nai

Dam Nai layout	Elevation (m)	Hub height (WTG) (m)	Number of WTGs	Type of WTG	WTG rating (MW)	Installed capacity (MW)	P50 total energy (GWh/year)	Capacity factor (CF) (%)	P50 net energy (GWh/year)	Net capacity factor (NCF) (%)	Net energy per WTG (GWh/year)	WTG details
Layout of 3 RevK wind turbines + 12 RevS turbines	10	80	3+12	G114–2.0 & 2.5 (pc@1.15 kg/m^3)	2.0 (Phase 1) & 2.5 (Phase 2)	36	129.6	41.2	116.4	36.9	7.76	RevK and RevS turbines

Table 8.17 Analysis of uncertainty sources and impact on energy production

Uncertainty sources	Wind Speed Percentage			Energy production (GWh/year)
	(%)	(m/s)	(%)	
Instruments	2.0	0.14	4.2	4.9
Long-term wind representation	1.5	0.10	3.1	3.6
Correlation coefficient	1.0	0.07	2.1	2.4
Extrapolation to turbine hub height	0.0	0.00	0.0	0.0
Seasonal uncertainty	0.0	0.00	0.0	0.0
Wind speed during	2.7	0.18	5.6	6.6
Measurement period			0.0	0.0
Terrain simulation wake loss simulation			0.7	0.7
Horizontal simulation			3.0	3.4
Substation measurements			0.3	0.3
Energy uncertainty over measurement period			6.6	7.6
Future wind variability (1 year)	7.0	0.48	14.7	17.1
Future wind variability (10 years)	2.2	0.15	4.6	5.4
Future wind variability (20 years)	1.6	0.11	3.3	3.8
Total energy uncertainty (1 year)			17.1	19.9
Total energy uncertainty (10 years)			9.8	11.4
Total energy uncertainty (20 years)			9.3	10.8

periods, extrapolation for hub height, seasonal bias), or from the simulation itself (e.g., horizontal simulation—distance from turbines to wind measurement masts—or wake loss simulation). Depending on the timeframe, the "future wind variability" has been reduced due to a decrease in long-term wind variability. All these uncertainties are used to calculate the P75.

The uncertainty calculated for the layout of 15 Gamesa G114 turbines at a hub height of 80 m indicates that the future uncertainty over 10 years will only be 9.9%, ensuring that the P50/P75 limits (with a deviation of around 5%) are within acceptable range, which is generally considered sufficient for financing purposes (Table 8.17).

The uncertainty for the 15 wind turbines is quite low due to a clear correlation with the long-term wind dataset and the turbines being located near the wind measurement mast (average distance ∼1 km). These uncertainties lead to a low 10-year uncertainty (9.9%)—the 10% value on P90 and 5% on P75 is generally considered sufficient for financial provision purposes.

8.6.7 Conclusion and recommendations

- The 24-month wind data collected from May 24, 2015 to May 23, 2017 at the 100 m wind mast in Dam Nai, Ninh Thuan Province, Vietnam was analyzed.
- The average wind speeds at heights of 100, 80, and 50 m were 6.53, 6.37, and 5.97 m/s, respectively, over the measured periods.
- The wind rose diagram corresponds to the predicted wind pattern in the area, with the stronger northeast monsoon winds during winter months, likely influenced by the topography of the surrounding region.
- **Area affected strongly by the monsoon**
  - **Summer monsoon period (April to September)**: Winds primarily come from the southeast and southwest, with low shear winds. Wind speeds peak during the day and decrease in the evening until mid-morning (type of light/thermal sea breeze).
  - **Winter monsoon period (mid-October to mid-March)**: Winds come from the north-northeast with significantly higher wind speeds, remaining quite strong throughout the day.
- The total shear measured was relatively low, with a shear coefficient of 0.129 according to the power law. This is consistent with the flat, less rugged terrain expected in the project area (rice fields).
- The calculated air density for the area is approximately 1.15 kg/m^3, which is within the expected range for the region.
- Initial turbulence and extreme wind analyzes indicate relatively low conditions for the project area (suggesting a Class III-C area), although the turbulence and extreme wind analysis based only on the 24-month measurement data might lead to an underestimation of the regional classification.
- Based on the 24-month data, the long-term wind speed has slightly decreased and is expected to be 7.00 m/s at 100 m above ground level and 6.83 m/s at 80 m above ground level at the wind mast location. These figures are similar to the estimates in the previous wind data report (less than a 1% difference). Additional data collection at the wind mast is needed to confirm these forecasts.
- The **Gamesa G114** (2.0 and 2.5 MW—Class II turbines) has been selected as the best option in terms of cost and availability on the market, and is highly suitable for the area. The estimated energy production based on the layout of 15 turbines within the project boundary, with a spacing of approximately 7.5D along the wind direction and 2D crosswind to ensure optimal energy production and minimize the occupation of rice fields, is as follows: The layout includes 3 ReV wind turbines for Phase 1 (limited to 6 MW), which will be built from mid-to-late 2017, and 12 RevS wind turbines for Phase 2 (limited to 30 MW), which will be built from mid-2018. The estimated AEP for the 15 Gamesa G114 turbines at 80 m height is 116.4 GWh/year, with a net capacity factor of 36.9%.

Further reading

[1] IRENA (International Renewable Energy Agency). *Data and visualisations.* Vietnam Ministry of Industry and Trade. *Power Development Plan VIII (2023).*
[2] Lowe Chemie. *Mechanical and Chemical Process.*
[3] AE Battery Manufacturer. *Technical Documentation on Battery Production,* AE Group AG, Leopoldshöhe, Germany.
[4] Balat H, Kırtay E. (2010). Hydrogen from biomass – Present scenario and future prospects. *Int J Hydrogen Energy.* 35:7416–26.
[5] Bhanu Mahajan (2012). *Negative Environment Impact of Solar Energy. Environmental Science and Policy Course.* CEPT University, India. Energy Policy, 33.
[6] Damon Turney and Vasilis Fthenakis (2011). Environmental impacts from the installation and operation of large-scale solar power plants. *Renewable and Sustainable Energy Reviews.* 15(6):3261–70.
[7] DOE (2011). *A National Offshore Wind Strategy: Creating an Offshore Wind Energy Industry in the United States.* 52 pp.
[8] GIZ (2018). Guidelines on Environmental and Social Impact Assessment for Wind Power Projects in Vietnam. Available at: http://gizenergy.org.vn/media/app/media/GIZ-ESP_ESIA%20Report_ENG_Final.pdf
[9] GSO (2006–2020). *Almanac Statistics of Vietnam Reports in 2006 – 2020.* General Statistics Office of Vietnam, in Vietnamese.
[10] IFC (2012). *Guidance Notes to Performance Standards on Environmental and Social Sustainability.* Washington, DC: International Finance Corporation. Available at: https://www.ifc.org/wps/wcm/connect/9fc3aaef-14c3-4489-acf1-a1c43d7f86ec/GN_English_2012_Full-Document_updated_June-14-2021.pdf?MOD=AJPERES&CVID=nXqnsJp
[11] International Energy Agency (2023). *Global EV Outlook 2023, Catching up with Climate Ambitions*; International Energy Agency: Paris, France.
[12] John Glasson, Riki Therivel and Andrew Chadwick (1999). *Introduction to Environmental Impact Assessment.* 2nd edition in 2005, UCL Press, the Taylor & Francis e-Library.
[13] Jose I. Bilbao, Garvin Heath, Alex Norgren, Marina M. Lunardi, Alberta Carpenter, Richard Corkish (2021). *PV Module Design for Recycling.* International Energy Agency (IEA), PVPS Task 12, Report T12-23:2021. Available at: https://iea-pvps.org/wp-content/uploads/2021/10/T12_2021_PV-Design-for-Recycling-Guidelines_Report.pdf
[14] Luca Ciacci and Fabrizio Passarini (2020). Life cycle assessment (LCA) of environmental and energy systems. *Energies.* 13(22):5892, https://doi.org/10.3390/en13225892
[15] Matt Rogers (2019). *Vietnam's renewable energy future.* Mckinsey Sustainability. Available at:https://www.mckinsey.com/capabilities/sustainability/our-insights/sustainability-blog/cop27-accelerating-decarbonization
[16] Md Shahariar Chowdhury, Kazi Sajedur Rahman, Tanjia Chowdhury, *et al.* (2020). An overview of solar photovoltaic panels' end-of-life material recycling. *Energy Strategy Reviews.* 27:100431.

[17] Ministry of Natural Resources and Environment (2010). *National Technical Regulation on Vibration*. Regulation No. QCVN 27:2010/BTNMT, signed on 16 December 2010. Available at: https://i.ndh.vn/attachment/2022/09/19/psi-bao-cao-nganh-dien-082022-pdf.pdf

[18] Ministry of Natural Resources and Environment (2015). *On Strategic Environmental Assessment, Environmental Impact Assessment and Environmental Protection Plan*. Circular No. 27/2015/TT-BTNMT, signed on 29 May 2015. Available at: https://datafiles.chinhphu.vn/cpp/files/vbpq/2015/06/27-tt-btnmt.signed.pdf

[19] Ministry of Natural Resources and Environment (2011). *Regulation on Strategic Environmental Assessment, Environmental Impact Assessment, and Environmental Protection Commitment*. Decree No. 29/2011/ND-CP, dated 18 April 2011. Available at: https://chinhphu.vn/default.aspx?pageid=27160&docid=100006

[20] National Institute of Environmental Health Sciences (1999). *Health Effects from Exposure to Power-Line Frequency Electric and Magnetic Fields*, Research Triangle Park, NC.

[21] Nguyen Minh Quang (2022). *Electricity Industry Update Report*. Oil and Gas Securities, August 2022. Available at: https://i.ndh.vn/attachment/2022/09/19/psi-bao-cao-nganh-dien-082022-pdf.pdf

[22] Pablo Hevia-Koch and Henrik Klinge Jacobsen (2019). Comparing offshore and onshore wind development considering acceptance costs. *Energy Policy*. 125:9–19. Available at: https://www.sciencedirect.com/science/article/abs/pii/S030142151830675X

[23] Prime Minister (2016). *Approving the Adjustment of the National Power Development Plan for the Period 2011–2020 with a Vision to 2030*. Decision No. 428/QD-TTg, signed and issued on 18 March 2016. Available at: https://datafiles.chinhphu.vn/cpp/files/vbpq/2016/03/428.signed.pdf

[24] Prime Minister (2015). *Regulations on Environmental Protection Planning, Strategic Environmental Assessment, Environmental Impact Assessment, and Environmental Protection Plans*. Decree No. 18/2015/ND-CP, signed and issued on 14 February 2015. Available at: https://datafiles.chinhphu.vn/cpp/files/vbpq/2015/02/18.signed_01.pdf

[25] SRV (2022). *Nationally Determined Contribution (NDC report)* (updated in 2022). *Socialist Republic of Viet Nam*, Ha Noi, 38p.

[26] Srinivasan, M., A. Velu, and B. Madhubabu (2019). Potential environmental impacts of solar energy technologies. *International Journal of Science and Research (IJSR)*. 9(5):792–5.

[27] Theocharis D Tsoutsos, Niki Frantzeskaki and Vassilis Gekas (2005). Environmental impact assessment of solar energy systems. *Energy Policy*. 33(3):289–96. Available at: https://doi.org/10.1016/S0301-4215(03)00241-6

[28] Tran Thien Khanh (2021). *Orientation for the Development of Clean Hydrogen Energy: Analysis of the Current Situation and Evaluation of Industrial Hydrogen Production Methods*. Available at: https://iced.org.vn/dinh-huong-phat-trien-nang-luong-sach-hydro-phan-tich-thuc-trang-va-danh-gia-cac-phuong-phap-san-xuat-hydro-trong-cong-nghiep/

[29] US. Embassy in Vietnam (2022). *International agreement to support Vietnam's climate and energy goals*. Available at: https://en.baoquocte.vn/international-agreement-to-support-vietnams-climate-and-energy-goals-209960.html
[30] VCBS (2023). *Electricity Industry Outlook Report 1H.2023: Electricity Consumption Demand to Grow Steadily in the Long Term*. Available at: https://finance.vietstock.vn/bao-cao-phan-tich/10735/bao-cao-trien-vong-1h-2023-nganh-dien-nhu-cau-tieu-thu-dien-tang-truong-on-dinh-trong-dai-han.htm
[31] World Bank (2015). *Environmental, Health, and Safety (EHS) Guidelines*. IFC E&S Washington, D.C. Available at: https://documents.worldbank.org/en/publication/documents-reports/documentdetail/157871484635724258/environmental-health-and-safety-general-guideline
[32] World Health Organization (2007). *Extremely Low Frequency Fields*. Environmental Health Criteria Monograph No. 238, Geneva.

Chapter 9
Circular economic solutions—effective management of "solar photovoltaic panel waste"

9.1 Introduction to circular economy scenarios for renewable energy development: case studies in Vietnam

9.1.1 Overview of circular economy and renewable energy

The circular economy (CE) is an economic model aimed at optimizing the use of resources, minimizing waste, and reducing negative environmental impacts. Instead of the traditional linear economy model of extraction, production, consumption, and disposal, the CE promotes the reuse of resources and the transformation of waste into inputs for new production cycles.

In the context of Vietnam, facing significant challenges such as resource depletion, environmental pollution, and climate change, developing renewable energy through the lens of the CE is an inevitable path forward. Vietnam has been importing energy since 2015 and is projected to need to import 100 mn tons of coal by 2030. To address these challenges, applying CE principles to renewable energy development will not only ensure energy security but also reduce pressure on resources and the environment.

9.1.2 Circular economy scenarios in renewable energy development

9.1.2.1 Using waste as raw material for renewable energy production

- **Waste-to-energy plants**: Converting municipal and industrial waste into energy sources. This model helps reduce landfill waste while harnessing heat from incineration to generate electricity. For example, waste-to-energy plants in Ho Chi Minh City have demonstrated the feasibility of this model.
- **Reusing industrial by-products**: Nestlé Vietnam has recycled boiler waste into non-fired bricks and produced fertilizer from non-hazardous sludge, contributing to reduced greenhouse gas emissions.

9.1.2.2 Using renewable energy in industrial production

- **Heineken Vietnam**: Nearly 99% of the company's waste and by-products are reused or recycled. Four out of six of Heineken's breweries use renewable energy and carbon-neutral fuels.
- **Unilever Vietnam**: Unilever implements a program to collect and recycle plastic packaging, helping reduce plastic waste in the environment.

9.1.2.3 Developing solar and wind energy

- **Installing rooftop solar systems**: Industrial parks and businesses have utilized rooftops to install solar panels, reducing energy costs and easing pressure on the national grid. For example, industrial parks in Ninh Binh and Da Nang have installed these systems, significantly saving on energy costs.
- **Developing wind farms**: Wind energy projects in Binh Thuan and Ninh Thuan not only generate clean energy but also promote the CE by reusing equipment and managing by-products from production.

9.1.2.4 Recycling alliances and business cooperation

- **Vietnam packaging recycling alliance (PRO)**: Established in 2019, PRO Vietnam involves nine pioneering businesses and focuses on recycling plastic packaging, contributing to reducing plastic waste in the environment and creating new business opportunities in waste management.

9.1.3 Case studies in Vietnam

9.1.3.1 Eco-industrial parks

Eco-industrial parks in Ninh Binh province, Can Tho city, and Da Nang city have successfully applied CE, saving USD 6.5 mn per year through water, thermal, and by-product reuse. These parks are leading models in renewable energy adoption and effective waste management.

9.1.3.2 Waste-to-energy conversion

Waste-to-energy plants in Soc Son (Hanoi city) and Binh Duong province not only alleviate waste disposal pressures but also contribute a stable power source to the national grid.

9.1.3.3 Pioneering foreign direct investment companies

Nestlé and Heineken are foreign companies investing in Vietnam with green production models, waste reuse, and renewable energy use. Their initiatives help spread awareness of CE within the Vietnamese business community.

9.1.4 Conclusion

Applying the CE in renewable energy development is not only a global trend but also an inevitable path for Vietnam to address resource, environmental, and energy challenges. Models such as eco-industrial parks, packaging recycling, and renewable energy in production are laying solid foundations for sustainable development. To be effective, there must be collaboration between the government, businesses, and communities in shifting mindsets, policies, and actions toward a greener, sustainable Vietnam.

9.2 Circular economy approach after the end-of-life of materials in solar and wind energy equipment

The CE is not only applied in production and use but also plays a crucial role in the end-of-life phase of renewable energy equipment such as solar panels and wind turbines. Effectively managing post-use materials will reduce environmental impacts, make use of recycled resources, and support sustainable growth in the renewable energy industry.

9.2.1 Challenges of solar and wind energy equipment after end-of-life

- **Solar panels**: Solar panels typically have a lifespan of 20–30 years. After reaching the end of their lifecycle, they become electronic waste in large quantities. Components like glass, silicon, silver, and aluminum are valuable for recycling but require complex separation processes and advanced technologies.
- **Wind turbines**: Wind turbines consist of parts such as blades (made from composite materials), shafts, and generators. Composite blades are durable but difficult to recycle, while metals like steel, copper, and rare earth elements in the shafts and generators can be reused.
- **Pollution and recycling costs**: High recycling costs and underdeveloped technologies mean some countries still resort to landfilling or incineration, leading to resource waste and increased environmental pollution.

9.2.2 Circular economy approaches after end-of-life use
9.2.2.1 Recycling and reusing materials

- **Solar panels**
  - **Material separation**: Solar panels are disassembled to separate glass, silicon, and metals like silver and aluminum. These materials are then recycled into raw materials or new products.
  - **Advanced recycling technologies**: Chemical and pyrolysis technologies are applied to treat silicon and silver, improving recovery rates and reducing waste.
- **Wind turbines**
  - **Reusing metals**: Steel, copper, and aluminum from wind turbines can be recycled into new products.

- **Composite materials**: Grinding and chemical technologies can be used to recycle composite materials from turbine blades into building materials such as cement or reinforced fibers.

9.2.2.2 Sustainable design

- **Design for easy disassembly**: Renewable energy equipment should be designed for easy disassembly and recycling at the end of their useful life.
- **Use of recycled materials**: Enhancing the use of recycled materials in manufacturing new equipment, creating a closed-loop material cycle.

9.2.2.3 Sharing economy and direct reuse

- **Reusing solar panels**: Damaged or underperforming solar panels can be repaired and reused in smaller projects or for applications that do not require high efficiency.
- **Reusing wind turbine blades**: Some creative projects use wind turbine blades for construction purposes, such as bridges, shelters, or infrastructure.

9.2.2.4 Waste collection and management systems

- **Developing specialized waste collection and management systems**: Establishing systems for collecting and managing waste from renewable energy equipment, especially in areas with large solar and wind energy projects.
- **Collaborating with recycling companies**: Partnering with recycling businesses to develop more efficient recycling solutions.

9.2.3 Case studies in Vietnam and globally

- **Vietnam case studies**
  - Several companies in Vietnam have initiated solar panel recycling projects, focusing on recovering silicon and aluminum.
  - Eco-industrial parks like those in Can Tho city and Da Nang city are applying CE, including recycling industrial waste, including renewable energy materials.
- **Global experience**
  - **PV cycle alliance (Europe)**: Implemented a solar panel recycling system with a material recovery rate of over 90%.
  - **Re-wind project (USA and Ireland)**: Reusing wind turbine blades for construction materials and infrastructure.

Adopting a CE approach after the end-of-life of materials in solar and wind energy equipment not only reduces waste but also creates opportunities for developing the recycling industry in Vietnam. The integration of modern technology, supportive policies, and community awareness will ensure effective management of these devices, contributing to a sustainable and environmentally friendly economy.

9.3 Solar photovoltaic panels: understanding the structure of solar panels to open up research on solar panel waste treatment

9.3.1 Structure of a solar photovoltaic panel

A solar photovoltaic (PV) panel is made up of several layers of different materials, each with a specific function. The combination of these components not only ensures operational efficiency but also presents challenges when the panel reaches the end of its lifecycle. The main components include:

- **Aluminum frame**
  - Role: Ensures durability, providing a solid structure for the panel and protecting it from external impacts.
  - Recycling: Nearly 100% recyclable.

- **Tempered glass**
  - Role: Protects the PV cells from weather conditions and impacts.
  - Recycling: Can be recycled up to 95% through standard processing methods.

- **Ethylene vinyl acetate (EVA) layer**
  - Role: Acts as an adhesive and protects the PV cells from vibration, dust, and moisture.
  - Challenge: This layer is difficult to recycle due to its complex polymer structure, requiring special methods to separate it from other layers.

- **Solar Cells**
  - Role: The main component that generates electricity, typically made from mono or polycrystalline silicon.
  - Recycling: Silicon can be recycled into new wafers, achieving an efficiency rate of 85%.

- **Backsheet**
  - Role: Provides insulation and mechanical protection.
  - Challenge: Made from polymers like polypropylene (PP), polyvinyl fluoride (PVF), or polyethylene terephthalate (PET), which are not biodegradable and require specialized technologies for recycling.

- **Junction box and cables**
  - Role: Collects and transmits electrical energy.
  - Recycling: The plastic and metal components of the junction box and cables can be recycled.

9.3.2 Challenges in solar panel waste treatment

- **Complexity of multilayer materials**: Separating the components for recycling requires advanced technologies and optimized processes.
- **Environmental impact**: If not properly managed, materials like polymers and EVA plastic can cause soil and water pollution.
- **Growing waste volume**: It is forecasted that solar panel waste in Vietnam could reach 3.25 mn tons by 2065, placing significant pressure on recycling systems.

9.3.3 Solar photovoltaic panels: researching the structure for waste treatment

9.3.3.1 Structure of a solar photovoltaic panel

A solar PV panel consists of multiple layers of materials, each with a specific function. The combination of these components ensures the panel's operational efficiency but also poses challenges when the panel reaches the end of its useful life. The main components include:

- **Aluminum frame**
  - **Role**: Ensures durability, providing a solid structure for the panel and protecting it from external impacts.
  - **Recycling**: Nearly 100% recyclable.

- **Tempered glass**
  - **Role**: Protects the PV cells from weather and impact.
  - **Recycling**: Can be recycled up to 95% through standard recycling processes.

- **EVA layer**
  - **Role**: Acts as an adhesive and protects PV cells from vibration, dust, and moisture.
  - **Challenge**: This layer is difficult to recycle due to its complex polymer structure, requiring specialized methods for separation from other layers.

- **Solar cells**
  - **Role**: The primary component generating electricity, typically made from mono- or polycrystalline silicon.
  - **Recycling**: Silicon can be recycled into new wafers with an efficiency rate of 85%.

- **Backsheet**
  - **Role**: Provides electrical insulation and mechanical protection.
  - **Challenge**: Made from polymers such as PP, PVF, or PET, which are not biodegradable and require specialized technology for recycling.

- **Junction box and cables**
  - Role: Collects and transmits electrical energy.

9.3.3.2 Challenges in solar panel waste treatment

- **Complexity of multilayer materials**: Separating the components for recycling requires advanced technology and optimized processes.
- **Environmental impact**: If not properly managed, materials like polymers and EVA plastic can cause soil and water pollution.
- **Increasing waste volume**: It is forecasted that solar panel waste in Vietnam could reach 3.25 mn tons by 2065, placing significant pressure on recycling systems.

9.3.3.3 Opportunities from waste treatment research

- **Material reuse**
  - Silicon and glass can be recycled with high efficiency, providing raw materials for new production.
  - The aluminum frame and metals in the junction box can be almost entirely reused.

- **Recycling technology development**
  - The process of treating silicon at 500 °C to separate plastics and silicon crystals has achieved an 85% recycling efficiency.
  - Thin-film panels can reach up to 95% recycling efficiency through grinding, centrifugation, and metal filtration processes.

- **Sustainability enhancement**
  - Efficient recycling of solar panels helps conserve resources and reduces greenhouse gas emissions.
  - It improves the renewable energy supply chain and supports the development of the green energy industry.

Research on the structure of solar PV panels is a crucial foundation for developing effective waste treatment processes. Recycling not only minimizes environmental pollution but also creates sustainable raw materials for the renewable energy industry, contributing to the development of a CE in the future.

9.4 "Solar panel waste": introducing the issue of solar panel waste and environmental challenges in the renewable energy sector

Global and Vietnam solar waste forecasts by 2050

The development of renewable energy, including solar energy, is an inevitable trend.

9.4.1 Solar energy development worldwide – current status and forecast

9.4.1.1 Solar energy waste in Vietnam

The projected solar energy capacity in Vietnam by 2045 is shown in Table 9.1. Based on this, the estimated amount of solar energy waste from 2040 to 2065 is also presented (Figure 9.1).

Solar energy waste in Vietnam will primarily arise from 2040 onwards (Figure 9.2). According to the 8th Power Development Plan, no new solar energy systems are expected to be installed during the 2020–2030 period, as the focus will be on building electricity transmission infrastructure. After 2030, the installation of new solar energy systems is expected to increase significantly, leading to an estimated solar energy waste of 3.25 mn tons by 2065.

9.4.1.2 Proportion of material weight and recycling potential of solar panel materials

- **Main material components by weight**:
  Materials such as glass, aluminum (Al), and silicon (Si) account for 87.85% of the total weight of a solar panel, while precious metals make up only 1%.

Table 9.1 Estimated solar energy waste and key recyclable materials in Vietnam

Year	2020	2030	2045
Total solar power capacity, MW (power development plan 8)	16,650	20,100	71,900
Solar power waste generated starting from 2040			
Year	2040	2050	2065
Total solar power waste, mn tons	0.67	0.19	3.25

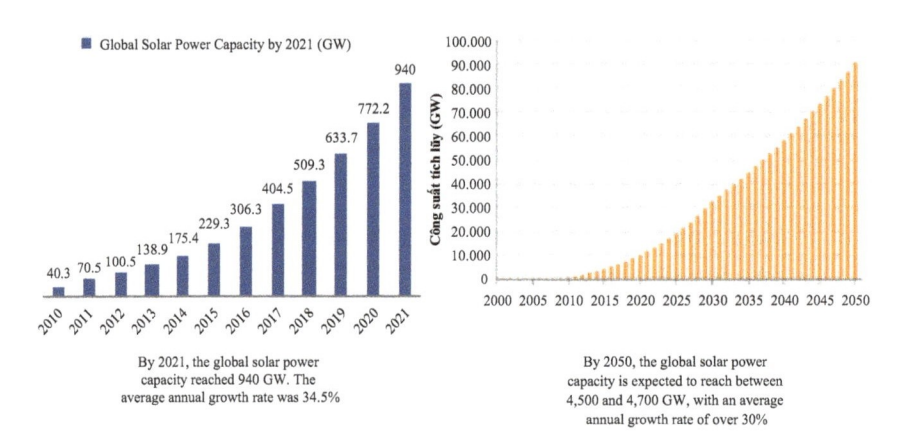

By 2021, the global solar power capacity reached 940 GW. The average annual growth rate was 34.5%

By 2050, the global solar power capacity is expected to reach between 4,500 and 4,700 GW, with an average annual growth rate of over 30%

Figure 9.1 Forecast of solar panel waste worldwide and in Vietnam by 2050 [1]

Circular economic solutions 393

Solar power capacity in Vietnam (MW)

Installed capacity (MW)

16,650

2018 2019 2020

By 2020, the solar power capacity connected to the grid in Vietnam reached 16,650 MW, providing an average of over 22 billion kWh per year (HSCS = 15.5%)

Projected solar power in Vietnam by 2045 (MW)

Projected solar power in Vietnam by 2045 (MW)

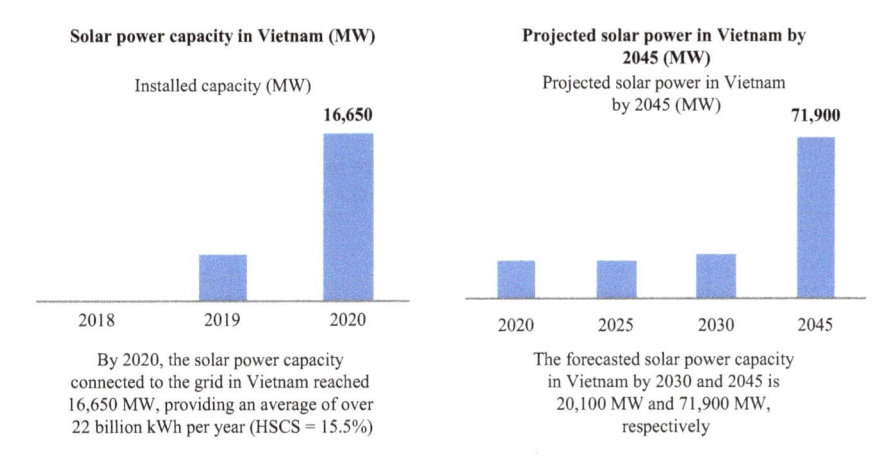

71,900

2020 2025 2030 2045

The forecasted solar power capacity in Vietnam by 2030 and 2045 is 20,100 MW and 71,900 MW, respectively

Figure 9.2   Solar power development in Vietnam (Courtesy: Electricity Planning 8)

- **Recycling rates for main materials**:
  – **Glass**: 95%
  – **Aluminum (Al)**: Nearly 100%
  – **Silicon (Si)**: 81%
  – **Initial polymers (e.g., EVA, Tedlar)**: These are often discarded and not recycled due to their low economic value.

9.4.1.3   Estimated amount of materials recovered from recycling

Based on installed solar energy capacity and global capacity forecasts, and using the recovery rates provided in Table 9.2, the amount of materials recovered from recycling solar energy waste worldwide can be estimated.

- **Potential benefits**

Recycling helps to maximize the use of valuable resources while reducing waste and environmental impact. With the increasing amount of solar energy waste expected from 2040 onwards, applying effective recycling solutions will play a crucial role in sustainable development.

9.4.1.4   Material recycling rates and potential

- **Proportion of main materials by weight**
  – **Glass, aluminum (Al), and silicon (Si)**: These make up 87.85% of the total weight of the solar panel.
  – **Precious metals**: Only 1%.
- **Recycling rates for main materials**
  – **Glass**: 95%.
  – **Aluminum (Al)**: Nearly 100%.
  – **Silicon (Si)**: 81%.
  – **Base polymers (e.g., EVA, Tedlar)**: Often discarded and not recycled due to their low economic value.

Table 9.2 Material recovery rates from solar panels after processing and recycling

Material	Material weight in a 20 kg crystalline silicon solar panel (kg)	Material percentage in solar panel (%)	Recovered material after recycling (kg)	Recovery rate (%)
Glass	14.8	74	14.06	95
Aluminum	2.06	10.3	2.06	100
Adhesive	1.31	6.55	–	–
Tedlar	0.72	3.6	–	–
Adhesive	0.232	1.16	–	–
Silicon (Si)	0.67	3.55	0.543	81
Silver (Ag)	0.034	0.17	0.017	50
Tin (Sn)	0.024	0.12	0.024	100
Zinc (Zn)	0.014	0.07	0.014	100
Copper (Cu)	0.114	0.57	0.114	100
Lead (Pb)	0.012	0.06	0.012	100
Other Materials	0.01	0.01	–	–
Total	**20.0**	**100**	**16.844**	**84.22**

Table 9.3 Forecast of solar panel waste and recyclable material recovery in Vietnam (2030–2050)

Year	2030	2040	2050
Solar panel waste (thousand tons)	8000	32,000	78,000
Recoverable materials after recycling (thousand tons)			
Glass	5,624	22,496	54,834
Aluminum	824	3,296	8,034
Silicon	217.07	868.3	2116.5
Silver	6.8	27.2	66.3
Tin	9.6	38.4	93.6
Zinc	5.6	22.4	54.6
Copper	45.6	182.4	444.6
Lead	4.8	19.2	46.8

9.4.1.5 Estimated recovered materials from recycling

Using data on installed solar energy capacity and global forecasts, combined with the recycling rates in Table 9.3, it can be predicted:

- A significant amount of material can be recycled from solar energy waste globally.
- This helps:
    - Optimize resource utilization.
    - Reduce environmental waste.

9.4.1.6 Benefits of recovering and reusing materials from solar panels

- **Environmental protection**: Recycling solar panels helps prevent waste from being discharged into the environment. It also reduces the need for raw material extraction, which helps protect ecosystems and limit pollution.
- **Conservation of natural resources**: Recycling reduces the need for new raw materials, thereby helping to protect non-renewable natural resources, especially rare minerals.
- **Energy savings**: Recycling solar panels saves energy by avoiding the need to process raw materials from scratch, thus reducing the energy consumption involved in manufacturing.
- **Job creation**: The collection and recycling of solar energy waste create numerous job opportunities, providing income for workers and contributing to the local economic development.

9.5 Solutions for solar panel waste: current solutions and future proposals

9.5.1 Solar energy waste recycling technology

Solar energy recycling technologies include methods for processing solar panels that have reached the end of their lifespan to recover and reuse as many materials as possible. The goal is to recycle these materials for the production of new panels or use them for other purposes, thereby optimizing product life cycles and reducing negative environmental impacts.

9.5.2 General cycle of solar energy waste recycling technology

The recycling technology for solar panel waste involves multiple stages to recover valuable materials from expired panels. This process not only brings environmental benefits but also has positive economic and social impacts. Below is the general cycle of the recycling process:

- **Collection**: Solar panels that have reached the end of their lifespan are collected from residential, commercial, and industrial systems.
- **Disassembly**: The panels are carefully disassembled to separate materials such as glass, aluminum, silicon, and polymers.
- **Material recovery**: Valuable materials like glass, aluminum, and silicon are recovered through processes like shredding, heating, and chemical treatments.
- **Reprocessing**: The recovered materials are reprocessed into usable forms for manufacturing new panels or for other applications.
- **Reuse**: Recycled materials are incorporated into the production of new solar panels or used in other industries, contributing to resource efficiency and sustainability.

396   *Clean energy in South-East Asia*

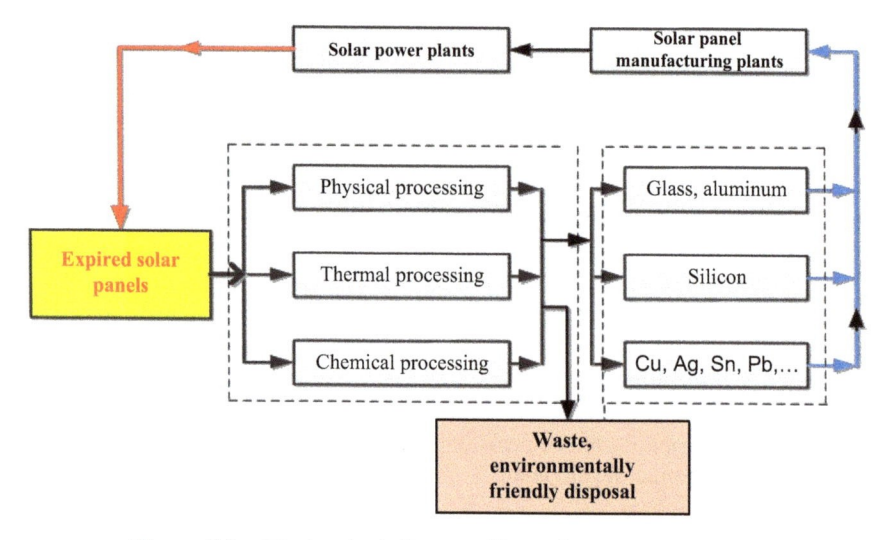

Figure 9.3   Mechanical disassembly and separation process

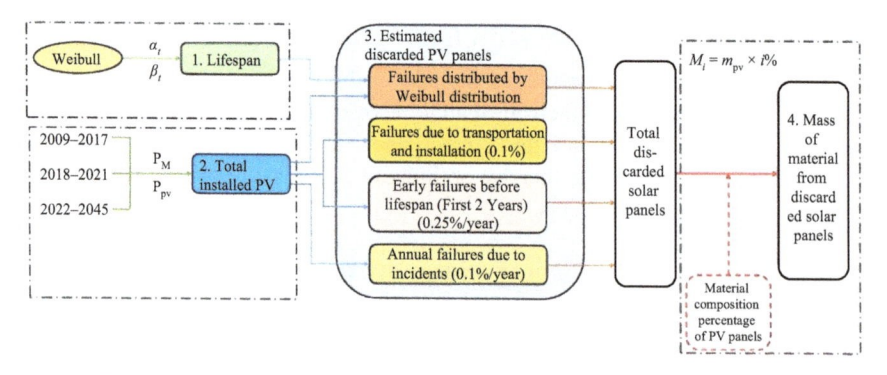

Figure 9.4   Estimated material flow from discarded solar panels

This cycle ensures that solar panel waste is efficiently managed, promoting a CE in the renewable energy sector (Figures 9.3 and 9.4).

9.5.3   Study on the waste flow of solar panels from solar power plants

- **Volume of solar panel waste**
  - **2019–2030**: The waste volume ranges from 2.4 thousand to 27.5 thousand tons.
  - **2030–2045**: The waste volume increases significantly, ranging from 27.5 thousand to 746.5 thousand tons.

- **Main causes of solar panel waste**
  - **Transportation and installation issues**: Damage or inefficiency during the transportation and installation of solar panels.
  - **Premature failures**: Panels may fail within the first two years of operation, leading to waste.
  - **Annual failures**: Panels can experience breakdowns or faults over time, contributing to waste.
  - **Weibull distribution of failures**: The failure rate of panels follows a Weibull distribution, meaning that failures are more likely as the panels age.

- **Composition of solar panel waste**
  - **Primary materials in the waste**
    - Glass
    - Aluminum frames
    - Copper wiring
    - EVA layers (ethylene vinyl acetate)

  - **Estimated waste from 2019 to 2045**
    - **Glass**: 1.5–3,100 thousand tons
    - **Aluminum frames**: 487.4 thousand tons
    - **Copper wiring**: 350.8 thousand tons
    - **EVA layers**: 378 thousand tons

- **Current Status of solar panel and wind turbine blade installation and operation in Vietnam**
  - **Field survey**:
    - **Provinces involved**: Bac Lieu and Dak Lak

  - **Survey participants**:
    - 8 large solar power plants
    - 7 project developers
    - 102 households
    - 6 buildings and agricultural farms with rooftop solar systems
    - Local power companies
    - Major equipment manufacturers

  - **Technology development (average solar panel capacity)**:
    - **Annual efficiency decline rate**: 0.5–1.1%
    - **Average solar panel efficiency**: 17–21.3%
    - **Average solar panel weight**: 22.5 kg

  - **Estimates for 2020**:
    - **Large solar power plants**: Approximately 28 mn solar panels
    - **Rooftop solar installations**: Approximately 17.5–23.5 mn solar panels

Proportion of solar panel types used

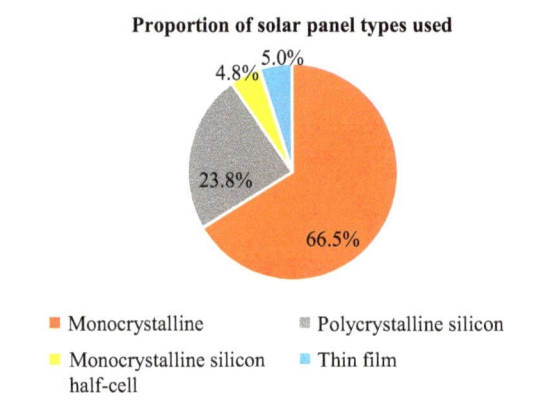

- Monocrystalline: 66.5%
- Polycrystalline silicon: 23.8%
- Monocrystalline silicon half-cell: 4.8%
- Thin film: 5.0%

Figure 9.5   Proportion of solar panel types in use

- **Development trends**:
    - **By 2030**: Around 37–50 mn solar panels are expected to be installed.
    - **By 2050**: The number of installed solar panels will reach between 219 and 293 mn.

This study highlights the growing volume of solar panel waste as installations increase and the long-term need for recycling and waste management technologies to address the environmental impact. The expected increase in installations underscores the importance of developing efficient recycling systems to manage both the waste from used panels and the resources that can be recovered (Figure 9.5).

9.5.4   Solar panel recycling technology worldwide

9.5.4.1   PV cycle technology

PV cycle is a nonprofit organization based in Europe that focuses on the management, collection, processing, and recycling of PV waste. They have tested and implemented several technologies to handle solar panel waste effectively:

- **Dismantling of electrical junction boxes and aluminum frames**
    - The process begins by removing the electrical junction boxes and aluminum frames from the solar panels. This step is essential to separate recyclable parts from non-recyclable materials and prepare the panels for further processing.

- **Crushing and sorting**
    - After dismantling, the panels are crushed into small fragments. These fragments are then sorted using advanced technologies, including **laser technology** and **screening methods**, to differentiate between various materials (such as glass, aluminum, silicon, and others).

Circular economic solutions 399

- **Separation and recycling of materials**
  - Once the materials are sorted, the individual components (e.g., glass, aluminum, silicon) are sent to specialized recycling companies. These companies are equipped to process the materials and recycle them into usable forms, often for the creation of new solar panels or for other industrial uses.

This process ensures that a significant portion of the materials from end-of-life solar panels can be recovered and reused, reducing waste and minimizing the environmental impact of discarded solar panels. PV cycle's approach demonstrates the potential for efficient and sustainable recycling of solar panel components, contributing to the CE in the solar energy sector (Figure 9.6).

9.5.4.2 Solar world technology

Since 2003, **SolarWorld** has been developing a pilot recycling line for PV module waste. This process involves several steps designed to recover valuable materials from old or decommissioned solar panels:

- **Dismantling**
  - The first step involves removing the electrical junction boxes and aluminum frames from the PV modules. This ensures that the components are separated before they undergo further processing.

- **Thermal treatment**
  - The PV modules, after the junction boxes and aluminum frames are removed, are heated to temperatures exceeding **600 °C**. At this high temperature, the **EVA (ethylene vinyl acetate)** layer, which is used to encapsulate the PV cells, is burned away. This process leaves behind the valuable materials such as **PV cells, glass, and metals**.

- **Manual sorting**
  - After the thermal treatment, the remaining materials are manually sorted. This step helps to further separate materials such as glass, metals (e.g., aluminum), and silicon cells.

- **Material transfer**
  - The separated materials are then transferred to specialized recycling facilities, where they are processed further and can be repurposed for new solar modules or other industrial applications.

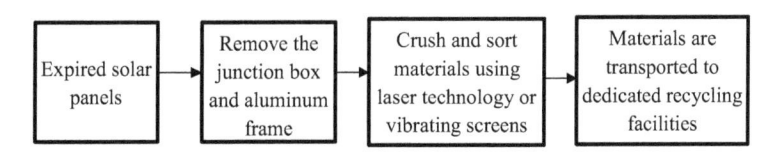

Expired solar panels → Remove the junction box and aluminum frame → Crush and sort materials using laser technology or vibrating screens → Materials are transported to dedicated recycling facilities

Figure 9.6 Integrated multi-method recycling process for expired solar panels

Remove the electrical junction box and aluminum frame → Incinerate (pyro-metallurgical process) → Separate PV panels, glass, and metals → Transfer the separated materials to other recycling facilities

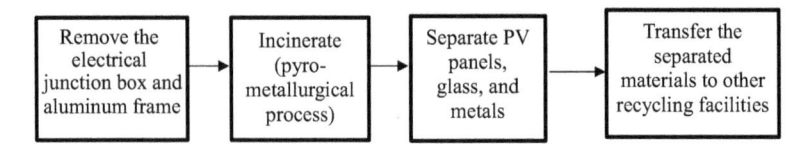

Figure 9.7 Thermal and material separation recycling process for expired solar panels

This approach by SolarWorld focuses on maximizing material recovery through a combination of mechanical, thermal, and manual processing methods. The goal is to make the recycling of solar panels more efficient, environmentally friendly, and cost-effective, supporting the sustainable lifecycle of solar energy systems (Figure 9.7).

9.5.4.3 Lowe Chemie technology – combined mechanical and chemical process

The recycling technology developed by **Lowe Chemie** involves a combination of mechanical and chemical processes to efficiently recover materials from used PV modules. The key steps are as follows:

- **Mechanical dismantling**
  - **Goal**: To remove the electrical junction boxes, glass, and aluminum frames from the PV modules.
  - **Method**: Mechanical techniques are used to disassemble and separate these components, ensuring that each material can be processed further in subsequent stages.

- **Mechanical grinding and sorting**
  - **Goal**: To process the remaining parts of the PV modules after dismantling.
  - **Method**: The remaining materials are ground into smaller pieces, and the materials are then sorted using mechanical techniques to separate different types of materials, such as metals, silicon, and glass.

- **Chemical processing**
  - **Goal**: To recover metals and silicon from the processed materials.
  - **Method**: Chemical treatments are applied to extract valuable metals (e.g., silver, copper) and silicon from the remaining materials. This step is essential for retrieving materials that are harder to separate using mechanical methods alone.

- **Solid-liquid separation**
  - **Goal**: To further refine the separation of materials.
  - **Method**: Metallurgical techniques are employed to separate solid materials from liquids, further enhancing the purity of the recovered materials.

- **Glass cleaning and aluminum refining**
  - **Goal**: To clean and refine the recovered glass and aluminum.

Circular economic solutions 401

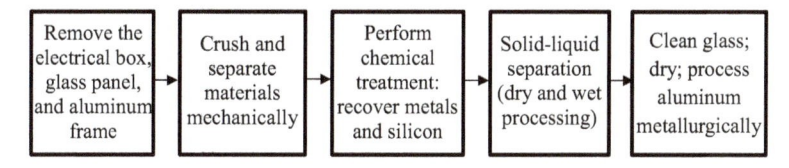

```
Remove the      Crush and      Perform        Solid-liquid    Clean glass;
electrical box, separate       chemical       separation      dry; process
glass panel,    materials   →  treatment:  →  (dry and wet  → aluminum
and aluminum →  mechanically   recover metals processing)     metallurgically
frame                          and silicon
```

Figure 9.8 Solar panel recycling process using the mechano-chemical method

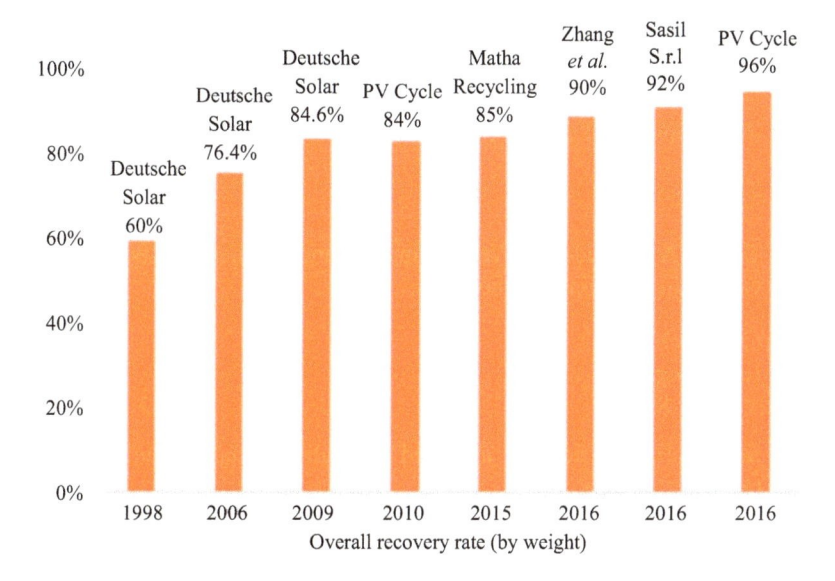

Figure 9.9 Advancements in silicon solar panel recycling efficiency from 1998 to 2016 (Each column represents the name of the company performing recycling and the corresponding recycling efficiency.) [3]

- **Method**: The glass is cleaned, and the aluminum is processed to remove impurities, making the materials suitable for reuse in new products or for manufacturing new solar panels.
- This combined mechanical and chemical approach aims to maximize the recovery of valuable materials, while ensuring that the process is efficient and environmentally sustainable. It reflects an advanced method for recycling PV waste, focusing on both the mechanical dismantling of modules and the chemical extraction of valuable components (Figures 9.8 and 9.9).

9.5.4.4 Experimental recycling technology by Lowe Chemie – mechanical and chemical process

- **Recycling efficiency**
  - Recycling efficiency is defined as the ratio of the total weight of materials recovered after the recycling process compared to the total weight of the

PV module before processing and recycling. Below are the performance trends of various companies involved in PV module recycling:

- **Deutsche solar**
  - Recycling efficiency over time:
    - 1998: 60%
    - 2006: 76.4%
    - 2009: 84.6%

- **Trend**: There is a significant increase in recycling efficiency over the years, reflecting advancements in recycling technology and processes.
- **PV cycle**
  - Recycling efficiency over time:
    - 2010: 84%
    - 2016: 96%

  - **Trend**: A marked improvement in efficiency, indicating successful innovations in recycling technologies.

- **Matha recycling**
  - Recycling efficiency:
    - 2015: 85%

  - **Status**: High efficiency achieved through the application of advanced recycling methods.

- **Zhang**
  - Recycling efficiency:
    - 2016: 90%

  - **Status**: High efficiency, reflecting the effectiveness of recycling practices and technologies.

- **Sasil S.r.l**
  - Recycling efficiency:
    - 2016: 92%

  - **Status**: One of the highest recorded efficiencies, demonstrating the effectiveness of their recycling process.

The recycling efficiency of PV module processing has significantly improved over the years, thanks to advancements in technology and recycling methods. The data shows that modern recycling technologies can achieve high recovery rates, with some companies reaching efficiencies as high as 96% or more. This progress is crucial for managing the growing volume of solar panel waste and maximizing resource recovery.

9.5.5 Experimental research on processing, recycling, and material recovery from crystalline silicon PV modules in Vietnam

9.5.5.1 Scope of research

This study is part of a scientific project assigned by the Ministry of Industry and Trade, conducted by the Vietnam Association for Science and Technology of Energy Saving and Efficiency (VASEP) during the 2021–2022 period.

Research content

- **Experimental technology**
  - The study involves testing technologies for processing, recycling, and material recovery from used crystalline silicon PV (Si—the solar panel) modules.

- **Initial technology selection**
  - The initial selection focuses on identifying suitable technologies for processing and recycling to recover materials such as glass, aluminum, and silicon, which together account for 85% of the total weight of the PV modules.

- **Feasibility assessment/**
  - The study evaluates the feasibility of the selected technologies to determine their effectiveness in recycling and material recovery.

Objective

The goal of the research is to develop and evaluate technologies for processing and recycling crystalline silicon PV modules, with an emphasis on recovering key materials such as glass, aluminum, and silicon. The project includes selecting suitable technologies and assessing their feasibility, contributing to the promotion of sustainable recycling methods in Vietnam.

9.5.5.2 Experimental research process

Phase 1: Survey and sample collection
- **Objective**
  - Collect used crystalline silicon PV modules (Si—the solar panel) from various sources, such as solar power plants and rooftop solar systems.
  - Assess the physical and chemical condition of the modules before processing.
- **Main activities**
  - Determine the damage rate and types of materials that need recycling.
  - Plan sample processing to ensure representativeness.

Phase 2: Technology selection
- **Objective**
  - Develop the experimental procedure based on globally tested and developed recycling technologies.

- **Main activities**
  - Test mechanical technology: dismantling aluminum frames, junction boxes, and glass layers.
  - Assess chemical technology: processing silicon and recovering metals from modules.
  - Evaluate thermal technology: handling EVA polymer films and removing impurities.

Phase 3: Processing and recycling experimentation
- **Objective**
  - Test the recycling efficiency of the selected technologies.
  - Recover key materials such as
    - **Glass**: Assess purity and recyclability.
    - **Aluminum**: Analyze purity after recycling.
    - **Silicon**: Evaluate reusability for new applications.
  - **Main activities**
    - Perform disassembly, grinding, separation, and thermal processing procedures.
    - Recover and classify materials.

Phase 4: Evaluation of efficiency and feasibility
- **Objective**
  - Evaluate the material recovery rate.
  - Assess technological and economic feasibility.
- **Main activities**
  - Compare recycling efficiency across technologies.
  - Propose improvements to the recycling process suited to conditions in Vietnam.

Phase 5: Reporting and proposals
- **Objective**
  - Complete the scientific report, presenting research results and specific recommendations.
- **Report contents**
  - Summarize data from the experimental process.
  - Propose a sustainable recycling model for PV modules in Vietnam (Figure 9.10).

9.5.5.3 Removing aluminum frame and junction box

- **Description of the process**
  - This is the first step in the recycling process of crystalline silicon PV modules (Si—the solar panel).
  - **Requirements**
    - Avoid cracking or breaking the solar panel during the disassembly process.
    - Ensure that the operation is simple and efficient.

Circular economic solutions 405

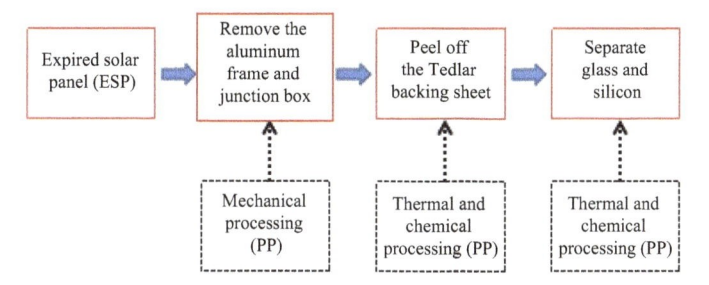

```
Expired solar → Remove the        → Peel off      → Separate
panel (ESP)     aluminum            the Tedlar      glass and
                frame and           backing sheet   silicon
                junction box
                     ↑                    ↑                ↑
                Mechanical          Thermal and     Thermal and
                processing          chemical        chemical
                (PP)                processing (PP) processing (PP)
```

Figure 9.10 Research and testing process

- **Methodology**
  - **Removing the aluminum frame**
    ○ Remove the aluminum frame surrounding the module using mechanical techniques without causing damage.

- **Removing the junction box**
  ○ Carefully detach the junction box to protect the internal electrical components.

9.5.5.4 Design and manufacture of aluminum frame and junction box removal machine

- **Objective**
  - Simplify the process of removing the aluminum frame and junction box.
  - Increase productivity while ensuring the integrity of the module after disassembly.

- **Key features of the machine**
  - **Automatic separation mechanism**
    ○ The machine is equipped with a mechanism to remove the aluminum frame and junction box using moderate mechanical pressure.

  - **Protection system**
    ○ Protects the surface of the module to avoid cracks or breaks during operation.

  - **User-friendly design**
    ○ Easy to operate and maintain, suitable for recycling facilities in Vietnam.

- **Expected results**
  - **Reduced disassembly time**
    ○ Shorten the time required for disassembly.

  - **High material recovery rate**
    ○ Ensure the recovery of materials (aluminum frame and electronic components) at the highest possible rate.

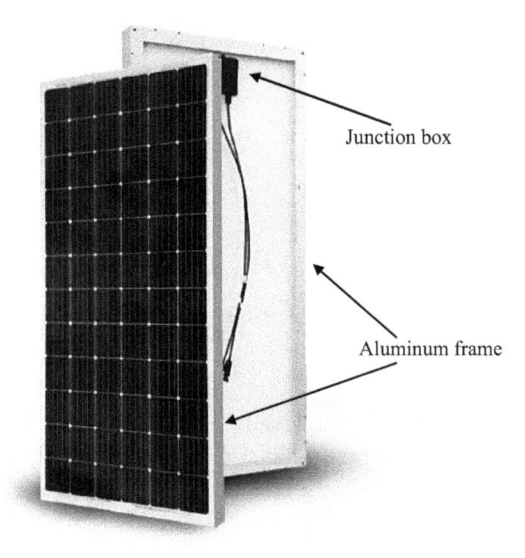

Figure 9.11 Structure of a solar panel with junction box and aluminum frame [4]

- **Sustainable and efficient recycling**
    - Meet the requirements of sustainable and effective recycling practices (Figure 9.11).

9.5.5.5 Position of the aluminum frame and junction box on the PV module

In crystalline silicon solar modules (Si—the solar panel), the aluminum frame and junction box are installed at specific locations to protect the internal components and support the module's structure. These components must be carefully disassembled to avoid damaging the solar panel.

- **Aluminum frame**
    - The frame is mounted around the module to protect and reinforce the module's structure. It provides stability and facilitates the installation of the module.
- **Junction box**
    - The junction box houses electrical components such as cables and connectors. It serves as the connection point between the modules and the electrical system and must be handled carefully to avoid damaging the electronic connections.

9.5.5.6 Machine for separating aluminum frame and junction box on PV module (the solar panel)

The development of a specialized machine for separating the aluminum frame and junction box from the solar module is crucial for enhancing the efficiency of the recycling process. The machine's design must ensure the integrity of the module while facilitating the removal of these components.

- **Key features**
    - **Automated separation mechanism**
        ○ The machine utilizes moderate mechanical pressure to separate the aluminum frame and junction box efficiently without causing harm to the module.
    - **Protection system**
        ○ Equipped with protective mechanisms to safeguard the module from cracks or other damage during the disassembly process.
- **Expected benefits**
    - **Improved efficiency**
        ○ Faster and more precise disassembly, reducing labor time and increasing throughput.
    - **Enhanced material recovery**
        ○ The design aims to recover a high percentage of valuable materials like aluminum and electronic components, making the recycling process more cost-effective.
    - **Sustainability**
        ○ The machine is intended to promote sustainable recycling practices by ensuring minimal damage to the module during disassembly, supporting overall material recovery (Figure 9.12).

Structure and function of components

- **Operational surface on the machine frame**
    - This is the primary interface where the operator interacts with the machine. It ensures that the operational processes are carried out smoothly and efficiently.

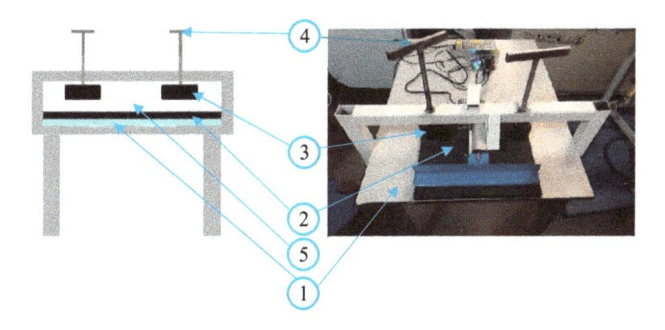

Figure 9.12 Structure of the clamp assembly for holding the solar panel in the separation device: (1) material panel (e.g., solar panel, polymer sheet, etc.), (2) rubber pad layer, (3) panel holding frame, (4) compression lever/force applicator, and (5) support base/frame.

- **Cushioning pad under the solar panel module**
  - The cushioning pad provides support and padding for the solar panel module during the separation process, helping to prevent damage to the solar panel.
- **Rubber clamp above the solar panel module**
  - This component holds the solar panel module securely in place from above, ensuring stability throughout the operation.
- **Screw mechanism with crank handle for height adjustment and rubber clamp fixation**
  - This mechanism allows for precise height adjustment to accommodate different panel sizes and ensures the rubber clamp is tightly secured.
- **Space for placing the solar panel module (thickness range from 0 mm to 60 mm)**
  - This area is designed to accommodate solar panel modules of varying thicknesses, allowing for flexible operation (Figure 9.13).

Control circuit

The control circuit is responsible for adjusting the speed, force, and other parameters of the piston/cylinder to push and lift the aluminum frame and junction box, ensuring that the separation process is carried out safely. Adjusting these parameters helps ensure that the separation process is smooth, preventing damage to the module components.

Separation process

- **Separation time**: The process of separating the aluminum frame and junction box occurs safely and relatively quickly, taking approximately 20 minutes or less for each solar module.

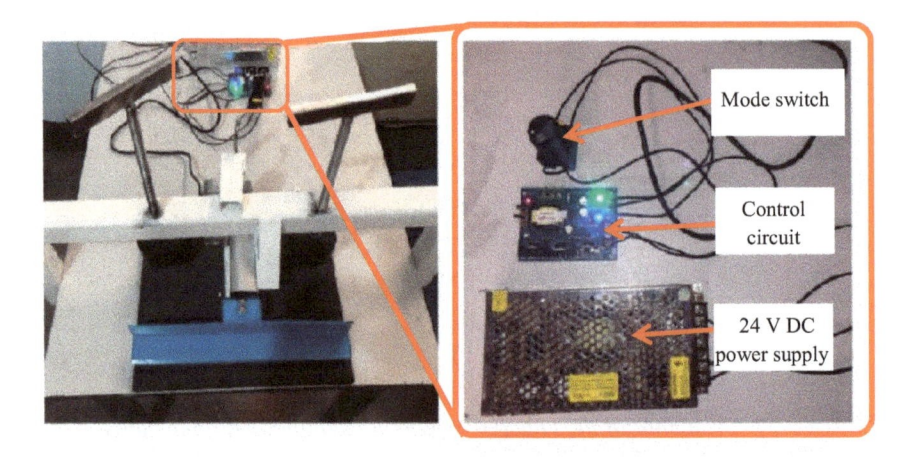

Figure 9.13 Structure and function: switch, control circuit, 24 VDC power supply

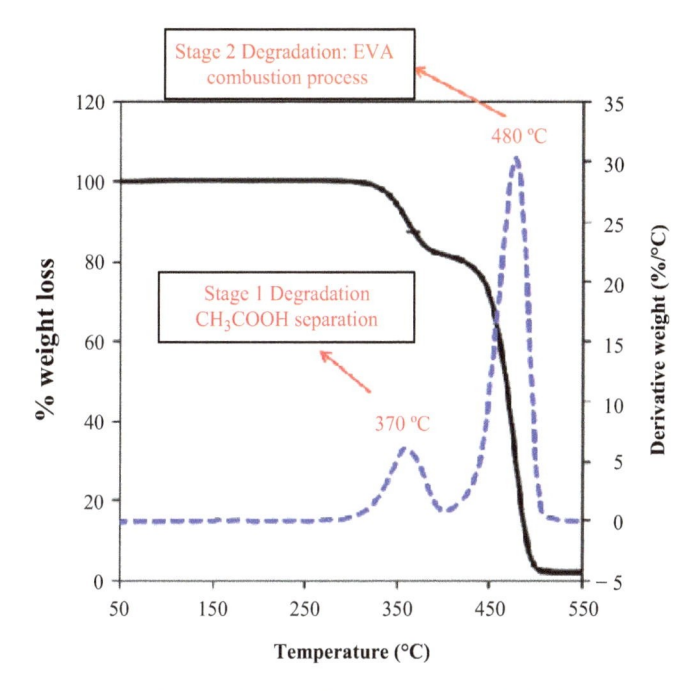

Figure 9.14   Thermal degradation process of EVA in two stages

- **Thermal method**: The thermal method utilizes heat to perform processes such as welding, separation, or altering the material properties. In this method, heat is applied to materials or components to change their physical properties, making them easier to handle or process. The thermal method is commonly used in industries for tasks such as welding, cutting, material separation, and in processes like heat treatment, where temperature changes are crucial to achieving the desired outcome.
- **Thermogravimetric analysis of EVA adhesive in new solar modules**: The thermogravimetric analysis (TGA) diagram of the EVA adhesive in new solar modules provides detailed information about the mass change of the EVA adhesive when exposed to varying temperatures. This process helps determine the temperature at which the EVA adhesive begins to decompose or change its properties, which is essential in selecting appropriate thermal conditions for processing or separating the adhesive from other components of the solar module (Figure 9.14).

9.5.5.7   Stages of degradation of EVA adhesive in solar modules

- **Stage 1**: **Acetic acid separation (370 °C)**: At a temperature of 370 °C, acetic acid is released, resulting in a mass loss of approximately 40% compared to the original mass due to the partial volatilization of the material through the following reaction:

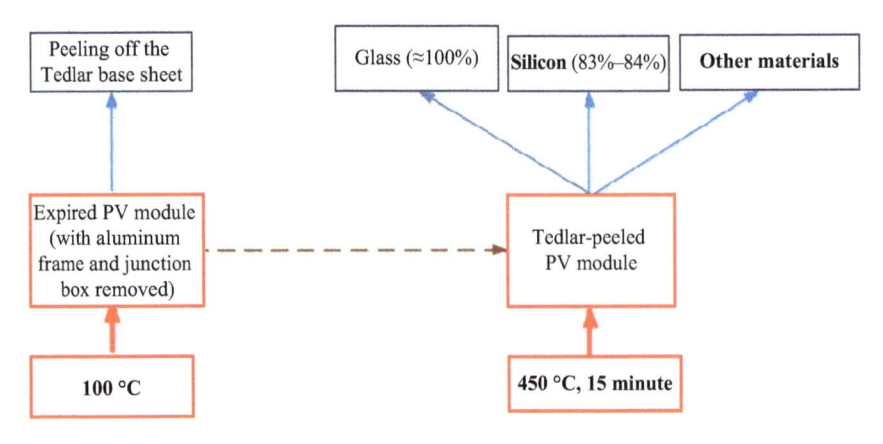

Figure 9.15 Carbonization process and carbon structure degradation (480 °C)

$$(CH2CH2)_x[CH2CH(OCOCH3)]_\gamma \rightarrow$$
$$(CH2CH2)_x[CH=CH]_\gamma + CH3COOH \text{ (axita cetic)}$$
$$(CH_2CH_2)_x[CH_2CH(OCOCH_3)]_\gamma \rightarrow$$
$$(CH_2CH_2)_x[CH=CH]_\gamma + CH_3COOH \text{(axita cetic)}$$
$$(CH2CH2)_x[CH2CH(OCOCH3)]_\gamma \rightarrow$$
$$(CH2CH2)_x[CH=CH]_\gamma + CH3COOH \text{(axit acetic)}$$

This reaction indicates the degradation process of the EVA adhesive, where acetic acid is formed as a volatile byproduct.

- **Stage 2: Carbonization and decomposition of carbon structure (480 °C):** At 480 °C, the EVA adhesive undergoes further intense degradation, leading to a significant reduction in mass and complete decomposition. During this stage, the EVA adhesive experiences carbonization and transforms into final products, including CO_2 and H_2O. Once the temperature exceeds 500 °C, the material is fully decomposed and ceases to exist (Figure 9.15).

9.5.6 Chemical method: using chemicals to separate materials in solar modules

Choosing solvent to dissolve EVA adhesive

Purpose: The aim is to use an appropriate solvent to dissolve the EVA adhesive in the solar module. EVA is a key component in connecting the layers of solar panels, particularly the glass and silicon PV cells. However, removing EVA is a critical step in the process of recycling solar modules.

Solvent selection: The choice of solvent plays a crucial role in the efficiency of the EVA separation process. The solvent must be capable of dissolving EVA without damaging other materials such as silicon, glass, or metal.

Some solvents that have been studied and tested include:

1. **Toluene**: This solvent effectively dissolves EVA, allowing for the removal of the adhesive without damaging other materials in the module.
2. **N-hexane**: Another organic solvent used to dissolve EVA, helping to separate EVA effectively at lower temperatures.
3. **Acetone**: Acetone can be used in certain cases to dissolve EVA, although it may not be as effective as toluene or hexane in fully removing the adhesive.
4. **Methanol**: Also an organic solvent, methanol can be used to separate EVA in some applications.

- **Advantages of the chemical method**
  - **High efficiency**: Capable of completely separating EVA from other module components.
  - **Time-saving**: The dissolution process occurs faster than mechanical or thermal methods.
  - **Protection of other materials**: The solvent can target only the EVA without harming other critical components like silicon, electrodes, or glass.
- **Issues to consider**
  - **Cost and feasibility**: The use of solvents may be costly, and the solvent needs to be recycled after use to prevent pollution.
  - **Safety and environmental impact**: Some solvents, such as toluene and acetone, can be hazardous if not used properly, and the disposal of chemical residues must be managed carefully.
  - **Emission of gaseous pollutants**: The use of solvents may result in harmful emissions if not properly controlled.

The chemical method, particularly the use of solvents to separate EVA, offers high efficiency in the recycling process of solar modules. However, attention must be paid to the cost, safety, and environmental impact of the process.

9.6 Approaches to managing "waste from renewable energy" for a circular economy and sustainable development

With the rapid development of renewable energy sources such as solar, wind, and biomass, managing and recycling "waste" from these sources is becoming a critical issue in building a CE and promoting sustainable development. While renewable energy sources offer numerous environmental and economic benefits, they also generate waste and materials that need to be properly managed to ensure long-term sustainability.

9.6.1 Managing waste from renewable energy

Waste from renewable energy primarily comes from systems like solar panels, wind turbines, and bioenergy devices. These wastes cannot be ignored and must be recycled or processed in a proper manner to minimize environmental impact. Materials such as glass, silicon, aluminum, copper, and metals in solar panels, or plastics and metals in wind turbines, can be recycled using modern processing and recycling technologies.

9.6.2 Developing recycling and processing technologies

A key solution for developing a CE from renewable energy is the advancement and application of advanced recycling technologies. These technologies can include:

- **Solar panel recycling**: After the end of their useful life, solar panels can be processed using mechanical, thermal, and chemical methods to recover materials like silicon, aluminum, glass, and precious metals. This recycling not only helps reduce waste but also regenerates valuable resources.
- **Wind turbine recycling**: Once wind turbines reach the end of their life cycle, they need to be dismantled and recycled. Components such as steel and copper can be recovered and reused, while other materials like carbon fiber or plastics need to be carefully processed to minimize environmental impact.
- **Biotechnology applications**: Using biotechnology to process waste from renewable energy sources, such as the biodegradation of organic materials from biomass energy, can play a vital role in recycling and reducing environmental impact.

9.6.3 Encouraging the adoption of the circular economy model

To successfully implement these solutions, it is essential to have encouragement from governments and organizations to promote the CE model. This includes:

- **Supportive policies**: Governments can offer policies and financial incentives for companies and organizations researching and developing recycling technologies for renewable energy.
- **Encouraging innovation**: Companies involved in the research and production of renewable energy devices should invest in the research and development of more efficient recycling technologies and the reuse of materials, helping to reduce waste and improve resource efficiency.
- **Public education and awareness**: Raising awareness in the community about the importance of recycling and managing waste from renewable energy will help promote environmental protection actions and support sustainable development.

9.6.4 Use of recycled materials in production

One of the key approaches in the CE is the use of recycled materials in production. Materials recovered from renewable energy sources can be repurposed to manufacture new products, such as:

- **Construction materials production**: Glass from solar panels can be used to produce building glass or soundproof glass.

- **Electronics and mechanical product manufacturing**: Precious metals like silver and copper from solar panels can be reused in the production of electronic and mechanical components, helping to reduce dependence on raw materials.

9.6.5 Creating value from waste

Waste from renewable energy is not just a problem to be managed; it can also become a valuable resource if recycled and utilized effectively. For instance, recovering metals from solar panel modules can help save on production costs, reduce the need for natural resource extraction, and minimize environmental pollution.

Developing and implementing recycling and waste management solutions from renewable energy sources will play a crucial role in promoting the CE and sustainable development. Governments, research organizations, and the business community need to collaborate in building effective recycling technologies while also implementing policies that reduce environmental impacts and improve resource efficiency.

Further reading

[1] IRENA (International Renewable Energy Agency). (2023). *Data and visualisations*. Vietnam Ministry of Industry and Trade. *Power Development Plan VIII*.
[2] Lowe Chemie. *Mechanical and Chemical Process*.
[3] AE Battery Manufacturer. *Technical Documentation on Battery Production*, AE Group AG, Leopoldshöhe, Germany.
[4] Balat H, Kırtay E. (2010). Hydrogen from biomass – Present scenario and future prospects. *Int J Hydrogen Energy*. 35:7416–26.
[5] Bhanu Mahajan (2012). *Negative Environment Impact of Solar Energy*. Environmental Science and Policy Course. CEPT University, India. Energy Policy, 33.
[6] Damon Turney and Vasilis Fthenakis (2011). Environmental impacts from the installation and operation of large-scale solar power plants. *Renewable and Sustainable Energy Reviews*. 15(6):3261–70.
[7] DOE (2011). *A National Offshore Wind Strategy: Creating an Offshore Wind Energy Industry in the United States*. 52 pp.
[8] GIZ (2018). *Guidelines on Environmental and Social Impact Assessment for Wind Power Projects in Vietnam*. Available at: http://gizenergy.org.vn/media/app/media/GIZ-ESP_ESIA%20Report_ENG_Final.pdf
[9] GSO (2006–2020). *Almanac Statistics of Vietnam Reports in 2006–2020*. General Statistics Office of Vietnam, in Vietnamese.
[10] IFC (2012). *Guidance Notes to Performance Standards on Environmental and Social Sustainability*. Washington, DC: International Finance Corporation.

Available at: https://www.ifc.org/wps/wcm/connect/9fc3aaef-14c3-4489-acf1-a1c43d7f86ec/GN_English_2012_Full-Document_updated_June-14-2021.pdf?MOD=AJPERES&CVID=nXqnsJp

[11] International Energy Agency (2023). *Global EV Outlook 2023, Catching up with Climate Ambitions*; International Energy Agency: Paris, France.

[12] John Glasson, Riki Therivel and Andrew Chadwick (1999). *Introduction to Environmental Impact Assessment*. 2nd edition in 2005, UCL Press, the Taylor & Francis e-Library.

[13] Jose I. Bilbao, Garvin Heath, Alex Norgren, Marina M. Lunardi, Alberta Carpenter, Richard Corkish (2021). *PV Module Design for Recycling*. International Energy Agency (IEA), PVPS Task 12, Report T12-23:2021. Available at: https://iea-pvps.org/wp-content/uploads/2021/10/T12_2021_PV-Design-for-Recycling-Guidelines_Report.pdf

[14] Luca Ciacci and Fabrizio Passarini (2020). Life cycle assessment (LCA) of environmental and energy systems. *Energies*. 13(22):5892, https://doi.org/10.3390/en13225892

[15] Matt Rogers (2019). *Vietnam's renewable energy future. McKinsey Sustainability*. Available at: https://www.mckinsey.com/capabilities/sustainability/our-insights/sustainability-blog/cop27-accelerating-decarbonization

[16] Md Shahariar Chowdhury, Kazi Sajedur Rahman, Tanjia Chowdhury, et al. (2020). An overview of solar photovoltaic panels' end-of-life material recycling. *Energy Strategy Reviews*. 27:100431.

[17] Ministry of Natural Resources and Environment (2010). *National Technical Regulation on Vibration*. Regulation No. QCVN 27:2010/BTNMT, signed on 16 December 2010. Available at: https://i.ndh.vn/attachment/2022/09/19/psi-bao-cao-nganh-dien-082022-pdf.pdf

[18] Ministry of Natural Resources and Environment (2015). *On Strategic Environmental Assessment, Environmental Impact Assessment and Environmental Protection Plan*. Circular No. 27/2015/TT-BTNMT, signed on 29 May 2015. Available at: https://datafiles.chinhphu.vn/cpp/files/vbpq/2015/06/27-tt-btnmt.signed.pdf

[19] Ministry of Natural Resources and Environment (2011). *Regulation on Strategic Environmental Assessment, Environmental Impact Assessment, and Environmental Protection Commitment*. Decree No. 29/2011/ND-CP, dated 18 April 2011. Available at: https://chinhphu.vn/default.aspx?pageid=27160&docid=100006

[20] National Institute of Environmental Health Sciences (1999). *Health Effects from Exposure to Power-Line Frequency Electric and Magnetic Fields*, Research Triangle Park, NC.

[21] Nguyen Minh Quang (2022). *Electricity Industry Update Report*. Oil and Gas Securities, August 2022. Available at: https://i.ndh.vn/attachment/2022/09/19/psi-bao-cao-nganh-dien-082022-pdf.pdf

[22] Pablo Hevia-Koch and Henrik Klinge Jacobsen (2019). Comparing offshore and onshore wind development considering acceptance costs. *Energy Policy*.

125:9–19. Available at: https://www.sciencedirect.com/science/article/abs/pii/S030142151830675X

[23] Prime Minister (2016). *Approving the Adjustment of the National Power Development Plan for the Period 2011–2020 with a Vision to 2030*. Decision No. 428/QD-TTg, signed and issued on 18 March 2016. Available at: https://datafiles.chinhphu.vn/cpp/files/vbpq/2016/03/428.signed.pdf

[24] Prime Minister (2015). *Regulations on Environmental Protection Planning, Strategic Environmental Assessment, Environmental Impact Assessment, and Environmental Protection Plans*. Decree No. 18/2015/ND-CP, signed and issued on 14 February 2015. Available at: https://datafiles.chinhphu.vn/cpp/files/vbpq/2015/02/18.signed_01.pdf

[25] SRV (2022). *Nationally Determined Contribution (NDC report)* (updated in 2022). Socialist Republic of Viet Nam, Ha Noi, 38 p.

[26] Srinivasan, M., A. Velu, and B. Madhubabu (2019). Potential Environmental Impacts of Solar Energy Technologies. *International Journal of Science and Research (IJSR)*. 9(5):792–5.

[27] Theocharis D Tsoutsos, Niki Frantzeskaki and Vassilis Gekas (2005). Environmental impact assessment of solar energy systems. *Energy Policy*. 33 (3):289–96. Available at: https://doi.org/10.1016/S0301-4215(03)00241-6

[28] Tran Thien Khanh (2021). *Orientation for the Development of Clean Hydrogen Energy: Analysis of the Current Situation and Evaluation of Industrial Hydrogen Production Methods*. Available at: https://iced.org.vn/dinh-huong-phat-trien-nang-luong-sach-hydro-phan-tich-thuc-trang-va-danh-gia-cac-phuong-phap-san-xuat-hydro-trong-cong-nghiep/

[29] US. Embassy in Vietnam (2022). *International agreement to support Vietnam's climate and energy goals*. Available at: https://en.baoquocte.vn/international-agreement-to-support-vietnams-climate-and-energy-goals-209960.html

[30] VCBS (2023). *Electricity Industry Outlook Report 1H.2023: Electricity Consumption Demand to Grow Steadily in the Long Term*. Available at: https://finance.vietstock.vn/bao-cao-phan-tich/10735/bao-cao-trien-vong-1h-2023-nganh-dien-nhu-cau-tieu-thu-dien-tang-truong-on-dinh-trong-dai-han.htm

[31] World Bank (2015). *Environmental, Health, and Safety (EHS) Guidelines*. IFC E&S Washington, D.C. Available at: https://documents.worldbank.org/en/publication/documents-reports/documentdetail/157871484635724258/environmental-health-and-safety-general-guideline

[32] World Health Organization (2007). *Extremely Low Frequency Fields*. Environmental Health Criteria Monograph No. 238, Geneva.

Chapter 10

Social impacts and trends in renewable energy development in developing countries: solar energy, wind energy, and beyond

10.1 Environmental impact assessment of solar and wind energy projects in Vietnam

10.1.1 Overview

Vietnam is a country located entirely within the tropical monsoon zone, with a land area of 331,212 km². It is geographically bounded by the northernmost point at 23°23′ N, 105°20′ E, the southernmost point at 8°34′ N, and the westernmost point at 22°22′ N, 102°10′ E, and the easternmost point at 12°40′ N, 109°24′ E. The East Sea (South China Sea) area is three times larger than the land area, approximately 1,000,000 km², with a coastline that stretches over 3,260 km from north to south. Vietnam, particularly in the central southern region, Central Highlands, Eastern and Western Mekong Delta, enjoys high solar radiation for several months each year, making it highly suitable for solar energy exploitation. On average, Vietnam has vast natural resource potential: 4–5 kWh/m²/day for solar energy and 3,000 km of coastline with stable wind speeds ranging from 5.5 to 7.3 m/s (Matt Rogers, 2019). With such geographic conditions, Vietnam has considerable advantages and potential for exploiting solar and wind energy, including wind in mountainous regions, coastal plains, near-shore areas, and even offshore. Several projects are awaiting approval for combined wind and solar energy farms.

Despite challenges related to capital mobilization policies and electricity pricing, most experts believe that renewable energy projects will continue to maintain their capacity for development and production in the coming years (Nguyen Minh Quang, 2022). An analysis by VCBS (2023) confirms that the average cost of energy (LCOE) in renewable energy has sharply decreased, especially regarding investment, with solar energy now being the renewable energy source (RES) with the lowest production cost. Recently, an international partnership group committed to mobilizing USD 15.5 bn from both public and private financial sources over 3–5 years to support Vietnam in achieving net-zero emissions by 2050 (US Embassy in Vietnam, 2022). One of the four significant goals in this commitment is to accelerate the deployment of renewable energy projects, aiming for renewable energy to account for 47% of total electricity generation by 2030, compared to the current plan of 36% by the end of the 2030s.

With the high potential for renewable energy development, particularly solar and wind farm projects, these are expected to explode in the coming years. This presents an opportunity to foster economic and social development for the country along the path of green energy. However, like any development project, aside from the positive aspects related to the correlation between the economy, environment, and society, there are also neutral impacts (minor, negligible impacts) and negative environmental and social impacts that investors must pay attention to and mitigate. If possible, it is advisable to compare the negative impacts of renewable energy projects from solar, wind, or tidal sources with those of existing traditional energy projects such as coal, oil, or gas power plants, hydropower plants, or nuclear power plants. This is reflected in the Vietnamese Government's policy (2016): "Prioritize the development of RES, creating breakthroughs in ensuring energy security, contributing to the conservation of energy resources, and reducing environmental pollution." The implementation of environmental impact reports, including social factors, aims to anticipate adverse impacts on ecosystems, the environment, landscape, biodiversity, as well as risks related to industrial safety or changes that may affect livelihoods, health, and other social issues. Conducting impact assessments aims to minimize negative factors, avoid operational costs, maintenance, or compensation expenses in the future. Community consultation will also foster societal consensus and better align with international criteria.

Any project or construction will have impacts on nature and society. Environmental impact assessments (EIAs) help us better understand the adverse changes in space and time (Figure 10.1), specifically through indicators such as air quality, water quality, noise, vegetation, habitats, biodiversity, community livelihoods, and safety for operational workers. Conducting environmental impact studies is a legal requirement under Article 28, Appendix II of Decree No. 18/2015/

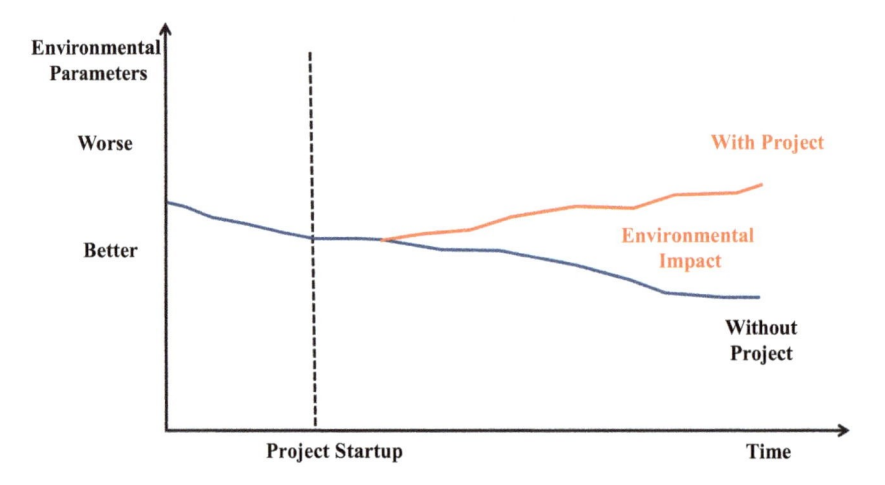

Figure 10.1 Natural changes with environmental impact

ND-CP issued by the Government (2015), which mandates that all solar and wind power plant projects covering an area greater than 100 ha or transmission lines above 110 kV or power stations with a capacity of 500 kV or more must submit an environmental impact report. In cases where wind energy projects cover an area smaller than 100 ha, environmental and social impact assessments may be considered within the project's feasibility study report. According to Decree No. 18/2015/ND-CP, the Ministry of Natural Resources and Environment is responsible for evaluating and approving EIAs for wind energy projects with a capacity of 50 MW or more, or for offshore wind projects covering 20 ha or more, or located in areas with protected forests, special-use forests, or natural forests exceeding 100 ha, or in rice-growing land of 10 ha or more. For projects falling below these thresholds, the local People's Committees are responsible for approval.

Any project or construction will have impacts on nature or society, and conducting EIA will allow us to better understand potential adverse changes over time and space (Figure 10.1), specifically regarding indicators such as air quality, water quality, noise, vegetation, habitats, biodiversity, community livelihoods, and worker safety. EIA studies are a legal requirement according to Section 28, Appendix II of Decree No. 18/2015/ND-CP (2015), which mandates that all projects involving solar and wind power plants with an area over 100 ha or transmission lines of 110 kV or substations with a capacity of 500 kV or more must conduct an environmental impact report. For wind energy projects under 100 ha, environmental and social impacts may be included in the project's feasibility study report. According to Decree No. 18/2015/ND-CP, the Ministry of Natural Resources and Environment is responsible for reviewing and approving EIAs for wind energy projects with a capacity of 50 MW or more, or nearshore wind projects with an area of seabed reclamation over 20 ha, or projects in protected forests or special-use forests over 20 ha, or natural forests over 100 ha, or agricultural land over 10 ha. Projects below these thresholds are approved by the provincial People's Committee where the project is located.

10.1.2 Environmental impact assessment for solar energy projects

The EIA report for solar energy projects can refer to the structure outlined in Appendix 1. Solar energy projects can take various forms, from installing solar panels to convert light energy into electricity, to installing solar concentrator mirrors to convert sunlight into thermal energy (heat insulation solar glass). Below are the environmental impacts that solar energy projects, including energy transmission systems, need to consider, especially for large-scale solar power plants. The benefits of solar energy projects can be summarized as listed in Table 10.1.

Most solar power projects require large areas of land or water, sometimes extending to several hundred or even thousands of hectares of natural land. The change in land use conditions can raise significant issues related to land rights, human health, livelihoods, vegetation and wildlife, hydrogeological resources, and climate change (Damon and Vasilis, 2011). The impacts and changes in the

Table 10.1 Benefits of solar energy projects

Environmental and climate change benefits	Economic and social benefits
• Significant reduction in greenhouse gas emissions, mainly CO_2, NO_x, CH_4, SO_2, and particulate matter. • Reduces surface water evaporation or surface moisture loss. • Limits the development of algae in reservoirs with solar panels. • Can rehabilitate barren desert land, when combined with grazing or crop cultivation under solar panels. • Can improve habitats for certain insects or underground animals that help aerate the soil, such as earthworms, crickets, and other microorganisms. • Converts previously inefficiently used land (barren land) into valuable and productive land.	• Provides local decentralized electricity, reducing the need for expanded transmission systems. Increases regional energy supply independence. • Creates new jobs for local labor and technicians. • Diversifies energy sources and enhances energy security. • Increases rural electrification rates. • Reduces power supply challenges and stress at the regional and national level. • Generates annual tax revenue, which increases the tax base. • Increases the government's reputation through policies that prioritize clean energy development. • Potential for carbon tax reduction on exported goods.

landscape can occur during the construction phase, such as large-scale excavation, piling, and the movement of construction vehicles (Tsoutsos et al., 2005). One of the most critical issues to limit is the deforestation for clearing land to install solar panels. If forest land must be sacrificed, it is necessary to choose degraded forests with poor biomass and assess the trade-off between the CO_2 absorption loss from the forest and the greenhouse gas (GHG) emissions from fossil fuel-based power plants, through life-cycle assessment (Luca and Fabrizio, 2020). The selected site for the solar project must carefully consider whether the vegetation cover, sensitive ecosystems, and biodiversity in the area will be significantly altered in a harmful way.

Like other infrastructure projects, during the construction of solar power plants, ecological impacts and changes in the landscape and noise levels may occur, such as large-scale earthworks, piling foundations, and the movement of construction machinery (Tsoutsos et al., 2005). This requires measuring and calculating the distance from the construction area center to residential areas, commercial zones, and limiting impact on tourism areas, natural scenic spots, and wildlife conservation zones. Investors need to quantify potential pollution sources and mitigation measures, including solid waste (construction debris, packaging for equipment, workers' domestic waste on-site), liquid waste (cleaning chemicals, vehicle oil, rust water, sewage from workers' activities, wash water for equipment and solar panels), and gas emissions (exhaust from construction vehicles, combusted materials, dust from vehicle operation, thermal mass from cooling equipment).

Additionally, the livelihoods and farming activities of local communities may be impacted by the project. It is crucial to consider the social aspect, as when people lose land for an energy project, even with compensation and land swaps, the emotional attachment to the original land is a negative impact. Therefore, public consultation meetings should be held early in the planning process to ensure the involvement of the public and relevant organizations to secure public acceptance. Workers at solar power construction sites may face heat stress, dehydration, glare, and accidents, which necessitate proper safety training and protective measures.

Currently, there is no clear evidence of the impact of solar panels on human health, including GHG emissions (National Institute of Environmental Health Sciences, 1999; World Health Organisation, 2007). Solar panels themselves do not produce pollutants during installation and operation. However, there may be chemical pollution risks in case of fires or during decommissioning after their operational lifespan, with the chemicals in the panels potentially leaking into the soil and groundwater, affecting the environment (Bhanu Mahajan, 2012). The disposal of solar panels after use should also be addressed in the environmental report (Md Shahariar Chowdhury *et al.*, 2020), with reference to the guidelines provided by the International Energy Agency (IEA) (Jose *et al.*, 2021). The potential risk of electric shock from the direct current generated by solar systems must also be considered, particularly for untrained workers who may not be fully aware of electrical safety measures (Srinivasan *et al.*, 2018).

10.1.3 Environmental impacts of wind power projects

Wind energy, harnessed from the movement of air masses, is recognized as a promising RES with relatively minimal impacts on the environment and biological resources compared to other energy sources. Wind energy projects, or wind farms, can be developed onshore (land-based) or offshore (including near-shore and offshore wind energy). Each type of wind energy installation, whether on land or at sea, has distinct economic and environmental impacts, which can be compared in Table 10.2. Offshore wind energy projects can be categorized based on the sea depth of the project area, referring to guidelines provided by the Department of Energy (DOE, 2011).

The environmental factors that need to be considered in wind energy projects can vary depending on whether the installation is onshore or offshore. It is crucial to take note of these differences when evaluating the potential impacts. Currently, most wind energy projects are financed, invested in, or technically supported by foreign entities. Therefore, in addition to national regulations, international standards (such as those by the IFC, 2012, and the World Bank, 2015) must be taken into account. The GIZ (2018) organization has published a guide on EIA for wind energy projects in Vietnam.

In general, according to international standards, there are eight performance standards that must be met for wind energy projects, which are summarized in Table 10.3.

Table 10.2 Comparison of key considerations for onshore and offshore wind energy projects

Onshore wind energy projects	Offshore wind energy projects
• Faster construction and installation time. • Significantly lower costs due to ease of transportation, construction, and management. • Easier integration with the power transmission system compared to offshore and less voltage drop between turbines and consumption points. • Onshore wind turbine towers have longer lifespans because they are less exposed to seawater corrosion and easier to maintain. • Wind speeds onshore are typically lower and less stable than offshore winds, so electricity generation efficiency is lower. • Proximity to residential areas may cause inconvenience due to noise or shadow flicker effects. • Compensation may be required for land use on agricultural or forested land. • Potential impact on migratory bird flight paths.	• Offshore winds are generally stronger and more stable than onshore winds, resulting in better power generation efficiency. • Can accommodate various tower heights and rotor diameters. • Higher construction costs due to more complex construction, better materials required, and higher resistance to saltwater corrosion. Offshore towers must withstand annual sea storms • More complex and costly transmission system and maintenance. • Nearly no costs for occupying marine areas. • Minimal impact on local residents in terms of noise, shadow flicker, or nearby landscape. • Can impact coral reefs, seagrass beds, and marine species' breeding areas • Navigation routes for maritime traffic must be considered.

Table 10.3 IFC performance standards (GIZ, 2018)

Performance standard	Description
1	Environmental and social risk and impact assessment and management
2	Labor and working conditions
3	Labor and working conditions
4	Community health, safety, and security
5	Land acquisition and involuntary resettlement
6	Biodiversity conservation and sustainable management of living natural resources
7	Indigenous peoples
8	Cultural heritage

Environmental impact assessment for onshore and offshore wind energy projects

For onshore wind energy projects

In the construction phase, the installation of foundations, turbine towers, blades, and power transmission systems requires attention to noise caused by drilling, heavy machinery, and crane operations. This noise may affect the health of nearby residents and workers on-site. Additionally, the excavation work, road construction, and land clearing may impact water flow, aquifers, deforestation, and the geological landscape. These activities can disrupt wildlife habitats or pasturelands. Furthermore, construction-related activities can lead to water pollution due to the discharge of oils and lubricants unless appropriate measures for waste management are implemented, as well as potential issues with worker sanitation facilities on the construction site.

For high-altitude construction work, worker health and safety are critical, especially to prevent accidents and fatalities. Specialized health monitoring for cardiovascular and neurological health, as well as overall safety measures, must be enforced.

The land acquisition for onshore wind farms should include adequate compensation for residents who lose their homes, agricultural land, or grazing areas. If necessary, relocation of graves or cultural relics should be considered. For displaced residents, suitable land for resettlement and the creation of new livelihood opportunities must be planned. Public consultation is essential for ensuring fair negotiations and community involvement.

During operation, the noise from turbines and blades can continue to be a concern if the installation is not located at an appropriate distance from residential areas or if there are no forest buffers to reduce noise propagation. In the event of high winds, tornadoes, or storms, turbines must have a safety mechanism to stop operation to prevent damage or tipping. Although the risk of blade failure due to extreme weather or material fatigue is low, it must be addressed with preventive measures. Emergency response plans should be prepared for such risks.

Shadow flicker (the phenomenon of blades casting moving shadows during daylight) can be bothersome to nearby residents, potentially causing visual discomfort and dizziness. This can be mitigated by avoiding the use of reflective paint, placing turbines far from residential areas, and planting green buffer zones.

Wind turbines or transmission systems can cause harm to birds, bats, or migratory species. A study of local wildlife behavior and biodiversity monitoring should be part of the planning process.

High-altitude wind projects, particularly those near airports or radar systems, should undergo assessments to determine any risks to aviation operations. The input of aviation experts may be needed to avoid accidents or interference with radar systems.

For nearshore wind energy projects

Offshore wind farms located near the shore (tidal flats and estuarine areas) must undergo similar environmental assessments as onshore projects, but with additional considerations for coastal ecosystems. The construction of foundations in these areas may alter water flow and increase the risk of coastal erosion. The corrosion of equipment due to saltwater exposure must also be taken into account.

Moreover, the livelihoods of local communities—such as fishing, aquaculture (shrimp, crab, shellfish), and salt farming—may be disrupted by the construction and operation of wind farms. Consideration must be given to whether relocation or compensation is necessary for affected areas. The impact on shipping routes and local tourism, including beaches and resorts, should also be evaluated. If the area is rich in mineral resources or is part of an offshore oil and gas reserve, these potential conflicts must be carefully considered.

For offshore wind energy projects

Offshore wind farms located far from the shore are subject to unique environmental considerations. The safety of structures against marine natural disasters (e.g., storms, tidal waves) and the potential impacts on marine life (e.g., fish, coral reefs, seagrass beds) must be assessed. The project's location along major shipping lanes or near marine protected areas requires careful evaluation of maritime safety. National security concerns regarding territorial waters also need to be considered.

10.1.4 Mitigation measures

Not all of the negative environmental and social impacts of renewable energy projects can be completely mitigated. However, we can implement measures to reduce these negative impacts without changing the energy supply option or incurring significant additional costs. Most of these mitigation solutions are related to the installation site, adjustments in the project's scale or nature, and changes to its characteristics (Tsoutsos et al., 2005). Suggested adjustments that can be applied include:

- Avoid installing solar or wind farms in areas with high population density, industrial zones, service infrastructure, or regions with scenic natural landscapes, tourist areas, resorts, or biodiversity reserves and historical-cultural sites.
- Implement occupational safety measures and machinery operation protocols during installation to minimize impacts on workers. Limit the use of harmful chemicals and biodegradable substances. Provide training for workers to familiarize them with the system and installation processes. Ensure proper safety gear and equipment for workers to prevent accidents and injuries.
- Conserve water resources and treat wastewater after use. Minimize deforestation, shrub clearance, and vegetation removal. Waste from construction and daily activities should be collected, sorted, and disposed of properly. After construction, return the land to its original condition or re-establish it in the best possible state.
- Estimate GHG emissions and propose maximization of reduction measures, including minimizing fuel consumption in operations and daily activities.
- Use new technologies that reduce environmental impacts (e.g., lower noise levels, reduced water use, less raw material demand, and reduced construction site area).

10.2 Raising environmental and social awareness in renewable energy projects

The implementation of renewable energy projects, especially solar and wind energy, is becoming an essential part of Vietnam's sustainable development strategy. However, alongside the obvious economic and environmental benefits, these projects may also have negative environmental and social impacts that need to be clearly understood and managed appropriately.

10.2.1 The importance of environmental and social awareness

Environmental and social awareness in renewable energy projects is crucial not only to minimize negative impacts on ecosystems, land, and water resources but also to safeguard the well-being of local communities. Renewable energy projects can affect landscapes, biodiversity, public health, as well as factors like employment and changes in local livelihoods. Particularly, these projects can directly impact people living near the installation sites, especially in densely populated areas or traditional craft villages.

10.2.2 Steps for implementation and impact assessment

To minimize negative impacts, the implementation of renewable energy projects must fully comply with EIA procedures. The EIA is a vital tool that helps assess potential impacts of a project on the natural environment and society. This step is essential for investors and regulatory bodies to identify potential issues and implement measures to control and mitigate negative impacts. EIA must be comprehensive, covering air quality, water quality, noise, biodiversity, labor safety, and public health.

10.2.3 Strengthening community and stakeholder involvement

Environmental and social awareness cannot solely rely on government agencies but should also be raised within the community and among stakeholders, especially those living in the project areas. One key factor in raising awareness is enhancing community participation throughout the implementation process. Public consultations should be regularly held to gather feedback and address local concerns, building social consensus. At the same time, stakeholders, including social organizations, businesses, and regulatory bodies, should take responsibility for supporting, providing information, and cooperating in environmental protection activities.

10.2.4 Training and capacity building for stakeholders

To develop environmental and social awareness, a structured training strategy is needed for all involved groups. Training on environmental management, social impacts of projects, and mitigation measures is essential. Particularly, for investors, local authorities, and local communities, a clear understanding of renewable energy issues will help them make informed decisions and minimize undesirable impacts.

10.2.5 Government policies and commitments

The Vietnamese government's policies on renewable energy, such as prioritizing the development of clean energy sources, play an important role in fostering awareness and actions for environmental protection. Specifically, targets for renewable energy development and commitments to reduce GHG emissions by 2050 will strongly drive these projects. However, for these goals to be successfully implemented, strict regulations, monitoring mechanisms, and environmental protection measures must be enforced.

Raising environmental and social awareness in the implementation of renewable energy projects is crucial for ensuring their sustainability and long-term effectiveness. By combining economic development with environmental protection, Vietnam can not only achieve its clean energy goals but also contribute to ecosystem preservation and improve the quality of life for its people.

10.3 Methods for environmental impact assessment of renewable energy sources to identify effective mitigation and control solutions

Methods for EIA of RES to identify effective mitigation and control solutions can be carried out through various steps and tools. This process involves analyzing the impact of renewable energy projects such as solar and wind energy on environmental, social, and human health factors.

First, EIA must be conducted from the early stages of the project, including site selection, scale, and design of renewable energy installations. The assessment methods should cover the effects on factors such as landscape, biodiversity, land, water, climate, and the economic and social activities of the surrounding community.

During construction, impacts such as noise from drilling, machinery, and vehicles, as well as construction activities, may affect the health of workers and nearby communities. In particular, excavation, road construction, and land leveling may alter water flow, affecting vegetation and ecosystems. To mitigate these impacts, protection plans should be implemented, including waste treatment and collection, controlling wastewater from machinery, and ensuring sanitary conditions for workers on-site.

Additionally, impacts during the operation phase of renewable energy projects, such as noise from wind turbines, "shadow flicker" effects, or the risk of animal collisions with equipment, should be thoroughly assessed. Mitigation solutions could include selecting sites further away from residential areas, using noise-reducing materials, or establishing green corridors to minimize impacts on wildlife.

Finally, effective mitigation solutions include applying new technologies to reduce noise, protect water resources, conserve fuel, and reduce GHG emissions. Adhering to labor safety standards, environmental protection regulations, and using advanced technologies will help minimize negative impacts while maintaining the effectiveness and sustainability of renewable energy projects.

In summary, implementing detailed and comprehensive EIA methods is essential to identify and implement effective mitigation solutions, in order to protect the environment and public health in renewable energy projects.

10.4 Example of environmental impact assessment in central Vietnam

10.4.1 Project introduction

The investment in the development of a large-scale solar energy project in Vietnam with a capacity of 50 MWp solar farm is essential for the following reasons and objectives:

General objective:
The development of renewable energy in general, and solar energy in particular, is an inevitable trend in global renewable energy development policies, especially as the investment cost for solar projects has decreased rapidly in recent years.

Specific objectives:

- The construction of this project will lay the foundation for mastering the technology, building, installing, and operating a large-scale solar power plant. This will serve as a model for expanding the application of this electricity supply model to other regions.
- The location of this project is on land used for annual crops and degraded hilly land, which is inefficient for agricultural and forestry development. Additionally, this location is not involved in the planning of other projects and is not inhabited. Therefore, priority should be given to developing solar power plants in these areas to maximize land use efficiency.
- The natural characteristics, the ability to easily connect to the power grid, existing transportation infrastructure, and the local land use planning are crucial factors for the feasibility and high efficiency of the project.
- This is a renewable and clean energy project that does not emit GHG or cause environmental pollution. It is applied in an area with significant solar energy potential, ensuring high feasibility. The project contributes to affirming Vietnam's international responsibility in environmental protection and fulfilling international commitments to reduce emissions.
- The development of solar energy in this project area will supplement local power supply for the province and neighboring provinces, meeting the 10% annual electricity consumption growth demand, reducing transmission losses, and avoiding the costs of building power lines from other areas.
- The solar electricity from the project will contribute to national energy security, gradually increasing the share of RES, ensuring energy security, mitigating climate change, protecting the environment, and promoting sustainable socio-economic development in line with Vietnam's renewable energy development strategy up to 2030, with a vision to 2050, and the national power development plan for the 2011–2020 period, considering the year 2030. Notably, the strategy emphasizes accelerating solar energy development so that solar energy will contribute approximately 0.5% of the total electricity production by 2020, 6% by 2030, and about 20% by 2050.

10.4.1.1 Assessment and forecasting of environmental impacts

The evaluation and forecasting of environmental impacts of the project on the environment and socio-economic conditions of the region are specified for each source of impact and each affected party. These impacts are considered in three stages:

- Project preparation phase
- Construction phase
- Operational phase
- Solar energy is considered an environmentally friendly energy source with minimal negative effects on society and the environment. It is a clean energy source that helps reduce dependence on fossil fuels when effectively harnessed. The use of solar energy does not emit GHGs such as CO_2, SO_2, or NO_2, contributing to environmental protection.
- Solar power development projects bring significant benefits, not only promoting economic and social development but also ensuring energy security both nationally and locally. However, the implementation of the project may cause environmental impacts in the project area, including both direct and indirect effects at varying levels. Some impacts may be temporary and negligible, while others could occur frequently throughout the project's implementation.
- Therefore, identifying and classifying the impacts associated with the project is essential. This process helps forecast, maintain, and enhance positive impacts while preventing and minimizing negative effects in a coordinated and effective manner throughout the project's execution.

10.4.2 Assessment and forecasting of impacts during the project preparation phase

The environmental impacts of the project during the preparation phase are summarized in Table 10.4.

Project objectives and environmental impacts

The project aims to minimize encroachment on residential land and historical and cultural sites such as temples, pagodas, and shrines. The surveyed area primarily consists of forest land, land used for perennial crops, and agricultural land cultivated by local residents.

The project will acquire approximately 70 ha of land to build a 50 MWp solar power plant and the foundation pylons for the 110 kV transmission line, along with 1.16 ha of temporary land for a safety corridor along the 13.782 km of transmission line (with a 7.5-m radius). The clearing of forest, trees, and crops will result in the loss of income from agricultural and forestry activities, impacting the quality of life and making it difficult for local residents to find suitable new occupations.

To mitigate these impacts, it is essential to involve the community and local authorities and ensure adequate compensation for affected households.

Impact duration: The impact duration typically exceeds the compensation and land clearance phase.

Table 10.4 Environmental impacts during project preparation phase

No.	Source of impact	Direct/indirect affected objects	Impacts	Impact duration
1	Land acquisition for the construction of 110 kV transmission line foundations and solar power plant	Households losing land, crops, and vegetation. Local authorities.	Land occupation for construction of the plant and foundations. Change of land use: from forest and agricultural land to specialized land (plant and foundations).	Long-term
2	Impact within the transmission line corridor	Land, crops, and vegetation of local residents.	Impact on daily activities, causing disruption for people losing their land. Limiting land use under the safety corridor: from perennial crops to short-height crops (6 m) or annual crops. Restricting the use of houses/buildings under the safety corridor. Cutting trees during the construction of the line and removing overgrown trees.	Temporary (during land clearance)

3	Demolition and vegetation clearing	Local environment (air quality).	• Air pollution and dust:
			  – Caused by demolition and vegetation clearing activities.
			  – From machinery used in demolition and clearing activities.
				Temporary (during land acquisition and clearing)

		Local environment (air quality).	• Air pollution and dust:
			  – Caused by demolition and vegetation clearing activities.
			  – From machinery used in demolition and clearing activities.
				Temporary (during land acquisition and clearing)

		Soil and water environment
		Residents living in the surrounding areas
		Direct workers
			• Noise:
			  – Caused by machinery operation at the construction site.
			  – From demolition activities.
			• Safety risks:
			  – Accidents due to falling debris, getting entangled in electrical systems.
				Temporary (during land acquisition and clearing)
				Temporary (during land acquisition and clearing)

4	Transportation and disposal of construction waste	Local air quality around the project area	• Air pollution and dust:
			  – Dust from transportation.
			  – Emissions from fuel-powered vehicles.
				Temporary (during transportation phase)

		Local residents and road users	• Noise:
			  – From transportation vehicles.
			• Traffic safety:
			  – Increased vehicles leading to higher accident risk.
				Temporary (during transport hours)

5	Workers' living conditions	Soil and air environment	• Excavated earth from digging. • Explosion risk from unexploded ordnance (UXO). • Generation of soil due to excavation.	Temporary (during transport hours)
6	Mine clearance	Soil and air environment	• Explosions caused by UXO (in cases where mines or ordnance remain).	Temporary (during transport hours)

Impact level: Medium, as all affected households are engaged in agricultural production.

10.4.3 Impact on air quality

Sources of emissions and dust:

Emissions from machinery operations during demolition and vegetation clearing activities.

Emissions from transportation vehicles carrying waste from land clearance.

- **Key pollutants**:
  - Dust, soil, rocks, and gases such as SO_2, NO_2, CO, and CO_2 from machinery and transportation vehicles.
- **Affected parties and impact scope**:
  - Air Pollution: Impacting air quality, workers, and local fauna and flora. The impact is significant within the project area but negligible in surrounding areas.
  - Dust: Affecting air quality, workers, and local fauna and flora. The impact is limited to the construction phase, primarily within the project site.
- **Impact assessment**:
  - Significant dust pollution: During the land clearance phase, dust pollution is significant but mainly contained within the project boundaries.
- **Vegetation clearing activities**:
  - Tasks: Demolition, clearing of trees and crops.
  - Machinery used: Bulldozers, excavators, graders.
  - Waste generated: Hundreds of tons of waste to be managed.
- **Dust and emissions generated**:
  - Source: Machinery operations, fuel combustion.
  - Pollutants: Dust, NO_x, SO_2, CO, etc.
  - Dependence: On the number, capacity, and condition of the equipment.

10.4.4 Impact on water quality

- **Sources of wastewater**:
  - Main sources: Rainwater runoff and domestic wastewater from the construction site.
- **Affected parties and impact scope**:
  - Targeted areas: Surface water bodies, aquatic organisms, and local residents along the route.
  - Impact scope: Rainwater runoff can carry soil, debris, and waste, leading to waterbody blockages, eutrophication, and increased turbidity.
- **Impact assessment**:
  - Rainwater runoff: Can carry waste materials, leading to sedimentation, flow blockage, and negative effects on low-lying areas and households if no proper drainage system is in place.

Table 10.5 Vibration levels of selected construction equipment (dB)

No.	Construction equipment	Vibration level at 10 m	Vibration level at 30 m	Vibration level at 30 m
1	Heavy-duty transport vehicle	74	64	54
2	Excavator	80	71	60
3	Bulldozer	79	69	59
4	Winch, light crane	77	67	57
5	Concrete mixer	76	66	56
6	Concrete vibrator, needle vibrator	82	72	62
7	Truck	74	64	54

- Mitigation measures: Develop an effective drainage system to collect and divert runoff, minimizing the impact.

10.4.5 Impact assessment and forecasting during the construction phase

Air quality impact assessment

- **Sources of emissions**
  - Emissions and dust from excavation, leveling, and transportation of materials and waste.
  - Dust from unloading, stockpiling materials, and operation of construction vehicles.

- **Affected parties and impact scope**
  - Targeted Areas: Air quality, workers, local fauna and flora.
  - Impact Scope: The impact is confined to the construction phase and the project site.

- **Primary impacts**
  - Excavation, dredging, transportation, and construction activities will generate dust, emissions, noise, and vibration, thus deteriorating air quality (Table 10.5).

10.4.6 Non-waste-related sources of impact

Noise and vibration impacts

During the construction phase, noise and vibration primarily arise from:

- Operation of construction machinery (bulldozers, compactors, rollers, etc.).
- Trucks transporting materials and equipment.

Noise and vibration levels are significant and continuous during construction, primarily affecting the surrounding environment within a radius of 50–100 m from the source.

10.4.7 Impact assessment and forecasting during the operational phase

10.4.7.1 Environmental impacts of the solar power plant

- **Air quality impacts**

    Compared to fossil fuel-based power plants, solar power plants do not emit harmful gases, dust, or GHGs, contributing to climate change mitigation. With a capacity of 50 MW, this project is considered a "clean energy" source, offering economic and social benefits to the local community while conserving fossil fuel resources.

- **Water quality impacts**
    - Domestic wastewater: Generated from 15 operational staff members, with an estimated volume of 1.8 m^3/day, containing organic matter, nutrients, and microorganisms. This wastewater will be treated to meet local standards before being discharged into the area's irrigation canal system.
    - Rainwater runoff: Rainwater, which is free of harmful substances, will be directed into the general stormwater drainage system.
    - Firefighting water: Water used during fire drills or emergencies may contain oil and will be treated in accordance with regulations by relevant authorities.

- **Solid waste impacts**
    - Household waste: Approximately 3.75–4.5 kg of daily household waste, primarily organic waste and inorganic materials like plastics and glass. The waste will be collected and processed by the local waste management service.
    - Hazardous waste:
        ○ PV solar panels: After 20 years of use, the panels will be replaced and will be collected and disposed of by the supplier.
        ○ Used oils and rags: Estimated at 300 kg per maintenance cycle (every 4–5 years) during equipment maintenance.
        ○ Office waste: Items like fluorescent lamps and ink cartridges (3 kg/month), which will be disposed of according to regulations.

- **Non-waste-related impacts**
    - Noise: The noise from the inverter is negligible, being only audible within 1–2 m of the source. It will not impact local residents due to the project's location being far from residential areas.
    - Electromagnetic fields (EMF): The highest intensity of the EMF at the 110 kV transmission line is 3.7 kV/m. This is not expected to have significant health impacts as long as safety distance guidelines are followed (Table 10.6).

During the operational phase, the 110 kV transmission line may generate EMF, which could potentially impact the health of maintenance workers and residents within the right-of-way corridor. The intensity of the EMF of the 110 kV transmission line has been calculated using Japan's CRIMAG model.

Table 10.6 Summary of environmental impacts during project operation phase

No.	Source of impact	Impacts	Direct/ indirectly affected parties	Impact level
1	Activities of operating workers	• Domestic wastewater • Domestic solid waste (DSW) • Hazardous waste (HW)	• Domestic wastewater • DSW • HW	Minor, controllable
2	Operation of the plant and solar panels	• Stormwater runoff and water from solar panel cleaning • Noise and vibration from transformers • Waste from damaged electrical equipment and solar panels • Vibration, noise, and waste oil from transformers • Glare caused by sunlight reflection on solar panels • Heat emitted from solar panels	• Maintenance workers • Local residents	Mild, long-term
3	Maintenance and management of transmission line safety corridor	Maintenance and management of transmission line safety corridor	Maintenance workers • Local residents • Biological resources and biodiversity	Mild, long-term
4	Incidents	• Short circuits, fires, explosions, transformer oil spills • Collapse of power poles, snapped lines • Weather-related incidents (wind, lightning, storms, floods, etc.)	• Maintenance workers • Local residents	Moderate, mitigable

The calculation results indicate that the maximum EMF intensity at a point 1 m above ground level (directly beneath the suspension point of the outer phase conductors) reaches 3.7 kV/m. While this level may pose certain effects, it does not cause significant health impacts if safety distances and health protection regulations are adhered to. Compliance with safety distance requirements and protective measures will effectively mitigate the negative impacts of EMF (Figure 10.2).

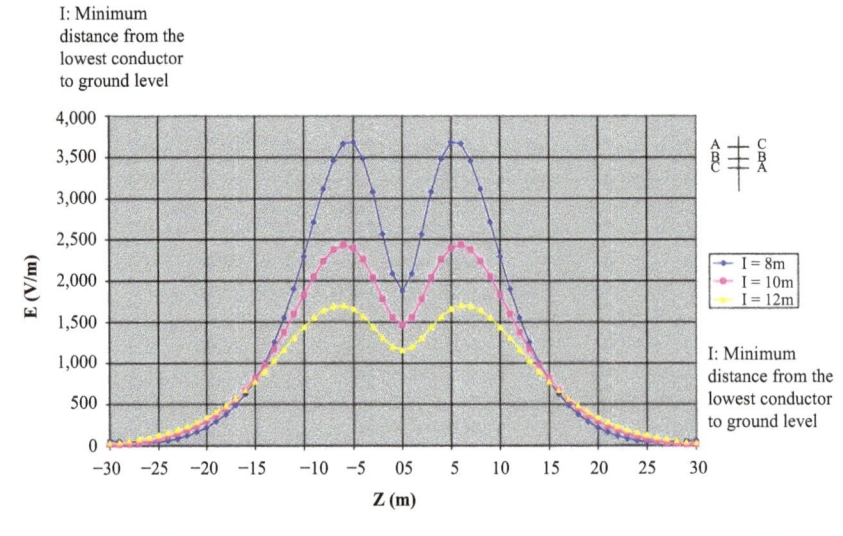

Figure 10.2 Electric field intensity at 1 m above the ground for a 110 kV transmission line

The solar power plant project ensures energy security, reduces emissions, and minimizes environmental pollution.

Environmental impact

- Preparation and construction phase: Dust, exhaust gases, noise, wastewater, solid waste, and hazardous waste will be generated.
- Operational phase: There will be almost no waste or noise, provided regulations are strictly adhered to.

Negative impacts are controlled through feasible mitigation measures, which are outlined in the EIA report.

Vietnam is considered to have significant potential for renewable energy development, with solar and wind energy attracting considerable investment. The exploitation of new energy sources and RES is crucial for ensuring energy security in the country and reducing GHG emissions. However, as with any project, there are negative impacts that must be mitigated throughout all stages, from initiation and construction to operation and decommissioning of equipment.

10.4.8 Recommendations

It is recommended that local authorities cooperate to support environmental protection efforts and that the Department of Natural Resources and Environment approves the EIA report promptly to ensure the project is implemented on schedule.

However, specific regulations and guidelines for assessing environmental impacts of solar and wind power projects are still lacking. Although some directives from the Ministry of Natural Resources and Environment exist, they remain somewhat general and lack specific details for different types of solar and wind power projects, as well as being incomplete and not fully aligned with international standards. Therefore, in the coming years, the Ministry of Natural Resources and Environment, the Ministry of Industry and Trade, and scientists should review policies, regulations, and standards in more detail. They should also publish comprehensive guidelines and provide training for EIA teams, while adding relevant content to training programs in environmental science, ecology, and energy fields.

10.4.9 Commitments

- Compliance with measures to mitigate negative impacts and environmental protection regulations.
- Strict monitoring of air emissions, wastewater, solid waste, and hazardous materials.
- Ensuring traffic safety and order within the project area.
- Remedying and compensating for damages in case of accidents.

The project will be implemented safely, efficiently, and sustainably, contributing to the socio-economic development of the region.

10.5 IoT solutions and smart lighting system, wireless charging stations for vehicles, and electric ferries using renewable energy

The Internet of Things (IoT) is no longer a foreign concept in today's advanced technological era. IoT is widely applied across various fields of life and production. To meet this trend and apply it in Vietnam, the study has focused on the theoretical foundations of hardware, software, and network protocols involved, from which a control model for a smart traffic light system was developed. This model allows for flexible control through a website and smartphone applications, ensuring energy efficiency and alignment with the practical conditions in Vietnam.

Current public lighting control systems
Currently, the control and monitoring of public lighting systems are still largely manual, leading to several issues such as:

- High operation time and costs.
- Safety concerns during extreme weather conditions, such as storms.
- Inability to update the status of damaged equipment promptly for replacement.

With the growing energy shortage, there is an urgent need to build an intelligent, automatic, and efficient control system. This system should optimize the on/off scheduling of lights to minimize energy waste.

Current issues

In Vietnam, some electrical device control models have been applied practically, but several limitations remain:

1. Inadequate scale: Most current systems are designed for large-scale applications, whereas the actual demand for small- and medium-sized organizations with fewer devices is very common.
2. High costs: The cost of existing control systems far exceeds the affordability of individuals and small organizations.
3. Poor security: Data security is not adequately ensured, leading to potential security risks.

IoT Application Solution

The use of wireless transmission methods in practice is still limited in Vietnam and has not been fully exploited. With the rapid development of the Internet and the ability to leverage available resources, the study proposes a solution for controlling public electrical devices via IoT.

This solution offers several benefits:

- User convenience: Remote control via the Internet.
- Operational efficiency: Saving time, reducing costs, and minimizing risks.
- Suitability for practical conditions: It is suitable for medium- and small-scale organizations and individuals.

The smart lighting system is developed based on the IoT technology and consists of the following main components:

- **Hardware**
  – The control model is designed to manage several public lighting devices.
  – These devices are directly connected to the Internet, forming the foundation for remote monitoring and control.

- **Management software**
  The software system includes:
  – **Website application**
  Allows users to access the system through a browser to monitor the on/off status and control the operation of the lights.
  – **Android application**
  Provides features for monitoring, controlling, displaying status, storing data, and tracking the operational status of the light poles.
  Control features: The system supports two flexible control modes:

- **Manual control**
  Users can turn the lights on/off directly through the app or website.
- **Automatic control**
  The system operates based on a pre-set schedule.

The automatic schedule can be easily changed and updated through the Android app, ensuring high flexibility.

- **Key advantages**
  - Enhances the ability to manage and monitor the public lighting system remotely.
  - Saves energy and reduces operational costs by optimizing the on/off timings.
  - User-friendly interface, easy to use on both the website and mobile devices.
  - Integrates data storage and analysis, helping track performance and ensuring timely equipment maintenance.

This system is not only suitable for the practical conditions in Vietnam but also contributes to modernizing public lighting management, moving toward smart and sustainable urban development.

10.5.1 Flowchart for central control unit algorithm

A safer lighting system for traffic. However, to fully harness these advantages, monitoring and controlling the lighting system becomes a crucial factor.

Operation process of the smart lighting control system

- **System initialization**

  When the program is activated, the system will automatically configure the necessary parameters to ensure stable operation.

- **Receiving data from the internet shield**

  The system receives signals from the Internet Shield. This data determines whether the lighting devices should be turned on or off.

- **Processing and controlling the devices**

Case 1 Signal requests to turn on all lights:

The central control unit analyzes the signal from the Internet Shield.

If the signal confirms the request to turn on all the lights is accurate, the system sends a control command through the Internet Shield to turn on all the lights.

If the signal is inaccurate, the command sent will be to turn off all the lights.

Case 2 Signal requests to turn on lights in an intermittent pattern:

The process is similar, where the central control unit analyzes the signal and sends a command via the Internet Shield to turn on the lights in the desired intermittent pattern or turn off unnecessary lights.

- **Automatic and synchronized operation**

The system operates flexibly and efficiently based on signals from the Internet Shield, ensuring that control requirements are executed accurately and synchronously.

This system not only optimizes lighting performance but also helps manage public lighting in a smart and energy-efficient way.

438 *Clean energy in South-East Asia*

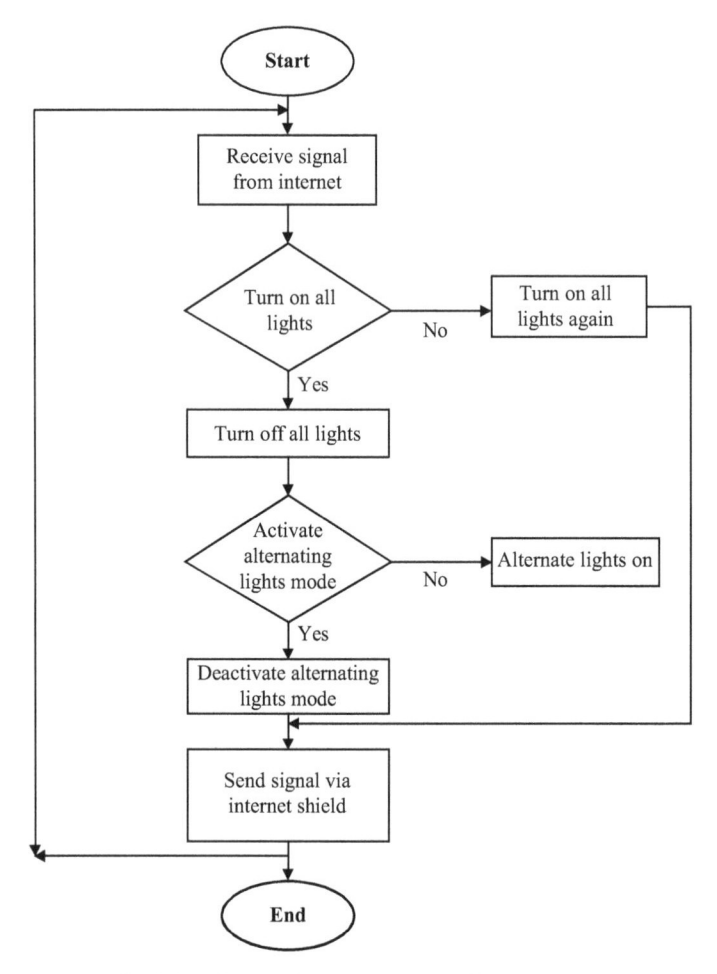

Figure 10.3 Flowchart of the microcontroller program algorithm

With the continuous development of LED technology and the growing demand for energy savings in the context of increasing urbanization, traffic lighting systems are undergoing a significant transformation with the adoption of LED lighting. LED technology not only saves electricity but also has a long lifespan, reduces light pollution, and lowers CO_2 emissions (Figure 10.3).

Flowchart representation
The smart lighting solution for traffic roads, using IoT technology combined with Long Range Radio (LoRa) networks, provides flexible and effective monitoring and control capabilities. This system allows the collection of data on the lighting status, temperature, humidity, brightness, current, and voltage consumption from LED lights. Based on this data, the system can automatically adjust the lighting power, turn lights on/off according to a schedule, and modify brightness levels to suit the specific conditions of each street area.

The application of LoRa technology in the smart lighting system ensures long-range communication, low energy consumption, and low deployment costs, making it highly suitable for large public lighting systems. LoRa allows LED lights to communicate with each other and with remote control stations without requiring a direct internet connection, enhancing management efficiency and reducing operational costs.

Additionally, the smart lighting system can integrate sensors to detect the presence of vehicles and pedestrians, automatically adjusting the light intensity to ensure traffic safety. For example, when a vehicle passes, the light intensity increases, and when there are no vehicles or pedestrians, the lights dim to save energy. The system can also trigger alarms when issues such as faulty lights or lost connections are detected, improving maintenance efficiency.

This solution not only helps save energy but also minimizes negative environmental impacts, assisting cities in reducing CO_2 emissions and light pollution. Furthermore, the system contributes to enhancing urban traffic quality and ensuring safety for road users at night.

10.5.2 *System test results*

The system was tested in a LAN environment using both Wi-Fi and ADSL modems, with the following results:

- In good connection conditions:
  - 10/10 successful presses on the internal website
  - 10/10 successful presses on the mobile software

- In poor connection conditions (approximately 4 s):
  - 10/10 successful presses on the internal website
  - 10/10 successful presses on the mobile software.

In conclusion, the smart lighting solution for traffic roads using LED technology combined with IoT and LoRa is an effective, flexible, and cost-efficient solution that optimizes energy usage and environmental protection, while enhancing the quality of life for residents and safety for road users.

Advantages
- Simple installation: The system is easily deployable on most existing Wi-Fi and ADSL modems.
- Fast processing performance: The average processing and response time is approximately 2 s under good network conditions, with a maximum of around 4 s under unstable connections.
- Device status identification: Ensures reliable device status monitoring through feedback signals.
- Scalability: Multiple devices can be controlled by adding relay modules or using expansion boards.
- Convenient control: Supports intelligent and user-friendly control modes through app 3.2.

Disadvantages
- Lack of security: The system currently does not integrate information security features.
- Signal delay: Signal delays depend largely on internet network quality, affecting operational efficiency.
- Suboptimal processing protocols: Protocols on mobile phones and websites are still basic, not fully optimized.
- Platform limitations: The system does not support multi-platform operations, reducing flexibility.
- Limited to Android OS: The system is currently only available for use on Android devices.

10.5.3 Development directions

With limited time and resources, the current system only addresses a small portion of a complete solution. To develop the system into a more practical and widely applicable solution, the following directions are proposed:

- **Improving mobile application**
  - Develop the application in the form of a service socket to handle signals in service format, reducing the resource load on mobile devices.
- **Develop hosting and server**
  - Rent or set up hosting/servers to enable remote device control anytime, anywhere, while ensuring stable connectivity.
- **Expand control scale**
  - Design the system to control additional devices, making it suitable for smart home applications.
- **Integrating smart home features**
  - The system should be compact, easy to deploy, and require minimal wiring, making it suitable for smart home setups.
- **Develop data collection network**
  - Build a remote data collection network for use in larger systems.
- **Use of embedded kits and cameras**
  - Integrate embedded kits to collect data from remote cameras, supporting surveillance applications.
- **Big data processing**
  - Transform microcontrollers into clients for data collection and upload to a server on a computer, linked to SQL for effective system monitoring.
- **Add sim900 module**
  - Integrate the Sim900 module to replace or support the Ethernet module in case of issues.
  - Combine the Arduino and Sim900 modules to control electrical devices via voice messages or SMS.

Social impacts and trends in RE development in developing countries 441

To meet the complex requirements of street lighting, more advanced solutions are needed, integrating real-time traffic data, adaptive control systems, and connecting to the smart city ecosystem. These advancements will not only help reduce energy waste but also maintain safety and enhance the overall effectiveness of urban infrastructure.

10.5.4 Wireless charging station for electric vehicles using renewable energy

- **Wireless charging solution for electric vehicles**
    - Today, environmental pollution is becoming more severe, and the depletion of fossil resources is driving the shift to green energy. Electric vehicles (EVs) are an effective green energy solution and are becoming increasingly popular worldwide. According to the IEA, EVs have developed rapidly over the past decade.
    - Figure 10.4 presents the IEA's statistics on the development of the EV market from 2013 to 2018. By 2018, the number of EVs had surpassed 5 mn, marking a 63% increase compared to 2017. It is estimated that demand for EVs will reach approximately 44 mn units annually by 2030.
    - In 2022, the Asia-Pacific (APAC) region accounted for 69.3% of sales, followed by Europe with 19% and the Americas with 10%. The dominant presence of China in the APAC region is a key factor for its leading market share. The APAC region is predicted to maintain its leading market share of 41.4% by 2035, followed by Europe at 31.6% and the Americas at 19.4%.

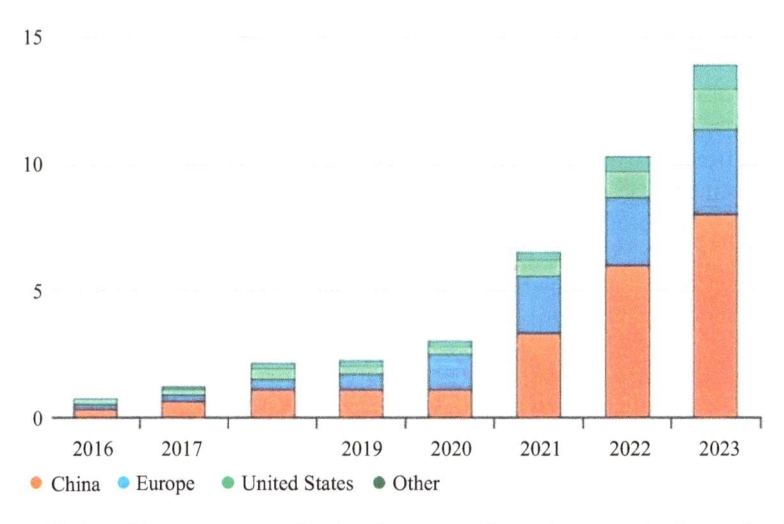

Figure 10.4 IEA statistics on the development of the electric vehicle market from 2016 to 2023

- Currently, EV chargers are being actively researched and developed. Wireless charging is a promising technology to replace plug-in chargers, eliminating the use of cables. Wireless charging for EVs offers more convenience and safety compared to traditional plug-in chargers.
- Wireless charging is one of the prominent applications of wireless power transfer (WPT) technology. WPT systems allow energy transfer over the air at distances ranging from a few millimeters to several hundred millimeters, with efficiencies exceeding 90%. WPT systems used in wireless charging for EVs are categorized into two types: static wireless charging and dynamic wireless charging.
- Static wireless charging
    - Requires the EV to park in the correct position relative to the transmitter to receive energy. Currently, static wireless chargers are already being marketed by major EV manufacturers like WiTricity and Qualcomm. Some EV manufacturers also offer wireless charging as an option when purchasing a vehicle. However, the disadvantages of static wireless chargers include long charging times, short driving range between charges, and large battery size and weight.
- Dynamic wireless charging
    - This solution addresses the disadvantages of static wireless charging. In a dynamic wireless charging system, the EV can charge while in motion. This system not only extends the EV's driving range but also significantly reduces the size and capacity of the battery. If 20% of the driving route is equipped with a 40 kW charging system, the EV's range can be extended by at least 80%. Energy transfer in dynamic wireless charging systems can be achieved by installing multiple transmitters under the road, similar to static wireless chargers. Instead of each EV having its own static charger, these chargers can be arranged and controlled to create a dynamic charging lane. Dynamic wireless charging systems can charge multiple EVs simultaneously and are compatible with various types of EVs, such as buses and cars, providing much higher efficiency than other charging systems. Currently, dynamic wireless charging systems are still in the research and testing phase, with power transmission up to 80 kW and transmission distance up to 500 mm. Recently, dynamic charging systems are being developed with various technologies and techniques. Additionally, many governments worldwide are implementing policies to promote the use of EVs for societal purposes.

10.5.5 Overview of the transportation sector in Vietnam
10.5.5.1 Transportation demand

Passenger transport demand: In recent years, economic development, urbanization, and population growth have driven an increase in transportation demand. Key highlights include:

- **Passenger transport growth**
  - From 2010 to 2019, the number of passenger trips increased by an average of 10% per year, rising from 97.93 bn passenger km in 2010 to 230.74 bn passenger km in 2019.
  - Due to the COVID-19 pandemic, passenger transport demand dropped by 32.8% in 2020.
- **Mode of transport breakdown (2019)**
  - Road: 75.4%
  - Air: 22%
  - Waterway: 1.6%
  - Rail: 0.97%

  Compared to 2010, road and air transport increased, while waterway and rail transport decreased. Specifically, air transport increased from 21.61 bn passenger km in 2010 to 77.40 bn passenger km in 2019, averaging 15.5% per year. Rail transport decreased by 3.6% per year.
- **Road passenger transport**
  - Includes both personal and public transport.
- **Personal transport**: Based on personal vehicles (bicycles, motorcycles, and cars). Motorcycles account for the largest share, with 72.06 mn motorcycles in 2020, averaging 0.74 per person.
- **Public transport**: Includes buses and intercity passenger vehicles, but still represents a smaller share compared to personal transport.
- **Current status of personal vehicle usage**
  - **Motorcycles**
    - Vietnam has one of the highest motorcycle usage rates in the world.
    - From 2014 to 2020, an average of 5.14 mn new motorcycles were added each year.
    - In 2020, there were approximately 237,000 electric motorcycles in use, with a strong trend of growth in urban areas.
  - **Cars (under nine seats)**
    - By the end of 2020, Vietnam had 2.43 mn cars.
    - Between 2014 and 2020, an average of 255,000 new cars were registered each year.

These figures highlight a significant shift in the transportation structure toward prioritizing modern vehicles and improving public transportation infrastructure, but they also create significant pressure on the transportation system and the environment (Tables 10.7 and 10.8).

Public passenger transport
In 2018, the total number of vehicles in the public passenger transport sector was 158,870, with 139,334 buses and 19,532 inter-provincial passenger vehicles. From 2014 to 2018, the number of buses increased by an average of 8,500 vehicles per year, while inter-provincial passenger vehicles grew by 9,760 vehicles annually. Regarding vehicle types, mini buses with fewer than 16 seats made up the largest

444 Clean energy in South-East Asia

Table 10.7 Number of registered personal road vehicles (units)

Year	2014	2015	2016	2017	2018	2019	2020
Cars	900,027	1,033,131	1,270,066	1,495,463	1,756,594	2,064,511	2,429,600
Motorcycles, mopeds	41,197,448	44,128,822	47,131,928	54,063,318	58,169,432	61,728,249	72,061,323
Electric motorcycles	NA	NA	501,400	728,451	1,075,630	1,250,298	1,449,379

Table 10.8 Household ownership rates of cars and motorcycles (%)

Year	Nationwide		Urban		Rural	
	Cars	Motorcycles	Cars	Motorcycles	Cars	Motorcycles
2006	0.1	52.7	0.4	71.5	0.0	45.5
2008	0.4	64.8	1.0	78.6	0.1	59.4
2010	1.2	71.9	2.8	80.4	0.5	68.2
2012	1.7	80.0	3.4	88.6	0.9	76.4
2014	2.0	82.8	4.1	89.6	1.1	79.7
2016	2.6	84.8	5.3	89.6	1.3	82.5
2018	3.2	86.7	6.1	90.5	1.7	84.7

Source: According to data from the General Statistics Office of Vietnam (2019), based on the 2018 Household Living Standards Survey.

share, accounting for over 62%, followed by medium-sized buses with a length of 9 m, representing 35.7%, and small buses with a length of 7 m, which made up 1.7%. In recent years, the trend has been the replacement of small buses with medium-sized ones. Notably, in 2017, the first bus rapid transit (BRT) line in Vietnam, with 35 buses, was launched in Hanoi.

10.5.5.2 Energy consumption and GHG emissions in the transportation sector

The rapid growth in transportation demand has led to a significant increase in energy consumption and negative environmental impacts. According to statistics, energy consumption in the transportation sector rose from 11,197 KTOE in 2014 to 14,238 KTOE in 2019, with an average annual growth rate of 4.9% during this period. This growth rate was higher than the overall growth rate of the energy sector, which was 3.4%, leading to an increase in the transportation sector's share of national energy consumption from 21.4% in 2014 to 23.0% in 2019.

Energy consumption in transportation includes fuels such as gasoline, diesel, fuel oil (FO), aviation gasoline, and electricity, with diesel accounting for the largest share (Table 10.9). Specifically:

- Gasoline: Mainly used in road transport vehicles such as motorcycles, passenger cars, and light service vehicles.
- Diesel: Used for road freight vehicles, railway transport, and inland waterway vessels.

Table 10.9 Fuel consumption

Fuel type	2014
Gasoline	4596.2
Diesel oil	5292.1
Fuel oil (FO)	234.0
Jet fuel	356.0
Electricity (GWh)	19.4

Source: According to statistics, energy consumption in the transport sector increased from 11,197 KTOE in 2014 to 14,238 KTOE in 2019.

Table 10.10 List of solutions for reducing greenhouse gas emissions in the transportation sector (updated NDC report)

No.	Solution name
1. Energy efficiency	
1.1	Fuel efficiency standards and emissions for new vehicles
1.2	Increase bus usage
1.3	Improve truck load factors
2. Shift in passenger transport modes from private vehicles	
2.1	Expand bus system
2.2	Expand bus rapid transit (BRT) system
2.3	Implement metro system
3. Shift in freight transport modes from road transport	
3.1	Shift from road to inland waterway transport
3.2	Shift from road to coastal transport
3.3	Shift from road to rail transport
4. Fuel substitution	
4.1	Encourage the use of biofuels (E5/E10)
4.2	Promote electric motorbikes
4.3	Introduce CNG buses
4.4	Encourage the use of electric cars and buses

Source: Ministry of Natural Resources and Environment (MONRE). (2020). Viet Nam's Updated Nationally Determined Contribution (NDC). Hanoi, Viet Nam.

- FO: Exclusively used for maritime transport.
- Aviation gasoline: Used solely in the aviation industry.
- Electricity: Currently limited to electric two wheelers and is expected to serve metro systems in the future.

This situation requires the implementation of energy optimization solutions and a shift to clean energy sources to mitigate environmental impacts (Table 10.10).

The EV market in Vietnam has not yet attracted as much attention compared to other countries in the region and globally. However, its development potential remains significant. EVs are an inevitable trend, aligning with the government's focus on using clean energy and environmental protection, especially following the commitments made at COP26. This presents an opportunity for investors to establish a foundation, including production facilities, supply chains, and human resource development, to prepare for the upcoming transition, as forecasted by Frost & Sullivan.

With a population of over 96 mn, Vietnam has a very low car ownership rate, only about 23 cars per 1,000 people in 2015, significantly lower than countries like Malaysia (439), Thailand (228), and Singapore (145). In contrast, motorcycles are the primary mode of transport, with more than half of the population owning one. In Hanoi, there are 6.6 mn vehicles, of which motorcycles account for 5.7 mn. In Ho Chi Minh City, the number rises to nearly 7.9 million vehicles, including more than 7.15 mn motorcycles by 2019.

The consequences of this reliance on motorcycles include severe traffic congestion in major cities like Hanoi city and Ho Chi Minh city, along with increasing air pollution. Air pollution, a serious issue in Vietnam, primarily originates from industrial emissions and vehicles, with motorcycles playing a significant role. Data shows that every year, Hanoi adds around 27,000 new vehicles, including 22,000 motorcycles and 5,000 cars, along with 1.2 mn vehicles from other provinces entering the city for commuting.

These issues highlight the urgent need for a shift toward environmentally friendly transportation, such as EVs, to reduce the negative environmental impacts and improve air quality.

10.5.5.3 Integration of renewable energy sources into electric vehicle systems

The development of EV charging points and the increasing demand for EV charging have prompted research on alleviating pressure on the power transmission system. A common approach is to integrate RES, such as wind and solar energy, into a smart grid model. These studies focus on balancing the benefits between RES-integrated power systems and the economics for EV users. In a smart grid, EVs can function as controllable loads or energy storage devices, supporting vehicle-to-grid technology, which allows EVs to provide energy back to the power grid.

Uncontrolled EV charging can put significant pressure on the distribution network. To address this, many studies have analyzed the coordination between the EV charging process and RES. EVs can also be used to provide ancillary services, helping to integrate RES more effectively into the power grid.

Additionally, WPT is an advanced method of transmitting energy through the air without physical connections. This technology is based on Maxwell's equations from 1862, which demonstrated radio waves and energy flow in Poynting's theorem (1884). Nikola Tesla explored wireless power transmission principles at the end of the nineteenth century, but initial experiments were not commercially successful due to safety issues, low efficiency, and high costs. Later, electromagnetic waves were primarily applied in wireless communication.

The integration of EVs, RES, and WPT holds the promise of breakthrough solutions for the energy and transportation industries in the future.

In dynamic WPT systems for EVs, the structure, coil design, and compensation circuits are key factors to ensure operational efficiency. The characteristic of charging EVs while in motion creates challenges related to power fluctuations, which affect transmission efficiency. This research introduces solutions to mitigate power fluctuations through optimal coil design using Ansys Maxwell simulation software.

In WPT systems, energy is transmitted through the air between the transmitting and receiving coils, leading to low coupling coefficients, high reactive inductance, large reactive power, and unstable efficiency. This study has developed and tested an load compensation circuit (LCC) design on both the primary and secondary sides, as follows:

- **LCC on the primary side**: Designed to:
  – Reduce reactive power.
  – Ensure soft switching (ZVS) conditions for the switches.
  – Optimize efficiency through adjusting the coupling between the transmitting coils.
- **LCC on the secondary side**: Optimized to:
  – Increase the efficiency of power transmission.
  – Ensure stable operation under varying system conditions.

The design solutions were validated through a series of simulations using software such as Ansys Maxwell, Ansys Electronics, and LTspice, as well as practical experiments on prototype models. The results indicate that this dynamic WPT system not only improves transmission efficiency but also minimizes power fluctuations, thereby enhancing its feasibility for widespread application in the future (Figure 10.5).

10.5.5.4 Design of a dynamic wireless charging system for electric vehicles

Overview of the structure

The dynamic WPT system, designed in a segmented manner, offers several benefits, including high efficiency, reduced electromagnetic interference, and compatibility with static charging systems. However, the major drawback is the complex structure, which requires multiple compensation circuits, power converters, and vehicle position detection systems, leading to high costs.

Proposed system structure:

- **Segmented and MODULAR design**:
  – The transmission path is divided into segment modules to optimize energy management.
  – Each primary side module includes three coils, with each coil equipped with its own compensation circuit.
- **Power supply**:
  – The transmission coils in each module are connected in parallel.
  – A single inverter powers the entire system, reducing the number of power converters, cutting costs, and optimizing efficiency.

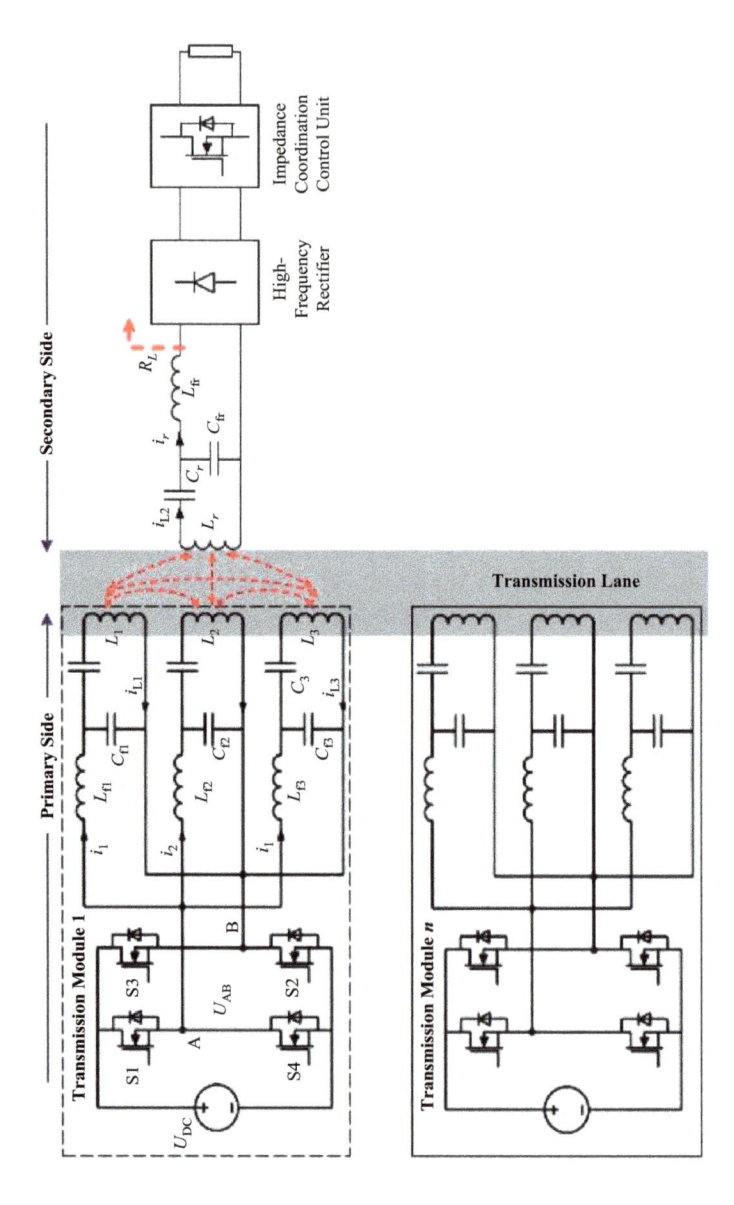

Figure 10.5 Block diagram of the system structure

- **Advantages of the primary side structure**:
  - High energy efficiency: Reduces energy loss through parallel connection and even distribution of power.
  - Simplified design: A single inverter per module reduces system control complexity.
  - Scalability: Modular design allows for easy adjustment of system size and power capacity based on requirements.
- **Secondary side**:
  - The receiving coils on the vehicle are optimized to ensure high transmission efficiency and minimize environmental influences.
  - The LCC is used on the secondary side to reduce reactive power and ensure conversion efficiency.

Illustration: The system structure is simulated (as shown in Figure 10.5) to visualize how the modules are connected and how the system operates during dynamic charging.

This proposed system has been verified through simulations and real-world testing using specialized software such as Ansys Maxwell, Ltspice, and other simulation tools. This ensures that the design meets both technical requirements and optimizes costs and operational performance.

- **Control of modules on/off**: The ability to control the on/off state of the modules based on the receiver's position improves efficiency and reduces electromagnetic interference.
- **Easy expansion**: The transmission path can be extended without requiring changes to the design.
- **Simple compensation design**: The segmented coils reduce peak voltage across each coil, making it easier to select the appropriate capacitors for testing systems, especially for high-power systems.

If the transmission coils are connected in series, only one compensation circuit is needed for each segment, simplifying the system structure. However, due to the high inductance of the transmission coils, capacitor design becomes challenging. Additionally, changing the length of the transmission path would require rewiring and redesigning the compensation circuit.

The number of coils in each transmission module is chosen to be three because the electromagnetic relationships between the coils are sufficient to reflect the characteristics of the dynamic charging path. Therefore, studying a single module provides a comprehensive understanding of the dynamic charging path characteristics.

In a WPT system, the entire section behind the power converter and controller (as shown in Figure 10.6) is replaced with an equivalent DC impedance, RLeq. Wireless energy transfer is considered from the high-frequency inverter input on the primary side to the output on the load impedance RLeq on the secondary side.

450 *Clean energy in South-East Asia*

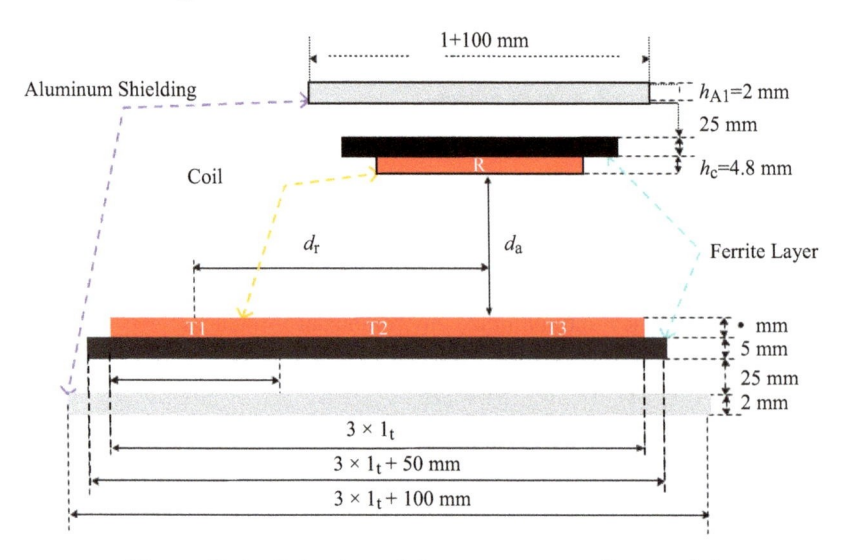

Figure 10.6   Side view of the magnetic coupling module

On the primary side, the DC input voltage (UDC) is converted into high-frequency AC voltage (uAB) by a single-phase inverter, passing through the LCC and transmitted from the primary side to the secondary side via a transformer link. On the secondary side, the AC voltage received from the coils is rectified into DC voltage by a high-frequency rectifier through the LCC and then converted by a control converter to match the equivalent impedance RLeq. The LCC on both sides are used to enhance transmission efficiency, reduce the rated power of devices, and provide soft switching conditions for power electronics. The receiving coil, Lr, is typically mounted under the vehicle, while the transmission coils L1, L2, and L3 are placed under the road. Multiple transmission coils placed next to each other form a dynamic charging lane for the vehicle. The electromagnetic coupling between the transmission coils and between the transmission and receiving coils is also illustrated in Figure 10.6.

10.5.5.5   Transformer design

An overview study shows that the segmented transmission system may encounter power spikes when the vehicle moves between adjacent transmitters, which impacts battery life. Current solutions to reduce power spikes are divided into three categories:

1. Compensation circuit design: This method is effective when the coupling factor changes within a narrow range.
2. Transmission current adjustment: Systems using DC/DC converters before the inverter to adjust the current, but this increases system costs and reduces overall efficiency. Additionally, this method requires knowledge of the mutual

inductance value, a two-way communication system, and is not suitable for dynamic systems with moving vehicles.
3. Coil Design: In coil design, power spikes are reduced by using three half-bridge inverters in parallel with an optimal receiver length. Uniform output is achieved only when the receiver length is 1.5 times the length of the transmitter. In, receivers are designed with multiple stacked coils to achieve uniform output, but this increases costs due to the additional coils, compensation circuits, and power converters. Power spikes are reduced by placing transmission coils next to each other and optimizing the receiver coil size. This structure is suitable for high-power applications.

Among these solutions, the third method is the most suitable for WPT systems used in dynamic charging, and it is actively being researched and developed by scientists. These studies focus on optimizing the sizes of the transmission and receiving coils to reduce power surge. However, results thus far have only been achieved when the vehicle is moving in a straight line, and the coupling characteristics when the vehicle is off-axis have not yet been analyzed.

This section presents a transformer design method that uses finite element analysis (FEA) simulations on Ansys Maxwell to reduce coupling fluctuations when the vehicle is in motion, thereby minimizing power surges. The coupling characteristics when the vehicle is off-axis are also analyzed, which is crucial for understanding the transmission characteristics and implementing advanced controls in the WPT system for dynamic wireless charging of EVs.

10.5.5.6 Transformer design structure

The transformer structure of the transmission and receiving modules is designed as shown in Figure 10.7. The transformer consists of three layers:

1. Layer one: A rectangular single-pole coil. In this design, multi-core copper wire is used to reduce high-frequency AC losses.
2. Layer two: Ferrite bars are placed underneath the coils to guide the magnetic field and enhance electromagnetic coupling between the transmission and receiving coils.

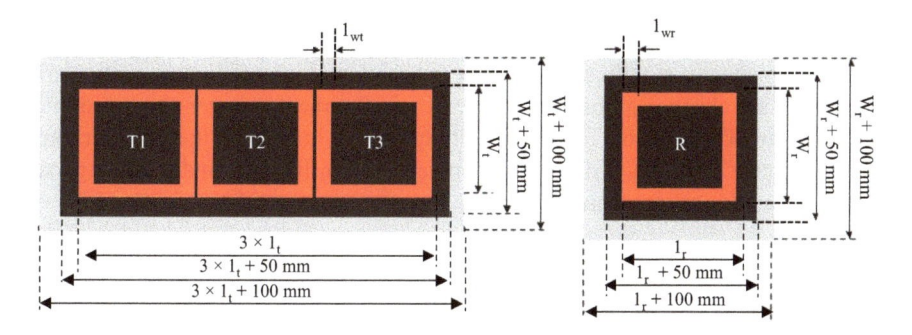

Figure 10.7 Top view of the transmission coil

3. Layer three: An aluminum shielding plate is used to prevent magnetic field leakage into the surrounding environment.

From a top-down view of the transmission coil, the following can typically be observed:

In this study, the theoretical foundations, design methods, and control techniques for the WPT system applied to dynamic wireless charging for EVs are presented in detail.

First, the system is designed with a modular structure on the transmission side, offering several advantages such as flexible transmission path scalability, easy capacitor design, and the ability to control the activation/deactivation of transmission modules based on the vehicle's position. This helps improve efficiency and reduce electromagnetic interference. The magnetic coupling is designed using FEA simulations on Ansys Maxwell to minimize power circuit fluctuations. The average magnetic coupling coefficient is 0.14 when the vehicle moves along the transmission path, with a fluctuation of ±6%. The coupling characteristics when the vehicle is off-axis are also studied to better understand the system's characteristics.

The LCC on both sides is optimized based on load values to maximize transmission efficiency while considering the effect of electromagnetic coupling between the transmission coils. The results show that the resonant frequency is unaffected by the coupling coefficient and load, the soft-switching condition (ZVS) for SiC MOSFETs is ensured, and high transmission efficiency is maintained across a wide frequency range. The output power ripple reaches 9.5%, and the average experimental efficiency of the system is 89.5%.

Based on the system design, advanced control methods to enhance efficiency have been implemented, including resonance tracking and optimal load tracking control. The resonance tracking method is applied when the coil parameters and compensation circuit change from the initial design, helping to narrow the soft-switching ZVS region of the inverter. This method achieves the desired efficiency when parameters change within ±7.5%, allowing the inverter to maintain high efficiency. Additionally, a new, simple, and effective method for estimating the coupling coefficient is introduced, enabling the output power to exceed 95%, and the overall system simulation efficiency surpasses 90%.

10.5.5.7 Optimizing performance in wireless power transfer

WPT systems face a significant challenge in achieving high efficiency, particularly when there are changes in load resistance or misalignment between the transmitter and receiver. This section focuses on studying the impact of these factors and proposes a new control algorithm to optimize system efficiency under varying load conditions and misalignments.

Detailed content

- **Current battery charging techniques**
  - Constant current-constant voltage (CC-CV) method: Widely used, but it has limitations such as longer charging times and reduced battery lifespan.

Social impacts and trends in RE development in developing countries 453

- – Pulse Charging: Helps distribute ions evenly in the battery electrolyte, reducing polarization and accelerating charging speed while improving battery life.
  – Sinusoidal Charging: Known for faster charging speed and lower heat generation. However, it presents challenges in impedance matching due to continuously changing impedance.
- **Challenges in WPT systems**
  – Efficiency drops when the load changes or there is misalignment between the transmitter and receiver.
  – The impedance at the load side (battery) changes continuously as the state of charge varies, complicating the task of optimizing impedance to maintain high efficiency.
- **Techniques to improve WPT efficiency**
  – Using rectifiers and DC–DC converters to adjust the load and source impedance for optimal performance. However, these solutions often require additional electronic components, increasing complexity and cost.
  – Some algorithms for maximum efficiency tracking (MET) adjust the phase or frequency of the inverter to optimize efficiency.
- **Proposed solution**
  – Hybrid control algorithm: Combines phase and frequency adjustment of the resonant inverter to reduce input power and optimize efficiency.
- **Key benefits**
  – Speeds up the battery charging process.
  – Optimizes efficiency in real-time when load or misalignment changes.
  – No need for additional conversion stages.
  – Maintains performance even when system parameters and positions change.

This work proposes a new control method that improves the efficiency of WPT systems under real-world conditions, minimizes costs, and enhances energy utilization efficiency.

10.5.5.8 Limitations and future research directions
Future research focus

Continue to expand transmission paths and implement control to switch transmission modules based on the position of the receiver to enhance efficiency.

Improve the structure of the optimal load tracking control converter. Currently, using diode rectifiers and boost converters reduces system efficiency. It is proposed to replace them with synchronous rectifiers to enhance efficiency.

The optimal load tracking control method is applied when the coupling coefficient (krk_rkr) and load change as the vehicle moves, increasing transmission efficiency by 6% compared to using fixed impedance control.

Finally, propose a power control method to accommodate different charging power levels for EVs moving simultaneously on the transmission lane. This method only requires control on the primary side, ensuring flexibility and high efficiency.

10.5.5.9 Planning, implementation, and operation of electric vehicle systems and charging infrastructure

To effectively plan and operate EV systems on roads, particularly the charging infrastructure, understanding the current demand and forecasting future potential at different stages is crucial. This study aims to develop a method to calculate the charging demand of EVs and establish the process for setting up public charging stations.

Research focus

- What types of vehicles will be used, and how many?
- Where, when, and for how long will the vehicles be charged?
- What factors influence the deployment of charging stations?

To support the electromobility system, a survey is necessary to accurately determine current and future charging demands. This helps identify the optimal locations for public charging stations.

Proposed calculation method: The introduced method is modular, allowing application to any type of EV or fleet and adaptable to any territory and time. This method helps determine:

- The frequency of charging,
- The duration of charging,
- The utilization rate of the chargers.

The results of the calculations can be visualized on a geographic map, aiding decision-making at various levels, from urban planning and EV service development to the operation of the power grid.

Future research directions

- Expectations for charging infrastructure,
- Requirements for construction and installation,
- Relevant legal frameworks.

10.5.6 Application of adaptive control research in energy management systems at Vietnamese ports

10.5.6.1 Model application

This section would introduce a model for applying adaptive control to energy management systems at ports in Vietnam. The adaptive control approach is designed to enhance the efficiency of energy use in port operations, accounting for variables such as port activity, energy demand, and available RES. Future research in this area will continue to develop optimized energy management strategies for improving sustainability and reducing costs (Figure 10.8).

The study focuses on the adaptive droop control method in DC microgrid systems, aiming to:

- Efficiently share power between RES at the port and vessels.
- Maintain DC voltage stability, particularly in environments with significant load fluctuations, such as ports and boat docking areas.

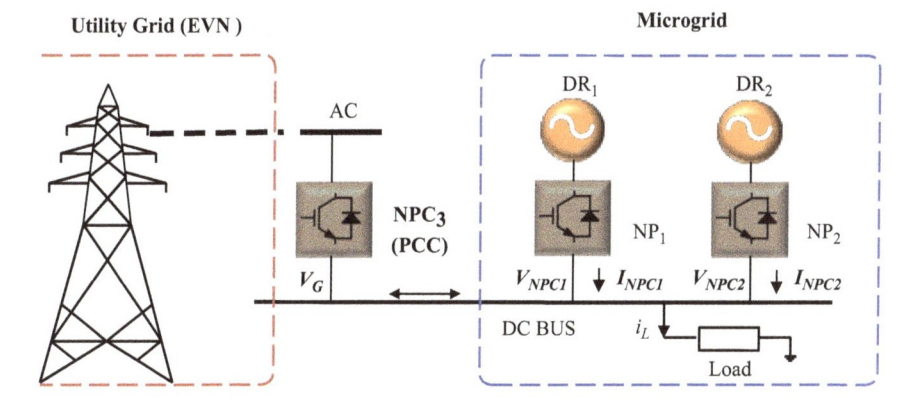

Social impacts and trends in RE development in developing countries 455

Figure 10.8 Grid-connected operation

This method is suitable for:

- Vessels integrated with renewable energy: Fishing boats, cruise ships, and cargo vessels with solar, wind, or storage battery systems. Sharing energy between sources on the vessel ensures optimized usage and reduces dependence on fossil fuels.
- Smart port systems: Ports need to manage energy from multiple vessels and provide a stable power grid for service operations, such as crane operation, electric boat charging, and logistics support equipment.

10.5.6.2 Port conditions in Vietnam and applications

- **Actual conditions at Vietnamese ports**
   - Large seaports: Vietnam has many large seaports such as Cai Mep–Thi Vai, Hai Phong city, and Da Nang city, with high energy demands to support logistics operations and docked vessels.
   - Small fishing and cruise vessels: Vietnam has thousands of fishing boats and cruise ships with integrated renewable energy, but energy management capabilities are limited. Energy support from the port could optimize the operational efficiency of these vessels.
   - Limited communication systems: Some ports are not equipped with advanced communication systems for managing distributed energy, making it difficult to coordinate power sources and loads.
- **Application of the research to port operations**
   - Energy sharing between vessels and ports
      o Real-world case: At a port, some vessels with excess energy from solar panels can transfer surplus electricity to the port, supplying power to port equipment or charging other vessels.
      o Benefits: Maximizing renewable energy use, minimizing waste, and reducing reliance on the national grid.

- DC voltage stability in microgrid systems:
  - Real-world case: Port operations, such as crane operation and electric boat charging, can cause significant load fluctuations. The adaptive control method ensures stable DC voltage, preventing imbalance when load devices change suddenly.
- Distributed energy management:
  - Real-world case: Each vessel and port area has independent energy sources. The adaptive control method can use network communication to manage these sources effectively, ensuring proper power sharing and system stability.

10.5.6.3 Challenges and recommendations

Challenges

- **High initial implementation costs**
  - Investment in distributed control systems, communication devices, and renewable energy infrastructure requires significant capital.
- **Limited communication infrastructure**
  - Many ports in Vietnam do not have advanced communication systems to manage energy in real-time.

Recommendations

- **Start with small-scale trials**
  - Begin with medium and small ports (e.g., cruise ship ports or fishing ports) to assess the effectiveness of the system.
- **Integrate modern communication technology**
  - Develop communication systems through Ethernet or wireless networks to support distributed control.
- **Combine with other control technologies**
  - Develop advanced adaptive control methods, such as composite adaptive control, to improve efficiency in fluctuating conditions.

The study on adaptive control holds great potential for energy management at Vietnamese ports. This system not only addresses the challenge of energy sharing and stability between vessels and ports but also supports the development of smart ports, optimizing operational performance and reducing dependence on fossil fuels. Implementation should be tested and adjusted according to real-world conditions to ensure effectiveness and feasibility.

10.6 Green hydrogen from wind and solar power via water electrolysis: an advanced solution for sustainable energy

For decades, fossil fuels have played a crucial role in energy supply, becoming the key resource for production and daily life. Currently, energy sources such as coal, oil, and natural gas account for nearly 80% of the global energy demand. However, according to the Fourth Assessment Report (AR4) of the Intergovernmental Panel on Climate Change, GHG emissions caused by human activity have been the primary driver of global warming since the mid-twentieth century. Long-term dependence on fossil fuels not only severely harms the environment but also negatively affects human health and ecosystems.

Moreover, with current consumption rates, global oil reserves are predicted to be depleted within the next 50 years. In response to this challenge, hydrogen has emerged as a promising energy source with the potential to meet sustainable energy needs and reduce harmful emissions through advanced production technologies.

One prominent approach is the development of solar-powered water electrolysis technology to produce green hydrogen. This solution aligns with the goals of the Paris Agreement, which aims to limit the global temperature increase to below 2.0°C by 2030 and 1.5°C by 2050. The European Green Deal also targets a reduction of at least 55% in GHG emissions compared to 1990 levels by 2030, requiring a significant increase in the use of RES such as solar and wind power.

However, due to the intermittent nature of renewable energy, effective storage solutions must be developed to prevent waste. One creative approach is to directly integrate low-cost photovoltaic modules with electrolysis devices. This technology, tested both in laboratory and real-world conditions, has shown high efficiency in converting solar energy into hydrogen (StH), promising widespread application and reasonable costs, especially in regions with mild climates.

The water electrolysis reaction occurs through two processes: the hydrogen evolution reaction and the oxygen evolution reaction. The general equation is:

$$2H_2O \rightarrow 2H_{2(gas)} + O_{2(gas)} 2H_2O \rightarrow 2H_{2(gas)} + O_{2(gas)} 2H_2O$$
$$\rightarrow 2H_{2(gas)} + O_{2(gas)}$$

The minimum voltage required for the process is 1.23 V under standard conditions, but a higher voltage is needed in practice to overcome energy losses.

The project has designed photovoltaic modules to ensure sufficient voltage for water splitting without external assistance, while also developing a model to optimize hydrogen yield based on real-world data and climate conditions (Figure 10.9).

10.6.1 Current hydrogen production methods

Water electrolysis for hydrogen production: Hydrogen can be produced through water electrolysis, where water is split into hydrogen and oxygen. This hydrogen can be utilized in combustion processes or hydrogen fuel cells (HFCs) to generate electricity. Since the cost of water is almost negligible, the overall production cost primarily depends on the cost of electricity. To enhance system efficiency, catalysts are dissolved in the water to improve electrical conductivity, which in turn increases hydrogen production. Although hydrogen production through water electrolysis has great potential, this technology still faces several challenges.

458   *Clean energy in South-East Asia*

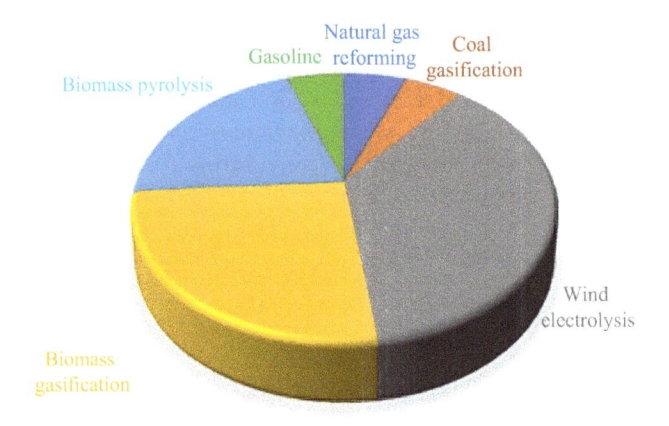

Figure 10.9   Overview of hydrogen production costs by different methods

- **Wind energy-based production**
  A more environmentally friendly method of producing hydrogen is using wind energy (WTH) combined with electrolysis. Excess electricity from wind can be stored as hydrogen, which can be converted back into electricity when wind speeds are low or when the grid is congested. In addition to its storage capacity, the WTH system has significant potential for use in light vehicles as a clean fuel source, helping to reduce GHG emissions.
- **Solar energy-based production**
  Solar energy is the largest energy source on the planet; however, less than 0.06% of global electricity demand is currently met by solar power. To address future energy shortages and increase solar energy usage, a proposed solution is the solar hydrogen system (SHS). However, the SHS system faces challenges, including suboptimal efficiency (only 8–14% of the energy from photovoltaic panels is used for water electrolysis) and high installation costs for photovoltaic systems, making it not yet feasible for commercial-scale deployment.
- **Biological hydrogen production**
  Agricultural and industrial waste, food waste, wastewater, and household solid waste are major global issues. To address these waste problems, biohydrogen technology has been implemented. This technology uses anaerobic microorganisms, such as fermenting bacteria, anaerobic bacteria, and cyanobacteria, to process environmental waste and produce biohydrogen. The amount of hydrogen produced depends largely on the input material and the fermentation pathways of the bacteria. Biohydrogen production is considered environmentally friendly and can be applied to various organic waste types. However, the hydrogen yield from this process remains limited.
- **Fast pyrolysis-based production**
  Fast pyrolysis occurs at high temperatures (300–700°C) with rapid heating rates (10–200°C per second) and short residence times, making it a preferred

method for hydrogen production. In this process, biomass is rapidly heated in an oxygen-free environment to produce vapor, which then condenses into dark brown bio-oil. The products of fast pyrolysis include gases such as CH_4, H_2, CO, CO_2, and others depending on the organic nature of the biomass. To enhance hydrogen yield, a product separation mechanism is applied, dividing the product into two types: water-soluble products used for hydrogen production and water-insoluble products used to make binders.
- **Supercritical water gasification (SCWG)**
Under normal conditions, water cannot dissolve biomass materials. Therefore, in this method, water is converted into a supercritical state. SCWG plays a crucial role in chemical reactions during the gasification process by enhancing the reactivity between organic materials. Superheated water molecules participate in basic reaction steps, acting as reactants or catalysts. When superheated water reacts with biomass, it produces a mixture of CO and H_2 in the gasification process.

10.6.2 The role of hydrogen in renewable energy

10.6.2.1 Hydrogen as a clean energy source

Hydrogen is considered a clean energy source because when used in fuel cells or combustion processes, the only byproduct is water. This helps reduce GHG emissions and air pollution, contributing to environmental protection. Hydrogen can be produced from RES like solar and wind through water electrolysis, converting renewable energy into a storable and usable form.

Combining hydrogen with other renewable energy sources (wind and solar): Hydrogen can be used to store energy from renewable sources like wind and solar, which are variable over time. When renewable energy is abundant, it can be converted into hydrogen through electrolysis. The hydrogen can then be stored and used when needed, helping to balance energy supply and demand. Additionally, hydrogen can be used in fuel cells to generate electricity, providing a stable and reliable energy source from renewables. This combination creates a sustainable energy system, reducing dependence on fossil fuels and minimizing environmental impact.

10.6.2.2 Hydrogen production methods

- **Hydrogen production from water (electrolysis)**
Water electrolysis is the process of using electricity to split water (H_2O) into hydrogen (H_2) and oxygen (O_2). This occurs in a device called an electrolyzer. When electricity passes through water, water molecules are split into hydrogen and oxygen ions, and the hydrogen ions combine to form hydrogen gas. This method is clean if the electricity source is from renewable energy.
- **Hydrogen production from natural gas**
The most common method today, which accounts for about 95% of global hydrogen production, is steam methane reforming. In this process, natural gas (mainly methane, CH_4) reacts with steam at high temperatures to produce

hydrogen, carbon monoxide, and a small amount of carbon dioxide. While this method is efficient and cost-effective, it emits CO_2 and is not entirely environmentally friendly.

- **Hydrogen production from biomass**
Biomass, such as crops or organic waste, can be used to produce hydrogen through processes like gasification or biological decomposition. Gasification converts biomass into syngas (synthesis gas), which contains hydrogen, carbon monoxide, and carbon dioxide. This syngas is then processed to extract hydrogen. Biological decomposition uses microorganisms to break down biomass into methane, which is then reformed into hydrogen. This method has great potential as it uses renewable raw materials.

10.6.2.3 Applications of hydrogen in renewable energy systems

- **Energy storage systems**
Hydrogen can store energy from renewable sources. When renewable energy is abundant, such as on sunny or windy days, it is used to produce hydrogen through electrolysis. The hydrogen can be stored and used when needed, helping to balance supply and demand and ensuring continuous energy availability.
- **Hydrogen fuel cells**
HFCs convert the chemical energy of hydrogen into electricity through an electrochemical reaction. These cells are not only efficient but also clean, with water as the only byproduct. Fuel cells can be used for small electronic devices, transportation such as cars and buses, and even power plants.
- **Hydrogen in electricity production**
Hydrogen can be used in gas turbines to generate electricity, either entirely with hydrogen or in a blend with natural gas. This helps reduce GHG emissions and air pollution while increasing the flexibility and reliability of the power system.

10.6.2.4 Benefits of using hydrogen

- **Reducing carbon emissions**
When hydrogen is burned or used in fuel cells, the only byproduct is water, reducing CO_2 and other GHG emissions. Using hydrogen in industry and transportation can significantly reduce harmful emissions, contributing to environmental protection and improving quality of life.
- **Enhancing energy efficiency**
Hydrogen has a high energy storage capacity, allowing efficient use of RES like wind and solar. When renewable energy is converted into hydrogen and stored, it can be used at any time, ensuring stable and continuous energy supply, optimizing energy system performance.
- **Renewability and reusability**
Hydrogen can be produced from various sources like water and biomass, and the production process is potentially limitless. This makes hydrogen a

sustainable energy source that will not deplete like fossil fuels. The byproducts of hydrogen production and use, primarily water, can be reused, conserving natural resources and minimizing environmental impact.

10.6.3 Groundbreaking ceremony for the first green hydrogen production Plant in Tra Vinh—March 30, 2023

On the morning of March 30, 2023, TGS Tra Vinh green hydrogen joint stock company held a groundbreaking ceremony for the construction of its green hydrogen production plant in Dong Hai commune, Duyen Hai district, Tra Vinh province. The ceremony was attended by local leaders, representatives from ministries, and energy and industrial experts.

The Tra Vinh Green Hydrogen Production Plant is being developed at the K8 mudflat area, Dong Thanh hamlet, Dong Hai commune, Duyen Hai district, covering more than 20.7 ha, with a total investment of up to 7,856 bn VND. This is one of the key projects in Vietnam's renewable energy sector and the country's first plant to produce green hydrogen using RES.

The plant will use seawater electrolysis technology, with electricity from renewable sources like solar and wind energy used to split water into hydrogen and oxygen. With a design capacity of 24,000 tons of hydrogen and 195,000 tons of oxygen annually, the plant is expected to begin operation in the first quarter of 2024. The hydrogen produced will initially be exported to international markets, as the domestic hydrogen market is not yet fully developed.

At the ceremony, Deputy Minister of Industry and Trade Nguyen Sinh Nhat Tan emphasized that the production of green hydrogen from renewable energy is part of Vietnam's strategy to develop the chemical industry. In recent years, green hydrogen has become a strategic energy source, expected to be a key element in the global energy transition toward sustainability, environmental friendliness, and reduced GHG emissions to combat climate change.

Moreover, the establishment of the green hydrogen plant in Tra Vinh will contribute to the local economy and help Tra Vinh become one of the prominent centers of green energy in the Mekong Delta and nationwide. The plant will not only produce hydrogen but also generate employment, promote research and development in renewable energy, and support the growth of industry and infrastructure in Tra Vinh.

Achieving these goals requires strong collaboration between the government, businesses, and the community. There is a need to invest in hydrogen technology research and development, build infrastructure, and create supportive policies. Together, we can drive the use of renewable energy and build a sustainable future for future generations.

The use of hydrogen in renewable energy systems offers both environmental benefits and the potential for sustainable development. With technological advancements and policy support, hydrogen is poised to become an integral part of the clean energy solution, contributing to the protection of the planet.

10.6.4 Conclusion

While the world is facing significant challenges due to the COVID-19 pandemic, addressing energy issues to ensure a stable foundation for the fight against the pandemic is crucial and urgent. Vietnam needs to make appropriate development decisions during this phase of renewal to build a national energy network based on its geographic advantages, strong agriculture, and recent innovations. Raising awareness of a new, versatile, low-emission energy source will create a solid foundation for the country's sustainable development. Recent resolutions from the 13th Party Congress emphasize the adoption of a circular economy in national development, laying the groundwork for Vietnam to become a country focused on green, clean, and sustainable growth.

Further reading

[1] Balat H, Kırtay E. (2010). Hydrogen from biomass – Present scenario and future prospects. *Int J Hydrogen Energy*, 35, 7416–7426.

[2] Bhanu Mahajan (2012). *Negative Environment Impact of Solar Energy. Environmental Science and Policy Course. CEPT University*, India. Energy Policy, 33.

[3] Damon Turney and Vasilis Fthenakis (2011). Environmental impacts from the installation and operation of large-scale solar power plants. *Renewable and Sustainable Energy Reviews*. 15(6), 3261–3270.

[4] DOE (2011). *A National Offshore Wind Strategy: Creating an Offshore Wind Energy Industry in the United States*. 52 pp.

[5] GIZ (2018). Guidelines on Environmental and Social Impact Assessment for Wind Power Projects in Vietnam. Available at: http://gizenergy.org.vn/media/app/media/GIZ-ESP_ESIA%20Report_ENG_Final.pdf

[6] GSO (2006–2020). *Almanac Statistics of Vietnam Reports in 2006 – 2020*. General Statistics Office of Vietnam, in Vietnamese.

[7] IFC (2012). *Guidance Notes to Performance Standards on Environmental and Social Sustainability*. Washington, DC: International Finance Corporation. Available at: https://www.ifc.org/wps/wcm/connect/9fc3aaef-14c3-4489-acf1-a1c43d7f86ec/GN_English_2012_Full-Document_updated_June-14-2021.pdf?MOD=AJPERES&CVID=nXqnsJp

[8] International Energy Agency (2023). *Global EV Outlook 2023, Catching up with Climate Ambitions*; International Energy Agency: Paris, France, 2023.

[9] John Glasson, Riki Therivel and Andrew Chadwick (1999). *Introduction to Environmental Impact Assessment*. 2nd edition in 2005, UCL Press, the Taylor & Francis e-Library.

[10] Jose I. Bilbao, Garvin Heath, Alex Norgren, Marina M. Lunardi, Alberta Carpenter, Richard Corkish (2021). *PV Module Design for Recycling*. International Energy Agency (IEA), PVPS Task 12, Report T12-23:2021. Available at: https://iea-pvps.org/wp-content/uploads/2021/10/T12_2021_PV-Design-for-Recycling-Guidelines_Report.pdf

[11] Luca Ciacci and Fabrizio Passarini (2020). *Life Cycle Assessment (LCA) of Environmental and Energy Systems*. Energies, 13(22), 5892; https://doi.org/10.3390/en13225892

[12] Matt Rogers (2019). *Vietnam's renewable energy future*. McKinsey Sustainability. Available at: https://www.mckinsey.com/capabilities/sustainability/our-insights/sustainability-blog/cop27-accelerating-decarbonization

[13] Md Shahariar Chowdhury, Kazi Sajedur Rahman, Tanjia Chowdhury, *et al.* (2020). An overview of solar photovoltaic panels' end-of-life material recycling. *Energy Strategy Reviews*, 27, 100431.

[14] Ministry of Natural Resources and Environment (2010). *National Technical Regulation on Vibration*. Regulation No. QCVN 27:2010/BTNMT, signed on 16 December 2010.

[15] Ministry of Natural Resources and Environment (2015). *On Strategic Environmental Assessment, Environmental Impact Assessment, and Environmental Protection Plan*. Circular No. 27/2015/TT-BTNMT, signed on 29 May 2015. Available at: https://datafiles.chinhphu.vn/cpp/files/vbpq/2015/06/27-tt-btnmt.signed.pdf

[16] Ministry of Natural Resources and Environment (2011). Regulation on Strategic Environmental Assessment, Environmental Impact Assessment, and Environmental Protection Commitment. Decree No. 29/2011/ND-CP dated 18 April 2011. Available at: https://chinhphu.vn/default.aspx?pageid=27160&docid=100006

[17] National Institute of Environmental Health Sciences (1999). *Health Effects from Exposure to Power-Line Frequency Electric and Magnetic Fields*, Research Triangle Park, NC.

[18] Nguyen Minh Quang (2022). *Electricity Industry Update Report*. Oil and Gas Securities, August 2022. Available at: https://i.ndh.vn/attachment/2022/09/19/psi-bao-cao-nganh-dien-082022-pdf.pdf

[19] Pablo Hevia-Koch and Henrik Klinge Jacobsen (2019). Comparing offshore and onshore wind development considering acceptance costs. *Energy Policy* 125 (2019), p. 9–19. Available at: https://www.sciencedirect.com/science/article/abs/pii/S030142151830675X

[20] Prime Minister (2016). *Approving the Adjustment of the National Power Development Plan for the Period 2011–2020 with a Vision to 2030*. Decision No. 428/QD-TTg, signed and issued on 18 March 2016. Available at: https://datafiles.chinhphu.vn/cpp/files/vbpq/2016/03/428.signed.pdf

[21] Prime Minister (2015). *Regulations on Environmental Protection Planning, Strategic Environmental Assessment, Environmental Impact Assessment and Environmental Protection Plans*. Decree No. 18/2015/ND-CP, signed and issued on 14 February 2015. Available at: https://datafiles.chinhphu.vn/cpp/files/vbpq/2015/02/18.signed_01.pdf

[22] SRV (2022). *Nationally Determined Contribution (NDC report)* (updated in 2022). Socialist Republic of Viet Nam, Ha Noi, 38 p.

[23] Srinivasan, M., A. Velu, and B. Madhubabu (2019). Potential environmental impacts of solar energy technologies. *International Journal of Science and Research (IJSR)*, 9(5), 792–795.

[24] Theocharis D Tsoutsos, Niki Frantzeskaki and Vassilis Gekas (2005). Environmental impact assessment of solar energy systems. *Energy Policy*, 33 (3), 289–296. Available at: https://doi.org/10.1016/S0301-4215(03)00241-6

[25] Tran Thien Khanh (2021). Orientation for the Development of Clean Hydrogen Energy: Analysis of the Current Situation and Evaluation of Industrial Hydrogen Production Methods. Available at: https://iced.org.vn/dinh-huong-phat-trien-nang-luong-sach-hydro-phan-tich-thuc-trang-va-danh-gia-cac-phuong-phap-san-xuat-hydro-trong-cong-nghiep/

[26] US. Embassy in Vietnam (2022). *International agreement to support Vietnam's climate and energy goals*. Available at: https://en.baoquocte.vn/international-agreement-to-support-vietnams-climate-and-energy-goals-209960.html

[27] VCBS (2023). Electricity Industry Outlook Report 1H.2023: Electricity Consumption Demand to Grow Steadily in the Long Term. Available at: https://finance.vietstock.vn/bao-cao-phan-tich/10735/bao-cao-trien-vong-1h-2023-nganh-dien-nhu-cau-tieu-thu-dien-tang-truong-on-dinh-trong-dai-han.htm

[28] World Bank (2015). *Environmental, Health, and Safety (EHS) Guidelines*. IFC E&S Washington, D.C. Available at: https://documents.worldbank.org/en/publication/documents-reports/documentdetail/157871484635724258/environmental-health-and-safety-general-guideline

[29] World Health Organization (2007). *Extremely Low Frequency Fields*. Environmental Health Criteria Monograph No. 238, Geneva.

[30] Ngo Dang Luu, Nguyen Hung, Nguyen Anh Tam, and Long D. Nguyen, (2021). "Wireless Power Transfer simulation for EV", Ministry of Culture, Sports and Tourism.

[31] Ngo Dang Luu, Nguyen Hung, and Long D. Nguyen, (2021). "Power flow calculation of the electric power system by MATLAB", Ministry of Culture, Sports and Tourism.

[32] Ngo Dang Luu, Nguyen Hung, Nguyen Anh Tam, and Long D. Nguyen, (2022). "Distance Relay Protection in AC Microgrid", Ministry of Culture, Sports and Tourism.

[33] Ngo Dang Luu, Nguyen Hung, Nguyen Anh Tam, and Long D. Nguyen, (2022). "Analysis of Solar Photovoltaic System Sunlight Shadowing", Ministry of Culture, Sports and Tourism.

[34] Ngo Dang Luu, Nguyen Hung, Le Anh Duc and Long D. Nguyen, (2022), "Estimate Frequency Response at the Command Line", Ministry of Culture, Sports and Tourism.

Index

Adam optimizer 81
adaptive neuro-fuzzy inference system (ANFIS) 143
 identification 146–7
 online ANFIS intelligent controller 144
 online training of 147–8
 structure 144–6
Add sim900 module 440
advanced distribution engineering productivity tool (ADEPT) 261–2
advanced metering infrastructure (AMI) 157
aesthetic differences 36
AI-based techniques 74
air circuit breaker (ACB) 70
Airdolphin Mark-Zero wind turbine 5
air quality impacts 432
alternating current (AC) 10, 42, 94, 164
 transmission 87–8
amorphous silicon (a-Si) 36–7
 solar modules 38
anti-islanding 164
artificial intelligence (AI) 59, 213, 215
 and automation in smart grids on cloud computing platforms 113
  improving reliability index set 113–14
  load transfer and grid upgrades for DAS project 118–20
  recommendations 120
  reduction of outage duration 113
  structure of DAS with FDIR function 114–18
 and automation in smart grids on cloud platforms 120
  ANN structure for fault location 121–2
  common faults in transmission lines 120–1
  fault location challenges on lines with series compensation 123
  fault location procedure using ANN 121
  key data features 122
  multilayer perceptron technique 122
  performance evaluation 122
  research methods and techniques 121
  research needs and objectives 121
  role of transmission lines 120
  types of faults studied 121
 for calculating and estimating inertia constants of power system 59
  current status 62
  development of renewable energy and impact on inertia of power system in Vietnam 60–2
  overview of power system inertia 59–60
 and data analysis for fault detection and prediction in multi-source energy systems 59

artificial neural networks (ANN) 121, 215
　application of 121
　fault location procedure using 121
　structure for fault location 121–2
avoided cost method 319

backpropagation algorithm 148
backstepping techniques 167
Bac Lieu WPP Bac Lieu Province 5
battery storage technologies 71–2
Bayesian neural network (BNN) 215
bell-shaped membership function 147
big data processing 440
Binh Dinh Province 5
biological hydrogen production 458
biomass energy systems 163
biotechnology applications 412
Bluetooth 219

cadmium telluride (CdTe) 34, 36
centralized FDIR structure 114–15
central processing unit (CPU) 137
chemical processing 400
circular economy (CE) 385
　approach after end-of-life of materials in solar and wind energy equipment 387
　approaches after end-of-life use 387–8
　case studies in Vietnam and globally 388–9
　challenges of solar and wind energy equipment after end-of-life 387
　approaches to managing waste from renewable energy for circular economy and sustainable development 411
　creating value from waste 413
　developing recycling and processing technologies 412
　encouraging adoption of circular economy model 412
　managing waste from renewable energy 411–12
　use of recycled materials in production 412–13
　case studies in Vietnam 386
　eco-industrial parks 386
　pioneering foreign direct investment companies 386
　waste-to-energy conversion 386
　in renewable energy development 385
　developing solar and wind energy 386
　recycling alliances and business cooperation 386
　using renewable energy in industrial production 386
　using waste as raw material for renewable energy production 385
　solar panel waste 391
　benefits of recovering and reusing materials from solar panels 395
　estimated amount of materials recovered from recycling 393
　estimated recovered materials from recycling 394
　material recycling rates and potential 393–4
　proportion of material weight and recycling potential of solar panel materials 392–3
　in Vietnam 392
　solar photovoltaic panels 389
　challenges in solar panel waste treatment 390

Index 467

researching structure for waste treatment 390–1
structure of 389
solutions for solar panel waste 395
  chemical method 410–11
  design and manufacture of aluminum frame and junction box removal machine 405–6
  experimental research process 403–4
  general cycle of solar energy waste recycling technology 395–6
  machine for separating aluminum frame and junction box on PV module 406–9
  position of aluminum frame and junction box on PV module 406
  removing aluminum frame and junction box 404–5
  scope of research 403
  solar energy waste recycling technology 395
  solar panel recycling technology worldwide 398–402
  stages of degradation of EVA adhesive in solar modules 409–10
  study on waste flow of solar panels from solar power plants 396–8
climate data analysis 373
cloud computing 170–2
cloud management 172
cloud technology 170–2
communication infrastructure 164
compensators 22
competitive bidding 341
composite materials 388
computer software 238–9
connection power 259–61
constant current-constant voltage (CC-CV) method 452

construction costs 368
construction investment consulting costs 369
contingency costs 369
controlled valve 88
conventional coordination FDIR structure 114
convolutional neural network (CNN) 215
copper indium gallium selenide (CIGS) 37–8
copper indium selenium (CIS) solar cells 37–8
crystalline silicon (c-Si) technology 38
Cu Lao Re 193
current fault location process 124
cybersecurity 157, 159

Dam An Khe Solar Power Plant 263, 266
Dam Nai Wind Energy Project 372
Dam Tra O Solar Power Plant 34
Da Nang city power grid
  DAS project with FDIR functionality at 120
  hardware configuration of 117
data analytics 157
data and research methodology 75
  research data 75–6
  research methodology 76
    Holt–Winters model 76–7
    N-BEATS block 79
    N-BEATS model 79
    Prophet model 78
    SARIMA model 77–8
databases (DBs) 216
data fitting 231
data linking software 109
data processing 75–6
Dau Tieng Solar Power Plant 32–3

DC–DC converters 453
decision trees (DT) 215–16
deep learning 79
demand response programs 159
demand-side management 333
Deutsche solar 402
device DB 235
diesel generator and transformer parameters 200–1
differential protection 129
Dinh Binh Hydroelectric Power Plant 284
direct current (DC) 10, 42, 164
distance protection (F21) 128
distributed energy resources (DERs) 163
distributed FDIR structure 114
distributed STATCOM (D-STATCOM) 143
Distribution Automation System (DAS) 107
  architecture model of 108
  definition and function of 111
  fault location 110
  load transfer and grid upgrades for 118–20
  model 111
  operating principle 111
  operational experience with 117
  software 109
  transfer method 110
  transfer operation–semi auto 110
distribution grid 112
distribution management system (DMS) 109
distribution reliability analysis (DRA) 261
domestic wastewater 432
doubly fed induction generators (DFIG) 167

dual-glass solar panels 40–1
dual-layer or triple-layer a-Si cells 38
Duc Pho District distribution grid 266
Duc Pho District electrical grid 264
Duc–Thu Duc North–Intel line 150–1
dynamic wireless charging 442

economic cost optimization function 186
economic efficiency 346
economic impact 330
  attracting investment and development 330
  job creation 330
electrical grid system 195
electricity forecasting
  long-term load forecasting 73
  medium-term load forecasting 73
  overall comparison 84–5
  real-time operation 73
  short-term load forecasting 73
  South Korea 83–4
  Spain 83
  Thailand 82–3
  Vietnam 81–2
electricity law 296, 297
Electricity of Vietnam (EVN) 337
electricity operation license 296, 297
Electricity Regulatory Authority of Vietnam (ERAV) 253, 338
Electric Reliability Council of Texas (ERCOT) 257
electric vehicles (EV) 185
  charging algorithm 189
  power balance graph with 191
  wireless charging solution for 441
electromagnetic clutch (EMC) 166
electromagnetic fields (EMF) 432
energy efficiency 346

energy estimation 378
energy management system (EMS) 163, 164, 188–9
energy savings 395
energy storage 44
  and management 346
  optimization 333
energy storage systems (ESS) 71, 163, 184, 460
energy storage technologies 71
  battery storage technologies 71–2
  classification based on application 71
  due to increasing variability from the integration of large-scale 71
environmental impact 330, 434
  protecting natural resources 330
  reducing pollution 330
environmental pollution 169
environmental protection 395
  policies 342
environmental sensors 218
equipment costs 368–9
ethylene vinyl acetate (EVA) layer 389
Euclidean neural network (ENN) 215
evaluation metrics 80
exponential smoothing (ETS) 74

fast pyrolysis-based production 458–9
fault current limiters (FCLs) 87
fault detection, isolation, and restoration (FDIR) 112
fault detection module 59, 138
fault diagnosis process 243–5
fault location 110
fault location, isolation, and service restoration (FLISR) function 109
fault prediction 59
fault-tolerant control (FTC) 22
feasibility assessment 403

feed-forward networks 215
Feed-in tariffs (FiT) 318
  benefits of competitive bidding 326
  challenges and solutions 326
  definition of 318
  policy recommendations 327
  pricing method 319
    existing methods and their advantages and disadvantages 319
    FiT price calculation process 321–3
    objective function for FiT price calculation 319–20
    simulation of FiT calculation process considering specific changing factors 320–1
  pricing policy 325
    for solar power 326
    for wind power 325–6
  results 323–5
financial mobilization 345
flexible alternating current transmission system (FACTS) 86, 141
flexible electricity pricing 341
fog computing 172
foreign direct investment (FDI) 337
fossil fuels 7
Fourth Assessment Report (AR4) 457
Fourth Industrial Revolution 213
frequency regulation 71
frequency response analysis 64
fuzzy logic (FL) 143, 215–16, 347

Gamesa G114 373, 381
Gaussian membership function variables 147
GE F650 relay 137–41
generalized high-level disturbance observer (GHODO) 22–3
Global Horizontal Irradiation (GHI) 32

gradient descent method 147
graph DB 235
grid-connected mode 164
grid regulations 296–8
grid-tied rooftop solar power system 47
grid-tied solar power system 43–4

half-cell solar panels 39
　dual-glass solar panels 40–1
　increased lifespan and durability 40
　optimal performance under shaded conditions 40
　reduced current intensity on busbars 39–40
　smaller busbar size increases light absorption efficiency 40
　structure 40–1
half-cell technology 38
　benefits of 39
Hamiltonian methods 167
hazardous waste 432
Heineken Vietnam 386
high-temperature superconducting (HTS) technology 86
high-voltage direct current (HVDC) transmission 86
　key technical requirements for 88
　potential integration of HVDC into existing power grid systems and real-world HVDC transmission projects 88
　assumptions used in calculations 88
　economic assumptions 89–92
　technical assumptions 88–9
　recent achievements in transmission technology 86
　efforts in ultra-high-voltage direct current and alternating current transmission 87–8

ETO thyristor applications in FACTS and HVDC 87
　fault current limiters 87
　superconducting technology 86
　trend toward smaller power systems 86–7
　variable frequency transformers 87
high-voltage ride-through (HVRT) 257
Holt–Winters model 73, 76–7, 80
horizontal-axis wind turbines (HAWT) 9
household waste 432
hybrid grid-tied solar power systems 44–5
hybrid wind–diesel power systems 169
hydroelectric power plants (HEPPs) 282
hydrogen
　benefits of using 460–1
　in electricity production 460
　production from biomass 460
　production from natural gas 459–60
　production from water 459
　production methods 459–60
　in renewable energy systems 460
hydrogen fuel cells (HFCs) 460

information DB 235
International Electrotechnical Commission (IEC) 88
Internet of Things (IoT) 213, 222–5, 435
　application of MLT in monitoring and analyzing operational status of photovoltaic panels 215–16
　application of wireless technology and sensor systems in monitoring and predicting performance of renewable energy sources 220
　on monitoring and analyzing operational status of photovoltaic panels 214–15

research content 216
research objective 216
in solar power monitoring 214
types of sensors and communication technologies for monitoring and managing renewable energy systems 218
  benefits of sensor and communication systems in renewable energy management 219–20
  communication technologies in renewable energy management 219
  sensors in renewable energy 218
interquartile range (IQR) 75
inverter DC/AC 49
investment policies 337
  capital mobilization policies and electricity pricing 341
    characteristics and policy requirements for renewable energy development 341
    developing human resources 342
    environmental protection policies 342
    factors affecting household investment intentions in rooftop solar energy sector 342–5
    mechanisms and policies to support investment 341
    preferential policies 342
    promoting development of independent renewable energy systems 341
  estimating economic and financial indicators of project and identifying funding sources for solar and wind energy systems 349
    calculating cost and benefit components of rooftop grid-connected solar systems 351
    calculating economic effectiveness of rooftop grid-connected solar system 351–3
    calculating effectiveness of the project 356
    calculating financial and economic effectiveness of rooftop grid-connected solar systems 353–4
    calculation of costs and benefits of grid-connected rooftop solar power systems 354
    calculation of costs and benefits of grid-connected wind power systems 355–6
    calculation of financial economic efficiency of wind power systems 354–5
    case studies in Vietnam 363–6
    combined approach between grid-connected solar energy and independent investment optimization 356–62
    hybrid solar-BESS system based on business models 362–3
  mechanisms, and competition among power producers in Vietnam 337
    enhancing national load dispatch center (A0) independence 338–9
    establishing transparent competitive bidding framework 339
    key policy recommendations 338
    promoting private investment in transmission infrastructure 339
    solutions for creating level playing field in clean energy infrastructure sector in Vietnam 339–40
    strengthening independence of ERAV 338
  practical example for some solar and wind power projects in Vietnam 367

conclusion and recommendations 381
estimate content 368–9
estimated cost 369–72
legal basis for estimate 367–8
overview of project characteristics 367
preliminary energy estimates 375–80
project information summary 372–5
procedure for building and developing renewable energy projects in Vietnam 345
technical criteria in construction and development of renewable energy projects 346
application of fuzzy logic algorithm for supplier selection of solar equipment 347
calculating connection to electrical system 348–9
power calculation 347–8
site selection, survey, land clearance, solar radiation calculation, and climate conditions 346–7
system testing and project handover 349
IoT-based solar energy monitoring system 233
achievements 247
API functions for data communication 236
challenges and limitations 247
design of monitoring and data collection unit 233–5
fault classification results 245
fault diagnosis process 243–5
integration of machine learning models into 242–3

key improvements 245
mobile application 237–8
mobile application and computer software interface 238–42
proposed IoT monitoring system 233
recommendations and development directions 247–8
server 235
Island mode 164

Japan's CRIMAG model 432
JSON string 236

KEPServerEX software 187, 192

lead-acid batteries 71–2
least squares linear (LLS) algorithm 375
levelized cost of energy (LCOE) 20
lightning protection system 50
Li-Ion batteries 72
Linear Quadratic Regulator (LQR) controller 26
Load Break Switches (LBS) 107
load shifting 71
load systems 164
logarithmic law 3
logistic regression 74
Long Range Radio (LoRa) networks 438
long short-term memory (LSTM) 74
long-term storage 71
Lowe Chemie–mechanical and chemical process 401–2
Lowe Chemie technology 400–1
low-voltage ride-through (LVRT) 136, 257
Ly Son island
power line system on 196
power supply options for 197

power supply system on 193–7
transformer substation system on 196
Ly Son power grid 207

machine learning (ML) 213, 215
  comparison of two models 232–3
  K-means algorithm model 229–30
  SVM algorithm model 230–1
machine learning-based techniques (MLT) 214–15
MATLAB/Simulink software 26, 192
MAX44009 sensor 235
maximum efficiency tracking (MET) 453
maximum power point (MPP) 53
maximum power point tracking (MPPT) controller 53, 132
mean absolute error (MAE) 80
mean absolute percentage error (MAPE) 73
mean squared error (MSE) 122
mechanical dismantling 400
medium-term storage 71
medium-voltage solar power plants 253
  harmonic distortion and voltage imbalance 254
  reactive power and voltage adjustment capability 253–4
  requirements for short-circuit current and fault clearing time 255
  voltage flicker requirements 254–5
  voltage range compliance 254
  voltage requirements for solar power plants 254
metal oxide varistors (MOV) 123
microgrid (MG) systems 163
  applications and examples of microgrid power grid systems deployment 173

analysis and proposed solutions to enhance stability 179–80
comparison with reality and other solutions 180–1
control of wind–diesel hybrid power system 173–4
general control algorithm for wind–diesel hybrid power generation system on Phu Quy Island 180
operational constraints 181–3
operation of wind–diesel hybrid power system 174–8
results from steady-state mode survey 178
wind–diesel hybrid power system in Phu Quy 173
definition and objectives of 163
development of microgrid model for Phu Quoc Island 183–7
impact of wind power plants on reliability of distribution grid 193
power supply system on Ly Son island 193–7
wind power potential in Ly Son 193
key components of 163–5
in renewable energy sources and cloud-based systems 165
cloud computing 170–2
problem statement 169
research on optimal hybrid wind–diesel power systems 166
research on stable operation of hybrid wind–diesel systems 166
research status in Vietnam 167–9
standalone wind power generation system 169–70
study of solutions to increase wind power penetration rate 166

study on improving power quality of hybrid wind–diesel power systems 166–7
wind–diesel hybrid power system 170
simulation results 190
electric vehicle charging scenario simulation 190–2
energy management system test 190
smart energy management algorithm 187
electric vehicle charging algorithm 189
energy management system algorithm 188–9
MATLAB-PLC communication module 187
Mini-SCADA 110
mobile application 216, 236
application interface flowchart 237–8
monitoring information interface 238
user access rights 236
mobile networks 219
model-based diagnostic methods (MBDM) 214
model diversity 84
model selection 84
molded case circuit breaker (MCCB) 70
momentary average interruption frequency index (MAIFI) 112
MongoDB software 235
monocrystalline cells 35
monocrystalline silicon (m-Si) 34, 42
multi-layer perceptron (MLP) 122, 146

National Power Development Plan (NPDP) 296
National system and market operator (NSMO) 291

NBeatsBlock class 81
N-BEATS model 73, 76, 79, 81–2
NBeatsNet class 81
network security 159
neuro-fuzzy systems 143
nickel-cadmium (NiCd) 72
nickel-metal hydride (NiMH) 72
node pricing method 318
non-waste-related impacts 432
North American Electric Reliability Corporation (NERC) 257

offset payments 341
offshore wind power technology 10
open-circuit voltage 221
operating characteristics in steady-state mode 178
operating principle 111
operating system 109
operational constraints 181
analysis of stable operating conditions 182–3
application of solutions to Phu Quoc system 183
modeling wind–diesel hybrid power system 182
overview and analysis of wind–diesel hybrid power system 182
proposed algorithm and program to identify appropriate wind station for isolated grids 183
recommendations and future research directions 183
optimize overcurrent relay (OCR) 127
output signal analysis (OSA) 214–15

particle swarm optimization (PSO) algorithm 127
PD170-6 wind generator 4
performance evaluation 122, 345

performance index 147
performance metrics 81
permanent magnet synchronous generators (PMSG) 167
PH500 wind generator 5
photovoltaic (PV) panels 34, 69, 163, 213
  failure cases in 221–2
  Internet of Things 222–5
  I–V and P–V characteristic curves of 221
  machine learning model for fault classification 225
    development of IoT-based solar energy monitoring system 233–6
    fault simulation 226–9
    machine learning model training results 229–33
    steps to build the classification model 225–6
  structure of 220
photovoltaic (PV) sources
  comprehensive genetic algorithm method 127
  differential protection and setting issues 127
  disconnection solutions according to IEEE 1547 standards 127
  optimizing directional overcurrent settings 127
  relay performance testing in high-PV penetration networks 127
photovoltaic (PV) systems
  faults in 132–5
  grounded PV system 139
  solutions for rapid detection of short circuit current in 135–7
PhuMy Solar Power Plant 34
Phu Quoc archipelago 184

Phu Quoc Island
  development of microgrid model for 183
    building power grid model in MATLAB 186–7
    electrical grid model 183–4
    power balance 185–6
    wind turbine model 184–5
  general control algorithm for wind–diesel hybrid power generation system on 180
Phu Quy
  operating conditions at 180
  wind–diesel hybrid power system 173
Phu Quy Island 208
plasma-enhanced chemical vapor deposition 37
pollution 387
polycrystalline cells 35–6
polycrystalline silicon (p-Si) 35, 42
polyethylene terephthalate (PET) 389
polynomial framework 26
polynomials 79
polypropylene (PP) 389
polyvinyl fluoride (PVF) 389
potential DC transmission projects 101
  transmission of electricity from south central coast to southeast region 101–2
power balance 185–6
power calculation 347–8
power electronics converters 164
power fluctuations 174
power grid (PG) 62
power purchase agreements (PPAs) 338
power quality 127, 168
power supply sources 193
power system (PS) 59

calculation of current cost
  adjustments when input factors
  change 92
calculation of present value 94–9
power system simulation in PSS/E
  92–3
transmission distance of 270 km
  93–4
transmission distance of 450 km
  99–101
challenges from renewable
  energy 60
energy storage technologies for
  renewable energy systems and
  load forecasting in solar and
  wind power systems 71
  energy storage technologies
    71–2
  research on application of seasonal
    time series models for monthly
    electricity production
    forecasting 72–85
inertia in traditional power systems
  60
real-time calculation and monitoring
  of 61–2
research on methodology for
  calculating and measuring PS
  inertia 62
  observations 65–6
  operation during lunar new year
    2021 66
  practical implications 63–4
  results of calculations and
    recommendations for power
    system operation 64–5
  theoretical basis 62–3
power system protection devices
  evaluation of harmonic analysis and
    equipment loss calculation using
    specialized software and
    effective selection of 66

rooftop solar project on three
  factory roofs of company 68–9
transformer structure 67–8
power system simulator (PSS) 261–2
preferential policies 342
probabilistic neural network (PNN)
  215
programmable logic controller (PLC)
  141
project management costs 369
Prophet model 73, 78, 80–2
proportional–integral (P–I) controller
  143
protection setting management
  function 109
protection system 164
provincial People's Committees
  (PPCs) 1
PSCAD software 277
  comments 282
  current status of transmission
    network in Binh Dinh Province
    285–6
  impact of Fujiwara Solar Power
    Plant on power grid of Binh
    Dinh Province 282–5
  modeling power losses on 280–2
  overview and modeling of power
    loss problem using 280
public–private partnerships (PPP) 20,
  329, 337
PV cycle technology 398–9

Quang Ngai Province electrical grid 263
Quota mechanism 341

real-time diagnostic methods (RDM)
  214
real-time inertia monitoring 61
  real-time and short-term grid
    dispatch 61

support for system operational
  planning and analysis 62
real-time monitoring and control 159
reasonable selection 70
recurrent neural networks (RNN) 74
recycling costs 387
reliability 172, 346
reliability indicator 201–7
renewable energy (RE) 60, 163, 213, 337
  communication technologies in 219
    IoT and specialized communication protocols 219
    WiFi and mobile networks 219
    Zigbee and Bluetooth 219
  development of FiT in Vietnam 318–27
  on environmental, economic, and social policies in Vietnam 329
    economic impact 330
    environmental impact 330
    social impact 331
  environmental impact assessment of solar and wind energy projects in Vietnam 417
    environmental impact assessment for solar energy projects 419–21
    environmental impacts of wind power projects 421–4
    mitigation measures 424
    overview 417–19
  example of environmental impact assessment in central Vietnam 427
    application of adaptive control research in energy management systems at Vietnamese ports 454–6
    assessment and forecasting of environmental impacts 428
    assessment and forecasting of impacts during project preparation phase 428–30
    commitments 435
    design of dynamic wireless charging system for electric vehicles 447–50
    energy consumption and GHG emissions in transportation sector 444–6
    impact assessment and forecasting during construction phase 431
    impact assessment and forecasting during operational phase 432–4
    impact on air quality 430
    impact on water quality 430–1
    integration of renewable energy sources into electric vehicle systems 446–7
    limitations and future research directions 453
    non-waste-related sources of impact 431
    optimizing performance in wireless power transfer 452–3
    planning, implementation, and operation of electric vehicle systems and charging infrastructure 454
    recommendations 434–5
    transformer design 450–1
    transformer design structure 451–2
    transportation demand 442–4
  green hydrogen from wind and solar power via water electrolysis 456
    current hydrogen production methods 457–9
    groundbreaking ceremony for the first green hydrogen production plant in Tra Vinh 461
    role of hydrogen in renewable energy 459–61

growth prospects for 61
impact of renewable energy systems on operating parameters of neighboring grid at connection point 261
   assessment of impact of solar power plant on voltage losses 266
   calculation of optimal reactive power compensation for 22 kV distribution line 471GC using CAPO compensation model 270–7
   current situation 266–70
   distribution grid on PSS/ADEPT software 262
   Duc Pho District electrical grid 264
   power grid in area before and after Dam An Khe Solar Power Plant 264–5
   PSS/ADEPT software 261–2
   Quang Ngai Province electrical grid 263
   reactive power compensation to reduce losses for Go Cong town 22 kV distribution network 266
   results of power grid simulation in project area 265–6
   simulating real project 262–3
   Tuy Phong Wind Power Plant project 277
impact on power system inertia 61
integration 71, 159
IoT solutions and smart lighting system, wireless charging stations for vehicles, and electric ferries using 435
   development directions 440–1
   flowchart for central control unit algorithm 437–9

   system test results 439–40
   wireless charging station for electric vehicles using 441–2
methods for environmental impact assessment of renewable energy sources to identify effective mitigation and control solutions 426
PSCAD software 277
   comments 282
   current status of transmission network in Binh Dinh Province 285–6
   impact of Fujiwara Solar Power Plant on power grid of Binh Dinh Province 282–5
   modeling power losses on 280–2
   overview and modeling of power loss problem using 280
raising environmental and social awareness in renewable energy projects 424
   government policies and commitments 425–6
   importance of environmental and social awareness 425
   steps for implementation and impact assessment 425
   strengthening community and stakeholder involvement 425
   training and capacity building for stakeholders 425
regulations on connecting wind and solar power projects to grid 252
   analysis of HVRT and LVRT by ERCOT and NERC 257–8
   analysis of LVRT and HVRT in Vietnam 258–9
   solar power 253
   solar power systems connected to low-voltage distribution grids 259–61

technical requirements for
    medium-voltage solar power
    plants 253–5
  technical requirements for solar
    power plants connected to
    transmission grid 255–7
  wind power 252–3
  sensors in 218
simulation of voltage drop and short
    circuit at rooftop solar power
    plant in Long An Province 286
  evaluation of selection of DC
    cables 289
  selection of DC wiring 286
  voltage drop check for DC cables
    286–9
solutions for renewable energy
    development in Vietnam 331–3
subsidy mechanism 327
  impact analysis 328
  limitations of current subsidy
    mechanisms 327–8
  proposed solutions 329
  relevant legal documents 328
Vietnamese electricity market 289
  analysis of simulation results 318
  current transmission pricing
    method in Vietnam 298–303
  problem statement 289–90
  transmission electricity price
    model in Vietnam's competitive
    wholesale electricity market
    303–17
  Vietnam's wholesale electricity
    market 290–8
Vietnam's power system 251
  challenges in ensuring electricity
    supply 252
  solar power 251
  strategic solutions 252
  structure of 251

renewable energy sources 163
renewable energy systems 220
research system diagram 136
reusing industrial by-products 385
rooftop solar power systems
  grid connection capacity of 51
  power generation of 51
  for self-generation and self-
    consumption 45
    flexible and modern connection
      solution 46
    general regulations 46
    proposed expansions 46–7
    specific regulations by installation
      capacity 46
root mean square error (RMSE) 73

Saigon Hi-Tech park power grid 149
SARIMA model 73, 75, 77–8,
    80, 82
SCADA software 110
sensors
  in renewable energy 218
    environmental sensors 218
    solar energy sensors 218
    wind sensors 218
short-circuit analysis method 141
short-circuit current 70
short-term storage 71
signal processing algorithms 215
silicon ribbon 42
simulation method 226–9
smart energy management algorithm
    187
  electric vehicle charging algorithm
    189
  energy management system
    algorithm 188–9
  MATLAB-PLC communication
    module 187

480 Clean energy in South-East Asia

smart grids 160
   artificial intelligence and automation in smart grids on cloud computing platforms 113
      improving reliability index set 113–14
      load transfer and grid upgrades for DAS project 118–20
      recommendations 120
      reduction of outage duration 113
      structure of DAS with FDIR function 114–18
   artificial intelligence and automation in smart grids on cloud platforms 120
      ANN structure for fault location 121–2
      common faults in transmission lines 120–1
      fault location challenges on lines with series compensation 123
      fault location procedure using ANN 121
      key data features 122
      multilayer perceptron technique 122
      performance evaluation 122
      research methods and techniques 121
      research needs and objectives 121
      role of transmission lines 120
      types of faults studied 121
   challenges and opportunities in implementing smart grids 156
      benefits and opportunities 159–60
      key challenges in implementation 157–9
   development journey and technology integration for 157
   distance protection for grid-connected PV sources 128
      changes with grid-connected PV systems 131
      differential protection for PV sources 128–30
      distance protection (F21) 128
      faults in PV systems 132–5
      GE F650 relay 137–41
      solutions for rapid detection of short circuit current in PV systems 135–7
      typical faults in PV systems 131–2
   fault location process and issues in power transmission lines in Vietnam 123
      current fault location process in Vietnamese power transmission 123–4
      current process 125
      existing issues 125
      fault location in integrated renewable energy systems 125
      future research directions 126
      issues in current fault location process 124
      new mathematical models 126
      technical causes of errors 125
      technical causes of fault location error 124–5
   importance in improving performance and network management 107
      required input data for DAS application 109
      software requirements 109–10
   integrating renewable energy into 333
   opportunities and challenges 156
   protection solutions for PV sources 126
      power quality and islanding issues 127

practices in Vietnam 127–8
protection strategies for PV sources 127
STATCOM for dynamic voltage stabilization in power systems 141
  design of online ANFIS intelligent controller 143–56
  objectives of research 143
technology and advances in 110
  application of distribution grid automation in enhancing power supply reliability at EVNCPC 111–12
  DAS 110
  definition and function of DMS 111
  DMS model 111
  Mini-SCADA 110
  operating principle 111
smart monitoring device (SMD) 214
social impact 331
  improving quality of life 331
  raising public awareness 331
solar energy 42
  efficient solar power system models 42
    grid-tied solar power system 43–4
    hybrid grid-tied solar power systems 44–5
    rooftop solar power systems for self-generation and self-consumption 45–7
    stand-alone solar power systems 42–3
  potential and development prospects of 31
  solar energy projects in Vietnam 32–4
  solar power with grid-tied rooftop systems 47–55

solar radiation coefficients in high-potential areas 31–2
structure and technical specifications of main components in solar power systems 42
traditional solar panel technology 34
  amorphous silicon 37
  amorphous silicon solar modules 38
  copper indium selenium solar cells 38
  monocrystalline cells 35
  monocrystalline silicon 34
  next-generation solar technology 38–41
  polycrystalline cells 35–6
  polycrystalline silicon 35
  thin-film solar panels 36–7
solar energy-based production 458
solar energy sensors 218
solar hydrogen system (SHS) 458
solar panels 35, 348, 387
  array-inverter layout 50
  operating efficiency of 51–2
  recycling 412
solar panel waste 391
  benefits of recovering and reusing materials from solar panels 395
  estimated amount of materials recovered from recycling 393
  estimated recovered materials from recycling 394
  material recycling rates and potential 393–4
  proportion of material weight and recycling potential of solar panel materials 392–3
  in Vietnam 392
solar photovoltaic panels 389
  challenges in solar panel waste treatment 390

researching structure for waste treatment 390–1
structure of 389
solar power 251, 253
   with grid-tied rooftop systems 47
      annual power generation chart 51
      connection plan 47–8
      design of solar-powered water pumping project 54–5
      factors affecting the curve 52–3
      grounding system 50
      inverter DC/AC 49
      key operating parameters for rooftop grid-connected solar power system 51
      lightning protection system 50
      operating efficiency of solar panels 51–2
      project introduction 47
      project layout solution 49–50
      project power generation simulation results 51
      solar panel array-inverter layout 50
      solar-powered automatic water pumping system 54
      solar power system 48
      solar tracking systems 53–4
      surge protection devices 50–1
   systems 42
      applications 42
      operating principle 42
solar-powered automatic water pumping system 54
solar-powered water pumping project 54–5
solar power plants 255
   negative sequence components 256
   reactive power and voltage regulation capability 255–6

voltage flicker requirements 256–7
solar radiation 163, 214, 347, 348
   coefficients in high-potential areas 31–2
   levels by region 33
solar tracking systems 53
   dual-axis solar tracking system 53–4
   single-axis solar tracking system 53
solar world technology 399–400
solid-liquid separation 400
solid waste impacts 432
space-saving 35
spinning reserve 71
squirrel cage induction generators (SCIG) 167
standard test conditions (STC) 221
STATCOM model 148–9
state of charge (SOC) 188
state-owned enterprises (SOEs) 337
   privatization and improved governance of 340
static wireless charging 442
sum of squares (SOS) technique 21
supercapacitors 164
superconducting technology 86
supercritical water gasification (SCWG) 459
supervisory control and data acquisition (SCADA) system 107
support vector machines (SVM) 215
surge protection devices (SPDs) 50
synchronous compensator (STATCOM) 87, 141
synchronous generator 63
system average interruption duration index (SAIDI) 112–13
system average interruption frequency index (SAIFI) 112

Index

Takagi–Sugeno–Kahn rule 216
Tedlar–PET–Tedlar (TPT) backsheet 40
temperature coefficient 221
thermal method 409
thermal treatment 399
thermogravimetric analysis (TGA) 409
thin-film solar panels (thin-film PV modules) 36–7
ThingSpeak application 214
threshold values 216
time–current characteristic (TCC) curve 131
time dial setting (TDS) 131
time series-based methods 74
total harmonic distortion of current (THDi) 67
total harmonic distortion of voltage (THDu) 67
tracking error 21–2
training data 215
transfer learning 79
transfer method 110
transformation methods 214
transformer design 450–1
transformer structure (MBA) 67–8
transmission lines
  common faults in 120–1
  role of 120
Tra Xom Hydroelectric Power Plant 284
trend toward smaller power systems 86–7
trigonometric functions 79
tripping time 70
turbine layout 377–8
Tuy Phong Wind Power Plant project 277

ultra-high-voltage direct current 87–8
uncertainty 378–80
uncontrolled valve 88
Unilever Vietnam 386
user DB 235
U.S. National Oceanic and Atmospheric Administration 7

variable frequency transformers (VFTs) 87
vertical-axis wind turbines (VAWT) 9
Vietnam
  current transmission pricing method in 298
    comparison and analysis of two methods 302–3
    electricity cost and transmission revenue differential 299–300
    group of electricity costs to determine total capital electricity costs 300
    group of electricity costs to determine total O&M electricity costs 300–1
    new investment grid 301–2
    total capital electricity costs 299
    total electricity delivered and received 299
    total transmission revenue 298–9
    transmission price 298
  practices in 127–8
  research status in 167
    study on selecting optimal wind–diesel hybrid power system 167
    study on stable operation and enhancing wind power penetration in hybrid wind–diesel power systems 167–9
  selected regions with highest wind energy potential in 2
  survey of installation and operating costs for wind power projects globally and in 19

offshore wind power projects 20
onshore wind power projects 19–20
wind energy potential as surveyed by World Bank 4
wind energy studies in 4–6
wind power investment projects in 6
Vietnamese government 28
Vietnam packaging recycling alliance (PRO) 386
Vietnam's competitive generation market (VCGM) 290
Vietnam's retail electricity market (VREM) 290
Vietnam's wholesale electricity market 290
  day-ahead market 291–3
  legal framework analysis in electricity market development 293–5
  member units participating in 292
  objectives of 290
  participants in 291
  principles for building VWEM 290–1
  structure of 291–2
  suitability of transmission pricing method for electricity market in 297–8
  suitability of transmission pricing method for electricity market in Vietnam 296–7
Vietnam's wholesale electricity market (VWEM) 290
Vinh Son Hydroelectric Power Plant 282
Voltage Ride-Through (VRT) capability 257
voltage support 71
voltage transformer (VT) 114

waste collection and management systems 388
waste-to-energy plants 385
water quality impacts 432
wavelet transformations 215
WiFi 214, 219
wind data 373
wind–diesel hybrid power system 170
  control of 173–4
  operation of 174–8
  in Phu Quy 173
wind–diesel power generation system 195
wind direction assessment 373
wind energy (WE) 71
  calculation of wind power generation 16–19
  conversion to electrical energy in wind power generation system 12
  disadvantages of 7–9
  environmental issues 6–7
  history of wind turbines development 9–10
  source 184
  studies in Vietnam 4–6
  types of generators used in wind power systems 10–14
  typical connection parameters for wind power projects selection of connection circuit parameters 14
    conductor cross-sect 15
    connection distance 14–15
    connection power capacity 15–16
    connection voltage 16
wind energy-based production 458
wind energy conversion systems (WECS) 10
  observer-based controller for 23
  parameters 22

wind farms 7–8
wind flower 17
wind power 28, 252
  draft regulations on wind power projects 253
  grid connection procedure 253
  investment projects in Vietnam 6
wind power plants (WPPs) 5, 102
  on distribution grid 102–3
  on power grid and statistical analysis of wind turbine failures at Tuy Phong Wind Power Plant—Binh Thuan 103
    failure parameters of wind turbines at 103–4
    power exchange characteristics between WPP and power grid 103
    practical operation experience 104
    significance of 104–5
wind resource grid (WRG) 375
wind sensors 218
wind speed 16, 199, 373

wind turbine model 184–5
wind turbines (WTs) 7, 163, 387–8
  development process of 11
  from leading manufacturers worldwide 13
  optimizing power calculation through nonlinear control and sensor fault monitoring for 20
  observer-based controller and fault-tolerant control design 21–2
  observer design for generalized high-level disturbances 22–6
  recycling 412
  structural safety of 9
wireless power transfer (WPT) 442
wireless sensor system 214
wire-to-wire short circuit 245
World Energy Council 7

Zigbee 214, 219
zonal division based on reliability indicators 201